MATH/STAT
LIBRARY

Statistics and Computing

Series Editors:
J. Chambers
W. Eddy
W. Härdle
S. Sheather
L. Tierney

Springer
Berlin
Heidelberg
New York
Hong Kong
London
Milan
Paris
Tokyo

Statistics and Computing

Dalgaard: Introductory Statistics with R.
Gentle: Elements of Computational Statistics.
Gentle: Numerical Linear Algebra for Applications in Statistics.
Gentle: Random Number Generation and Monte Carlo Methods, 2nd Edition.
Härdle/Klinke/Turlach: XploRe: An Interactive Statistical Computing Environment.
Hörmann/Leydold/Derflinger: Automatic Nonuniform Random Variate Generation
Krause/Olson: The Basics of S and S-Plus, 3rd Edition.
Lange: Numerical Analysis for Statisticians.
Loader: Local Regression and Likelihood.
Ó Ruanaidh/Fitzgerald: Numerical Bayesian Methods Applied to Signal Processing.
Pannatier: VARIOWIN: Software for Spatial Data Analysis in 2D.
Pinheiro/Bates: Mixed-Effects Models in S and S-Plus.
Venables/Ripley: Modern Applied Statistics with S, 4th Edition.
Venables/Ripley: S Programming.
Wilkinson: The Grammar of Graphics.

Wolfgang Hörmann
Josef Leydold
Gerhard Derflinger

Automatic Nonuniform Random Variate Generation

 Springer

Wolfgang Hörmann
Bogazici University
Dept. of Industrial Engineering
80815 Istanbul
Turkey
hormannw@boun.edu.tr

Josef Leydold
Gerhard Derflinger
University of Economics and Business Administration
Dept. for Applied Statistics & Data Processing
Augasse 2-6
1090 Wien, Austria
Josef.Leydold@statistik.wu-wien.ac.at
Gerhard.Derflinger@statistik.wu-wien.ac.at

Series Editors:

J. Chambers
Bell Labs. Lucent Technologies
600 Mountain Ave.
Murray Hill, NJ 07974
USA

W. Eddy
Department of Statistics
Carnegie Mellon University
Pittsburgh, PA 15213
USA

W. Härdle
Institut für Statistik und
Ökonometrie
Humboldt-Universität zu Berlin
Spandauer Straße 1
10178 Berlin
Germany

S. Sheather
Australien Graduate
School of Management
University of New South Wales
Sydney, NSW 2052
Australia

L. Tierney
Department of Statistics and Actuarial Science
University of Iowa
Iowa City, IA 52242-1419
USA

Cataloging-in-Publication Data applied for

A catalog record for this book is available from the Library of Congress.

Bibliographic information published by Die Deutsche Bibliothek
Die Deutsche Bibliothek lists this publication in the Deutsche Nationalbibliografie;
detailed bibliographic data is available in the Internet at <http://dnb.ddb.de>.

Mathematics Subject Classification (2000): 65C10, 65-02, 62-02, 65C05, 68U20, 62H12, 62G05, 65C20

ISBN 3-540-40652-2 Springer-Verlag Berlin Heidelberg New York

This work is subject to copyright. All rights are reserved, whether the whole or part of the material is concerned, specifically the rights of translation, reprinting, reuse of illustrations, recitation, broadcasting, reproduction on microfilm or in any other way, and storage in data banks. Duplication of this publication or parts thereof is permitted only under the provisions of the German Copyright Law of September 9, 1965, in its current version, and permission for use must always be obtained from Springer-Verlag. Violations are liable for prosecution under the German Copyright Law.

Springer-Verlag Berlin Heidelberg New York
a member of BertelsmannSpringer Science+Business Media GmbH

springeronline.com

© Springer-Verlag Berlin Heidelberg 2004
Printed in Germany

The use of general descriptive names, registered names, trademarks etc. in this publication does not imply, even in the absence of a specific statement, that such names are exempt from the relevant protective laws and regulations and therefore free for general use.

Cover design: *design & production,* Heidelberg
Typesetting by the authors using a Springer LATEX macro package
Printed on acid-free paper   40/3142ck-5 4 3 2 1 0

Preface

This book deals with non-uniform random variate generation algorithms. It is readily accepted that these algorithms are a keystone to stochastic simulation. They are used in all fields of simulation application reaching from computer science, engineering and finance to Bayesian statistics. Even experts in the field of mathematical statistics and probability theory rely on these methods for checking their theoretical results. Nevertheless, the last monographs on random variate generation were published about 15 years ago. But it is not true that research in this area has ceased in recent years. Thus one objective of this text is to demonstrate that – unified under the concept of automatic generators – this field comprises lots of recent ideas and developments. As a second objective we want to show that these ideas, little known to most users of simulation till now, are not only based on a nice theory. They also lead to new algorithms that have practical and theoretical advantages in simulation practice. So we tried to find a balance between the necessary mathematical theory and the more applied description of the new algorithms. We also share with the readers our experience concerning the implementation and the numerical problems of the new algorithms.

The application of these new algorithms is not difficult due to the freely available UNU.RAN library and the "code generator" that can be tested on our web server. This book provides the theoretical background to fully appreciate the advantages of this new automatic approach to random variate generation. Please check our webpage

$$\text{http://statistik.wu-wien.ac.at/arvag/}$$

that also contains the C code of most of the algorithms of this book and some additional material.

The first basis of this book is the work of many researchers in the field of random variate generation, mainly between 1960 and the late eighties of the last century. At that time the authors have started to work together in Vienna. Our entrance into the world of random variate generation was facilitated by the book of Luc Devroye and by the personal discussions, help and encouragement of Ulrich Dieter and Ernst Stadlober. We know how much we have profited from their help. It is a good place here to express our gratitude to these three colleagues as we cannot imagine that our research in this field could have started without them. Our work on automatic random variate generation algorithms started about ten years ago. The acceptance of research projects on automatic random variate generation increased our motivation to start with the development of the UNU.RAN software library and to start with a unified description of our research results in a monograph. About five years have passed since we started those research projects. The book is considerably larger than planned originally as we tried to include recent and applied fields like Markov Chain Monte Carlo, perfect sampling, generation of copula with specified marginals, and resampling methods. We also included a chapter on simulation applications (in finance and Bayesian statistics) to demonstrate how automatic random variate generation algorithms can aid simulation practice.

We want to thank our colleagues at the department in Vienna for their cooperation, in particular Michael Hauser (who also contributed Chap. 13), Günter Tirler und Erich Janka for their work on our UNU.RAN library. The first author knows that this work would not have been possible without the understanding and support of all colleagues at the department in Istanbul, especially the dean of the Engineering Faculty, Ali Riza Kaylan, and the department head, Ilhan Or. He is also grateful to the students in his courses on random variate generation and to his master thesis students who helped to correct and improve many of the algorithm descriptions.

We also thank Pierre L'Ecuyer and the members of the p-lab project in Salzburg, especially Peter Hellekalek, for their help in questions regarding uniform random number generation. We are especially grateful to James Falk who – during his stay in Istanbul – patiently read through the whole manuscript to correct our English. (Of course he is not responsible for any errors that we have introduced during the final stage of preparation.) We gratefully acknowledge the support of our research by the APART Scholar ship of the Austrian Academy of Science and by the Austrian *Fonds zur Förderung der Wissenschaftlichen Forschung*, Project No. 12805-MAT.

Vienna
Istanbul
July 24, 2003

Wolfgang Hörmann
Josef Leydold
Gerhard Derflinger

Contents

Part I Preliminaries

1 Introduction 3

2 General Principles in Random Variate Generation 13
   2.1 The Inversion Method 13
   2.2 The Rejection Method 16
   2.3 Composition 26
   2.4 The Ratio-of-Uniforms Method (RoU) 32
   2.5 Almost-Exact Inversion 35
   2.6 Using Special Properties of the Distribution 37
   2.7 Exercises 39

3 General Principles for Discrete Distributions 43
   3.1 Inversion 43
   3.2 The Alias Method 47
   3.3 Discrete Rejection 50
   3.4 Exercises 52

Part II Continuous Univariate Distributions

4 Transformed Density Rejection (TDR) 55
   4.1 The Main Idea 55
   4.2 The Class T_c of Transformations 59
   4.3 T_c-Concave Distributions 63
   4.4 Construction Points 69
   4.5 Algorithms and Variants of Transformed Density Rejection ... 88
   4.6 Other Transformations 95
   4.7 Generalizations of Transformed Density Rejection 97
   4.8 Automatic Ratio-of-Uniforms Method 103

VIII Contents

    4.9 Exercises 109

5 Strip Methods 113
    5.1 Staircase-Shaped Hat Functions ("Ahrens Method") 114
    5.2 Horizontal Strips 123
    5.3 Exercises 124

6 Methods Based on General Inequalities 125
    6.1 Monotone Densities 126
    6.2 Lipschitz Densities 134
    6.3 Generators for $T_{-1/2}$-Concave Densities 135
    6.4 Generators for T_c-Concave Densities 142
    6.5 Exercises 152

7 Numerical Inversion 155
    7.1 Search Algorithms Without Tables 156
    7.2 Fast Numerical Inversion 158
    7.3 Exercises 164

8 Comparison and General Considerations 165
    8.1 The UNU.RAN Library 166
    8.2 Timing Results 169
    8.3 Quality of Generated Samples 173
    8.4 Special Applications 184
    8.5 Summary 188

9 Distributions Where the Density Is Not Known Explicitly . 193
    9.1 Known Hazard-Rate 193
    9.2 The Series Method 201
    9.3 Known Fourier Coefficients 204
    9.4 Known Characteristic Function 206
    9.5 Exercises 210

Part III Discrete Univariate Distributions

10 Discrete Distributions 215
    10.1 Guide Table Method for Unbounded Domains 216
    10.2 Transformed Probability Rejection (TPR) 219
    10.3 Short Algorithms Based on General Inequalities 227
    10.4 Distributions Where the Probabilities Are Not Known
          Explicitly 232
    10.5 Computational Experience 236
    10.6 Summary 239
    10.7 Exercises 241

Part IV Random Vectors

11 Multivariate Distributions 245
  11.1 General Principles for Generating Random Vectors 246
  11.2 Uniformly Distributed Random Vectors 252
  11.3 Multivariate Transformed Density Rejection 264
  11.4 Orthomonotone Densities................................. 280
  11.5 Computational Experience................................ 294
  11.6 Multivariate Discrete Distributions 295
  11.7 Exercises 300

Part V Implicit Modeling

12 Combination of Generation and Modeling 305
  12.1 Generalizing a Sample 306
  12.2 Generalizing a Vector-Sample 319
  12.3 Modeling of Distributions with Limited Information 322
  12.4 Distribution with Known Moments 323
  12.5 Generation of Random Vectors where only Correlation and
       Marginal Distributions are Known 328
  12.6 Exercises 343

**13 Time Series (Authors Michael Hauser and Wolfgang
    Hörmann)** 345
  13.1 Stationary Gaussian Time Series 347
  13.2 Non-Gaussian Time Series 356
  13.3 Exercises 362

14 Markov Chain Monte Carlo Methods 363
  14.1 Markov Chain Sampling Algorithms 364
  14.2 Perfect Sampling for Markov Chain Monte Carlo 372
  14.3 Markov Chain Monte Carlo Methods for Random Vectors 376
  14.4 Exercises 385

15 Some Simulation Examples 387
  15.1 Financial Simulation 387
  15.2 Bayesian Statistics...................................... 400
  15.3 Exercises 409

List of Algorithms 411

References... 415

Author index 429

Selected Notation 433

Subject Index and Glossary 435

Part I

Preliminaries

1
Introduction

Non-uniform random variate generation is a small field of research somewhere between mathematics, statistics and computer science. It started in the fifties of the last century in the "stone-age" of computers. Since then its development has mainly been driven by the wish of researchers to solve generation problems necessary to run their simulation models. Also the need for fast generators has been an important driving force. The main mathematical problems that have to be solved concern the distribution of transformed random variates and finding tight inequalities. Also implementing and testing the proposed algorithms has been an important part of the research work. A large number of research papers in this field has been published in the seventies and early eighties. The main bibliographical landmark of this development is the book of Devroye (1986a), that is commonly addressed as the "bible" of random variate generation. We can certainly say that random variate generation has become an accepted research area considered as a subarea of statistical computing and simulation methodology. Practically all text-books on discrete event simulation or Monte Carlo methods include at least one chapter on random variate generation; within simulation courses it is taught even to undergraduate students.

More important is the fact that random variate generation is used by lots of more or less educated users of stochastic simulation. Random variate generation code is found in spreadsheets and in expensive discrete event simulation software and of course in a variety of programming languages. Probably many of these users do not bother about the methods and ideas of the generation algorithms. They just want to generate random variates with the desired distribution. "The problems of random variate generation are solved" these people may say. And they are right as long as they are only interested in popular standard distributions like normal, gamma, beta, or Weibull distributions.

Why Universal Random Variate Generation?

The situation changes considerably if the user is interested in non standard distributions. For example, she wants to simulate models that include the generalized inverse Gaussian distribution, the hyperbolic distribution, or any other distribution that is less common or even newly defined for her purpose. Then the user had two possibilities: She could either find some code (or a paper) dealing with the generation of random variates from her distribution, or she needed some knowledge on random variate generation (or find an expert) to design her own algorithm. Today the user has a third possibility: She can find a universal generator suitable for her distribution, perhaps in our C library UNU.RAN (Universal Non-Uniform RANdom variate generation). Then she can generate variates without designing a new generator; a function that evaluates e.g. the density of the distribution or the hazard rate is sufficient.

This book is concentrating on the third possibility. We present ideas and unifying concepts of universal random variate generation, and demonstrate how they can be used to obtain fast and robust algorithms. The book is presenting the first step of random variate generation, the design of the algorithms. The second step, i.e. the implementation of most of the presented algorithms, can be found in our C library UNU.RAN (Universal Non-Uniform RANdom number generators, Leydold, Hörmann, Janka, and Tirler, 2002). As some of these algorithms are rather long it is not possible, and probably not desirable, to present all implementation details in this book as they would hide the main ideas.

What Is Non-Uniform Random Variate Generation?

Usually random variates are generated by transforming a sequence of independent uniform random numbers on $(0,1)$ into a sequence of independent random variates of the desired distribution. This transformation needs not be one-to-one. We assume here, as it is generally done in the literature, that we have an ideal source of uniform random numbers available. An assumption which is not too unrealistic if we think of fast, modern uniform random number generators that have cycle lengths in the order 2 raised to the power of several hundreds or even thousands and equidistribution property up to several hundred dimensions, e.g., Matsumoto and Nishimura (1998) or L'Ecuyer (1999); see also the pLab website maintained by Hellekalek (2002) for further links. A collection of many published uniform random number generators – good ones and bad ones – is compiled by Entacher (2000).

Given that (ideal) source of uniform random numbers, the well known inversion, (acceptance-) rejection and decomposition methods can be used to obtain exact random variate generation algorithms for standard distributions. We do not want to give a historical overview here but it is remarkable that the rejection method dates back to von Neumann (1951). Later refinements of the general methods were developed mainly to design fast algorithms, if possible with short code and small memory requirements. For the normal distribution compare e.g. Box and Muller (1958), Marsaglia, MacLaren, and Bray (1964), Ahrens and Dieter (1972, 1973, 1988), and Kinderman and Ramage (1976). For the gamma distribution see e.g. Cheng (1977), Schmeiser and Lal (1980), Cheng and Feast (1980), and Ahrens and Dieter (1982). In the books of Devroye (1986a) and Dagpunar (1988) you can find an impressive number of references for articles dealing with random variate generation for standard distributions. The techniques and general methods we describe in this book are very closely related to those developed for standard distributions. We know what we owe to these "pioneers" in random variate generation. Even more as we have started our research in this field with generators for standard distributions as well (see e.g. Hörmann and Derflinger, 1990). However, in this book we try to demonstrate that several of these main ideas can be applied to build universal generators for fairly large distribution families. Thus we do not concentrate on standard distributions but try to identify large classes of distributions that allow for universal generation. Ideally these classes should contain most important standard distributions. So the reader will come across quite a few standard distributions as we use them as examples for figures and timings. They are best suited for this purpose as the reader is familiar with their properties. We could have included fairly exotic distributions as well, as we have done it in some of the empirical comparisons.

What Is a Universal Generator?

A *universal* (also called *automatic* or *black-box*) generator is a computer program that can sample from a large family of distributions. The distributions are characterized by a program that evaluates (e.g.) the density, the cumulative distribution function, or the hazard rate; often some other information like the mode of the distribution is required. A universal generator typically starts with a *setup* that computes all constants necessary for the generation, e.g. the hat function for a rejection algorithm. In the *sampling* part of the program these constants are used to generate random variates. If we want to generate a large sample from a single distribution, the setup part is executed only once whereas the sampling is repeated very often. The average execution time of the sampling algorithm to generate one random variate (without taking the setup into account) is called *marginal execution time*. Clearly the *setup time* is less important than the marginal execution time if we want to generate a large sample from a single distribution.

Fast universal generators for discrete distributions are well known and frequently used. The indexed search method (Chen and Asau, 1974) and the alias method (Walker, 1974, 1977) can generate from any discrete distribution with known probability vector and bounded domain. For continuous distributions the design of universal generators started in the eighties and is mainly linked with the name of Luc Devroye. During the last decade research in this direction was intensified (see e.g. Gilks and Wild, 1992; Hörmann, 1995; Ahrens, 1995; Leydold, 2000a) and generalized to random vectors (see e.g. Devroye, 1997a; Leydold and Hörmann, 1998). However it seems that these developments are little known and hardly used. When we have started our UNU.RAN project in 1999 we checked several well known scientific libraries: IMSL, Cern, NAG, Crand, Ranlib, Numerical recipes, and GSL (Gnu Scientific Library). Although all of them include quite a few random variate generation algorithms for continuous standard distributions we were not able to find a single universal generator for continuous distributions in any of them. And the situation is similar for random-vector generation methods. This has been a main motivation for us to start our project with the aim to write both a book and a C library to describe and realize theory and practice of universal random variate and random vector generation.

Why We Have Written this Book?

We are convinced that universal random variate generation is a concept of greatest practical importance. It also leads to nice mathematical theory and results. Neither the mathematics nor the implemented algorithms have been easily accessible up to now, as there is – up to our knowledge – no book and no software library available yet that includes the main concepts of universal random variate generation.

Implementation of Universal Algorithms

One main problem when implementing universal algorithms for continuous distributions in a software library is certainly the application programming interface (API). Considering generators for a fixed distribution everything is simple. Any programmer can easily guess that the following C statements are assigning a realization from a uniform, a standard normal, and a gamma(5) random variate to the respective variables.

```
x  = random();
xn = randnormal();
xg = randgamma(5.);
```

For a universal method the situation is clearly more complicated. The setup often is very expensive compared to the marginal generation time and thus has to be separated from the sampling part of a universal algorithm. Moreover, to run such a setup routine we need a lot more parameters. In a routine like randgamma(5.) all required information is used to build the algorithm and is thus contained implicitly in the algorithm. For black-box algorithms we of course have to provide this information explicitly. This may include – depending on the chosen algorithm – the density, its derivative and its mode, its cumulative distribution function, its hazard rate, or similar data and functions to describe the desired distribution. Furthermore, all the blackbox algorithms have their own parameters to adjust the algorithm to the given distribution. All of this information is needed in the setup routine to construct a generator for the distribution and to store all necessary constants. The sampling program then uses these constants to generate random variates.

There are two very different general solutions for the implementation of automatic algorithms. First, we can make a library that contains both a setup routine and a sampling routine using an object-oriented design. The setup routine creates an instance of a *generator object* that contains all necessary constants. The sampling routine is using this generator object for generation. If we want to sample from several different distributions in one simulation we can create instances of such generator objects for all of them and can use them for sampling as required. In our library UNU.RAN we have implemented this idea. For further details see Sect. 8.1.

Our experiences with the implementation of universal algorithms in a flexible, reliable, and robust way results in rather large computer code. As the reader will find out herself, the complexity of such a library arises from the setup step, from parts performing adaptive steps, and (especially) from checking the data given by the user, since not every method can be used for every distribution. The sampling routine itself, however, is very simple and consists only of a few lines of code. Installing and using such a library might seem too tedious for "just a random number generator" at a first glance, especially when only a generator for a particular distribution is required. As a solution to this problem we can use universal methods to realize the concept of an *automatic code generator for random variate generation*. In this second approach we use the constants that are computed in the setup to produce a single piece of code in a high level language for a generator of the desired distribution. Such a code generator has the advantage that it is also comparatively simple to generate code for different programming languages. Moreover, we can use a graphical user interface (GUI) to simplify the task of obtaining a generator for the desired distribution for a practitioner or researcher with little background in random variate generation. We also have implemented a proof of concept study using a web based interface. It can be found at http://statistik.wu-wien.ac.at/anuran/. Currently, program code in C, FORTRAN, and Java can be generated and downloaded. For more details see Leydold, Derflinger, Tirler, and Hörmann (2003).

It should be clear that the algorithm design of the setup and the sampling routine remain the same for both possible implementation concepts. Thus we will not consider the differences between these two approaches throughout the book. The only important difference to remember is that the speed of the setup is of little or no concern for the code generator whereas it may become important if we use generator objects.

Theoretical Concepts in the Book

We have spoken quite a lot about algorithms and implementation so far but most of the pages of the book are devoted to the theoretical concepts we need for designing random variate generation algorithms. Typically we are concerned with the following problems:

- We have to show that the algorithms generate variates from the desired distribution.
- For rejection algorithms we use inequalities to design upper and lower bounds for densities.
- The automatic construction of hat functions for rejection methods requires design points. Hence we also need algorithms to find such points. These algorithms are either simple and fast, or provide optimal solutions (or both).
- We need properties of the distribution families associated with the universal algorithms.
- We need simple (sufficient) conditions for the (large) class of distributions for which an algorithm is applicable. When the condition is not easily computable either for the routines in the library or for the user of such a library, a black-box algorithm is of little practical use.
- The complexity of an algorithm is the number of operations it requires for generating a random variate. For most algorithms the complexity is a random variate itself that depends on the number of iterations I till an algorithm terminates. For many universal algorithms we can obtain bounds for the expected number of iterations $\mathrm{E}(I)$.
- We want to have some estimates on the "quality" of the generated random variates. In simulation studies streams of *pseudo*-random numbers are used, i.e. streams of numbers that cannot be distinguished from a stream of (real) random numbers by means of some statistical tests. Such streams always have internal structures (see e.g. L'Ecuyer (1998) for a short review) and we should take care that the transformation of the uniform *pseudo*-random numbers into non-uniform *pseudo*-random variates do not interfere with the structures of this stream.

Chapters of the Book

Chapter 2 presents the most important basic concepts of random variate generation: *Inversion*, *rejection*, and *composition* for continuous random variates. Thus it is crucial to the rest of the book as practical all of the algorithms presented in the book depend on one or several of these principles. Chapter 3 continues with the basic methods for generating from discrete distributions, among them inversion by sequential search, the indexed search and the alias method.

Part II of the book deals with continuous univariate distributions. Chapters 4, 5, and 6 present three quite different approaches to design universal algorithms by utilizing the rejection method. Chapter 7 realizes the same task using numerical inversion. Chapter 8 compares different aspects of the universal algorithms presented so far including our computational experiences when using the UNU.RAN library. It also describes the main design of the UNU.RAN programming interface that can be used for generating variates from discrete distributions and from random vectors as well. Part II closes with Chap. 9 that collects different special algorithms for the case that the density or cumulative distribution function of the distribution is not known.

Part III consists only of Chap. 10. It explains recent universal algorithms for discrete distributions, among them the indexed search method for distributions with unbounded domain and different universal rejection methods.

Part IV, that only consists of Chap. 11, presents general methods to generate random vectors. It also demonstrates how the rejection method can be utilized to obtain universal algorithms for multivariate distributions. The practical application of these methods are restricted to dimensions up to about ten.

Part V contains different methods and applications that are closely related to random variate generation. Chapter 12 collects random variate generation procedures for different situations where no full characterization of the distribution is available. Thus the decision about the generation procedures is implicitly also including a modeling decision. Chapter 13 (co-authored by M. Hauser) presents very efficient algorithms to sample Gaussian time series and time series with non-Gaussian one-dimensional marginals. They work for time series of length up to one million. Markov Chain Monte Carlo (MCMC) algorithms have become frequently used in the last years. Chapter 14 gives a short introduction into these methods. It compares them with the random vector generation algorithms of Chap. 11 and discusses how MCMC can be used to generate iid. random vectors. The final Chap. 15 presents some simulation examples for financial engineering and Bayesian statistics to demonstrate, at least briefly, how some of the algorithms presented in the book can be used in practice.

Reader Guidelines

This book is a research monograph as we tried to cover – at least shortly – all relevant universal random variate generation methods found in the literature. On the other hand the necessary mathematical and statistical tools are fairly basic which should make most of the concepts and algorithms accessible for all graduate students with sound background in calculus and probability theory. The book mainly describes the mathematical ideas and concepts together with the algorithms; it is not closely related to any programming language and is generally not discussing technical details of the algorithm that may depend on the implementation. The interested reader can use the source code of UNU.RAN to see possible solutions to the implementation details for most of the important algorithms. Of course the book can also be seen as the "documentation" explaining the deeper mathematics behind the UNU.RAN algorithms. Thus it should also be useful for all users of simulations who want to gain more insight into the way random variate generation works.

In general the chapters collect algorithms that can be applied to similar random variate generation problems. Only the generation of continuous one-dimensional random variates with known density contains so much material that is was partitioned into the Chaps. 4, 5, and 6. Several of the chapters are relatively self contained. Nevertheless, there are some dependencies that lead to the following suggestions: For readers not too familiar with random variate generation we recommend to read Chap. 2 and Chap. 3 before continuing with any of the other chapters. Chapters 4 to 7 may be read in an arbitrary order; some readers may want to consult Chap. 8 first to decide which of the methods is most useful for them. Chapters 9, 12, and 13 are self contained, whereas some sections of Chaps. 10 and 11 are are generalizations of ideas presented in Chaps. 4 and 6. Chapter 14 is self contained, but to appreciate its developments we recommend to have a look at Chap. 11 first. In Chap. 15 algorithms of Chaps. 2, 4, 11 and 12 are used for the different simulations.

Course Outlines

We have included exercises at the end of most of the chapters as we used parts of the book also for teaching courses in simulation and random variate generation. It is obvious that the random variate generation part of a simulation course will mainly use Chaps. 2 and 3. Therefore we tried to give a broad and simple development of the main ideas there. It is possible to add the main idea of universal random variate generation by including for example Sect. 4.1. Selected parts of Chap. 15 are useful for a simulation course as well.

A special course on random variate generation should start with most of the material of Chaps. 2 and 3. It can continue with selected sections of Chap. 4. Then the instructor may choose chapters freely, according to her or the students' preferences: For example, for a course with special emphasis on multivariate simulation she could continue with Chaps. 11 and 14; for a course concentrating on the fast generation of univariate continuous distributions with Chaps. 5, 7, and 8. Parts of Chap. 15 can be included to demonstrate possible applications of the presented algorithms.

What Is New?

The main new point in this book is that we use the paradigm of universal random variate generation throughout the book. Looking at the details we may say that the rigorous treatment of all conditions and all implications of transformed density rejection in Chap. 4 is probably the most important contribution. The second is our detailed discussion of the universal generation of random vectors. The discussion of Markov chain Monte Carlo methods for generating iid. random vectors and its comparison with standard rejection algorithms is new.

Throughout the book we present new improved versions of algorithms: Among them are variants of transformed density rejection, fast numerical inversion, universal methods for increasing hazard rate, indexed search for discrete distributions with unbounded domain, improved universal rejection algorithms for orthomonotone densities, and exact universal algorithms based on MCMC and perfect sampling.

Practical Assumptions

All of the algorithms of this book were designed for practical use in simulation. Therefore we need the simplifying assumption of above that we have a source of truly uniform and iid. (independent and identically distributed) random variates available. The second problem we have to consider is related to the fact that a computer cannot store and manipulate real numbers.

Devroye (1986a) assumes an idealized numerical model of arbitrary precision. Using this model inversion requires arbitrary precision as well and is therefore impossible in finite time unless we are given the inverse cumulative distribution function. On the other hand there are generation algorithms (like the series method) that require the evaluation of sums with large summands and alternating signs, which are exact in the idealized numerical model of arbitrary precision.

If we consider the inversion and the series method implemented in a modern computing environment (e.g. compliant with the IEEE floating point standard) then – using bisection – inversion may be slow but it is certainly possible to implement it with working precision close to machine precision. On the other hand, the series method may not work as – due to extinction – an alternating series might "numerically" converge to wrong values. We do not know the future development of floating point arithmetic but we do not want to use an idealized model that is so different from the behavior of today's standard computing environments. Therefore we decided to include numerical inversion algorithms in this book as long as we can easily reach error bounds that are close to machine precision (i.e. about 10^{-10} or 10^{-12}). We also include the series method but in the algorithm descriptions we clearly state warnings about the possible numerical errors.

The speed of random variate generation procedures is still of some importance as faster algorithms allow for larger sample sizes and thus for shorter confidence intervals for the simulation results. In our timing experiments we have experienced a great variability of the results. They depended not only on the computing environment, the compiler, and the uniform generator but also on coding details like stack variables versus heap for storing the constants etc. To decrease this variability we decided to define the *relative generation time* of an algorithm as the generation time divided by the generation time for the exponential distribution using inversion which is done by

$X \leftarrow -\log(1 - \text{random}())$. Of course this time has to be taken in exactly the same programming environment, using the same type of function call etc. The relative generation time is still influenced by many factors and we should not consider differences of less than 25 %. Nevertheless, it can give us a crude idea about the speed of a certain random variate generation method.

We conclude this introduction with a statement about the speed of our universal generators. The relative generation time for the fastest algorithms of the Chaps. 4, 5 and 7 are not depending on the desired density and are close to one; i.e. these methods are about as fast as the inversion method for exponential variates and they can sample at that speed from many different distributions.

2

General Principles in Random Variate Generation

This chapter is collecting the basic ideas and methods for the generation of continuous non-uniform random variates. They serve as the building blocks of universal algorithms. Most of these methods can also be used for discrete distributions which will be presented in Chap. 3. These methods can also be used as starting points for compiling algorithms to generate random vectors, which will be done in Chap. 11.

Almost all methods that are used in random variate generation can be classified into the following main groups: inversion methods, rejection methods, composition, and algorithms that use special properties of the distribution. As this book is dealing with universal methods we will concentrate on the first three groups here. We present the inversion (Sect. 2.1), the rejection (Sect. 2.2) and the composition principle (Sect. 2.3) as well as the ratio-of-uniforms method (Sect. 2.4) and almost-exact inversion (Sect. 2.5), which can be both seen as variants of the rejection method. In Sect. 2.6 we demonstrate how special properties of the distribution can be utilized to sample from a triangular density and from the normal distribution. There exist many other specialized algorithms for random variate generation like, for example, the acceptance-complement method and the Forsythe-von Neumann method. We refer the interested reader to the book of Devroye (1986a) who gives an excellent and detailed description. Dagpunar (1988) also gives a clear presentation of many specialized methods. General descriptions of random variate generation are also included in many books on simulation. Recent books with a long discussion of random variate generation include Gentle (1998) and Fishman (1996). Combined with Chap. 3 this chapter is well suited for the random variate generation part of a course on simulation.

2.1 The Inversion Method

The *inversion method* is based on the following simple theorem.

2 General Principles in Random Variate Generation

Theorem 2.1. *Let $F(x)$ be a continuous cumulative distribution function (cdf) and U a uniform $U(0,1)$ random number. Then the random variate $X = F^{-1}(U)$ has the cdf F. Furthermore, if X has cdf F, then $F(X)$ is uniformly distributed.*

Proof. Due to the fact that the cdf is monotonically increasing we get:

$$\mathrm{Prob}(X \leq x) = \mathrm{Prob}(F^{-1}(U) \leq x) = \mathrm{Prob}(U \leq F(x)) = F(x)$$

and this precisely means that $F(x)$ is the cdf of X.
The second statement immediately follows from

$$\mathrm{Prob}(F(X) \leq u) = \mathrm{Prob}(X \leq F^{-1}(u)) = F(F^{-1}(u)) = u \ . \quad \square$$

There are problems with the definition of the inverse of the cdf if the cdf is constant over an interval. It is not difficult to prove that for such cdf both statements remain correct if we define the inverse cdf F^{-1} by

$$F^{-1}(u) = \inf\{x \colon F(x) \geq u\}, \quad \text{for } 0 < u < 1 \ .$$

Applying this principle leads to the very simple and short Algorithm 2.1.

Algorithm 2.1 Inversion

Require: Inverse of the cumulative distribution function $F^{-1}(u)$.
Output: Random variate X with cdf F.
1: Generate $U \sim U(0,1)$.
2: Set $X \leftarrow F^{-1}(U)$.
3: **return** X.

Example 2.2. Figure 2.1 illustrates the inversion principle for the standard exponential distribution with cdf and inverse of the cdf

$$F(x) = 1 - \exp(-x), \qquad F^{-1}(u) = -\log(1-u) \ .$$

It is indicated how the uniform variates $U = 0.4$ and $U = 0.95$ are transformed into values close to 0.5 and 3, respectively.

Inversion is clearly a very simple and short method. It is easily applicable, provided that the inverse cdf is known in closed form and routines are available in program libraries that compute it at acceptable costs. Table 2.1 gives some examples for such distributions.

The inversion method also has special advantages that make it attractive for simulation purposes. It preserves the structural properties of the underlying uniform pseudo-random number generator. Consequently it can be used, e.g., for variance reduction techniques, it is easy to sample from truncated

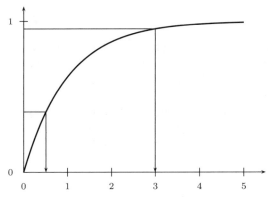

Fig. 2.1. The cdf of the exponential distribution with the inversion method for $U = 0.4$ and $U = 0.95$

Table 2.1. Distributions for which inversion is possible in closed form

distribution	density $f(x)$	cdf $F(x)$	inverse $F^{-1}(u)$
Uniform ($a \leq x \leq b$)	$1/(b-a)$	$(x-a)/(b-a)$	$a + u(b-a)$
Triangular ($0 \leq x \leq a$)	$(1-x/a)(2/a)$	$(x - x^2/(2a))(2/a)$	$a(1 - \sqrt{1-u})$
Exponential ($x \geq 0$)	e^{-x}	$1 - e^{-x}$	$-\log(1-u)$
Cauchy	$1/(\pi(1+x^2))$	$1/2 + \arctan(x)/\pi$	$\tan(\pi(u - 1/2))$
Pareto ($x \geq b > 0$)	$a b^a x^{-a-1}$	$1 - (b/x)^a$	$b/(1-u)^{1/a}$
Weibull ($x \geq 0$)	$b x^{b-1} e^{-x^b}$	$1 - e^{-x^b}$	$(-\log(1-u))^{1/b}$
Logistic	$1/(2 + e^x + e^{-x})$	$1/(1+e^{-x})$	$\log(u/(1-u))$
Hyperbolic secant	See Exercise 2.2		

distributions, from marginal distributions, and order statistics. This is important in the framework of quasi-Monte Carlo computing, since there no other methods that can be used for importance sampling have been developed, yet. Moreover the quality of the generated random variables depends only on the underlying uniform (pseudo-) random number generator. Another important advantage of the inversion method is that we can easily characterize its performance. To generate one random variate we always need exactly one uniform variate and one evaluation of the inverse cdf. So its speed mainly depends on the costs for evaluating the inverse cdf. Hence inversion is often considered as the method of choice in the simulation literature, see e.g. Bratley, Fox, and Schrage (1983).

Unfortunately computing the inverse cdf is, for many important standard distributions (e.g. for normal, student, gamma, and beta-distributions), comparatively expensive. Often no such routines are available in standard program libraries. Then numerical methods for inverting the cdf are necessary, e.g. Newton's method. Such procedures, however, have the disadvantage that they may be slow or not exact, i.e. they compute approximate values. Although, in theory numerical methods work with all distributions where the cdf is available, in practice technical problems have to be considered. Sometimes numerical inversion can only be implemented by a "brute force" method using huge tables. We discuss universal algorithms based on numerical inversion in Sects. 7.1 and 7.2. Sometimes neither the cdf nor its inverse are available and alternative methods are required.

Despite the serious numerical problems the excellent structural properties of inversion serve as a model for more sophisticated methods of non-uniform random variate generation, that are faster and that do not share its technical difficulties.

2.2 The Rejection Method

2.2.1 The Basic Principle

The *rejection method*, often called *acceptance-rejection method*, was suggested by John von Neumann (1951). Since then it has proven to be the most flexible and most efficient method to generate variates from continuous distributions. It is based upon the knowledge of the density function of the distribution. It uses the fundamental property that if a random point (X, Y) is uniformly distributed in the region $\mathcal{G}_f$ between the graph of the density function f and the x-axis then X has density f. We will restate this observation in a more formal language later and start to illustrate this idea with the following example.

Example 2.3. We want to sample random variates from the distribution with density
$$f(x) = \tfrac{1}{2} \sin(x), \quad \text{for } 0 \leq x \leq \pi \,.$$
For this purpose we generate random points that are uniformly distributed in the region between the graph of $f(x)$ and the x-axis, i.e., the shaded region in Fig. 2.2. In general this is not a trivial problem but in this example we can use a simple trick: Sample a random point (X, Y) uniformly in the bounding rectangle $(0, \pi) \times (0, \tfrac{1}{2})$. This is easy since each coordinate can be sampled independently from the respective uniform distributions $U(0, \pi)$ and $U(0, \tfrac{1}{2})$. Whenever the point falls into the shaded region below the graph (indicated by • in Fig. 2.2), i.e. when $Y \leq \tfrac{1}{2} \sin(X)$, we accept it and return X as a random variate from the distribution with density $f(x) = \tfrac{1}{2} \sin(x)$. Otherwise we have to reject the point (indicated by ○), and try again.

It is obvious that this idea works for every distribution with a bounded density on a bounded domain. Moreover it is quite clear that we can use this procedure with any multiple of the density. We call such a function a *quasi-density*. E.g., we can equally well use the quasi-density $f(x) = \sin(x)$ on $[0, \pi]$ in the above example. Algorithm 2.2 (Rejection-from-Uniform-Hat) gives the details.

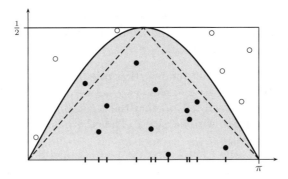

Fig. 2.2. Rejection from a constant hat: density $f(x) = \frac{1}{2}\sin(x)$ on $[0, \pi]$ (solid line) and squeeze $s(x)$ (dashed line). 18 points are generated, 7 are rejected ($\circ$), 11 are accepted ($\bullet$) and their x-coordinates are stored for the output (indicated on x-axis)

Algorithm 2.2 Rejection-from-Uniform-Hat

Require: Quasi-density $f(x)$ on bounded domain (b_l, b_r), upper bound $M \geq f(x)$.
Output: Random variate X with density prop. to f.
1: **repeat**
2:     Generate $U \sim U(0,1)$.
3:     $X \leftarrow b_l + (b_r - b_l)\,U$
4:     Generate $V \sim U(0,1)$.
5:     $Y \leftarrow M\,V$
6: **until** $Y \leq f(X)$.
7: **return** X.

The theoretical foundation for this algorithm is given by the following theorems.

Theorem 2.4. *Let $f(x)$ be an arbitrary density and α some positive constant. If the random pair (X, Y) is uniformly distributed on the set*

$$\mathcal{G}_{\alpha f} = \{(x,y): 0 < y \leq \alpha\,f(x)\}$$

then X is a random variate with density $f(x)$.

Proof. The random pair (X, Y) has by assumption the joint density: $f(x, y) = \frac{1}{\alpha}$ for $(x, y) \in \mathcal{G}_{\alpha f}$ and zero elsewhere. Then the marginal density $f_X(x)$ of X is given by

$$f_X(x) = \int_0^{\alpha f(x)} \frac{1}{\alpha}\,\mathrm{d}t = f(x)$$

which completes the proof. □

Theorem 2.5. *Let (X_1, Y_1), (X_2, Y_2), (X_3, Y_3), ..., be a sequence of iid. random pairs uniformly distributed in a set A. Then for a (measurable) subset $B \subset A$ with $\mathrm{Prob}((X, Y) \in B) > 0$ the subsequence of all pairs $(X_i, Y_i) \in B$ is a sequence of iid. pairs uniformly distributed in B.*

Proof. It is quite obvious that this is a sequence of iid. pairs of random numbers. For the exact formal proof of the uniformity in B some measure theory is required (see Devroye, 1986a). □

Theorem 2.5 states that by the rejection principle all the accepted pairs are uniformly distributed over the region $\mathcal{G}_f$ between (quasi-) density f and the x-axis. Theorem 2.4 then tells us that the x-coordinate of these accepted pairs follows a distribution with density f. Thus the validity of rejection from a constant hat is proven.

From Fig. 2.2 we easily can see that the performance of the algorithm depends heavily on the area of the enveloping rectangle. Moreover the method does not work if the target distribution has infinite tails. Hence non-rectangular shaped regions for the envelopes are important and we have to solve the problem of sampling uniformly from such domains. Looking again at Example 2.3 we notice that the x-coordinate of the random point (X, Y) was sampled by inversion from the uniform distribution on the domain of the given density. This motivates us to replace the quasi-density of the uniform distribution by the quasi-density $h(x)$ of some other appropriate distribution. We only have to take care that it is chosen such that it is always an upper bound, i.e.,

$$h(x) \geq f(x), \quad \text{for all } x \in D \, (\subseteq \mathbb{R}) \, .$$

The following theorem verifies this idea. It shows the converse direction of Thm. 2.4.

Theorem 2.6. *If X is a random variate with density $g(x)$ and Y a uniform variate on $(0, \alpha\, g(X))$, for some constant $\alpha > 0$, then the random pair (X, Y) is uniformly distributed on the set*

$$\mathcal{G}_{\alpha\, g} = \{(x, y) : 0 < y \leq \alpha\, g(x)\} \, .$$

Proof. By assumption the pair (X, Y) has the marginal distribution $h_X(x) = g(x)$ and the conditional distribution of Y given X is uniform on $[0, \alpha g(X)]$. Thus the two-dimensional distribution of (X, Y) is uniquely determined and it can only be the uniform distribution on $\mathcal{G}_{\alpha\, g}$ due to Thm. 2.4. □

Now we are ready for the standard rejection algorithm. We need a density $f(x)$ and a *dominating function* or *hat function* $h(x) \geq f(x)$. Algorithm 2.3 summarizes the idea.

A common method to find a hat h for a quasi-density f is by choosing some density g and finding a constant $\beta > 0$ such that $\beta\, g(x) \geq f(x)$ for all x in the domain of f. We then set $h(x) = \beta\, g(x)$ for the hat. Clearly the rejection algorithm will only result in a simplification if generation from the quasi-density $h(x)$ is easier than that from $f(x)$. For simplicity we will call the distribution with density proportional to the hat function the *hat distribution*. In Chaps. 4 and 6 we will introduce methods for constructing such hat functions h automatically. In almost all cases $h(x)$ is selected such

Algorithm 2.3 Rejection (Basic method)

Require: Quasi-density $f(x)$, hat function $h(x) \geq f(x)$, method to sample from hat.
Output: Random variate X with density prop. to f.
1: **repeat**
2:     Generate X with density prop. to h.
3:     Generate $V \sim U(0,1)$.
4:     $Y \leftarrow V\,h(X)$.
5: **until** $Y \leq f(X)$.
6: **return** X.

---

that we can generate from its distribution by inversion (Algorithm 2.4). If this is the case we can easily generate X with density $h(x)$ and a $U(0,1)$ uniform random number U. By Thm. 2.6 the random pair $(X, V\,h(X))$ is then uniformly distributed below the graph of $h(x)$ and X is accepted as a sample from f if $(X, V\,h(X))$ is below $f(X)$, i.e., if the acceptance condition $V\,h(X) \leq f(X)$ is fulfilled.

Algorithm 2.4 Rejection-with-Inversion

Require: Quasi-density $f(x)$, hat function $h(x)$, inverse cdf $H^{-1}(x)$ for hat.
Output: Random variate X with density prop. to f.
1: **repeat**
2:     Generate $U \sim U(0,1)$.
3:     $X \leftarrow H^{-1}(U)$.
4:     Generate $V \sim U(0,1)$.
5:     $Y \leftarrow V\,h(X)$.
6: **until** $Y \leq f(X)$.
7: **return** X.

Example 2.7. We consider the half-normal distributions with quasi-density

$$f(x) = \exp(-x^2/2), \quad \text{for } x \geq 0\,.$$

Figure 2.3 shows the density of this distribution together with an exponential hat function. To be precise we use the exponential distribution with expectation one as the hat distribution. So we have $g(x) = \exp(-x)$ and $h(x) = \beta \exp(-x)$. We try to choose β such that $f(x) \leq h(x)$. It is intuitively clear that we should use the smallest possible β which we can find by using

$$\beta = \sup_{x \geq 0} \left(\frac{f(x)}{g(x)} \right)\,.$$

By standard calculus we thus get for our example $\beta = \sqrt{e}$ and therefore $h(x) = \sqrt{e}\exp(-x)$. To speed up the algorithm the general acceptance condition

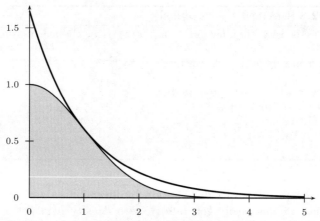

Fig. 2.3. Rejection for the half-normal distribution using an exponential hat function

$V h(X) \leq f(X)$ can be rewritten as $\log V \leq X - X^2/2 - 1/2$. Thus applying rejection with inversion we arrive at the following algorithm:

Output: Half-normal distributed random variate X.
1: **repeat**
2:     Generate $U \sim U(0,1)$.
3:     $X \leftarrow -\log(1-U)$. /* Generate from exponential hat */
4:     Generate $V \sim U(0,1)$.
5: **until** $\log V \leq X - X^2/2 - 1/2$.
6: **return** X.

2.2.2 The Squeeze Principle

If we have a closer look at Algorithm 2.3 (Rejection) it is obvious that evaluating the acceptance condition is a central and, for many distributions, an expensive step. Therefore we can speed up the rejection algorithm considerably if we find easy to evaluate lower bounds $s(x)$ (so called *squeezes*) for the density $f(x)$, i.e.,

$$s(x) \leq f(x), \quad \text{for all } x \in D \, (\subseteq \mathbb{R}) \, .$$

If the pair $(X, V h(X))$ is below the squeeze $s(x)$ we can accept X without evaluating the density. Algorithm 2.5 contains all details.

Example 2.8. A simple example for a squeeze for the quasi-density $f(x) = \sin(x)$ on $[0, \pi]$ in Example 2.3 we can use (see dotted lines in Fig. 2.2)

$$s(x) = \tfrac{1}{2} - \left| \tfrac{1}{2} - \tfrac{x}{\pi} \right|, \quad \text{for } 0 \leq x \leq \pi \, .$$

2.2 The Rejection Method

Algorithm 2.5 Rejection-with-Squeeze

Require: Quasi-density $f(x)$, hat function $h(x)$, squeeze function $s(x)$,
    method to sample from hat.
Output: Random variate X with density prop. to f.
1: **loop**
2:    Generate X with density prop. to h.
3:    Generate $V \sim U(0, 1)$.
4:    $Y \leftarrow V\,h(X)$.
5:    **If** $Y \leq s(X)$ **then return** X. /* evaluate squeeze */
6:    **If** $Y \leq f(X)$ **then return** X. /* evaluate density */

In the literature on random variate generation algorithms, easy to evaluate upper bounds for the density are called squeezes as well. As it is better to use good-fitting hat functions instead we do not explain the use of such upper bounds here.

2.2.3 Performance Characteristics

It is not difficult to analyze the rejection algorithm: As we will prove in the below Thm. 2.9 the key parameter is the *acceptance probability* $\frac{1}{\alpha}$ or its reciprocal value α called the *rejection constant*. It is calculated as the ratio $\alpha = A_h/A_f$ of the area below the hat and the area below the density. As we can see from Algorithm 2.3 (Rejection) the area A_f below the density f is not required to generate random variates by rejection. A_f is not equal to one for quasi-densities and might be unknown. Thus we introduce the ratio $\varrho_{hs} = A_h/A_s$ of the area below the hat and the area below the squeeze as convenient upper bound to the rejection constant, which can be used when A_f is not known. As we will see in Chap. 4 it can be used as a parameter to control the performance of black-box algorithms. It is not difficult to prove the following proposition.

Theorem 2.9. *For a quasi-density f and the rejection Algorithm 2.5 (Rejection-with-Squeeze) denote the areas below the squeeze, below the quasi-density and below the hat by A_s, A_f, and A_h, respectively. Set*

$$\alpha = \frac{A_h}{A_f} \quad \text{and} \quad \varrho_{hs} = \frac{A_h}{A_s}$$

and denote the number of iterations till acceptance with I. Then we have

$$\text{Prob}(I = i) = \tfrac{1}{\alpha}\left(1 - \tfrac{1}{\alpha}\right)^{i-1}, \qquad \text{E}(I) = \alpha \leq \varrho_{hs}, \qquad \text{Var}(I) = \alpha(\alpha - 1)\,.$$

Denote the number of evaluations of the density f by $\#f$. Then the expected number $\text{E}(\#U)$ of uniform variates needed to generate one non-uniform variate and the expected number $\text{E}(\#f)$ of evaluations of the density is given by

$$\mathrm{E}(\#U) = 2\,\alpha, \qquad \mathrm{E}(\#f) = \frac{A_{h-s}}{A_f} \leq \varrho_{hs} - 1$$

where $A_{h-s} = A_h - A_s$.

Proof. For the rejection algorithm the probability that the variate is generated in one trial is called the acceptance probability. As the generated pairs (X, Y) are uniformly distributed on $\mathcal{G}_h$, the region below the hat, the probability of acceptance (i.e. the probability that the pair is below the density) is equal to the ratio of the areas A_f/A_h which is equal to $\frac{1}{\alpha}$ by the definition of α. As the uniform random numbers used are assumed to be independent we can see that I follows a geometric distribution with parameter $1/\alpha$; thus the expectation and variance of I are calculated from the respective formulas for the geometric distribution. The inequality $\alpha \leq \varrho_{hs}$ follows immediately from the fact that $A_s \leq A_f$.

The result for $\mathrm{E}(\#U)$ is now trivial as we need two uniform variates in every iteration of the loop. For the last equation notice that the the number of the evaluations of the density (i.e. $\#f$) is $I-1$ with probability A_s/A_f (this is the case when the finally accepted pair is below the squeeze $s(x)$); otherwise (i.e. with probability $1 - A_s/A_f$) we have $\#f = I$ (the case where the finally accepted pair is above $s(x)$). So $\#f$ is a mixture of I and $I-1$. We find

$$\mathrm{E}(\#f) = (\mathrm{E}(I) - 1) A_s/A_f + \mathrm{E}(I)(1 - A_s/A_f)$$

and replacing $\mathrm{E}(I)$ with A_h/A_f completes the proof. □

2.2.4 Finding (Good) Hat Functions

A hat function must be an upper bound for the density. In practice it is only useful if it is possible to generate random variates from its corresponding distribution by inversion (if a fast implementation of the inverse cdf is available). Thus the distributions listed in Table 2.1 are the first candidates. For a given density f on domain D we can construct a hat function by means of a density g by setting

$$h(x) = \beta\, g(x)$$

where

$$\beta \geq \sup_{x \in D} \frac{f(x)}{g(x)}.$$

Such a constant $\beta > 0$ ensures that this h is a dominating function for the density f. Notice that β is identical to the ratio α of Thm. 2.9 when f and g are densities. For bounded densities on bounded domains (intervals) any quasi-density g that is bounded away from 0 could be used. Otherwise some difficulties may arise. For densities with unbounded domains we have to choose a density g that has tails as least as heavy as the density f of the target distribution. If $f(x)$ is unbounded then g must have its "infinite peaks" at the same location as f and they must not be as "sharp" as those of f.

Among these possible hat functions we should select one that has an acceptance probability $\frac{1}{\alpha}$ close to one. Devroye (1986a) classifies the approaches to construct hat functions into two main groups: one is based on inspecting the shape of the density and applying analytical devices to obtain inequalities; the second approach starts from a parametric family of possible hat functions that have about the same shape as the density and optimizes the parameters such that the acceptance probability is maximized.

As methods for constructing hat functions are introduced in Chaps. 4 and 6 we will demonstrate these ideas here only for a few examples. First we will illustrate the latter method.

Optimizing the Parameters of a Hat Function

To find the optimal parameters for the hat function it may be necessary to solve difficult optimization problems. We look here at one simple example that is of little practical value but nicely illustrates the principle.

Example 2.10. The density of the gamma(2) distribution is given by

$$f(x) = x \exp(-x), \quad \text{for } x \geq 0 \ .$$

We use the family of exponential distributions with quasi-densities $g_\lambda(x) = \exp(-\lambda x)$ to construct a suitable hat function. We first have to find the value of β and the acceptance probability for an arbitrary fixed parameter $\lambda > 0$ and get

$$\beta = \sup_{x \geq 0} \left(\frac{x e^{-x}}{e^{-\lambda x}} \right) = \sup_{x \geq 0} \left(x e^{x(\lambda - 1)} \right)$$

which is bounded for $0 < \lambda < 1$. Taking the logarithm and very simple calculus shows us that a maximum is attained for $x = \frac{1}{1-\lambda}$ and we find

$$\beta = \frac{1}{e(1-\lambda)} \quad \text{and} \quad h(x) = \frac{e^{-\lambda x}}{e(1-\lambda)} \ .$$

It is then easy to compute the rejection constant α depending on λ as

$$\alpha(\lambda) = \frac{A_h}{A_f} = \frac{1}{e\,\lambda(1-\lambda)} \ .$$

To minimize $\alpha(\lambda)$ we vary λ and find: $\lambda = \frac{1}{2}$ and $\alpha = \frac{4}{e}$. Figure 2.4 shows the density and the optimal exponential hat.

Using Inequalities to Construct Hat Functions

There are lots of examples of this method in the literature, especially in the book of Devroye (1986a). The automatic construction of hat functions as described in Chaps. 4 and 6 would also fall into this subsection. We demonstrate this method here for the same density as above.

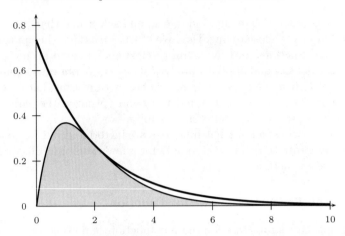

Fig. 2.4. Optimal exponential hat for the gamma(2) distribution with density $f(x) = x \exp(-x)$

Example 2.11. We want to fit an exponential hat $h(x) = \exp(a + bx)$, $b < 0$ that touches the density $f(x) = x \exp(-x)$, $x \geq 0$, of the gamma(2) distribution in a point x_0. Hence we have to solve the two equations

$$h(x_0) = f(x_0) \text{ and } h'(x_0) = f'(x_0).$$

Taking the logarithm for both of them we get

$$a + b x_0 = \log x_0 - x_0 \text{ and } \log b + a + b x_0 = -x_0 + \log(1 - x_0)$$

with the solution $b = (1-x_0)/x_0$ and $a = \log(x_0)-1$. Thus for a given point of contact x_0 we have constructed the exponential function $h(x) = \exp(a + b x)$ that touches f. What remains, however, is to show that this function is a hat function, i.e., an upper bound for the density. For fixed x_0 we could easily check this by plotting the graph of the density and of the hat. For a general exact proof recall that $x \exp(-x)$ is log-concave, i.e., its logarithm is a concave function. This implies that an exponential hat touching in a single point is an upper bound for all x. To find the x_0 minimizing the rejection constant α we compute

$$\alpha = \frac{A_h}{A_f} = \frac{e^a}{-b} = \frac{x_0^2}{e(x_0 - 1)}.$$

For the optimal hat we then have $x_0 = 2$ and $\alpha = \frac{4}{e}$, which is the same as the optimal exponential hat we have constructed in the above example (cf. Fig. 2.4).

Up to now it is a matter of taste which of the two methods to construct the hat function is preferred. We will see in Chap. 4 below that the second method is closely linked with transformed density rejection.

2.2 The Rejection Method

Example 2.12. Consider the distribution with density proportional to $f(x) = 1/x$ for $1 \leq x \leq 3$. This function is monotonically decreasing and thus it is easy to construct a stair-case shaped hat function h with $h(x) = f(1) = 1$ for $1 \leq x \leq x_0$ and $h(x) = f(x_0) = 1/x_0$ for $x_0 \leq x \leq 3$. Then $A_h = (x_0 - 1) + (3 - x_0)/x_0$ and it is no problem to find that A_h (and therefore α) is minimized for $x_0 = \sqrt{3}$. This is the optimal hat if we minimize α and thus the expected number of iterations of the rejection algorithm.

If we add the trivial squeeze $s(x) = f(x_0) = 1/x_0$ for $1 \leq x \leq x_0$ and $s(x) = f(3) = 1/3$ for $x_0 \leq x \leq 3$ it may be interesting to minimize the expected number of evaluations of f. Theorem 2.9 implies that we then have to minimize the area between hat and squeeze A_{h-s} instead of A_h. For our example we can easily compute

$$A_{h-s} = (1 - \tfrac{1}{x_0})(x_0 - 1) + (\tfrac{1}{x_0} - \tfrac{1}{3})(3 - x_0)$$

which is minimized for $x_0 = \sqrt{3}$. Hence for $f(x) = 1/x$ minimizing A_h and A_{h-s} leads to the same result. But this is only due to the density we have chosen. Figure 2.5 (l.h.s.) shows the density of this example with optimal hat and squeeze.

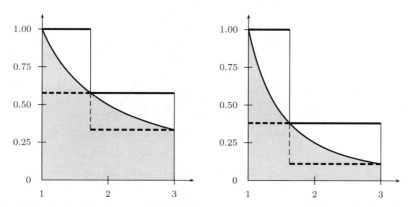

Fig. 2.5. Optimal stair-case shaped hat and squeeze for $f(x) = 1/x$ (l.h.s.) and $f(x) = 1/x^2$ (r.h.s.)

Example 2.13. Now consider the distribution with density proportional to $f(x) = 1/x^2$ for $1 \leq x \leq 3$. Again we want to use a stair-case function with two steps to construct a hat function. Using the same idea as in the above example we find $x_0 = 1.634\ldots$ if we minimize the rejection constant α numerically; whereas when minimizing the expected number of evaluations of the the density, i.e. A_{h-s}, we obtain $x_0 = 1.623\ldots$. Figure 2.5 (r.h.s.) shows the density of this example with hat and squeeze minimizing A_{h-s}.

Exercise 2.9 contains another example that shows the difference between optimizing A_h and A_{h-s}. If we decompose the domain of the distribution into more than two subintervals and use piece-wise constant hat and squeeze we get a fast automatic algorithm for monotone densities with bounded support called "Ahrens Method" (see Sect. 5.1).

2.3 Composition

The composition method is an important principle to facilitate and speed up random variate generation. The basic idea is simple and can be described by means of Fig. 2.6. To generate random variates with a given density we first split the shaded region below the density into three parts. Then select one of these randomly with probabilities given by their respective areas. Finally generate a random variate from the density of the selected part by inversion and return it as random variate of the full distribution.

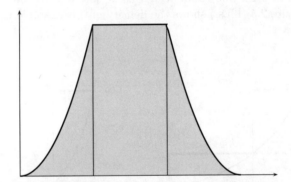

Fig. 2.6. The density is decomposed into three parts

We can restate this idea in a formal language. If our target density f can be written as a discrete finite mixture

$$f(x) = \sum_{i=1}^{n} w_i \, f_i(x)$$

where the f_i's are given density functions and the w_i's form a probability vector (i.e., $w_i > 0$ for all i and $\sum w_i = 1$), then random variates can be obtained by generating a discrete random variate I with probabilities w_i first to decide which part f_i should be chosen. Then a random variate with density f_I is generated and returned. Algorithm 2.6 (Composition) contains the details. Generators for discrete random variates with arbitrary probability vector are described in Sects. 3.1 and 3.2 below.

The composition method is especially useful when the distribution we are interested in is itself defined as a mixture, as it is the case for example for

Algorithm 2.6 Composition

Require: Decomposition of density $f(x) = \sum_{i=1}^{n} w_i f_i(x)$; generators for each f_i.
Output: Random variate X with density f.
1: Generate discrete random variate J with probability vector (w_i).
2: Generate X with density f_J.
3: **return** X.

---

normal mixtures (see Exercise 2.11). For other distributions it is rarely possible to find a simple discrete mixture that is exactly the same as the desired distribution. Nevertheless decomposition of the area below the density into triangles and rectangles has been used to design fast but complicated generators for special distributions. This technique is sometimes called *patchwork method*, see Kinderman and Ramage (1976) or Stadlober and Zechner (1999) for examples.

2.3.1 Composition-Rejection

A very powerful application of the composition principle, however, is partitioning the domain of the distribution into subintervals. To obtain the probabilities w_i we need to know the area below the density for these intervals and it is necessary to evaluate the cdf at the boundaries of the intervals which is often not possible. But it is easy for the hat distribution of a rejection algorithm since usually the cdf of the hat is known as the inversion method is used for sampling. Algorithm 2.7 (Composition-Rejection) gives the formal description of this idea which will be very important for universal methods. Here we use the inversion method to sample from the hat distribution. Examples 2.12 and 2.13 can be seen as simple examples of this idea.

Algorithm 2.7 Composition-Rejection

Require: Quasi-density $f(x)$ on domain (b_l, b_r);
   partition of domain into N intervals (b_{i-1}, b_i), $b_l = b_0 < b_{i-1} < b_i < b_N = b_r$;
   hat $h_i(x)$, inverse cdf of hat $H_i^{-1}(x)$, and probabilities w_i prop. to areas below h_i for each interval (b_{i-1}, b_i).
Output: Random variate X with density prop. to f.
1: **loop**
2:   Generate discrete random variate J with probability vector (w_i).
     /* Use Alg. 3.1 or 3.2 */
3:   Generate $U \sim U(0, 1)$.
4:   $X \leftarrow H_J^{-1}(U)$.
5:   Generate $Y \sim U(0, h_J(X))$.
6:   **if** $Y \leq f(X)$ **then**
7:     **return** X.

Example 2.14. (Continuation of Example 2.12)
Looking at Fig. 2.5 (l.h.s.) it is not difficult to collect all informations necessary to spell out the details of Algorithm 2.7 (Composition-Rejection) for this simple special case with constant hat. We also include the simple constant squeezes shown in Fig. 2.5.

Output: Generates variates with quasi-density $1/x$ on the interval $(1,3)$.
1: Set $b_0 \leftarrow 1$, $b_1 \leftarrow \sqrt{3}$, $b_2 \leftarrow 3$ and $\omega_1 \leftarrow \sqrt{3} - 1$, $\omega_2 \leftarrow (3 - \sqrt{3})/\sqrt{3}$.
   /* Notice that $\omega_1 = \omega_2$ */
2: Set $h_1 \leftarrow 1$, $h_2 \leftarrow 1/\sqrt{3}$ and $s_1 \leftarrow 1/\sqrt{3}$, $s_2 \leftarrow 1/3$.
3: **loop**
4:   Generate $U \sim U(0, \omega_1 + \omega_2)$.
5:   **if** $U < \omega_1$ **then**
6:     Set $J \leftarrow 1$.
7:   **else**
8:     $J \leftarrow 2$.
9:   Generate $U \sim U(0,1)$.
10: Set $X \leftarrow b_{J-1} + U(b_J - b_{J-1})$.
11: Generate $V \sim U(0,1)$.
12: $Y \leftarrow V h_J$.
13: **if** $Y \leq s_J$ **then** /* evaluate squeeze */
14:    **return** X.
15: **if** $Y \leq 1/X$ **then** /* evaluate density */
16:    **return** X.

2.3.2 Recycling Uniform Random Numbers

The advantage of Algorithm 2.7 (Composition-Rejection) is that we can use a simple local hat in every interval but still can expect a good fit and a small rejection constant α when we use sufficiently many intervals. The disadvantage of course is the additional discrete random variate we have to generate. Note that the generation can be done quite fast using the techniques explained in Sects. 3.1 and 3.2 below. When we use discrete inversion (Algorithms 3.1, Sequential-Search, and 3.2, Indexed-Search), I is returned if the uniform random number U generated in the first step of these algorithms falls into the interval $[F(I-1), F(I))$, where $F(i) = \sum_{k \leq i} w_i$ denotes the cdf of the discrete distribution. Since U is also uniformly distributed over this interval, we get a new $U(0,1)$ random variate, which is independent from I, by the linear transformation
$$U' = \frac{U - F(I-1)}{F(I) - F(I-1)}.$$

Thus we can save the call to the uniform random number generator in Step 3 by this technique. This method is called *recycling* of uniform random numbers in the literature. Since available pseudo-random numbers are not really continuous but have some "resolution" this technique should be used with care because of the lost precision in very short intervals.

In Algorithm 2.7 (Composition-Rejection) generation from the hat distribution in each interval is always realized by inversion. Thus considering the recycling principle leads to the following important observation:
Using inversion to generate the discrete random variate I and then the recycling principle to get an uniform random number U in Step 3 is equivalent to using the inversion method with the global hat function, that is defined as

$$h(x) = \sum_{i=1}^{N} h_i(x)\, \mathbf{1}_{(b_{i-1},b_i)} \,.$$

Instead of inversion to generate the discrete random variate we need a search procedure to find the correct interval of the hat-function. For this algorithm we do not need the composition method explicitly. This is the reason that we were able to introduce two applications of Algorithm 2.7 with constant hat in Examples 2.12 and 2.13. As a simple example and for the sake of completeness we spell out the details of the generator of Example 2.14 with recycling.

Example 2.15. Continuation of Example 2.14. See Fig. 2.5 (l.h.s.).
Output: Generates variates with quasi-density $1/x$ on the interval $(1,3)$.
1: Set $b_0 \leftarrow 1$, $b_1 \leftarrow \sqrt{3}$, $b_2 \leftarrow 3$ and $\omega_1 \leftarrow \sqrt{3} - 1$, $\omega_2 \leftarrow (3 - \sqrt{3})/\sqrt{3}$.
 /* Notice that $\omega_1 = \omega_2$ */
2: Set $h_1 \leftarrow 1$, $h_2 \leftarrow 1/\sqrt{3}$ and $s_1 \leftarrow 1/\sqrt{3}$, $s_2 \leftarrow 1/3$.
3: **loop**
4:   Generate $U \sim U(0, \omega_1 + \omega_2)$.
5:   **if** $U < \omega_1$ **then**
6:     Set $J \leftarrow 1$.
7:     Set $U \leftarrow U/\omega_1$. /* U is the recycled $U(0,1)$ variate. */
8:   **else**
9:     $J \leftarrow 2$.
10:    Set $U \leftarrow (U - \omega_1)/\omega_2$. /* U is the recycled $U(0,1)$ variate. */
11: Set $X \leftarrow b_{J-1} + U\,(b_J - b_{J-1})$.
12: Generate $V \sim U(0,1)$.
13: $Y \leftarrow V\,h_J$.
14: **if** $Y \leq s_J$ **then** /* evaluate squeeze */
15:    return X.
16: **if** $Y \leq 1/X$ **then** /* evaluate density */
17:    return X.

It is easy to see that for this algorithm $A_h = 1.4641$ and $A_s = 0.8453$. A simple integration yields $A_f = \log 3 = 1.0986$; we have $\alpha = 1.3327$ and $\varrho_{hs} = 1.7321$. Using Thm. 2.9 we can calculate the expected number of uniforms required $\mathrm{E}(\#U) = 2\,\alpha = 2.6654$ and the expected number of evaluations of the density $\mathrm{E}(\#f) = \frac{A_{h-s}}{A_f} = 0.5633$.

2.3.3 Immediate Acceptance

It is no problem to add squeezes to Algorithm 2.7 (Composition-Rejection). This will reduce the number of evaluations of the density f but leaves the expected number of used uniform random numbers unchanged. Adding a squeeze s_i in an interval (b_{i-1}, b_i) splits the region below the hat h_i into two parts and it is possible to decompose the density f further by $h_i(x) = s_i(x) + (h_i(x) - s_i(x))$. Figure 2.7 illustrates the situation. It is obvious that any random point (X, Y) generated by the algorithm that falls into the lower part below the squeeze can be accepted immediately without evaluating the density. If X is generated (by inversion) from the distribution with density proportional to the squeeze s_i, it is not even necessary to generate Y.

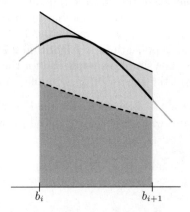

Fig. 2.7. The squeeze (dashed line) splits region below the hat into two parts. In the lower part a random point can be accepted immediately

A possible algorithm using this idea modifies Algorithm Composition-Rejection in the following way: make a refinement of the decomposition by splitting each of the regions in the intervals into the region above and below the squeeze. Use a discrete random variate to select one of the intervals and additionally select the upper or lower part. If the part below the squeeze has been chosen, sample from the distribution with density proportional to the squeeze and return the generated point immediately. If the part between hat and squeeze has been chosen, sample from the distribution with density proportional to $h(x) - s(x)$. Sample a uniform random number Y on $[s(X), h(X)]$ and accept if $Y \leq f(X)$, otherwise reject and try again with a new discrete random variate.

Unfortunately it is often very difficult to sample from the distribution with density proportional to $h(x) - s(x)$. The best and easiest solution to this problem is to use a squeeze that is proportional to the hat, i.e., in each interval (b_{i-1}, b_i) we set $s_i(x) = \nu_i h_i(x)$ for a suitable constant $0 \leq \nu_i \leq 1$. Algorithm 2.8 (Rejection-Immediate-Acceptance) shows the details.

Since it is not obvious that this algorithm works we include a formal proof for completeness.

2.3 Composition 31

Algorithm 2.8 Rejection-Immediate-Acceptance

Require: Quasi-density $f(x)$ on domain (b_l, b_r);
  partition of domain into N intervals (b_{i-1}, b_i), $b_l = b_0 < b_{i-1} < b_i < b_N = b_r$;
  hat $h_i(x)$, inverse cdf of hat $H_i^{-1}(x)$, and probabilities w_i prop. to areas below
  h_i for each interval (b_{i-1}, b_i);
  constants ν_i such that $s_i(x) = \nu_i h_i(x) \leq f(x)$ for $b_{i-1} \leq x \leq b_i$.
Output: Random variate X with density prop. to f.

1: **loop**
2:   Generate discrete random variate J with probability vector (w_i).
      /* Use Alg. 3.1 or 3.2 */
3:   Generate $U \sim U(0,1)$. /* use recycling technique */
4:   **if** $U \leq \nu_J$ **then** /* below squeeze, immediate acceptance */
5:     $X \leftarrow H_J^{-1}(U/\nu_J)$.
6:     **return** X.
7:   **else** /* rejection step */
8:     $X \leftarrow H_J^{-1}((U - \nu_J)/(1 - \nu_J))$.
9:     Generate $Y \sim U(\nu_J h_J(X), h_J(X))$.
10:    **if** $Y \leq f(X)$ **then**
11:      **return** X.

Theorem 2.16. *Algorithm 2.8 (Rejection-Immediate-Acceptance) works.*

Proof. $\hat{h}_i(x) = h_i(x)/w_i$ is a density function and we have a decomposition of the hat h given by $h(x) = \sum_{i=1}^{N} \left(\nu_i w_i \hat{h}_i(x) + (1 - \nu_i) w_i \hat{h}_i(x) \right)$. Since U/ν_J and $(U - \nu_J)/(1 - \nu_J)$ are $U(0,1)$ random numbers, the algorithm generates random variates from the corresponding hat distribution in the respective Steps 5 and 8. Thus if we add an additional Step 5a "Generate $Y \sim U(0, \nu_J h_J(X))$." we generate pairs (X, Y) uniformly distributed below h_i that are accepted if $Y \leq f(X)$. Hence this algorithm is correct as it is a rejection algorithm and omitting Step 5a does not change the output. □

At last we notice that this algorithm reduces the expected number of used uniform random numbers. If the ratio ϱ_{hs} is close to one, then this number is almost reduced to one half.

Theorem 2.17. *For Algorithm 2.8 (Rejection-Immediate-Acceptance) we denote the areas below the squeeze, below the density and below the hat by A_s, A_f, and A_h, respectively. Set $\alpha = \frac{A_h}{A_f}$ and $\varrho_{hs} = \frac{A_h}{A_s}$ and denote the number of evaluations of the density f by $\#f$. Then the expected number $\mathrm{E}(\#U)$ of uniform variates needed to generate one non-uniform variate and the expected number $\mathrm{E}(\#f)$ of evaluations of the density is given by*

$$\mathrm{E}(\#U) = 2\alpha - A_s/A_f \leq 2\varrho_{hs} - 1/\varrho_{hs}, \qquad \mathrm{E}(\#f) = \frac{A_{h-s}}{A_f} \leq \varrho_{hs} - 1 \ .$$

Proof. Denote the number of iterations till acceptance by I. Since this algorithm is just a sophisticated version of a rejection algorithm with squeeze (see proof of Thm. 2.16) we have by Thm. 2.9, $\mathrm{E}(I) = \alpha = A_h/A_f$ and $\mathrm{E}(\#f) = A_{h-s}/A_f$. Then the number of required uniform random numbers $\#U$ is given $(2I - 1)$ with probability A_s/A_f and $2I$ with probability $(1 - A_s/A_f)$. Hence using $\mathrm{E}(I) = A_h/A_f$ we find for the expectation $\mathrm{E}(\#U) = A_s/A_f\,(2\,\mathrm{E}(I) - 1) + (1 - A_s/A_f)\,2\,\mathrm{E}(I) = (2\,A_h - A_s)/A_f = 2A_h/A_f - A_s/A_f \leq 2A_h/A_s - A_s/A_h = 2\,\varrho_{hs} - 1/\varrho_{hs}$. □

We conclude this section with a simple example for Algorithm 2.8.

Example 2.18. Continuation of Example 2.15. See Figure 2.5 (l.h.s.).
Output: Generates variates with quasi-density $1/x$ on the interval $(1, 3)$.
1: Set $b_0 \leftarrow 1$, $b_1 \leftarrow \sqrt{3}$, $b_2 \leftarrow 3$ and $\omega_1 \leftarrow \sqrt{3} - 1$, $\omega_2 \leftarrow (3 - \sqrt{3})/\sqrt{3}$.
   /* Notice that $\omega_1 = \omega_2$ */
2: Set $h_1 \leftarrow 1$, $h_2 \leftarrow 1/\sqrt{3}$ and $\nu_1 \leftarrow 1/\sqrt{3}$, $\nu_2 \leftarrow 1/\sqrt{3}$.
3: **loop**
4:    Generate $U \sim U(0, \omega_1 + \omega_2)$.
5:    **if** $U < \omega_1$ **then**
6:      Set $J \leftarrow 1$.
7:      Set $U \leftarrow U/\omega_1$. /* U is the recycled $U(0,1)$ variate. */
8:    **else**
9:      $J \leftarrow 2$.
10:     Set $U \leftarrow (U - \omega_1)/\omega_2$. /* U is the recycled $U(0,1)$ variate. */
11:   **if** $U < \nu_J$ **then** /* below squeeze, immediate acceptance */
12:     Set $X \leftarrow b_{J-1} + U\,(b_J - b_{J-1})/\nu_J$.
13:     **return** X.
14:   **else**
15:     Set $X \leftarrow b_{J-1} + (b_J - b_{J-1})(U - \nu_J)/(1 - \nu_J)$.
16:     Generate $Y \sim U(\nu_J h_J, h_J)$.
17:     **if** $Y \leq 1/X$ **then** /* evaluate density */
18:       **return** X.

Again we have $A_h = 1.4641$, $A_s = 0.8453$, $A_f = \log 3 = 1.0986$ and $\alpha = 1.3327$. Using Thm. 2.17 we can calculate the expected number of uniforms required $\mathrm{E}(\#U) = 1.8959$ compared to 2.6654 without immediate acceptance; the expected number of evaluations of the density $\mathrm{E}(\#f) = 0.5633$ is unchanged.

2.4 The Ratio-of-Uniforms Method (RoU)

The construction of an appropriate hat function for the given density is the crucial step for constructing rejection algorithms. Equivalently we can try to find an appropriate envelope for the the region $\mathcal{G}_f$ between the graph of the density and the x-axis, such that we can easily sample uniformly distributed random points. This task could become easier if we can find transformations that map $\mathcal{G}_f$ into a region of more suitable shape.

Example 2.19. Consider the Cauchy distribution with density $f(x) = \frac{1}{\pi(1+x^2)}$. One can find the following simple method:

2.4 The Ratio-of-Uniforms Method (RoU)

Output: Cauchy distributed random variate X.
1: **loop**
2:    Generate $U \sim U(-1,1)$ and $V \sim U(0,1)$.
3:    **if** $U^2 + V^2 \leq 1$ **return** U/V.

To see that this algorithm works notice that it generates uniformly distributed points in a half-disc by rejection and that $U/V = x$ describes a straight line through the origin. The angle between this line and the v-axis is called $\theta = \arctan x$. Hence the probability that $X \leq x$ is proportional to the area in the half-disc between the negative part of the u-axis and this straight line. This area is given by $(\frac{\pi}{2} + \theta) = (\frac{\pi}{2} + \arctan x)$. Hence we find

$$\operatorname{Prob}(X \leq x) = \tfrac{1}{\pi}(\tfrac{\pi}{2} + \arctan x) = \int_{-\infty}^{x} \tfrac{1}{\pi(1+t^2)}\, dt = F(x)$$

where $F(x)$ denotes the cdf of the Cauchy distribution. Thus the algorithm generates Cauchy distributed random variates.

The fundamental principle behind this algorithm is the fact that the region below the density is transformed by the transformation

$$(X, Y) \mapsto (U, V) = (2\, X\sqrt{Y},\, 2\,\sqrt{Y})$$

into a half-disc in such a way that the ratio between the area of the image to the area of the preimage is constant (see the shaded regions in Fig. 2.8). This is due to the fact that that the Jacobian of this transformation is constant.

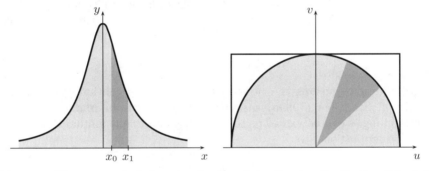

Fig. 2.8. The region $\mathcal{G}_f$ below the density of the Cauchy distribution (l.h.s.) is mapped onto the half-disc $\mathcal{A}_f$ (r.h.s.) such that the ratio between the area of the image and the area of the preimage (shaded region) is constant for all x_0 and x_1

This example is a special case of a more general principle, called the *Ratio-of-uniforms (RoU)* method. It is based on the following theorem.

Theorem 2.20 (Kinderman and Monahan 1977). *Let X be a random variable with quasi-density function $f(x)$ with support (b_l, b_r) not necessarily finite. Let μ be some constant. If (U, V) is uniformly distributed in*

$$\mathcal{A}_f = \{(u,v)\colon 0 < v \leq \sqrt{f(u/v + \mu)},\ b_l < u/v + \mu < b_r\},$$

then $X = U/V + \mu$ has probability density function $f(x)$.

Proof. Consider the transformation

$$\mathbb{R} \times (0, \infty) \to \mathbb{R} \times (0, \infty), \quad (U, V) \mapsto (X, Y) = (U/V + \mu, V^2). \tag{2.1}$$

It maps $\mathcal{A}_f$ one-to-one and onto the region $\mathcal{G}_f = \{(x, y) : 0 < y \leq f(x)\}$. Since the Jacobian of this transformation is 2 the joint density function of X and Y is given by $w(x, y) = 1/(2 |\mathcal{A}_f|)$, if $0 < y \leq f(x)$, and $w(x, y) = 0$ otherwise. (Here $|\mathcal{A}_f|$ denotes the area of $\mathcal{A}_f$.) So the proposition follows from Thm. 2.4. Note that the above arguments also imply $2|\mathcal{A}_f| = |\mathcal{G}_f| = A_f$. □

For most distributions it is best to set the constant μ equal to the mode of the distribution. For sampling random points uniformly distributed in $\mathcal{A}_f$ rejection from a convenient enveloping region is used, usually the *minimal bounding rectangle*, i.e. the smallest possible rectangle that contains $\mathcal{A}_f$. It is given by

$$\mathcal{R}_{\mathrm{mbr}} = \{(u, v) : u^- \leq u \leq u^+, 0 \leq v \leq v^+\}$$

where

$$\begin{aligned} v^+ &= \sup_{b_l < x < b_r} \sqrt{f(x)}, \\ u^- &= \inf_{b_l < x < b_r} (x - \mu)\sqrt{f(x)}, \\ u^+ &= \sup_{b_l < x < b_r} (x - \mu)\sqrt{f(x)}. \end{aligned} \tag{2.2}$$

Algorithm 2.9 (RoU) gives the details. Since this is a rejection algorithms we can use the note at the end of the proof of Thm. 2.20 to see that the rejection constant is equal to

$$\alpha = 2 v^+ (u^+ - u^-)/A_f. \tag{2.3}$$

In addition squeezes can be used to avoid the evaluation of the quasi-density f. Stadlober (1989b) has shown that the ratio-of-uniforms method with rectangles can be viewed as rejection from a table mountain shaped hat function proportional to

$$h(x) = \begin{cases} (u^-/x)^2 & \text{for } x < u^-/v^+, \\ (v^+)^2 & \text{for } u^-/v^+ \leq x \leq u^+/v^+, \\ (u^+/x)^2 & \text{for } x > u^+/v^+. \end{cases}$$

Here we have assumed for simplicity that $\mu = 0$.

To apply Algorithm 2.9 (RoU) to a distribution we can easily use (2.2) to determine the design constants u^-, u^+ and v^+. This simple algorithm works for all distributions with bounded densities that have subquadratic tails (i.e. tails like $1/x^2$ or lower) and for most standard distributions it has quite good rejection constants. (E.g. 1.3688 for the normal and 1.4715 for the exponential distribution.) The details of the RoU algorithm for the normal distribution are posed as Exercise 2.14. The details for the gamma(3) distribution are given in the next example.

Algorithm 2.9 RoU (Ratio-of-uniforms)

Require: Quasi-density $f(x)$ on domain (b_l, b_r);
   constant μ, (minimal) bounding rectangle $(u^-, u^+) \times (0, v^+)$.
Output: Random variate X with density prop. to f.
1: **loop**
2:   Generate $U \sim U(u^-, u^+)$ and $V \sim U(0, v^+)$.
3:   $X \leftarrow U/V + \mu$.
4:   **if** $V^2 \leq f(X)$ **return** X.

Example 2.21. The gamma(3) distribution has the quasi-density function $f(x) = x^2 \exp(-x)$ which has its mode at 2. So it is convenient to use $\mu = 2$. Using (2.2) and numeric optimization we easily find $u^- = -0.6449$ and $u^+ = 1.23602$. As we know that the mode of the density is 2 we get $v^+ = \sqrt{f(2)} = 0.73576$. Using (2.3) and $A_f = 2$ we find for the rejection constant $\alpha = 1.3836$.

It is possible to generalize the idea of Thm. 2.20 by using other transformations as the one in (2.1), see Wakefield, Gelfand, and Smith (1991), and Sect. 6.4.1.

2.5 Almost-Exact Inversion

When we use the rejection method to generate random numbers with density f as described in Sect. 2.2 we need a hat function h for the density. Then we generate a random number X from the hat distribution and a uniform $U(0,1)$ random number Y. If $Y h(X) \leq f(X)$ the point X is accepted, otherwise we have to try again. In almost all cases, however, $h(x)$ is selected such that we can generate from its distribution by inversion, see Algorithm 2.4 (Rejection-with-Inversion) on p. 19. Therefore we can equally well describe the hat distribution by its inverse cdf Ψ. As the density of the hat distribution is then $(\Psi^{-1})'(x)$ the acceptance condition reads

$$Y \beta (\Psi^{-1})'(X) \leq f(X)$$

where the constant β is chosen such that $\beta (\Psi^{-1})'(x)$ is a hat function to the density $f(x)$. Since $(\Psi^{-1})'(x) = 1/\Psi'(u)$ with $x = \Psi(u)$ this can be simplified to

$$Y \leq \tfrac{1}{\beta} f(\Psi(U)) \, \Psi'(U) \qquad (2.4)$$

where U is the uniform random number used to compute X by inversion. This means that it is possible to generate random variates without knowing the density or the cdf of the hat distribution. It is enough to know the inverse of the cdf Ψ of the hat distribution and its derivative. Thus this method allows for hat distributions that cannot be used for ordinary rejection algorithms.

Example 2.22. Wallace (1976) applied this idea to design a generator for the normal distribution. To demonstrate the main principle we restrict the presentation to the half-normal distribution with quasi-density $f(x) = \exp(-x^2/2)$ for $x \geq 0$ (see Fig. 2.9, l.h.s.). Using $\Psi(u) = b\,u\,(1 + c/(1-u))$ with $b = 1.22$ and $c = 0.14$ we arrive at the following algorithm.

Output: Half-normal distributed random variate X.
1: **repeat**
2:    Generate $U \sim U(0,1)$.
3:    $X \leftarrow 1.22\,U\,(1 + 0.14/(1-U))$. /* $\Psi(U)$ */
4:    Generate $Y \sim U(0,1)$.
5: **until** $Y \leq 0.8543034\,\exp(-X^2/2)\,(1 + 0.14/(1-U)^2)$. /* condition (2.4) */
6: **return** X.

It is obvious that this idea can be extended to a very short algorithm for the normal distribution (Exercise 2.16).

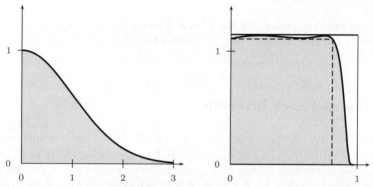

Fig. 2.9. Density of half-normal distribution (l.h.s.) and its conjugate density using $\Psi(u) = 1.22\,u\,(1 + 0.14/(1-u))$ (r.h.s.). The rejection algorithm for sampling from the conjugate density uses constant hat and squeeze

We can describe the same idea from a different point of view: A random variate can be generated as $F^{-1}(U)$ where U is a $U(0,1)$ random variate. If the inverse cdf F^{-1} is not feasible to compute, we could use an approximation Ψ instead. Of course, $\Psi(U)$ does not have the desired distribution. But this is true for $\Psi(Z)$ where Z has the property $\Psi(Z) = F^{-1}(U) = X$. Thus $Z = \Psi^{-1}(X)$ and using the transformation theorem (see Exercise 2.19) we can compute the density t of Z

$$t(z) = f(\Psi(z))\,\Psi'(z)\,. \tag{2.5}$$

We call $t(z)$ the *conjugate density* of f. Figure 2.9 (r.h.s.) shows the conjugate density for Example 2.22. Algorithm 2.10 (Almost-exact-Inversion) based on this principle is called *almost-exact inversion* by Devroye (1986a). A similar idea, called *exact-approximation method*, is proposed by Marsaglia (1984).

Algorithm 2.10 Almost-exact-Inversion

Require: Quasi-density $f(x)$; approximation $\Psi(z)$ to the inverse cdf.
Output: Random variate X with density prop. to f.
1: Generate Z with density $t(z) = f(\Psi(z))\,\Psi'(z)$.
2: $X \leftarrow \Psi(Z)$.
3: **return** X.

Generation from the conjugate density t can be done by a rejection method. Wallace (1976) called this method *transformed rejection*. It is obvious that Z is uniformly distributed if $\Psi = F^{-1}$. If Ψ is close to the inverse cdf then t is close to the uniform density. Thus fast algorithms can be compiled for many distributions using rejection from a constant hat. To avoid the expensive evaluation of the conjugate density t it is also possible to use a constant squeeze (see Hörmann and Derflinger (1994a) and Ahrens and Dieter (1988) for examples). Compiling such an algorithm is straightforward and is left to the reader (Exercise 2.15).

There is also a third point of view for this method. Consider the map

$$(X,Y) \mapsto (U,V) = (\Psi^{-1}(X), Y\, t(\Psi^{-1}(X))/f(X))$$

where f is the density of the desired distribution and where t is the conjugate density as defined in (2.5). It maps the region $\mathcal{G}_f$ between the graph of the density and the x-axis into the region $\mathcal{G}_t$ between the graph of the conjugate density and the u-axis (the shaded regions in Fig. 2.9). Its Jacobian is constant and equal to one (see Exercise 2.18). Hence the area of the image of any region equals to the area of its preimage. Thus almost-exact inversion can be seen as another method to transform the unbounded region between the graph of the density and the x-axis into some bounded convenient region for which it is easier to find hat function and squeeze. In fact this method together with the inversion method, the ratio-of-uniforms method and its generalizations can be described by a general framework using transformations with constant Jacobians (Jones and Lunn, 1996).

2.6 Using Special Properties of the Distribution

For many important distributions we can exploit special properties to obtain simple and elegant random variate generators. Obviously algorithms based on special properties will not lead to automatic algorithms for large families of distributions. Thus we only give two examples for this principle here.

The Symmetric Triangular Distribution.

The symmetric triangular distribution between 0 and 2 has density $f(x) = x$ on $(0,1)$ and $f(x) = 2 - x$ on $(1,2)$. We can generate it easily with inversion but have to use two different formulas for the inverse cdf for the two different regions. However, the generation algorithm becomes even simpler if we recall a property of the symmetric triangular distribution: It can be written as the sum of two independent $U(0,1)$ random variates. So it is enough to generate two uniform variates and to return their sum.

The Normal Distribution.

The great practical and theoretical importance of the normal distribution is well known. Unfortunately the cdf and the inverse cdf of the normal distribution are numerically unfriendly. Hence the inversion method is slow and also inaccurate. But due to Box and Muller (1958) we have a simple exact method available. It is based on the idea to generate the two-dimensional standard multinormal distribution (which has density $f(x,y) = \exp(-\frac{1}{2}(x^2+y^2))/(2\pi)$ by generating a random angle and a random radius. It is easy to see and well known that the multinormal distribution is radially symmetric. So the random angle is uniformly distributed on $(0, 2\pi)$. To find the distribution R of the radius we use the observation that the contour-lines of the density are all circles with the origin as center. Thus integrating the normal density along a circle with radius r gives $2r\pi f(0,r) = r\exp(-r^2/2)$ which is the density $f_R(r)$. It is not difficult to prove (see Exercise 2.19) that

$$R = \sqrt{2E} = \sqrt{-2\log(U)},$$

where E denotes a standard exponential and U a $U(0,1)$ random number. Hence we can easily generate the angle and the radius of the multinormal distribution and can use the sine and cosine function to compute the pair (X, Y), which are – due to well known properties of the multinormal distribution – two independent standard normal variates. Algorithm 2.11 contains the details of that algorithm.

Algorithm 2.11 Box-Muller

Output: Standard normal distributed variates.

1: stored ← false.
2: if stored == true then
3:    stored ← false.
4:    return X.
5: else
6:    Generate $U \sim U(0,1)$ and $V \sim U(0,1)$.
7:    $R \leftarrow \sqrt{-2\log(V)}$.
8:    $X \leftarrow R\cos(2\pi U)$.
9:    $Y \leftarrow R\sin(2\pi U)$.
10:   stored ← true.
11:   return Y.

It is a simple, fast and exact normal generator, probably the one that should be used if speed is not of primary importance. If a normal variate with mean μ and variance σ^2 is needed we generate the standard-normal variate Z and return $X \leftarrow \mu + \sigma Z$.

2.7 Exercises

Inversion

Exercise 2.1. Spell out the details of the inversion algorithm to generate from the staircase-shaped quasi-density, $f(x) = \lfloor x \rfloor$ for $1 \leq x < 5$ and zero elsewhere.

Exercise 2.2. Find the density and the inverse cdf for the hyperbolic secant distribution. It has cdf $F(x) = \frac{2}{\pi} \arctan(e^x)$.

Exercise 2.3. Spell out the details of the inversion algorithm to generate from the Laplace density, which is proportional to $e^{-|x|}$.

Rejection

Exercise 2.4. For $f(x) = \exp(-x^2/2)$ (Normal distribution) and $h(x) = \frac{c}{1+x^2}$ (Cauchy distribution) find the smallest possible c. Use the result to give the details of a random-variate generation algorithm for the normal distribution that uses rejection from the Cauchy distribution.

Exercise 2.5. Example 2.3 describes a rejection algorithm from a uniform hat to generate variates with density $f(x) = \frac{1}{2} \sin x$ for $0 \leq x \leq \pi$. Compile the details of this algorithm. Make all necessary changes to add the squeeze $s(x) = \frac{1}{2} - |\frac{1}{2} - x/\pi|$.

Exercise 2.6. Compute the expected number of uniforms $E(\#U)$ and the expected number of evaluations of the density $E(\#f)$ for the Algorithm in Exercise 2.5.

Exercise 2.7. Consider the class of hat-functions $\tilde{h}(x) = 1/(b+x)^2$ with $b > 0$. Construct the optimal hat for the exponential distribution $f(x) = \exp(-x)$. Compute the multiplicative constant β for general b first; then find the expression of $A_h = \alpha$ for general b and find the optimal b.

Exercise 2.8. We consider the same problem as for Exercise 2.7. But this time we start with constructing the hat $h(x) = a/(b+x)^2$, that touches in x_0. Then find the expression of $A_h = \alpha$ for general x_0 and find the optimal x_0. Check that the final optimal hat is the same as for Exercise 2.7.

Exercise 2.9. $f(x) = e^{-x^2/2}$ for $0 \leq x \leq 2$. Construct – in analogy to the Examples 2.12 and 2.13 – a hat function and a squeeze constant on two subintervals. Optimize the point x_0 which splits the two intervals with a numerical search procedure. Compute the expected number of uniforms required and the expected number of evaluations of f required. Compare these result for minimized A_h and minimized A_{h-s}.

Composition

Exercise 2.10. Spell out the details of a composition algorithm for sampling from the distribution with the quasi-density shown in Fig. 2.6. It is given by $f(x) = x^2$ for $x \in [0, 1)$, 1 for $x \in [1, 2)$, $(x-3)^2$ for $x \in [2, 3]$, and 0 otherwise.

Exercise 2.11. Use the composition method and spell out the details of a generator for a mixture of two normal distributions. Parameters of this mixture family are μ_1, σ_1, μ_2, σ_2 and the probability p that the first normal distribution is used. (Use the Box-Muller method to generate the required standard normal variates.)

Exercise 2.12. Spell out all the details of the composition-rejection algorithm using the optimal hat and squeeze constructed in Exercise 2.9.
(*Hint*: The algorithm is similar to the sequence of algorithms given in the Examples 2.14, 2.15 and 2.18).

The Ratio-of-Uniforms Method (RoU)

Exercise 2.13. Show that Example 2.19 is a special case of Algorithm 2.9 (RoU).

Exercise 2.14. Spell out all the details of an algorithm for generating normal random variates that is based on the ratio-of-uniforms method. Check that the rejection constant is really 1.3688.

Almost-Exact Inversion

Exercise 2.15. Compile an algorithm based on almost-exact inversion that uses a constant hat and a squeeze-rectangle for the conjugate density. As input of this algorithm use transformation Ψ, maximum of the conjugate density and squeeze rectangle $(0, u_0) \times (0, v_0)$.

Exercise 2.16. Use the idea of Example 2.22 for constructing a short and fast algorithm for the normal distribution. Add a rectangular squeeze to reduce the number of evaluations of the conjugate density. See Hörmann and Derflinger (1994a).

Exercise 2.17. Use the algorithm of Example 2.15 to compile the details for a generator for the Cauchy distribution with quasi-density $f(x) = 1/(1+x^2)$. Use transformation $\Psi(u) = b\,(u-\frac{1}{2})\,(1+c/(\frac{1}{2}-|u-\frac{1}{2}|))$. The squeeze rectangle should have maximal area. See Ahrens and Dieter (1988).

Exercise 2.18. Show by a straightforward computation that the transformation
$$(X,Y) \mapsto (U,V) = (\Psi^{-1}(X), Y\,t(\Psi^{-1}(X))/f(X))$$
has Jacobian 1. f denotes a density function and t is the conjugate density as defined in (2.5).

Using Special Properties of the Distribution

Exercise 2.19. For the validity of the Box-Muller method we have to show that a random variate R with density
$$f_R(r) = r\exp(-r^2/2)$$
can be generated by:
$$R = \sqrt{2\,E} = \sqrt{-2\log(U)}$$
where E denotes a standard exponential and U a $U(0,1)$ random variate. Use the transformation theorem for continuous random variates to show that the random variate $\sqrt{-2\log(U)}$ has really density $f_R(r)$.

Hint: The "Transformation Theorem" states that for a random variate X with density f and a monotone and smooth transform $y = R(x)$ the density of the transformed random variate $Y = R(X)$ is equal to $f(R^{-1}(y))|(R^{-1})'(y)|$.

3
General Principles for Discrete Distributions

A discrete random variable is a random variable taking only integer values. The distribution of a discrete random variable X is determined by its *probability mass function* (pmf), $p_k = \text{Prob}(X = k)$. It is also called *probability vector* if its support is bounded from below. In the latter case we assume without loss of generality that X is only taking values on the nonnegative integers. We then write $(p_0, p_1, \ldots)$ for the probability vector. To generate discrete random variates we can apply the inversion (Sect. 3.1) and the rejection principle (Sect. 3.3) that are also of major importance for continuous distributions. We also present the alias method (Sect. 3.2) which was developed especially for discrete distributions.

The algorithms presented in this chapter are not only important for discrete generation problems. They also serve as important building blocks for most universal generators for continuous distributions described in Chaps. 4, 5 and 7.

3.1 Inversion

3.1.1 Inversion by Sequential Search

We have already presented the idea of the inversion method to generate from continuous random variables. For a discrete random variable X we can write it mathematically in the same way:

$$X = F^{-1}(U) = \inf\{x\colon F(x) \geq U\}$$

where F is the cdf of the desired distribution and U is a uniform $(0, 1)$ random number. The difference compared to the continuous case is that F is now a step-function. Figure 3.1 illustrates the idea of discrete inversion for a simple distribution. To realize this idea on a computer we have to use a search algorithm. In Algorithm 3.1 we have stated the simplest version called

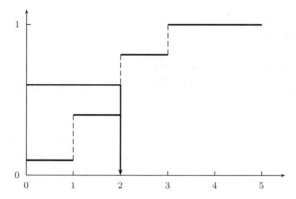

Fig. 3.1. Discrete inversion for a simple distribution

Sequential Search. Here the cdf is computed on-the-fly as sum of the probabilities p_k, since this is usually much cheaper than computing the cdf directly. It is obvious that the basic form of the search algorithm only works for discrete random variables with probability mass functions p_k with $k \geq 0$.

Algorithm 3.1 Sequential-Search

Require: Probability mass function p_k, $k \geq 0$.
Output: Random variate X with given probability vector.
1: Generate $U \sim U(0, 1)$.
2: Set $X \leftarrow 0$, $P \leftarrow p_0$.
3: **while** $U > P$ **do**
4:     Set $X \leftarrow X + 1$, $P \leftarrow P + p_X$.
5: **return** X.

Algorithm 3.1 (Sequential-Search) is a true black-box algorithm as it works in principle with any computable probability vector. But, with the exception of some very simple discrete distributions, sequential search algorithms become very slow as the while-loop has to be repeated very often. The number of iterations I, i.e. the number of comparisons in the while condition, is given by

$$\mathrm{E}(I) = \mathrm{E}(X) + 1$$

and can therefore become arbitrary large or even infinity if the tail of the distribution is very heavy. Another serious problem can be critical round-off errors due to summing up many probabilities p_k.

Devroye (1986a) explains binary search and other strategies to improve sequential search. The resulting algorithms are faster but more complicated then sequential search, whose main advantage is its unbeatable simplicity. On the other hand no search procedure is as fast as table aided search, which we describe in the next section.

3.1.2 Indexed Search (Guide Table Method)

The idea to speed up the sequential search algorithm is easy to understand. Instead of starting always at 0 we store a table of size C with starting points for our search. For this table we compute $F^{-1}(U)$ for C equidistributed values of U, i.e. for $u_i = i/C$, $i = 0, \ldots, C - 1$. Such a table is called *guide table* (or *hash table*) and its entries g_i are computed as

$$g_i = F^{-1}(u_i) = \inf\{j \colon F(j) \geq i/C\}\,.$$

Then we find for every $U \in (0,1)$, $g_{\lfloor U C \rfloor} = F^{-1}(\lfloor U C \rfloor/C) \leq F^{-1}(U)$. This shows that we can start our sequential search procedure from g_{k_0}, where the index k_0 of the correct table entry can be found rapidly by means of the truncation operation, i.e.

$$k_0 = \lfloor U C \rfloor\,.$$

This method is also known as *Indexed Search* and has first been suggested by Chen and Asau (1974). Algorithm 3.2 works out the details for finite probability vectors $(p_0, \ldots, p_{L-1})$, Figure 3.2 sketches the situation for a simple example.

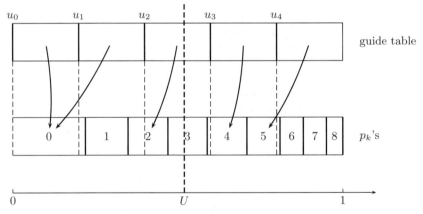

Fig. 3.2. Guide table of size $C = 5$ for a probability vector of length $L = 9$. The lower bar indicates the (cumulative) probabilities for the events 0, 1, ..., 8. The j-th table entry is created such that it points to the least k with $F(k) \geq u_j = j/C$. For the shown U we find $\lfloor U C \rfloor = u_2$. Since the table entry for u_2 is 2, i.e. $g_2 = 2$, we have to start sequential search at $k = 2$ and return $X = 3$

We can see that the two main differences to sequential search are that we start searching at the number determined by the guide table, and that we have to compute and store the cumulative probabilities P_i in the setup as we have to know the cumulative probability for the starting point of the search algorithm. The rounding problems that can occur in Algorithm 3.1

Algorithm 3.2 Indexed-Search

Require: Probability vector $(p_0, p_1, \ldots, p_{L-1})$; size C of guide table.
Output: Random variate X with given probability vector.
  /* Setup: Computes the guide table g_i for $i = 0, 1, \ldots, C-1$. */
1: Compute cumulative probabilities $P_i \leftarrow \sum_{j=0}^{i} p_j$.
2: Set $g_0 \leftarrow 0$, $i \leftarrow 0$.
3: **for** $j = 1$ to $C - 1$ **do**
4:    **while** $j/C > P_i$ **do**
5:       Set $i \leftarrow i + 1$.
6:    Set $g_j \leftarrow i$.
  /* Generator */
7: Generate $U \sim U(0, 1)$.
8: Set $X \leftarrow g_{\lfloor UC \rfloor}$.
9: **while** $U > P_X$ **do**
10:   Set $X \leftarrow X + 1$.
11: **return** X.

---

can occur here as well. We could try to add a sanity check to guarantee that all values in the guide table are smaller than L but it is not easy to compute the cumulative probabilities P_j such that we are really protected against all possible round-off errors. Compared with Algorithm 3.1 we have now the obvious drawback of a slow setup. The computation of the cumulative probabilities requires $O(L)$ operations whereas the computation of the guide table executes in $O(\max\{L, C\})$ time.

What we gain is really high speed as the marginal execution time of the sampling algorithm becomes very small. The expected number I of comparisons for the while-loop depends on the ratio between the length L of the probability vector and the size C of the guide table,

$$\mathrm{E}(I) = 1 + \mathrm{E}(\text{\# of } P_i \text{ in interval } [u_j, u_{j+1})) \leq 1 + \frac{L}{C}.$$

This consideration shows that there is a trade-off between speed and the size of the guide table. Cache-effects in modern computers will however slow down the speed-up for really large table sizes. Thus we recommend to use a guide table that is about two times larger than the probability vector to obtain optimal speed. But of course this number strongly depends on the length L, the computer, operating system and compiler used.

Remark. Indexed search is one of the key building blocks for fast automatic algorithm for sampling from continuous distributions (see e.g. Algorithm 4.1, TDR). Then it is possible to recycle the uniform random number U in Algorithm 3.2, since $U' = (U - P_{X-1})/(P_X - P_{X-1})$ is again a $U(0, 1)$ random number independent of U (and X). However, this procedure should be used with care since common sources of random numbers have only limited resolution.

3.2 The Alias Method

A second well known and ingenious method for fast generation from a discrete distribution with a finite number of different values is the *alias method* introduced by Walker (1974, 1977). It is based on the fact that any discrete probability vector of length L can be expressed as equi-probable mixture of L two-point distributions. To store these two-point distributions we need in principle three tables of size L: The table (i_k) containing the first value of all two-point distributions, the table (a_k) containing the second value of all two-point distributions and the value (q_k) containing the probabilities for the first values i_k. As we will see below it is possible to find two-point distributions such that we have $i_k = k$. This means that we only need the tables for the "aliases" a_k and for the probabilities q_k.

We can use such tables to compile a very fast and simple algorithm to sample from the given distribution, see Algorithm 3.3 (Alias-Sample) and Fig. 3.3. In the description of the algorithm we have included two variants for Step 2. The one in parenthesis is utilizing the idea of recycling to avoid the necessity of a second uniform random number. However, if L is very large this variant could be dangerous as it depends on the precision and the randomness of lower order bits of the uniform random number generator.

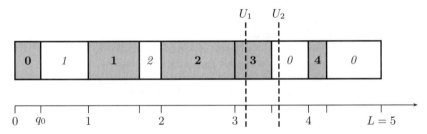

Fig. 3.3. Sampling from the discrete distribution of Fig. 3.4 using the alias method. The probabilities q_k are shown as gray bars, the probabilities of choosing an alias a_k as white bars. The figures in the bars indicate the random variate that is returned for a sampled uniform random number $U \sim U(0, L)$ when using the recycle principle. E.g., for U_1 we return 3, since $3 < U_1 - 3 \le q_3$; for U_2 we return the alias $a_3 = 0$

We can clearly see the strength of the alias method when looking at Algorithm 3.3. The sampling algorithm is very short and requires only one comparison and less than two table accesses. The price we have to pay is again a comparatively slow and complicated setup and the storage requirements of two tables of size L.

The setup Algorithm 3.4 (Alias-Setup) that computes the necessary tables (a_k) and (q_k) is due to Kronmal and Peterson (1979). Our goal is to transform the histogram of the given probabilities into a "squared histogram", by means of a strategy that Marsaglia calls the "Robin Hood Algorithm": take from the

Algorithm 3.3 Alias-Sample

Require: Probability vector $(p_0, p_1, \ldots, p_{L-1})$ of length L.
Output: Random variate X with given probability vector.
/* Setup */
1: Compute tables (a_k) and (q_k). /* call Algorithm 3.4 (Alias-Setup) */
/* Generator */
2: Generate $U \sim U(0, 1)$.
3: Set $X \leftarrow \lfloor LU \rfloor$.
4: Generate $V \sim U(0, 1)$. (*Or:* Set $V \leftarrow LU - X$.)
5: **if** $V < q_X$ **then**
6:    return X.
7: **else**
8:    return a_X.

Algorithm 3.4 Alias-Setup

Require: Probability vector $(p_0, p_1, \ldots, p_{L-1})$ of length L.
Output: Integer-table (a_k) and probabilities (q_k) (both tables have size L) necessary for Algorithm 3.3 (Alias-Sample).
1: **for** $l = 0$ to $L - 1$ **do**
2:    Set $q_l \leftarrow L p_l$.
3: Initialize the integer sets $Greater = \{l : q_l \geq 1\}$ and $Smaller = \{l : q_l < 1\}$.
4: **while** $Smaller$ and $Greater$ not empty **do**
5:    Choose (arbitrary) $k \in Greater$, and $l \in Smaller$.
   /* Take from the "rich" k and give to the "poor" l */
6:    Set $a_l \leftarrow k$. /* Store the donor. */
7:    Remove l from $Smaller$. /* q_l and a_l are now finalized. */
8:    $q_k \leftarrow q_k - (1 - q_l)$. /* As k gives to l reduce "probability" q_k */
9:    **if** $q_k < 1$ **then**
10:      Move k from $Greater$ to $Smaller$. /* k has given too much */

"rich" to bring the "poor" up to average. In each iteration of the while-loop we chose an entry where the probability is greater than the average ("rich", stored in $Greater$) and an entry where the probability is smaller than the average ("poor", stored in $Smaller$). Then the "rich" gives some of its probability to the "poor"; Figure 3.4 illustrates this idea. Notice that in each step the size of $Smaller$ goes down by one. Thus the while-loop is executed exactly L times and the total execution time of the algorithm is $O(L)$. After each step the total excess probability in $Greater$ equals the total shortage in $Smaller$, i.e.

$$\sum_{i \in Greater} q_i - 1 = \sum_{i \in Smaller} 1 - q_i \,.$$

Consequently $Greater$ and $Smaller$ can only become empty at the same time after the last step. The algorithm also implies that for a number i with probability p_i less than average, i.e. $p_i < 1/L$, no "alias" $a_j = i$ can be constructed. This shows that i will be generated with the correct probability. The reduction

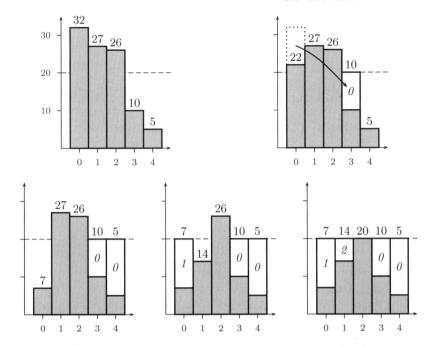

Fig. 3.4. Robin Hood algorithm for computing the table of "aliases" a_k and of the probabilities q_k. In the first step we choose bar 0 with probability greater than average (dashed line) and bar 3 with probability less than average. Then we split the "rich" bar 0 and give the upper part to the "poor" bar 3. We have to store the donor in a_3 and update the probability q_0. In the next iteration we proceed in the same way with bars 0 and 4 and continue until we end up with the squared histogram in the last figure. The alias (donors) a_k are printed into the upper (white) part of each bar, the probabilities q_k (in %) are printed on their tops

of q_k in Step 8 is the key to understand that all other probabilities are finally correct as well. Implementing sets *Greater* and *Smaller* can be accomplished in many ways, e.g. with a simple push/pop stack or with a linked list. Notice, that there also can occur numerical difficulties caused by round-off errors if the probabilities do not exactly sum to one. To avoid problems it is therefore better to test if either of the two sets is empty in Step 4 of Algorithm 3.4 (Alias-Setup).

The alias method has very fast marginal execution times. But it can be still speeded-up at the expense of larger tables and a slower setup. This variant called *alias-urn method* has been suggested by Peterson and Kronmal (1982). It is based on the idea that we can add $C - L$ probabilities 0 to the given probability vector of length L, thus obtaining a vector of length C ($> L$). If we look at the details of Algorithm 3.4 (Alias-Setup) we can see that $p_i = 0$ implies $q_i = 0$. This means that for $i \geq L$ we know that $q_i = 0$. So we can save one table access in that case and can immediately return a_i. Algorithm 3.5

Algorithm 3.5 Alias-Urn

Require: Probability vector $(p_0, p_1, \ldots, p_{L-1})$ of length L; total length C of tables.
Output: Random variate X with given probability vector.
/* Setup */
1: Compute tables (a_k) and (q_k) for vector $(p_0, \ldots, p_{L-1}, 0, \ldots, 0)$ of length $C \geq L$, using Algorithm 3.4 (Alias-Setup).
/* Generator */
2: Generate $U \sim U(0, 1)$.
3: Set $X \leftarrow \lfloor CU \rfloor$.
4: **if** $X \geq L$ **then**
5:     return a_X.
6: **else**
7:     Generate $V \sim U(0,1)$.   (*Or:* Set $V \leftarrow CU - X$.)
8:     **if** $V < q_X$ **then**
9:         return X.
10:    **else**
11:        return a_X.

contains the details of this method. It can be seen that for $C \to \infty$ the expected number of comparisons and the expected number of table accesses tend to one. As mentioned above for the guide table method the size of large tables has a negative influence on the speed of a program due to cache effects. Therefore we do not recommend to choose C too large. When using the recycling principle in Algorithm 3.3 we could not observe any difference in speed between Algorithms 3.3 and 3.5.

3.3 Discrete Rejection

Reading the above two sections it may look as if the alias method or indexed search are the final solution of the problem to sample from discrete distributions. However, this is not true. The main disadvantages of these algorithms are the slow setup, the size of the necessary tables and the problem with rounding errors in the setup. This can become prohibitive if the probability vector of the desired distribution is long (e.g. $L = 10^5$ or 10^6). We also see that the alias method described above only works for distributions that can take only finitely many values. If there is a tail it has to be cut off somewhere, where the probabilities are "almost 0" but this means that the alias method is not applicable to heavy-tailed distributions as the size of the tables and the execution time of the setup explodes for such distributions. Nevertheless, the alias method or guide table inversion remain the best choice for most every-day discrete generation problems. But for some applications universal rejection algorithms for discrete distributions are of practical importance; especially when the tails of the distribution are very heavy, the setup times and the memory requirements of the algorithm are of importance.

We have discussed several rejection algorithms for continuous distributions in Sect. 2.2. There are three general approaches to apply the idea of rejection to discrete distributions (Fig. 3.5):

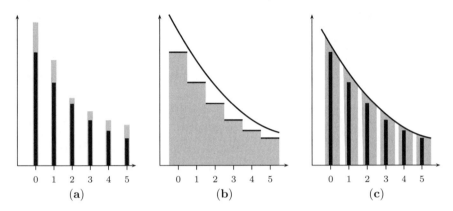

Fig. 3.5. Three approaches for using the rejection method for discrete distributions: (**a**) rejection from discrete dominating distribution; (**b**) use histogram function as density together with rejection from continuous hat function and return result rounded to next integer; (**c**) use continuous density to compute probabilities for discrete hat function

- *Use a discrete dominating distribution.*
  The advantage of this approach is the possibly very close fit of the dominating distribution. The disadvantage lies in the fact that, with the exception of the geometric distribution, there are not many useful discrete dominating distributions that can easily be generated by inversion.
- *Use the histogram of the discrete distribution as density for a continuous rejection algorithm.*
  By truncation (or rounding) we then obtain the desired discrete random variate. This method was used by many authors (including the first author of this book) to design Poisson and Binomial generators as it leads to uniformly fast generators (see Devroye (1986a) and references given there). The disadvantage of this method is the bad fit of the continuous hat if neighboring probabilities are not close together.
- *Combine these two methods.* This is a new variant of discrete rejection called *rejection-inversion* (Hörmann and Derflinger, 1996), that will be described in Sect. 10.2.1.

These concepts are used in Chap. 10 to develop automatic methods for discrete distributions. Since we are mainly interested in universal methods we do not give explicit examples here.

3.4 Exercises

Exercise 3.1. Construct "by hand" a guide table of size four for the probability vector $(0.1, 0.3, 0.4, 0.2)$. Compute the expected number of evaluations of the condition in the while-loop of Algorithm 3.2.

Exercise 3.2. Construct "by hand" the Tables a_i and q_i for the alias method for the probability vector $(0.1, 0.3, 0.4, 0.2)$.

Part II

Continuous Univariate Distributions

Part II

Continuous Univariate Distributions

4

Transformed Density Rejection (TDR)

The main problem when utilizing the acceptance-rejection method for designing a random variate generator is the choice of the hat function. In Sect. 2.2 we have demonstrated how we can find a hat function "by hand" for some simple examples. But for an automatic algorithm we need a design method that can be executed by the computer. It is possible to find a single hat that can be used for all distributions of fairly large families but it is obvious that the fit of such a "global hat" cannot be very close for all distributions. In Chap. 6 we are going to discuss how to find and use such global hat functions. In this chapter we are dealing with methods that can construct hat functions automatically around a given point locally, at least if the density satisfies some regularity conditions. *Transformed density rejection* is such a principle that can be applied to a large class of continuous distributions. A first variant has been published by Gilks and Wild (1992) under the name *adaptive rejection sampling*. The use of a general transformation and the name *transformed density rejection* was suggested by Hörmann (1995). In this presentation we try to develop the idea such that the reader can also see the (in our opinion) "beautiful" mathematical background (Sects. 4.2–4.4). The details and variants of the algorithms are presented in Sect. 4.4. Generalizations and special cases are discussed in Sects. 4.6 and 4.7. In Sect. 4.8 we develop the automatic ratio-of-uniforms method and show its equivalence to transformed density rejection. Transformed density rejection can also be applied to multivariate distributions (see Sect. 11.3).

4.1 The Main Idea

A suitable method for finding a hat function h for a given density f is to partition its domain into several intervals and construct the hat locally on each of these parts. Then the hat can be created by these components using the composition method (see Sect. 2.3). For example it is very easy to construct a hat for a concave density by using the (pointwise) minimum of several tangents. Figure 4.1 shows such an example for the quasi-density $f(x) = \sin(x)$ on $(0, \pi)$ with three tangents. It also shows that we can construct a squeeze s by means of secants between the construction points of the tangents. It is quite obvious that the ratio $\varrho_{hs} = A_h/A_s$ can be made smaller by inserting more and more tangents.

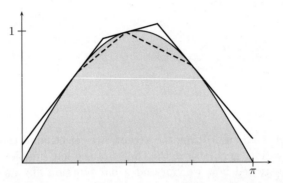

Fig. 4.1. Quasi-density $f(x) = \sin(x)$ on $(0, \pi)$ with the minimum of three tangents as hat. Secants between the construction points serve as squeeze

This idea is very simple but only of limited practical use since the family of concave distributions does not contain many prominent members. Nevertheless, it becomes a very powerful method if we use it together with the fact that many densities f can be "forced" to be concave by applying a strictly monotonically increasing transformation T such that $T(f(x))$ is concave. For example it is well known that the densities of many of the most important standard distributions are log-concave (i.e. the functions $\log(f(x))$ are concave). Thus log-concave densities serve as an example with $T = \log$. In analogy we call a density f, with $T(f(x))$ concave, T-*concave* (Hörmann, 1995). In abuse of language we call a distribution with a T-concave density, a T-*concave* distribution. It should be noted here that such densities are always unimodal. We will see in Thm. 4.5 that we can find transformations that make practically all unimodal densities T-concave.

To construct hat and squeeze functions for a T-concave distribution we can proceed in the following way. It is clear that this idea only works if T is differentiable and invertible. Let f be a T-concave quasi-density with domain (b_l, b_r), not necessarily finite, i.e. we may have $b_l = -\infty$ or $b_r = \infty$ or both.

(1) Transform the density into the concave function $\tilde{f}(x) = T(f(x))$.
(2) Find N suitable design points $b_l < p_1 < p_2 < \ldots < p_N < b_r$ (see Sect. 4.4 below).
    Notice: The theory works equally well if we set $p_1 = b_l$ or $p_N = b_r$ provided that the corresponding boundary is finite. Including this variant requires the distinction between several cases concerning the intervals

4.1 The Main Idea

at the boundaries and makes the presentation less clear. All necessary changes are straightforward and left to the reader.

(3) Construct tangents $\tilde{f}(p_i) + \tilde{f}'(p_i)(x - p_i)$ in all design points p_i. By the concavity of $\tilde{f}$ the function

$$\tilde{h}(x) = \min_{1 \leq i \leq N} \left(\tilde{f}(p_i) + \tilde{f}'(p_i)(x - p_i) \right)$$

is obviously a hat for the transformed density $\tilde{f}$, i.e., $\tilde{h}(x) \geq \tilde{f}(x)$.

(4) Compute the intersection points b_i of the respective tangents in p_i and p_{i+1}, and set $b_0 = b_l$ and $b_N = b_r$. Define

$$\tilde{h}_i(x) = (\tilde{f}(p_i) + \tilde{f}'(p_i)(x - p_i)) \, \mathbf{1}_{[b_{i-1}, b_i)}$$

where $\mathbf{1}_{[b_{i-1}, b_i)}$ denotes the indicator function of the set $[b_{i-1}, b_i)$. We then have $\tilde{h}(x) = \sum \tilde{h}_i(x)$.

(5) Transform $\tilde{h}(x)$ back into the original scale to get a hat h for the original density f:

$$h(x) = T^{-1}(\tilde{h}(x)) = \sum_{i=1}^{N} h_i(x)$$

where $h_i(x) = T^{-1}(\tilde{h}_i(x))$ for $x \in [b_{i-1}, b_i)$ and 0 otherwise.

(6) The transformed squeeze is constructed analogously as secants between the points of contact $(p_i, \tilde{f}(p_i))$ of the transformed density:

$$\tilde{s}_i(x) = \left(\tilde{f}(p_i) + \frac{\tilde{f}(p_{i+1}) - \tilde{f}(p_i)}{p_{i+1} - p_i}(x - p_i) \right) \mathbf{1}_{[p_i, p_{i+1})} \quad \text{for } 1 \leq i \leq N-1$$

and

$$s(x) = \sum_{i=1}^{N-1} s_i(x) \quad \text{with} \quad s_i(x) = T^{-1}(\tilde{s}_i(x)) \text{ for } x \in [p_i, p_{i+1}).$$

Figure 4.2 shows the construction of hat and squeeze in the transformed scale (l.h.s.) and the original scale (r.h.s.).

It is now necessary to identify transformations T that are useful for the procedure of constructing hat functions. Such transformations have to fulfill the below conditions. However, we have to distinguish between densities with bounded and unbounded domain.

(T1) $T: \mathbb{R}^+ \to \mathbb{R}$ is differentiable and $T'(x) > 0$.

For distributions with unbounded domain we further need

(T2) $\lim_{x \to 0} T(x) = -\infty$.

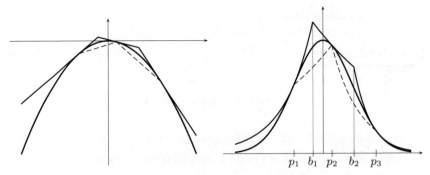

Fig. 4.2. Hat function and squeeze with three points of contact for the normal distribution and logarithm as transformation, design points at -1, 0.3, and 1.8. Transformed scale (l.h.s.) and original scale (r.h.s.)

(T3) The inverse transform $T^{-1}(x)$ has an integrable left tail, i.e., $\int_{-\infty}^{x} T^{-1}(t)\,dt$ is bounded for all fixed x in the image of T.

Condition (T1) simply implies that the inverse transform $T^{-1}(x) \geq 0$ exists and is monotonically increasing. (T2) is convenient since it is a necessary condition for the existence of T-concave distributions with unbounded support. Condition (T3) is required for the existence of a hat function with bounded integral

$$\int_{b_l}^{b_r} h(x)\,dx = \sum_{i=1}^{N} A_i < \infty$$

where

$$A_i = \int_{b_{i-1}}^{b_i} h_i(x)\,dx = \int_{b_{i-1}}^{b_i} T^{-1}(\tilde{f}(p_i) + \tilde{f}'(p_i)(x - p_i))\,dx$$

denotes the area below the hat in the i-th subinterval. Of course it is necessary that the set $\{\tilde{f}(p_i) + \tilde{f}'(p_i)(x - p_i): x \in (b_{i-1}, b_i)\}$ is a subset of the image of T.

For distributions with bounded domain (T2) and (T3) are not required. Due to (T1) T is continuous and invertible. If all points of the transformed hat are in the image of T (as stated above) this is enough to guarantee hats with bounded integral over the bounded domain of the distribution.

For sampling from the *hat distribution*, i.e., the distribution with quasi-density $h(x)$, the composition method (Sect. 2.3) is the obvious choice. Thus we define the "cdf" of the hat restricted to the various subintervals by

$$H_i(x) = \int_{b_{i-1}}^{x} h_i(t)\,dt = \int_{b_{i-1}}^{x} T^{-1}(\tilde{f}(p_i) + \tilde{f}'(p_i)(t - p_i))\,dt, \quad \text{for } b_{i-1} \leq x \leq b_i \tag{4.1}$$

and get

$$A_i = H_i(b_i) = \int_{b_{i-1}}^{b_i} h_i(t)\,dt \tag{4.2}$$

for the area below the hat in the i-th subinterval. Now let $F_T(x)$ denote an antiderivative of the inverse transformation T^{-1}. Especially if (T2) and (T3) hold we set

$$F_T(x) = \int_{-\infty}^{x} T^{-1}(t)\,dt\ .$$

If $\tilde{f}'(p_i) \neq 0$, (4.1) can be written as

$$H_i(x) = \frac{1}{\tilde{f}'(p_i)} \left(F_T(\tilde{f}(p_i) + \tilde{f}'(p_i)(x - p_i)) - F_T(\tilde{f}(p_i) + \tilde{f}'(p_i)(b_{i-1} - p_i)) \right). \tag{4.3}$$

To sample from the hat function in the different intervals we will use inversion so it is also necessary that we can invert the H_i, which is clearly possible if we have F_T^{-1} available. Conditions (T1) and (T3) imply that $F_T^{-1}(x)$ exists. So we can obtain the formula

$$H_i^{-1}(u) =$$
$$p_i + \frac{1}{\tilde{f}'(p_i)} \left(F_T^{-1}\left(\tilde{f}'(p_i)\,u + F_T(\tilde{f}(p_i) + \tilde{f}'(p_i)(b_{i-1} - p_i)) \right) - \tilde{f}(p_i) \right) \tag{4.4}$$
$$\text{for } 0 \leq u \leq A_i$$

provided that $\tilde{f}'(p_i) \neq 0$. Otherwise we simply have $H_i^{-1}(u) = b_{i-1} + u/f(p_i)$.

The basic algorithm for transformed density rejection is presented as Algorithm 4.1 (TDR). It requires a transformation T and design points. The choice of these two most important ingredients of transformed density rejection is discussed in the next sections. It should be noted here that Step 8 can be executed in constant time by means of *Indexed Search* (Sect. 3.1.2). Notice that then the random variate X is generated by inversion from the hat distribution, since the uniform random numbers are recycled in Step 9 (Sect. 2.3.2). For this reason we do not recommend the *Alias Method* (Sect. 3.2) in this case.

Remark. It should be noted here that we need the derivative of the density f only for constructing the tangent to the transformed density $\tilde{f}(x) = T(f(x))$. Indeed, we just need the slope of any tangent. Thus the algorithm works equally well if f' is a function that returns the slope of one possible "tangent" even if it is not unique. The quasi-density $f(x) = \max(1 - |x|, 0)$ serves as a simple example, since it is not differentiable in $x = 0$ but we can use $f'(0) = 0$ to construct a hat function.

4.2 The Class T_c of Transformations

In this section we want to find transformations that lead to fast and easy to handle automatic algorithms. To allow for unbounded domains, every such transformation must fulfill conditions (T1)–(T3) above, but which are the best candidates? Here is a list of some additional important criteria:

Algorithm 4.1 TDR (Transformed-Density-Rejection)

Require: T-concave quasi-density $f(x)$ with domain (b_l, b_r);
transformation $T(x)$, $\tilde{f}(x) = T(f(x))$, construction points $p_1 < \ldots < p_N$.
Output: Random variate X with density prop. to f.
/* Setup: Construct hat $h(x)$ and squeeze $s(x)$. */
1: Compute $\tilde{f}(p_i) = T(f(p_i))$ and $\tilde{f}'(p_i)$, for $i = 1, \ldots, N$.
2: Compute $\tilde{s}'(p_i) = (\tilde{f}(p_{i+1}) - \tilde{f}(p_i))/(p_{i+1} - p_i)$, for $i = 1, \ldots, N-1$.
3: Compute b_i as the intersection point of the tangents of $\tilde{f}(x)$ in the points p_i and p_{i+1}, for $i = 1, \ldots, N-1$.
   Set $b_0 \leftarrow b_l$ and $b_N \leftarrow b_r$.
4: Compute areas A_i below the hat $h(x)$ for each subinterval (b_{i-1}, b_i) using (4.2), for $i = 1, \ldots, N$.
5: Set $A \leftarrow \sum_{i=1}^{N} A_i$.
/* Generator */
6: **loop**
7:   Generate $V \sim U(0, A)$.
     /* Generate J with probability vector proportional to $(A_1, \ldots, A_N)$. */
8:   $J \leftarrow \min\{J: A_1 + \cdots + A_J \geq V\}$. /* use *Indexed Search* (Sect. 3.1.2) */
     /* Recycle uniform random number */
9:   $V \leftarrow V - (A_1 + \cdots + A_{J-1})$. /* $V \sim U(0, A_J)$. */
     /* Generate X with density proportional to h_J. */
10: Compute $X \leftarrow H_J^{-1}(V)$ using (4.4).
     /* Evaluate hat and squeeze */
11: Compute $h(X) \leftarrow T^{-1}(\tilde{f}(p_J) + \tilde{f}'(p_J)(X - p_J))$.
12: **if** $J > 1$ and $X < p_J$ **then**
13:    Compute $s(X) \leftarrow T^{-1}(\tilde{f}(p_{J-1}) + \tilde{s}'(p_{J-1})(X - p_{J-1}))$.
14: **else if** $J < N$ and $X > p_J$ **then**
15:    Compute $s(X) \leftarrow T^{-1}(\tilde{f}(p_J) + \tilde{s}'(p_J)(X - p_J))$.
16: **else**
17:    Set $s(X) \leftarrow 0$.
     /* Accept or reject */
18: Generate $U \sim U(0, 1)$.
19: **if** $U h(X) \leq s(X)$ **then** /* evaluate squeeze */
20:    **return** X.
21: **if** $U h(X) \leq f(X)$ **then** /* evaluate density */
22:    **return** X.

1. Simplicity of T, F_T and their inverse functions.
2. The transformation should lead to scale invariant hat functions:
   Denote by $h_{T,f,p}(x)$ the local hat constructed for the quasi-density f in the point p using the transformation T, i.e.,

$$h_{T,f,p}(x) = T^{-1}[T(f(p)) + T(f(p))' \cdot (x-p)] \;.$$

Then we say a hat is *scale invariant* if and only if

$$h_{T,\gamma f,p}(x) = \gamma\, h_{T,f,p}(x) \quad \text{for all } \gamma > 0.$$

3. A large class of densities should be T-concave.
4. T should lead to good fitting hat functions.

We will see that criteria 1 and 2 lead to the same class of transformations whereas criteria 3 and 4 are actually contradicting each other. Criterion 2 is important because often we only have given a quasi-density without the knowledge of the normalization constant but we do not want the performance of the algorithm to depend on the actual choice of the scale factor.

When we try to find transformations that fulfill criterion 1 we certainly will try the logarithm. Indeed, using this transformation was the starting point for the development of transformed density rejection (Gilks and Wild, 1992). The logarithm is the limiting case $\lambda \to 0$ of the class of Box-Cox transforms $T_\lambda(x) = \frac{x^\lambda - 1}{\lambda}$ that are frequently used in time-series analysis (cf. Box and Cox, 1964). We prefer a rescaled and translated version of that class. We define
$$T_c(x) = \text{sign}(c)\, x^c \quad \text{and} \quad T_0(x) = \log(x) \quad (4.5)$$
which leads to a class of simple transformations. Adding $\text{sign}(c)$ in this definition is necessary to make T_c increasing for all c. For designing simple and broadly applicable versions of Algorithm 4.1 (TDR) the most important members of this family of transformations are $T_0(x) = \log(x)$ and $T_{-1/2}(x) = -1/\sqrt{x}$. Table 4.1 gives all information necessary to use the class T_c. Notice that the transformations T_c are applicable for densities with unbounded domains only for $c \in (-1, 0]$.

Table 4.1. The class T_c of transformations

	$c > 0$	$c = 0$	$c < 0,\ c \neq -1$	$c = -1/2$
$T(x)$	$\mathbb{R}^+ \to \mathbb{R}^+ : x^c$	$\mathbb{R}^+ \to \mathbb{R} : \log(x)$	$\mathbb{R}^+ \to \mathbb{R}^- : -x^c$	$-x^{-1/2}$
$T^{-1}(x)$	$\mathbb{R}^+ \to \mathbb{R}^+ : x^{1/c}$	$\mathbb{R} \to \mathbb{R}^+ : e^x$	$\mathbb{R}^- \to \mathbb{R}^+ : (-x)^{1/c}$	x^{-2}
$F_T(x)$	$\mathbb{R}^+ \to \mathbb{R}^+ : \frac{x^{\frac{c+1}{c}}}{\frac{c+1}{c}}$	$\mathbb{R} \to \mathbb{R}^+ : e^x$	$\mathbb{R}^- \to \mathbb{R}^+ : \frac{-(-x)^{\frac{c+1}{c}}}{\frac{c+1}{c}}$	$-1/x$
$F_T^{-1}(x)$	$\mathbb{R}^+ \to \mathbb{R}^+ : \left(x \frac{c+1}{c}\right)^{\frac{c}{c+1}}$	$\mathbb{R}^+ \to \mathbb{R} : \log(x)$	$\mathbb{R}^+ \to \mathbb{R}^- : -\left(-x \frac{c+1}{c}\right)^{\frac{c}{c+1}}$	$-1/x$

This class and its rescaled and translated versions are obviously scale invariant. The following theorem shows that these are indeed the only transformations that fulfill the scale-invariance criterion 2 above.

Theorem 4.1. *The hat constructed by transformed density rejection is scale invariant (see criterion 2 above) for every (differentiable) density if and only if the transformation can be written as $T(x) = \alpha + \beta\, x^c$ or $T(x) = \alpha + \beta \log(x)$ with $\alpha, \beta, c \in \mathbb{R}$.*

Proof. We construct local hats $h_f(x)$ and $h_{\gamma f}(x)$ at an arbitrary point p_0 for the density $f(x)$ and the quasi-density $\gamma f(x)$, respectively. For the local hats the condition $h_{\gamma f}(x) = \gamma h_f(x)$ can be written as

$$T^{-1}[T(\gamma f(p_0)) + T(\gamma f(p_0))' \cdot (x - p_0)] = \gamma T^{-1}[T(f(p_0)) + T(f(p_0))' \cdot (x - p_0)]$$

or equivalently

$$T(\gamma T^{-1}[T(f(p_0)) + T(f(p_0))' \cdot (x - p_0)]) = T(\gamma f(p_0)) + T(\gamma f(p_0))' \cdot (x - p_0) \,, \tag{4.6}$$

i.e., $T(\gamma T^{-1}(u))$ maps one tangent line into the other one. Since $T'(x) > 0$ by condition (T1), $T(f(p_0))' = 0$ implies $f'(p_0) = 0$ and thus $T(\gamma f(p_0))' = 0$. In this case (4.6) reduces to the identity $T(\gamma f(p_0)) = T(\gamma f(p_0))$ which trivially holds for any transformation T. Now (4.6) must hold for every density and every tangent to its transformed density and we assume $T(f(p_0))' \neq 0$ and $T(\gamma f(p_0))' \neq 0$ in the sequel for verifying the necessity of the proposition. We write

$$z = T^{-1}[T(f(p_0)) + T(f(p_0))' \cdot (x - p_0)] \,.$$

Then $x - p_0 = (T(z) - T(f(p_0)))/T(f(p_0))'$ and inserting this into the right hand side of (4.6) we find

$$T(\gamma z) = T(\gamma f(p_0)) + T(\gamma f(p_0))'/T(f(p_0))' \cdot (T(z) - T(f(p_0))) \,.$$

By setting

$$a(\gamma) = T(\gamma f(p_0)) - T(f(p_0)) T(\gamma f(p_0))'/T(f(p_0))'$$
$$b(\gamma) = T(\gamma f(p_0))'/T(f(p_0))'$$

this can be written as

$$T(\gamma z) = a(\gamma) + b(\gamma) T(z) \,. \tag{4.7}$$

Notice that $b(\gamma) \neq 0$ for all $\gamma > 0$ by our assumptions. As T is differentiable by condition (T1), it follows that $a(\gamma)$ and $b(\gamma)$ are differentiable as well. Differentiating of Equation (4.7) with respect to z and γ results in

$$\gamma T'(\gamma z) = b(\gamma) T'(z) \quad \text{and} \quad z T'(\gamma z) = a'(\gamma) + b'(\gamma) T(z) \,.$$

Elimination of $T'(\gamma z)$ results in the differential equation

$$b(\gamma) z T'(z) = \gamma a'(\gamma) + \gamma b'(\gamma) T(z) \quad \text{for } z \neq 0.$$

Notice that there is at most one x where $z = z(x) = 0$, since $T(f(p_0))' \neq 0$ by assumption.

If $b'(\gamma) = 0$ this differential equation can be solved by the method of separation of variables. One gets the unique solution

$$T(z) = \frac{\gamma a'(\gamma)}{b(\gamma)} \log(z) + C(\gamma)$$

which is of the form $T(x) = \alpha + \beta \log(x)$ and belongs to the class T_c with $c = 0$. For $b'(\gamma) \neq 0$ solving the homogeneous equation and applying the method of variation of the constant leads to the unique solution

$$T(z) = C(\gamma)\, z^{\gamma b'(\gamma)/b(\gamma)} - \frac{a'(\gamma)}{b'(\gamma)}$$

which is of the form $T(x) = \alpha + \beta\, x^c$. This shows the necessity of the assumption. The sufficiency of the condition can be checked easily (see Exercise 4.1). □

Theorem 4.1 allows us to call the transformations T_c the (only) class of practically useful transformations for transformed density rejection as, for all other transformations, the fit of the hat function depends on the multiplicative constant used for f, which might even be unknown to the user of the algorithm. Nevertheless, other transformations may be of some interest, see Sect. 4.6.

Perhaps it is in place here to point out the nice special role of the logarithm (and its rescaled and translated versions) in the class T_c. As we can see from Table 4.1 the logarithm is the only member of the family that has as image all real numbers $\mathbb{R}$. All other members of the family have as image only a one sided interval. Therefore a problem can arise in Algorithm 4.1 (TDR) for $c < 0$. Since then $T^{-1}(x)$ is only defined for negative arguments we are in trouble when two tangents in the transformed scale intersect above the x-axis. In this case it is not possible to transform the tangents back into the original scale and thus we cannot construct a hat. It is not too difficult to avoid this case. The simplest way out is to use the mode of the density always as construction point of a horizontal tangent together with at least one construction point on either side of it.

4.3 T_c-Concave Distributions

We have just seen that it is reasonable to restrict our main attention to the family of transformations T_c, as they are the only transformations that leave the constructed hat independent from the chosen scaling constant of the density. Most importantly, the well known family of log-concave distributions is a special class of T_c-concave distributions. It includes many important standard distributions and has some interesting mathematical properties like being closed under convolution (see Devroye, 1986a, Sect. VII.2, and references given therein). If we try to characterize the shape of log-concave densities we can easily find a list of necessary conditions: All log-concave densities are continuous, unimodal and have exponential or sub-exponential tails. Using a verbal explanation we could add: And they must not be "too strongly convex".

For a better understanding of this statement it is instructive to examine the shape of the hats constructed by TDR. Around a design point p_i the hat has the form $T^{-1}(\tilde{f}(p_i) + \tilde{f}'(p_i)(x - p_i))$ which can be written as $(k\,x + d)^{1/c}$ for T_c and $\exp(k\,x + d)$ for T_0. We can compute the constants k and d immediately if we use the transformed density $\tilde{f}$, since the values of the density f and of its derivative at the design point p_i are equal to the respective values for the hat.

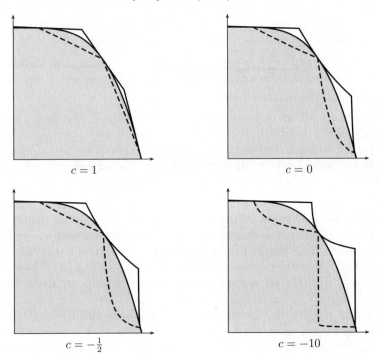

Fig. 4.3. Hat and squeeze for the quasi-density $f(x) = 1-x^4$, $0 \leq x \leq 1$ constructed for different values of c

As an illustration we use the quasi-density $f(x) = 1 - x^4$ for $0 \leq x \leq 1$. The sequence of pictures in Fig. 4.3 shows hat and squeeze for this density constructed by TDR with design points 0.2, 0.7, and 0.99 for different values of c (1, 0, -0.5, and -10). If we take a look at the behavior of the hat function around the (middle) design point at 0.7 we can see that in the first picture the hat is simply the tangent as T_1 is the identity function. In the other pictures, with c becoming smaller, the hats become stronger and stronger convex due to the stronger concavity of T. It is also obvious that the best fitting hat for our density is that with the largest c, i.e. with $c = 1$. On the other hand we know that this transformation is only useful for distributions that are concave. If the density were convex in a region around the design point the "hat" constructed by using T_1 (i.e., the tangent in the design point) is not above the density as it is not T-concave in that point. Thus we can easily see why there must be a trade off between the criteria 3 and 4 of Sect. 4.2. The best fitting hat only makes the current density T-concave but no density that is more convex (or less concave) than the given density. In Fig. 4.3 we can also see that the shapes of the squeeze (dashed lines) designed by different transformations T_c are analogous to the shapes of the hat functions.

Among the c's used in Fig. 4.3, $c = -10$ produces the worst fitting hat and squeeze but at the same time it can be apparently used for most densities, as its curvature around the design point is strongest. Bigger c's lead to hats and squeezes that approximate the density more closely than smaller c's. From Fig. 4.3 we can easily guess that hat function and squeeze constructed by transformed density rejection converges pointwise to step functions for $c \to -\infty$. This idea to use piecewise constant functions as hats for monotone densities seems to have been around quite a while (see e.g. Devroye, 1986a). It has been presented by Ahrens (1995) as a fast and relatively simple automatic algorithm, see Sect. 5.1. Unfortunately, we cannot write a meaningful transformation for the limiting case. So the view of transformed density rejection is not useful to construct the hat, which can be constructed easily anyway. Nevertheless, there are questions about the optimal placement of the design points where we can exploit the transformed density rejection interpretation of the step-function method.

In the light of the above discussion of the nature of the hat functions we can give the informal definition: A density is log-concave if a local approximation of the density by the exponential function $\exp(kx + d)$ is always above the density. If we restrict ourselves to two times differentiable densities we can easily write the necessary and sufficient condition for being log-concave as $\log(f(x))'' \leq 0$, or equivalently

$$f''(x) \leq \frac{f'(x)^2}{f(x)} .$$

We can formulate this condition for T_c-concave distribution as well.

Lemma 4.2. *A two times differentiable density $f(x)$ is T_c-concave if and only if*

$$f''(x) + (c-1)\frac{f'(x)^2}{f(x)} \leq 0 \quad \text{for all } x \text{ with } f(x) > 0. \tag{4.8}$$

Proof. For $c = 0$ we get the result by setting $\log(f(x))'' \leq 0$.
For $c \neq 0$ we have

$$T_c(f(x))'' = |c|f(x)^{c-2}\left((c-1)f'(x)^2 + f(x)f''(x)\right) \leq 0$$

for all x with $f(x) > 0$. □

An immediate consequence of Lemma 4.2 is the following theorem. As a corollary we have the important fact that any class of T_c-concave distributions with $c \leq 0$ includes all log-concave distributions.

Theorem 4.3. *Every T_{c_1}-concave density is also T_{c_2}-concave for any $c_2 < c_1$.*

Proof. As $\frac{f'(x)^2}{f(x)} \geq 0$ the theorem follows directly from Lemma 4.2. □

For $c > 0$ we can show the following theorem about the support of T_c-concave distributions. It again illustrates the above lemma that larger values of c lead to a less general class of T_c-concave distributions.

Theorem 4.4. *For $c > 0$ every T_c-concave distribution has bounded domain.*

Proof. For positive c the transformed density $T_c(f(x))$ is nonnegative. Therefore $T_c(f(x))$ can only be concave if its support is bounded. Thus $f(x)$ itself must have bounded support too. □

An informal description of T_c-concave distributions for $c < 0$ includes the following necessary conditions: f is continuous (on its domain) and unimodal, the tails of f go to zero at least as fast as the function $x^{1/c}$ (which means subquadratic tails for the most important special case $c = -1/2$). For a formal proof of this characterization see Exercise 4.2. As an immediate consequence we can see that for $c \leq -1$ there exist T_c-concave functions that are no longer densities as their integral is not bounded.

Considering condition (T3) of Sect. 4.1 ($T^{-1}(x)$ has an integrable left tail), that is necessary to obtain hat functions with bounded integral, we can make a similar observation. If the given density f has a bounded domain, we obtain hat functions with bounded integral for arbitrary c. For distributions with unbounded domain the tails of the hat must go faster to zero than $1/x$ for $x \to \infty$. As the hat function is of the form $(kx + d)^{1/c}$ it is obvious that for $c \leq -1$ condition (T3) is violated for distributions with unbounded domain. On the other hand for $-1 < c \leq 0$ condition (T3) is always fulfilled. So for these c we obtain a hat with bounded integral provided that the hat tends to zero when x tends to $\pm\infty$. Thus if we are interested in distributions with unbounded domain transformations T_c are only useful for $-1 < c \leq 0$. That is the reason why only these c's were considered by Hörmann (1995), where this class of transformations was introduced. The cases $c = 0$ and $c = -1/2$ are of major practical importance due to the simple forms of the required functions (cf. Table 4.1). c in the interval $(-1, -0.5)$ is of interest, if we want to generate from heavy tailed distributions.

Schmetterer (1994) raised the question if it is possible to give a general characterization of the family of all functions that can be transformed into concave functions. To answer this question it is instructive to introduce the function $\mathrm{lc}_f(x)$ that computes the maximal c that is necessary to make the density f, T_c-concave in the point x. We get this c by regarding the inequality (4.8) as an equality and expressing c therein. So we arrive at the definition of the *local concavity*:

$$\mathrm{lc}_f(x) = 1 - \frac{f''(x)\,f(x)}{f'(x)^2}\,. \tag{4.9}$$

We can use this new function to check that for example the quasi-density $f(x) = 1 + x^3$ on some interval around 0 is not T_c-concave for any c, because

$\mathrm{lc}_f(x) = 1 - 2(1+x^3)/3\,x^3$ has a negative pole at 0. A second example is given in Exercise 4.3. The following theorem partly answers the question of L. Schmetterer.

Theorem 4.5. *Let f be a bounded, unimodal, two-times differentiable density where for all x we have either $f'(x) \neq 0$ and $f''(x)$ bounded, or $f'(x) = 0$ and $f''(x) < 0$. Then there exists a c such that f is T_c-concave.*

Proof. From (4.9) we see that $\mathrm{lc}_f(x) > -\infty$ is certainly fulfilled if the conditions of the theorem are satisfied. □

We include a simple useful lemma together with a corollary that we will need in Chap. 11.

Lemma 4.6. *For all two times differentiable functions f, f_1, f_2 we have:*

$$\mathrm{lc}_{f_1 f_2}(x) = \frac{f_2(x)^2 f_1'(x)^2 \mathrm{lc}_{f_1}(x) + f_1(x)^2 f_2'(x)^2 \mathrm{lc}_{f_2}(x)}{(f_1'(x)f_2(x) + f_1(x)f_2'(x))^2}, \quad (4.10)$$

$$\mathrm{lc}_{f^a}(x) = \frac{\mathrm{lc}_f(x)}{a}. \quad (4.11)$$

Proof. Both results can be checked by straightforward computations. □

Corollary 4.7. *For any T_c-concave density f and real number $a > 0$ the function f^a is $T_{c/a}$-concave.*

Proof. This result follows directly from the pointwise result given in (4.11) above. □

If $X_1, X_2, \ldots, X_n$ are independent and identically distributed (iid.) random variables, then the order statistics for this sample are denoted by $X_{(1)}, X_{(2)}, \ldots, X_{(n)}$ where $X_{(1)} \leq X_{(2)} \leq \ldots \leq X_{(n)}$. It is well known that $U_{(i)}$, the i-th order statistic from a uniform sample of size n, has a beta distribution with parameters i and $n - i + 1$ and thus the density

$$f_{U_{(i)}}(x) = k\, x^{i-1}(1-x)^{n-i}$$

where $k = n \binom{n-1}{i-1}$ is some normalization constant. For an arbitrary continuous distribution with density f and cdf F we can easily see (by the transformation theorem) that

$$f_{X_{(i)}}(x) = k\, f(x) F(x)^{i-1}(1 - F(x))^{n-i}.$$

The following theorem is perhaps not too well known.

4 Transformed Density Rejection (TDR)

Theorem 4.8. *For a continuous, log-concave distribution, all order statistics have a log-concave distribution.*

Proof. The theorem can be shown using a result of Prekopa (1973) that states that all marginal distributions of a log-concave distribution are again log-concave. Together with the formula of the multidimensional distribution of order statistics this implies our theorem. For an elementary proof see Hörmann and Derflinger (2002). □

Hörmann and Derflinger (2002) prove that the maximum and the minimum order statistics of a T_c-concave distribution are T_c-concave for the same c as long as $-1/2 \leq c \leq 0$ but not for $c < -1/2$. In the same paper they conjecture, without a proof, that the same is true for all order statistics and $-1/2 \leq c \leq 0$. We are able to give here a proof for this conjecture for the case $c = -1/2$ which is most important in practice and not included in the paper of Hörmann and Derflinger (2002). For the proof for the maximum and the minimum for $c > -1/2$ we refer to the original paper. The proof for the case $c = -1/2$ is based on the following result.

Lemma 4.9. *We are given a random variate Z with density $f_Z(x)$ and a random variate X with density $f(x)$ and cdf $F(x)$.*
If X and Z are $T_{-1/2}$-concave this implies that the transformed variable $Y = F^{-1}(Z)$ with density
$$f_Y(x) = f_Z(F(x))f(x)$$
is $T_{-1/2}$-concave as well.

Proof. Using $f_Y(x) = f_Z(F(x))f(x)$, the definition of the local concavity and some simplifications we get writing F, f, f' and f'' for $F(x)$, $f(x)$, $f'(x)$ and $f''(x)$:

$$\mathrm{lc}_{f_Y}(x) = $$
$$\frac{f'_Z(F)^2 f^4 - f_Z(F)f^4 f''_Z(F) - f_Z(F)f'_Z(F)f^2 f' + f_Z(F)^2 f'^2 - f_Z(F)^2 f f''}{(f'_Z(F)f^2 + f_Z(F)f')^2}.$$

Using some algebra it can be verified that this is the same as:

$$\mathrm{lc}_{f_Y}(x) + \frac{1}{2} = \frac{f'_Z(F)^2 f^4 (\mathrm{lc}_{f_Z}(F) + 1/2) + f_Z(F)^2 f'^2 (\mathrm{lc}_f(x) + 1/2)}{(f'_Z(F)f^2 + f_Z(F)f')^2}.$$

So clearly the fact that $\mathrm{lc}_{f_Z}(F(x)) + 1/2 \geq 0$ and $\mathrm{lc}_f(x) + 1/2 \geq 0$ for arbitrary x imply that $\mathrm{lc}_{f_Y}(x) \geq -1/2$ which completes the proof. □

Theorem 4.10. *If a random variate X is $T_{-1/2}$-concave, then this implies that all order statistics of X are $T_{-1/2}$-concave as well.*

Table 4.2. T_c-concave distributions

Name of distribution	quasi-density	parameters	T-concave for
Normal	$e^{-\frac{x^2}{2}}$		logarithm
Exponential	$\lambda e^{-\lambda x}$	$\lambda > 0$	logarithm
Gamma	$x^{a-1} e^{-x}$	$a \geq 1$	logarithm
Beta	$x^{a-1}(1-x)^{b-1}$	$a, b \geq 1$	logarithm
Cauchy	$1/(1+x^2)$		$c = -1/2$
t-distribution	$\left(1 + \frac{x^2}{a}\right)^{-\frac{a+1}{2}}$	$a > 0$	$c = \frac{-1}{1+a}$
Pearson VI (or beta-prime)	$\frac{x^{a-1}}{(1+x)^{a+b}}$	$a \geq 1$	$c = \frac{-1}{1+b}$
Perks	$\frac{1}{e^x + e^{-x} + a}$	$a > -2$	logarithm
Generalized inverse Gaussian	$x^{\lambda-1} e^{-(\omega/2)(x+1/x)}$	$\lambda \geq 1, \omega > 0$ $\lambda > 0, \omega \geq 0.5$	logarithm $c = -1/2$
lognormal	$e^{-\ln(x-\mu)^2/(2\sigma^2)}/x$	$\sigma \leq \sqrt{2}$	$c = -1/2$
Weibull	$x^{a-1} \exp(-x^a)$	$a \geq 1$	logarithm
Planck	$x^a/(e^x - 1)$	$a \geq 1$	logarithm
Burr	$x^{a-1}/(1+x^a)^b$	$a \geq 1, b \geq 2$	$c = -1/2$
Snedecor's F	$\frac{x^{m/2-1}}{(1+m/n\, x)^{(m+n)/2}}$	$m, n \geq 2$	$c = -1/2$
Gumbel	$e^{-kx - \exp{-kx}}$	$k > 0$	logarithm
Exponential power	$e^{-|x|^a}$	$a \geq 1$	logarithm
Hyperbolic	$\exp(-\zeta\sqrt{1+x^2})$	$\zeta > 0$	logarithm

Proof. An order statistic $X_{(i)}$ can be written as

$$X_{(i)} = F^{-1}(U_{(i)})$$

where F denotes the cdf of X and $U_{(i)}$ the uniform order statistic. It is well known that uniform order statistics follow the beta-distribution and that the beta-distribution is log-concave and therefore also $T_{-1/2}$-concave. Thus the theorem follows directly from Lemma 4.9. □

We conclude this section with Table 4.2. It gives a list of important standard distributions (all distributions that are included as main chapters in the book of Johnson, Kotz, and Balakrishnan, 1994, 1995) and includes information about the c that leads to T_c-concavity for the given distribution family. Note that there are variants of TDR collected in Sect. 4.7 below that allow the generation of non-T_c-concave distributions.

4.4 Construction Points

Like the performance of every rejection algorithm the performance of Algorithm 4.1 (TDR) depends on the rejection constant $\alpha = A_h/A_f$; the number of evaluations of f, $\mathrm{E}(\#f)$, is also influenced by the size of A_s. If the area below the density is unknown the ratio $\varrho_{hs} = A_h/A_s$ (see Thm. 2.9 in Sect. 2.2.3) is a simple bound for α. Clearly ϱ_{hs} and α depend on the choice of the construction points for the tangents. Possible choices lie between two extremal positions:

(1) Use only a few constructions points with little storage requirement and fast setup but with moderate marginal generation time.
(2) Use many and if possible optimal construction points. By "optimal" we mean that either $\mathrm{E}(\#f)$ or $\mathrm{E}(\#U)$ or the ratio ϱ_{hs} is minimized for the given density and number of tangents. Then the marginal generation if very fast at the expense of a slower setup.

In the following we present several methods for different purposes, which are summarized in Sect. 4.4.8.

4.4.1 One Construction Point, Monotone Densities

Although some of the results can be generalized to non-monotone densities as well (see Derflinger and Hörmann, 2002) we restrict our attention here to the monotone case. So we assume throughout this section that T is a transformation fulfilling the conditions given in Sect. 4.1. The quasi-density $f(x)$ is a monotonically decreasing two times differentiable function on $[b_l, b_r)$ which is strictly T-concave with respect to the current transformation, i.e. $T(f(x))'' < 0$. We also consider the case where $b_r = \infty$.

As we have – for the moment – only one design point and thus one subinterval we can simplify the notation of Sect. 4.1 a bit. For the tangent of the transformed density in the point p we write

$$\tilde{h}(x) = \tilde{h}(p, x) = \tilde{f}(p) + \tilde{f}'(p)(x - p) \quad \text{for } b_l \leq x < b_r$$

where $\tilde{f}(x) = T(f(x))$ again denotes the transformed density. The hat function is therefore

$$h(x) = h(p, x) = T^{-1}(\tilde{h}(p, x)) \ .$$

For finding the best possible hat we try to minimize the area $A_h(p)$ below the hat function $h(p, x)$ to obtain the optimal design point p. We start with considerations that hold for a general transformation T. We can easily compute:

$$A_h(p) = \int_{b_l}^{b_r} h(p, x) \, \mathrm{d}x =$$
$$= \frac{1}{\tilde{f}'(p)} (F_T(\tilde{f}(p) + \tilde{f}'(p)(b_r - p)) - F_T(\tilde{f}(p) + \tilde{f}'(p)(b_l - p))) \ .$$

Differentiating with respect to p, and a slight rearrangement, yield

$$A'_h(p) = \frac{\tilde{f}(p)''}{\tilde{f}(p)'}(-A_h(p) + h(p, b_l)(p - b_l) + h(p, b_r)(b_r - p)) \,.$$

By the T-concavity and monotonicity of f we have $\tilde{f}(x)'' < 0$ and $\tilde{f}(x)' < 0$ for $x > b_l$. Hence we have the following simple condition for a stationary point $(p_0, A_h(p_0))$:

$$A_h(p_0) = h(p_0, b_l)(p_0 - b_l) + h(p_0, b_r)(b_r - p_0) \,. \tag{4.12}$$

For the case that $b_r = \infty$ we can obtain with similar arguments the condition

$$A_h(p_0) = h(p_0, b_l)(p_0 - b_l) \,. \tag{4.13}$$

So far we only have used simple analysis and no assumptions are needed about T but it is not clear yet that we only have a single stationary point. For the case of transformations T_c it has been proven by Derflinger and Hörmann (2002) that for any monotone, strictly T_c-concave density only one stationary point exists which must thus be the global minimum. As the case T_c contains all transformations with practical importance we restrict ourselves to this class here. The above result is surprising in two respects:

- The optimal point of contact $P_0 = (p_0, h(p_0, p_0))$ only depends on the hat function. It does not depend on the density $f(x)$. A given hat, with fixed $A_h(p_0)$, $h(p_0, b_l)$, and $h(p_0, b_r)$ has only one point with respective property (4.12) or (4.13). This is obvious as for a fixed hat $h(x)$ with fixed borders b_l and b_r, $h(b_l)$ and $h(b_r)$ are fixed as well. So these two equations are simple linear equations in p_0 and it is easy to show that they have a solution p_0 with $b_l \leq p_0 \leq b_r$. It follows that all quasi-densities that a hat $h(p_0, x)$ touches optimally, are touched in the same point P_0. We call this point P_0 the *characteristic point* of the hat.
- Equation (4.12) shows a simple geometric relation: The area under the hat is equal to the sum of the areas of two rectangles $(b_l, p_0) \times (0, h(b_l))$ and $(p_0, b_r) \times (0, h(b_r))$. This is shown in Fig. 4.4. For the case of unbounded domain this relation is even easier: The area under the hat is equal to the area of the rectangle $(b_l, p_0) \times (0, h(b_l))$, see Fig. 4.5.

Using (4.12) we can prove relations between the characteristic point and the function values of the optimal hat at the borders; we obtain e.g. for $-1 < c < 0$

$$h(p_0)^c = \frac{1}{1+c} \frac{h(b_l)^{1+c} - h(b_r)^{1+c}}{h(b_l) - h(b_r)} \,. \tag{4.14}$$

For $c = -1/2$ this simplifies to

$$\sqrt{h(p_0)} = \frac{\sqrt{h(b_l)} + \sqrt{h(b_r)}}{2} \,,$$

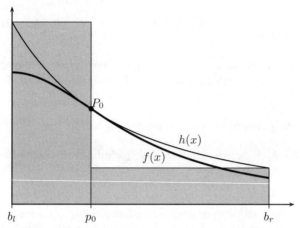

Fig. 4.4. The sum of the areas of the two shaded rectangles is equal to the area below the hat function $h(x)$

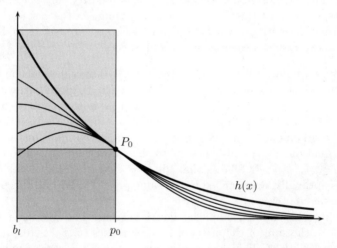

Fig. 4.5. The hat $h(x)$ touches several arbitrary selected quasi-densities in the same point P_0. The rectangle $(b_l, p_0) \times (0, h(p_0))$ is maximal for all hat functions and all quasi-densities. The area of the rectangle $(b_l, p_0) \times (0, h(b_l))$ is equal to the area below the hat

for $c = 0$ it is possible to prove

$$\log(h(p_0)) = \frac{h(b_l)\log(h(b_l)) - h(b_r)\log(h(b_r))}{h(b_l) - h(b_r)} - 1 \,. \tag{4.15}$$

Unfortunately all the conditions for p_0 above are in terms of values of the hat function but in practice we have to find p_0 for a given quasi-density $f(x)$. To do this we can use (4.12) for a numerical search procedure to find the optimal point p_0 for a given density f; but such a procedure is not really simpler than searching for p_0 by directly minimizing $A_h(p)$.

This situation changes for the important special case of unbounded domain ($b_r = \infty$). Due to $h(b_r) = 0$ Equations (4.14) and (4.15) simplify to

$$\frac{h(b_l)}{h(p_0)} = (1+c)^{1/c}, \text{ for } -1 < c < 0 \quad \text{and} \quad \frac{h(b_l)}{h(p_0)} = e, \text{ for } c = 0, \quad (4.16)$$

that can be reformulated for the transformed hat (i.e. the tangent) as

$$\frac{\tilde{h}(b_l)}{\tilde{h}(p_0)} = 1+c, \text{ for } -1 < c < 0 \quad \text{and} \quad \tilde{h}(b_l) - \tilde{h}(p_0) = 1, \text{ for } c = 0.$$

The last condition can be conveniently used for finding the optimal design point for given density on (b_l, b_r) in an iterative procedure.

To develop a second simple procedure for unbounded domain consider the rectangle $(0, z) \times (0, h(z))$ below a fixed hat function. We define the point z_0 with $b_l < z_0$ as the point where the area of the rectangle below h is maximized. It is not difficult to show (see Derflinger and Hörmann, 2002) that the point z_0 and the characteristic point p_0 coincide for transformations T_c. But as the hat and the density coincide for p_0, i.e. $h(p_0) = f(p_0)$ and $h'(p_0) = f'(p_0)$, this implies that the area of the rectangle below the density $(0, p) \times (0, f(p))$ is maximized for $p = p_0$.

Summarizing, the *characteristic point* $P_0 = (p_0, h(p_0))$ has in the case of unbounded domain and transformations T_c, $-1 < c \leq 0$, the following important properties. (Fig. 4.5 shows a hat touching optimally four quasi-densities in the same point P_0.)

- It is the unique point of optimal contact. (For non-monotone densities this has not been proven yet but we conjecture that it is true in the case of unbounded domain.)
- The area under the hat is equal to the rectangle area $(p_0 - b_l) \times h(b_l)$.
- The rectangle area $(p - b_l) \times h(p)$ below the hat is maximal for $p = p_0$.
- The rectangle area $(p - b_l) \times f(p)$ below any density curve touched in P_0 is maximal for $p = p_0$. (Remark: $f(p_0) = h(p_0)$)

From the last property it follows that a density has the same universal optimal point of contact for all T_c-hats independent of c, $-1 < c \leq 0$. The last property also leads to a simple procedure for finding this optimal point of contact: Find numerically the solution for $\max_p (p - b_l) f(p)$. This procedure, together with the theorem below, is due to Hörmann and Derflinger (1996).

Theorem 4.11. *For given, fixed c and monotone strictly T_c-concave quasi-density $f(x)$ on (b_l, ∞) the unique optimal design point p_0 has the property:*

$$\max_p (p - b_l) f(p) = (p_0 - b_l) f(p_0).$$

The rejection constant of the optimal hat with single design point p_0 is bounded by:

$$\alpha = \frac{A_h}{A_f} \leq (1+c)^{1/c} \text{ for } -1 < c < 0 \quad \text{and} \quad \frac{A_h}{A_f} \leq e \text{ for } c = 0.$$

Proof. The choice of p_0 was discussed above.

The area under the optimal hat is equal to $(p_0 - b_l)h(b_l)$. Among all monotone quasi-densities which are touched by the hat in $P_0 = (p_0, h(p_0))$ the uniform density $f(x) = h(p_0)$ for $b_l \leq x \leq p_0$ has the smallest area below its density curve. Thus we get (using (4.16)):

$$\frac{A_h}{A_f} \leq \frac{(p_0 - b_l)h(b_l)}{(p_0 - b_l)h(p_0)} = \frac{h(b_l)}{h(p_0)} = \begin{cases} (1+c)^{1/c} & \text{for } -1 < c < 0, \\ e & \text{for } c = 0. \end{cases}$$

and $\frac{A_h}{A_f} \leq e$ for $c = 0$. □

4.4.2 Two and Three Points of Contact, Domain $\mathbb{R}$

In the last subsection we have in some detail considered the mathematical properties of optimal hat functions for the case of only one point of contact. It turned out that they result in simple numerical procedures for the case of monotone densities on $[b_l, \infty)$. In this subsection we will see that the mathematical results of the last subsection can be used to find optimal hat functions (i.e. hats with minimal A_h) for the case of unimodal densities with domain $\mathbb{R}$ and $N = 2$ or 3 points of contact.

We start with two simple but crucial general observations valid for arbitrary $N > 1$. (Recall from Sect. 4.1 that the b_i divide the domain of the distribution into N subintervals.)

- If the global hat is optimal then the hat of every subinterval (b_i, b_{i+1}) is optimal as well.
- The optimal global hat function is continuous in all points b_i. (Otherwise the area below the hat can easily be reduced in the region of the discontinuity.)

Now we can easily proof:

Theorem 4.12. *For given, fixed c and strictly T_c-concave quasi-density $f(x)$ on $\mathbb{R}$ the optimal design points p_1 and p_2 have the property that $(p_1, p_2) \times (0, f(p_1))$ is the maximal rectangle below f. The rejection constant of that optimal hat is bounded by:*

$$\alpha = \frac{A_h}{A_f} \leq (1+c)^{1/c} \text{ for } -1 < c < 0 \quad \text{and} \quad \frac{A_h}{A_f} \leq e \text{ for } c = 0.$$

Proof. Eq. (4.16) implies that for an optimal hat the ratio of $h(b_1)/h(p_0) = h(b_1)/f(p_0)$ is fixed. Thus $f(p_1) = f(p_2)$. From Thm. 4.11 we know that we can find p_0 as the point that maximizes the area of the rectangle below f. Together with $f(p_1) = f(p_2)$ this directly implies that we can find a continuous hat that is optimal for both subintervals by choosing p_1 and p_2 as the borders of the maximal rectangle below f. As this is the unique hat that is optimal for both subintervals it must be the optimal hat, see also Fig. 4.6.
The global bound for α follows directly from applying Thm. 4.11 to both subintervals. □

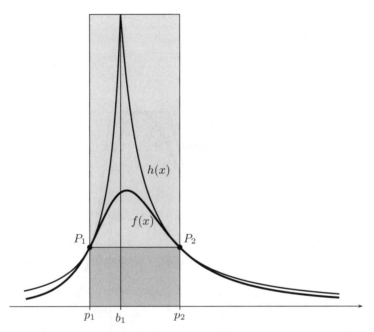

Fig. 4.6. Two construction points: The rectangle $(p_1, p_2) \times (0, h(p_1))$ is maximal below $h(x)$ and maximal below $f(x)$. The area of the rectangle $(p_1, p_2) \times (0, h(b_1))$ is equal to the area below the hat

Theorem 4.12 shows that the two-point case is very similar to the one-point case; again we have to find the maximal rectangle below the density and again the optimal design points are not influenced by c.

For the case $N = 3$ we restrict our attention to the case that the hat touches in the mode m, i.e. $p_2 = m$, and we thus have a horizontal tangent. In this case we can give simple conditions for optimal design points p_1 and p_3 that have been developed for $c = 0$ in (Devroye, 1986a, Sect. VII.2.6) and for general transformations T in Hörmann (1995).

Theorem 4.13. *Let f be a T_c-concave quasi-density on $\mathbb{R}$ with mode m and $p_2 = m$. Then the area below the hat is minimized when p_1 and p_3 fulfill the condition*

$$f(p_1) = f(p_3) = f(m)(1+c)^{-1/c} \text{ for } c < 0$$

and

$$f(p_1) = f(p_3) = f(m)/e \text{ for } c = 0,$$

see also Fig. 4.7. The area below the hat function is equal to $f(m)(p_3 - p_1)$ and $\alpha = A_h/A_f$ is bounded for all T_c-concave distributions by

$$\begin{aligned} t_o &= \frac{1}{1 - 1/(1+c)^{1+1/c}} \quad \text{for } c < 0 \quad \text{and} \\ t_o &= \frac{e}{e-1} = 1.582\ldots \quad \text{for } c = 0. \end{aligned} \quad (4.17)$$

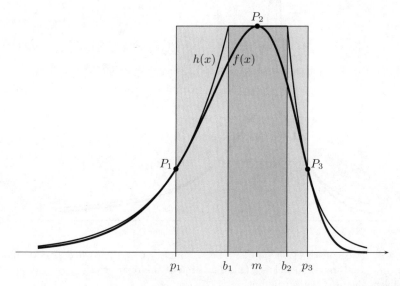

Fig. 4.7. Three construction points with one tangent in the mode m

Proof. The optimality of p_1 and p_3 follows directly from (4.16) together with the consideration that the optimal hat must be continuous and the fact that $p_2 = m$ and the hat function $h_2(x) = f(m)$.

Equation (4.13) implies that for the optimal hat the area below the hat for $x > b_2$ is $h(b_2)(p_3 - b_2) = f(m)(p_3 - b_2)$; the total area below the optimal hat is $A_h = f(m)(p_3 - p_1)$. It is easy to see that the T_c-concave quasi-density f with minimal A_f that touches the hat in all three design points has a transformed density $T(f(x))$ that is linear on (p_1, m) and linear on (m, p_3) and zero outside that interval. By integration we obtain $A_f = (p_3 - p_1)f(m)/t_o$, which proofs the bounds of α for the optimal hat. □

A Simple Choice for $N = 3$

We have seen in the discussion above that we can find convenient conditions for the optimal design points for the cases $N = 1, 2,$ and 3, and unbounded domain. But all these conditions lead to numerical search procedures and thus the setup of the resulting algorithms is not very simple and will include quite a few evaluations of the density. If A_f is known, a much simpler choice of the design points is possible for the case $N = 3$: a hat function with horizontal tangent in the mode and one back-transformed tangent for the left and for the right tail. In the case of a monotone density on (b_l, ∞) this hat reduces to the constant part and the right tail. We can prove the following theorem called the "minimax approach" by Devroye (1986a) who considered only the case $c = 0$.

Theorem 4.14. *We are given a monotone T_c-concave quasi-density f with domain (b_l, ∞) and $-1 < c \leq 0$. We construct a hat with horizontal tangent in $p_1 = b_l$ and a back-transformed tangent that touches in the second design-point p_2. For known A_f the very simple choice*

$$p_2 = b_l + \frac{t_o A_f}{f(b_l)} \quad \text{(for t_o see (4.17))} \tag{4.18}$$

guarantees that $\alpha = A_h/A_f$ is bounded by t_o.
For a T_c-concave quasi-density f with domain $\mathbb{R}$ and mode m we construct a similar hat with horizontal tangent in $p_2 = m$ and back-transformed tangents that touch in the design-points p_1 and p_3. The choice

$$p_1, p_3 = m \pm \frac{t_o A_f}{f(m)} \tag{4.19}$$

guarantees that $\alpha = A_h/A_f$ is bounded by $2t_o$ for arbitrary T_c-concave distributions on $\mathbb{R}$.

Remark. Note that for the case with domain $\mathbb{R}$ and known $F(m)$ (i.e. cdf at the mode), which includes all symmetric distributions, we can compute A_f for $x < m$ and $x > m$ separately. So we can use the formula of the monotone case for the left and the right tail and obtain $\alpha = A_h/A_f \leq t_o$ instead of $2t_o$.

Proof. We start with the case of monotone f. Without loss of generality we may assume $b_l = 0$, $f(b_l) = 1$ and $A_f = 1$; then $p_2 = t_o$. It is easy to show that in this class of densities the density $f_1(x)$, with $T_c(f_1(x))$ linear on $(0, t_o)$ and $f_1(x) = 0$ outside of that interval, fulfills the optimality condition of Thm. 4.13. On the other hand for all T_c-concave distributions with $b_l = 0$,

$f(b_l) = 1$ and $A_f = 1$ we have $f(t_o) \leq f_1(t_o)$. This implies that for all densities of this class the optimal point of contact is smaller than or equal to t_o. Thus (4.13) implies that for $p_2 = t_o$ we get $A_h \leq f(b_l)(t_o - b_l) = t_o$. The general result for monotone densities follows as a change of $f(b_l)$ and A_f is just a change of scale that dose not influence the ratio α.

For the domain $\mathbb{R}$ with unknown $F(m)$ we have to take the same formula for the design points and thus the area below the hat may be up to $2 t_o$ but clearly not more than that. □

4.4.3 Asymptotically Optimal Constructions Points

We have seen above that the computation of optimal construction points is not simple even for the case of few points; and for $N > 3$ there are no simple procedures available to obtain optimal points. But it is possible to find a rule for optimal design points in the asymptotic sense, i.e. for $N \to \infty$. Our extensive experiments with many different distributions have shown that these asymptotic rules lead to excellent results even for N around 10. Therefore we want to give a short overview of the main result and its practical potential in this section. The technical details are contained in Derflinger and Hörmann (2002) and will be published elsewhere.

To obtain an optimal hat for TDR we can try to minimize the expected number of iterations, $E(I)$, and therefore the area A_h below the hat or equivalently the area A_{h-f} between hat and density. It is also sensible to minimize the expected number of evaluations of f, $E(\#f)$, and consequently the area A_{h-s} between hat and squeeze. To be able to obtain an asymptotic theorem we have to restrict our attention to the case of bounded domains and $p_1 = b_l$, $p_N = b_r$. Note that we have exactly $N-1$ intervals. In order to describe the location of the design points for the asymptotic case we define a continuous "partitioning function" $v(x) > \epsilon$ with $\int_{b_l}^{b_r} v(x)\,dx = 1$ that defines the location of the design points p_i by the condition:

$$\int_{p_i}^{p_{i+1}} v(x)\,dx = \frac{1}{N-1}, \quad \text{for } i = 1, 2, \ldots, N-1.$$

To obtain an optimal partitioning function it is first necessary to develop the area $A_{h-f}(p_i, p_i + \Delta_i)$, with $\Delta_i = p_{i+1} - p_i$, between hat and density in the interval $(p_i, p_i + \Delta_i)$ into a MacLaurin series for Δ_i. For this we must suppose that $f(x)$ is five times differentiable. After some calculations and the continuous continuation of $A_{h-f}(p_i, p_i + \Delta_i)$ and its derivatives to $\Delta_i = 0$ we are able to obtain:

$$A_{h-f}(p_i, p_i + \Delta_i) = -\frac{T(f(p_i))''}{24 T'(f(p_i))} \Delta_i^3 + O(\Delta_i^4)$$

and $A_{h-s}(p_i, p_i + \Delta_i) = 3 A_{h-f}(p_i, p_i + \Delta_i)$. Using this Taylor series expansion it is possible to prove the following proposition.

Theorem 4.15. *We are given a four times differentiable density $f(x)$ on a bounded domain (b_l, b_r) with $T'(f(x))$ bounded and $T(f(x))'' < 0$. For $p_1 = b_l$, $p_N = b_r$ and $N \to \infty$ the uniquely determined partitioning function*

$$\tilde{v}(x) = \frac{\theta(x)^{1/3}}{\int_{b_l}^{b_r} \theta(x)^{1/3}\, dx} \quad \text{with} \quad \theta(x) = -\frac{T(f(x))''}{24 T'(f(x))}$$

is leading to the minimal value for A_{h-f} and A_{h-s} among all partitioning functions that have $\epsilon < v(x) < M$ for $b_l \le x \le b_r$. For the optimal partition we get

$$\lim_{N \to \infty} (N-1)^2 A_{h-f} = \left(\int_{b_l}^{b_r} \theta(x)^{1/3}\, dx \right)^3.$$

In addition we have $\lim_{N\to\infty} (N-1)^2 A_{h-s} = \lim_{N\to\infty} (N-1)^2\, 3 A_{h-f}$.

Proof. See Derflinger and Hörmann (2002). The idea of the proof is that using the assumptions we can show that

$$\lim_{N\to\infty} (A_{h-f}(N-1)^2) = \int_{b_l}^{b_r} \frac{\theta(x)}{v(x)^2}\, dx .$$

Then the result follows by minimizing the right hand side of the above equation under the constraint $\int_{b_l}^{b_r} v(x)\, dx = 1$. With the help of the calculus of variations it is possible to find the optimal partitioning function $\tilde{v}(x)$. □

For T_c-concave distributions we could use the above theorem directly for the case of bounded domain (b_l, b_r), and strictly concave transformed density. Unfortunately the theorem does not give us any hint how to place the points p_1 and p_N, neither for bounded nor for unbounded domain. But it is possible to obtain an excellent approximation for A_{h-s} given p_1, p_N using the asymptotic formula for the center and the area below the hat for the tail parts:

$$A_{h-s}(p_1, p_N) = \int_{x \le p_1} h(x)\, dx + \frac{3}{(N-1)^2} \left(\int_{p_1}^{p_N} \theta(x)^{1/3}\, dx \right)^3 + \int_{x \ge p_N} h(x)\, dx.$$

If we use numerical integration (for example trapezoidal rule with Euler-MacLaurin correction) for the integral of $f(x)$ over (p_1, p_n) we can develop an analogous formula for A_h: Then we can minimize the approximate $A_{h-s}(p_1, p_N)$ (or A_h) by searching for optimal values of p_1 and p_N. To find the points $p_2, p_3, \ldots, p_{N-1}$ we can use numeric integration and a search algorithm to guarantee

$$\left(\int_{p_i}^{p_{i+1}} \theta(x)^{1/3}\, dx \right) = \frac{\left(\int_{p_1}^{p_N} \theta(x)^{1/3}\, dx \right)}{N-1}.$$

To compute the points $p_1, \ldots, p_N$ following the above ideas we can either use a mathematical software package capable of doing symbolic computations or we use numerical approximations for computing the two derivatives necessary to evaluate $\theta(x)^{1/3}$. As the minima are extremely flat a very rough procedure is sufficient: We erect a grid of about 40 points (this number depends on the density and on the parameters set), evaluate $\theta(x)^{1/3}$ at these points and use instead of this function the approximation composed from the linear interpolations between neighboring points. Experience has shown that we loose practically nothing for the final values obtained for A_h and A_{h-s}. But the set-up, although somewhat complicated, becomes relatively fast. The design points can then be found by replacing $\theta(x)^{1/3}$ in the two formulas above by its linear interpolation. Our experience with various distributions shows that the minima are so flat that, using this crude numerical approximation, we loose only very little for the final value obtained for A_{h-s}.

4.4.4 Equiangular Points

The method presented in the above section delivers (almost) optimal construction points but is quite complicated. As a simple heuristic it is possible to choose points $p_1, \ldots, p_N$ with equidistributed angles. Let m be the mode or a point near the mode of the density. Then the following can be used as construction points:

$$p_i = m + \xi \tan(-\pi/2 + i\pi/(N+1)), \quad i = 1, \ldots, N, \quad (4.20)$$

where ξ is chosen according to the scale parameter of the distribution. For example ξ equal to half of the interquartile range is optimal for the Cauchy distribution and leads to acceptable results for many other distributions. $\xi = \sigma$ is a good choice for the normal distribution. If the density function has a bounded domain (b_l, b_r) this has to be modified to

$$p_i = m + \xi \tan(\theta_l + i(\theta_r - \theta_l)/(N+1)), \quad i = 1, \ldots, N \quad (4.21)$$

where $\theta_l = \arctan(b_l - m)$ and $\theta_r = \arctan(b_r - m)$. Notice that for p_i close to m this formula gives approximately the arithmetic mean of its neighbors; for very large values of p_i approximately is the harmonic mean of its neighbors. For the motivation of this idea see Sect. 4.8.

Numerical experiments with several density functions have shown that the simple rule of thumb with $\xi = 1$ results in an acceptable good hat function for many distributions with scale parameter not too different from one. Especially it works well when $\sup_{x \in (b_l, b_r)} |x - m|\sqrt{f(x)}$ is of the same order of magnitude as $\sup_{x \in (b_l, b_r)} \sqrt{f(x)}$. Indeed, if we set $u = \sup_{x \in (b_l, b_r)} |x - m|\sqrt{f(x)}$ and $v = \sup_{x \in (b_l, b_r)} \sqrt{f(x)}$ then we have the worst case bound (Leydold, 2000a)

$$\varrho_{hs} < 1 + \frac{u^2 + v^2}{uv}\left(\frac{\pi}{N} + O(N^{-2})\right).$$

In our numerical experiments we have observed much faster convergence for many important distributions, especially ϱ_{hs} mostly behaves like $1 + \text{const}/N^2$

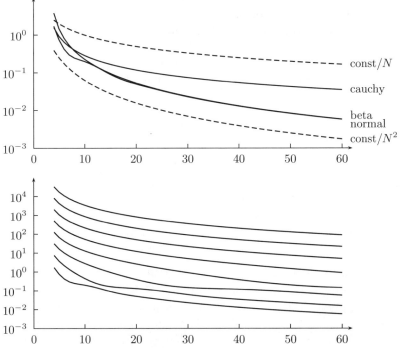

Fig. 4.8. $\varrho_{hs} - 1$ for different numbers of equiangular construction points using $T_{-1/2}$ and $\xi = 1$. First row: normal, beta(5,15) and Cauchy distribution compared to const/N and const/N^2. Second row: normal distribution with standard deviations $\sigma = 1, 2, 4, 8, 16, 32, 64,$ and 128 (from bottom to top)

for log-concave distributions. The constant factor, however, can be large, see Fig. 4.8. Using equiangular construction points is thus only applicable for "well shaped" densities and it is recommended to use them in combination with (derandomized) adaptive rejection sampling (see below).

Remark. Fig. 4.8 shows that for the Cauchy distribution we find $\varrho_{hs} \approx 1 + \text{const}/N$. Nevertheless, we even find in this example for the rejection constant $\alpha \approx 1 + \text{const}/N^2$. This differences between α and ϱ_{hs} is caused by the very heavy tails of this distribution and the fact that there is no squeeze left of p_1 and right of p_N.

4.4.5 Adaptive Rejection Sampling (ARS)

Gilks and Wild (1992) have introduced the ingenious concept of *adaptive rejection sampling* for the problem of finding appropriate construction points for tangents to the transformed density rejection method. It works in the following way:

1. Start with (at least) two points on either side of the mode or with three points: two on either side of the mode and a third point close to the mode. then sample points from the hat distribution.
2. If we have to evaluate the density f at a sampled point x, add x as a new construction point, until a certain stopping criterion is fulfilled, typically if the desired ratio ϱ_{hs} or the maximal number of construction points is reached.

Remark. Gilks and Wild (1992) have suggested this idea for the situation of Gibbs sampling (see Sect. 15.2.1). There they need only one random variate of a given density and the mode and A_f are unknown. So they start ARS and sample from the often very loosely fitting starting hat. If the generated variate is rejected they add it as a construction point. Then they repeat this procedure until the first variate is accepted and returned as the required single variate with the given density. If the densities are very expensive to evaluate, this procedure seems to be the fastest to generate a single variate, although its performance depends on the starting values.

To ensure that the starting hat is bounded, one construction point at (or at least close to) the mode should be used. Therefore it is better to start with three points instead of two. Points calculated by the equiangular rule of thumb are useful starting points for this algorithm. Figure 4.9 demonstrates the idea. Notice that the area between hat and squeeze decreases quickly in the adaptive steps.

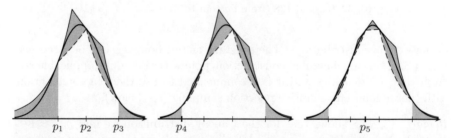

Fig. 4.9. Adaptive rejection sampling: Start with construction points p_1, p_2, and p_3. When p_4 is sampled from the hat distribution but rejected then add p_4 as new construction point. Continue and use the next rejected point p_5. The shaded region indicates the area between hat and squeeze

To see that this method works, notice that the new construction point is sampled from the quasi-density $h(x) - s(x)$. The probability that a particular interval (p_{i-1}, p_i) is split by a new construction point p is given by the relative area of the region between hat and squeeze in this interval compared to the total area $A_h - A_s$. The region in such an interval is triangular shaped and thus it is more likely that the interval is split near the center than near the

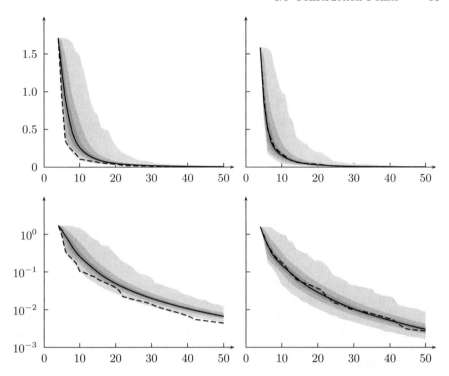

Fig. 4.10. Convergence of $\varrho_{hs} - 1$ when using adaptive rejection sampling for standard normal (l.h.s.) and Cauchy distribution (r.h.s.) (3 starting points using the equiangular rule; transformation $T_{-1/2}$). Range, 5% and 95%-percentiles, quartiles, and median are shown (100 000 samples). The dashed line gives the result using DARS with expected point as new construction point. First row: normal scale, second row: logarithmic scale

boundary. This also holds for the intervals (b_l, p_1) and (p_N, b_r) at the boundary of the domain.

Adaptive rejection sampling is the most flexible method for finding construction points. By a simple parameter, namely the ratio ϱ_{hs}, the performance of Algorithm 4.1 (TDR) can be easily controlled. It is applicable when we want to construct a hat with $\varrho_{hs} \approx 1$ as well as for sampling only a few random numbers from a particular distribution without having an expensive setup.

Figure 4.10 shows the result of a simulation for the standard normal and the Cauchy distribution with three starting points using the equiangular rule (100 000 samples). $\varrho_{hs} - 1$ is plotted against the number N of construction points. The range of ϱ_{hs} is given by the light shaded area, 90%- and 50%-tolerance intervals are shown by dark shaded areas, the median by the solid line.

Empirically we find $\varrho_{hs} = 1+O(N^{-2})$ for (almost) all samples. Simulations with other distributions and starting values show that convergence is even faster for other (non-normal) distributions. However, analytical investigations are of interest. Upper bounds for the expected value of $\varrho_{hs}(N)$ are an open problem.

It should also be noted here that ARS is the most robust method for finding construction points for densities with extreme shapes.

4.4.6 Derandomized Adaptive Rejection Sampling (DARS)

The setting of optimal construction points is the most expensive setup technique. Moreover it is not easy to estimate the necessary number of optimal construction points for the aimed ratio ϱ_{hs} in advance. The equiangular rule is cheap, but performs terrible for, e.g., the normal distribution with standard deviation 10^{-5}.

Adaptive rejection sampling, on the other hand, has a stochastic component that makes it difficult to predict the actual number of constructions points for a given target ratio ϱ_{hs}. It might be large in rare events. Moreover, for sampling efficiently from the vector $(A_1, \ldots, A_N)$ in Step 8 of Algorithm 4.1 (TDR) it is necessary to build a table for indexed search. It has to be recomputed in $O(N)$ time whenever a construction point is added. Furthermore there is no clear separation between the setup and the generation part.

These disadvantages can be avoided when we use the expected value of the distribution given by the quasi-density $h - s$. Applying this idea for *derandomized adaptive rejection sampling* results in the following method (Leydold, Janka, and Hörmann, 2002):

1. Start with a proper set of construction points as for adaptive rejection sampling (two arbitrary points on either side of the mode, optimal construction points, or the equiangular rule).
2. Each of the intervals (p_{i-1}, p_i), where the area between hat and squeeze is too large, is split by inserting a new construction point. (This includes the intervals $(p_0, p_1) = (b_l, p_1)$ and $(p_N, p_{N+1}) = (p_N, b_r)$.) As threshold values a constant fraction of the average area over all intervals is used. I.e., if $A_{i,h}$ and $A_{i,s}$ denote the respective areas below hat and squeeze in the interval (p_{i-1}, p_i), then this interval is split whenever

$$A_{i,h} - A_{i,s} \geq \kappa\,(A_{h-s})/(N+1)$$

for some constant κ. κ should be less than 1 to ensure that at least one interval is split. A good choice for this constant is $\kappa = 0.99$.

For finding splitting points in the intervals the following rules can be applied:

(a) the expected new construction point in adaptive rejection sampling, i.e.,

$$p_{\exp} = \frac{\int_{p_{i-1}}^{p_i} x\left(h(x) - s(x)\right) \mathrm{d}x}{\int_{p_{i-1}}^{p_i} h(x) - s(x) \, \mathrm{d}x}, \tag{4.22}$$

(b) the "arc-mean" of the boundary of the interval (p_{i-1}, p_i), i.e.,

$$p_{\mathrm{arc}} = \tan((\arctan(p_{i-1}) + \arctan(p_i))/2), \tag{4.23}$$

where $\arctan(\pm\infty)$ is set to $\pm\pi/2$.

Notice that rule (a) cannot be applied for unbounded intervals if $c \leq -1/2$. Then we have to switch to rule (b).

Another rule for inserting a new construction point might be the arithmetic mean of the boundary of an interval (p_{i-1}, p_i). However, this only works well if the density is close to constant in the interval. Another "obvious" choice for a splitting point are the intersection points b_i. However, this leads to less useful points if the slope of the transformed density is large.

3. If ϱ_{hs} is larger than the desired bound repeat Step 2 with updated threshold value. Otherwise we are done.

Figure 4.10 shows the performance of DARS compared to ARS for the standard normal and the Cauchy distributions. The difference between these two distributions is mainly due to the fact that the expected value $p_{\exp}$ does not exist for unbounded intervals for the Cauchy distribution and that the second rule using p_{arc} does not work very well for distributions with heavy tails.

Due to round-off errors it might happen that it is not possible to use the new construction point in Step 2. This happens for example if the density is extremely small, or if the new construction point is close to one of the old intervals. Then the implemented algorithm has to take care to avoid infinite loops. Additionally more complicated search methods like bisection have to be used. The simplest solution to this problem is to add some points using ARS and to return to DARS afterwards.

4.4.7 Convergence

The convergence of ϱ_{hs} towards 1 is important for setup time and storage requirements of the resulting algorithm when a small value of ϱ_{hs} should be reached. Computational experiences show that one finds for the above methods

$$\varrho_{hs} \approx 1 + \mathrm{const}/N^2 \,.$$

It is hard to give a proof for this observation especially for adaptive rejection sampling. The following theorem does not fully verify this empirical result but at least it gives us some confidence. Notice that for a large number N of construction points we find in each (bounded) interval (p_{i-1}, p_i), $\varrho_{hs} - 1 \approx 3(\alpha-1)$, see Sect. 4.4.3. Moreover, equidistributed construction points usually do not result in an optimal hat.

4 Transformed Density Rejection (TDR)

Theorem 4.16. *For a two times differentiable T-concave quasi-density function f with bounded domain, and N design-points placed in equal distances, the area between hat and density is $O(N^{-2})$. I.e., we find for the rejection constant α,*

$$\alpha = 1 + O(N^{-2}) \,.$$

Proof. In each interval h and f are both two times differentiable functions with the same first-order Taylor-expansion in the design point. Thus we have $|h(x) - f(x)| = O(r^2)$ around each design point p, where $r = |x - p|$. Since we have N design points with equal distances, we find for the average radius $r = O(N^{-1})$, which implies that the average distance $h(x) - f(x) = O(N^{-2})$. As we have assumed a bounded domain (b_l, b_r) we get $\int_{b_l}^{b_r} |h(x) - f(x)|\,\mathrm{d}x = O(N^{-2})$. □

4.4.8 Summary

Figure 4.11 gives an overview of the convergence for $\varrho_{hs} - 1$ for the described methods. The actual choice of the method for computing appropriate construction points for the hat depends, of course, on the particular application. We have two criteria: (1) storage requirements, and (2) setup time versus generation time. Of course the price of the setup depends on the total cost for generating all required random variates, i.e., on the total number of generated points. A guide for choosing a method is given in Table 4.3.

Table 4.3. Recommended method for crucial and non-crucial storage requirements and setup time. $\mathrm{Opt}_3 \ldots$ 3 optimal points (4.4.2), $\mathrm{Opt}_n \ldots n$ optimal points (4.4.3), $\mathrm{Eq}_n \ldots n$ equiangular points (4.4.4), (D)ARS $\ldots$ (derandomized) adaptive rejection sampling (4.4.5 and 4.4.6); $A\,\&\,B \ldots$ start with A continue with B

table setup	small	large
very fast	Eq_2 & ARS	Eq_2 & ARS
fast	Opt_3 & ARS	Eq_5 & ARS
slow	Opt_n	Eq_{10} & DARS or Opt_n

The actual choice of the number of construction points and the appropriate method for finding such points depends of course on the application. We know from the results of Chap. 2 (see Thm. 2.9) that the performance of a rejection algorithm is mainly described by

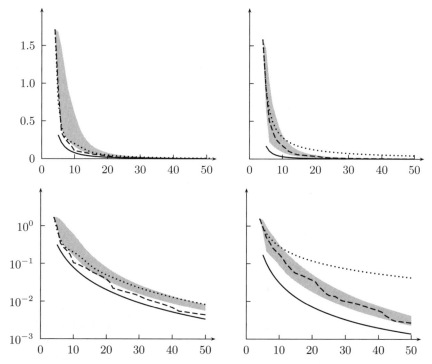

Fig. 4.11. Convergence of $\varrho_{hs} - 1$ with $T_{-1/2}$ for standard normal (l.h.s.) and Cauchy distribution (r.h.s.) using ARS (shaded region: 5% and 95%-percentiles), DARS (dashed line), equiangular points (dotted line), and (optimal) construction points using the method from Sect. 4.4.3 (solid line).
First row: standard scale, second row: logarithmic scale

$$\mathrm{E}(I) = \alpha = A_h/A_f \qquad \mathrm{E}(\#f) = \frac{A_{h-s}}{A_f} \ .$$

Theorem 4.16 above shows that we have

$$A_{h-s} = O(1/N^2) \ .$$

This result together with our experiments with many distributions show that even for a moderate number of design points (e.g. $N > 50$) we can expect that $\mathrm{E}(I)$ is close to one and $\mathrm{E}(\#f)$ close to zero. As A_f is not always available the performance of Algorithm 4.1 (TDR) can be approximately described by the more convenient ratio

$$\varrho_{hs} = \frac{A_h}{A_s}$$

which is an upper bound for the rejection constant α. The number $(\varrho_{hs} - 1)$ is an upper bound for the expected number of evaluations of the density function per generated random variate, $\mathrm{E}(\#f)$. If the user specifies the desired ϱ_{hs} we can start with a small number of arbitrary construction points on either side of the mode and then – using (derandomized) adaptive rejection sampling – add construction points till the desired value of ϱ_{hs} is reached. For ϱ_{hs} close to 1 Algorithm 4.1 (TDR) is fast and the marginal generation time does not depend on the given density f. Moreover, it is close to inversion from the cdf and thus inherits many of the desired properties of the inversion method (but is much faster for almost all distributions).

4.5 Algorithms and Variants of Transformed Density Rejection

In this section we describe variants of the basic method that make the universal algorithm faster and/or its user interface more comfortable. Additionally, we give some detailed ready to use algorithms for special cases.

4.5.1 Use Shifted Secants for Constructing Hat

It can be inconvenient, slow, or even almost impossible to code the derivative of the density f. To avoid this problem it is possible to replace the tangent of the transformed density $\tilde{f}$ at a point p by the line through the point $(p, \tilde{f}(p - \Delta))$ with the slope $(\tilde{f}(p - \Delta) - \tilde{f}(p))/\Delta$, where Δ must be chosen such that $p - \Delta$ lies between the mode m of the density and p. Due to the concavity of $\tilde{f}$ this line is always greater than or equal to $\tilde{f}$ for such a Δ, see Fig. 4.12. This constraint on Δ, rephrased as $m - p < \Delta < 0$ if $p < m$ and $0 < \Delta < p - m$ if $p > m$, also guarantees that $p - \Delta$ always lies in the support of the density f.

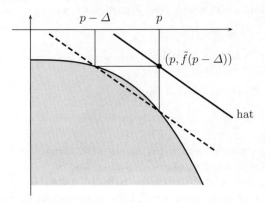

Fig. 4.12. The secant through $p - \Delta$ and p shifted into the point $(p, \tilde{f}(p - \Delta))$ is always above the transformed density $\tilde{f}$

For the choice of Δ we have to consider two antagonistic tendencies: (i) Larger values of $|\Delta|$ increase the area below the hat for a given set of construction points. Furthermore, the intersection points of the "tangents" at two closely neighbored construction points p_{i-1} and p_i now might not be in the interval (p_{i+1}, p_i). (ii) On the other hand, small values of $|\Delta|$ bear the danger of critical round-off errors.

An additional problem occurs when the mode m is not known. Then it can happen that incidentally we have $p - \Delta < m < p$ and consequently the shifted secant line might not be a hat function any more. Taking $|\Delta|$ too large increases the danger for this error, while small values for $|\Delta|$ make the possible error negligible compared to the accuracy of the floating point arithmetic.

In computing the slope of the shifted secant $(\tilde{f}(p-\Delta)-\tilde{f}(p))/\Delta$ there is the subtraction of two floating point numbers numbers of possibly very different size $(p - \Delta)$ and the subtraction of two floating point numbers of almost the same size $(\tilde{f}(p - \Delta) - \tilde{f}(p))$. Both operations may result in catastrophic cancellation errors when choosing Δ very small, e.g. 10^{-20}. To control this problem for the first operation it is enough to choose $|\Delta| = \epsilon |p|$. With $\epsilon = 10^{-5}$ this guarantees that we do not loose more than half of the digits if the precision of the used floating point numbers is at least 10 decimal places. (The precision is about 16 decimal places for double precision numbers on most modern computers). But what can we say about the second operation? If $\tilde{f}$ is very flat in that region the suggested choice of Δ may be too small. Clearly we would like to have $\Delta_y = |\tilde{f}(p - \Delta) - \tilde{f}(p)| \geq |\tilde{f}(p)| \epsilon$. Using $\Delta_y = \tilde{f}'(p) \Delta$ we get $\Delta \geq \frac{|\tilde{f}(p)|}{|\tilde{f}'(p)|} \epsilon$. As we do not know the derivative we can use the lower bound $|\tilde{f}'(p)| \geq |\tilde{f}(m) - \tilde{f}(p)|/|m - p|$ to obtain the formula:

$$\Delta = \text{sign}(m - p)\, \epsilon\, \max(|p|, |\tilde{f}(p)|\, |m - p|/|\tilde{f}(m) - \tilde{f}(p)|, \epsilon)\,.$$

The parameter ϵ within the maximum was added to avoid unnecessary small values of Δ for the case that both p and $|\tilde{f}(p)|$ are very close to 0. As stated above $\epsilon = 10^{-5}$ should be a sensible choice for most densities and computing environments.

Gilks (1992) explains another idea to obtain a derivative free version of TDR. To construct the hat he uses (non-shifted) secants in those regions where they are above the transformed density. The idea is useful for the situation of Gibbs sampling (see Sect. 15.2.1), especially when the densities are very expensive to evaluate. As the fit of such a hat is clearly worse than the hat explained above and the organization of the algorithm is more complicated we do not provide details here.

4.5.2 Proportional Squeeze

In the algorithm by Gilks and Wild (1992) (Algorithm 4.1, TDR) secants between the construction points p_i are used to construct the squeeze. Evans and Swartz (1998) have used secants between the boundary points b_i, instead. This reduces some programming complexity because then the squeeze is a linear function in all intervals (b_{i-1}, b_i). However, $N-1$ additional evaluations of the density function f (at all b_i) are necessary in the setup. This approach can be further simplified at the price of a slightly increased ϱ_{hs} by using a squeeze that is proportional to the hat h in the interval (b_{i-1}, b_i) (see Fig. 4.13). In detail we set $s(x) = s_i(x) = \nu_i h(x)$ for $b_{i-1} \leq x \leq b_i$ and a proper constant ν_i. Then we have a faster rejection step. No squeeze has to be evaluated, U has only to be compared to the constant ν_i.

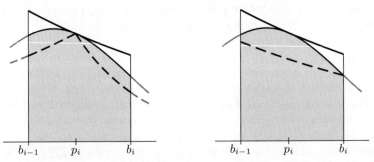

Fig. 4.13. Hat function and squeeze used in the original algorithm (l.h.s.) and the modified algorithm with proportional squeeze (r.h.s.)

Obviously the best choice for ν_i is given by $\nu_i = \min_{x \in (b_{i-1}, b_i)} f(x)/h_i(x)$. When using the transformation T_c, it is easy to find ν_i. In the transformed scale the squeeze $T_c(s_i(x)) = T_c(\nu_i h(x))$ is again a linear function. Thus by the concavity of the transformed density we only have to evaluate $f(x)/h(x)$ at the boundary points b_{i-1} and b_i of the interval. The value of ν_i is then the smaller of these two ratios. If the interval is not bounded we set $\nu_i = 0$. Algorithm 4.2 (TDR-Proportional-Squeeze) uses this new squeeze. Notice that, at the expense of Step 3, we neither have to evaluate the hat nor the squeeze in Step 12. So the generation part of this algorithm is considerably shorter and simpler than that of Algorithm 4.1 (TDR). This has also a positive influence on the marginal speed of the algorithm.

Due to the different construction of the squeeze we have to reconsider the choice of the construction points. When using adaptive rejection sampling for Algorithm 4.2 (TDR-Proportional-Squeeze) it is necessary to change the rule: Now we add a new construction point x whenever a point x is *rejected* as this leads to better fitting hats than adding a point whenever we have to evaluate the density function. Notice that derandomized rejection sampling as described in Sect. 4.4.6 cannot be used analogously. Asymptotically optimal construction points can be computed exactly with the same method as described in Sect. 4.4.3. In general we can say that construction points calculated for the original Algorithm 4.1 (TDR) work also very well for the new variant.

4.5 Algorithms and Variants of Transformed Density Rejection 91

Algorithm 4.2 TDR-Proportional-Squeeze

Require: T-concave quasi-density $f(x)$ with domain (b_l, b_r);
  transformation $T(x)$, $\tilde{f}(x) = T(f(x))$, construction points $p_1 < \ldots < p_N$.
Output: Random variate X with density prop. to f.
  /* Setup: Construct hat $h(x)$ and squeeze $s(x)$. */
1: Compute $\tilde{f}(p_i) = T(f(p_i))$ and $\tilde{f}'(p_i)$, for $i = 1, \ldots, N$.
2: Compute b_i as the intersection point of the tangents of $\tilde{f}(x)$ in the points p_i and
  p_{i+1}, for $i = 1, \ldots, N-1$.
  Set $b_0 \leftarrow b_l$ and $b_N \leftarrow b_r$.
3: Compute $\nu_i = \min\left(f(b_{i-1})/h(b_{i-1}), f(b_i)/h(b_i)\right)$, for $i = 1, \ldots, N$.
  Set $\nu_1 \leftarrow 0$ if $b_l = -\infty$, and $\nu_N \leftarrow 0$ if $b_r = \infty$.
4: Compute areas A_i below hat $h(x)$ for each subinterval (b_{i-1}, b_i) using (4.2), for
  $i = 1, \ldots, N$.
5: Set $A \leftarrow \sum_{i=1}^{N} A_i$.
  /* Generator */
6: **loop**
7:   Generate $V \sim U(0, A)$.
    /* Generate J with probability vector proportional to $(A_1, \ldots, A_N)$. */
8:   $J \leftarrow \min\{J : A_1 + \cdots + A_J \geq V\}$. /* use *Indexed Search* (Sect. 3.1.2) */
    /* Recycle uniform random number */
9:   $V \leftarrow V - (A_1 + \cdots + A_{J-1})$. /* $V \sim U(0, A_J)$. */
    /* Generate X with density proportional to h_J. */
10: Compute $X \leftarrow H_J^{-1}(V)$ using (4.4).
11: Generate $U \sim U(0, 1)$.
12: **if** $U \leq \nu_J$ **then** /* evaluate squeeze */
13:    **return** X.
14: Compute $h(X) \leftarrow T^{-1}(\tilde{f}(p_J) + \tilde{f}'(p_J)(X - p_J))$.
15: **if** $U h(X) \leq f(X)$ **then** /* evaluate density */
16:    **return** X.

4.5.3 Region of Immediate Acceptance

It may be important in certain situations to design a transformed density rejection algorithm where the expected number of uniform random numbers $E(\#U)$ is close to one. Such an algorithm requires sampling from the region between the hat and the squeeze, see Sect. 2.3.3. Leydold (2000b) suggested a quite complicated version that directly uses the hat and squeeze of Algorithm 4.1 (TDR). But when we use a squeeze that is proportional to the hat h as suggested above in Algorithm 4.2 (TDR-Proportional-Squeeze), the region between hat and squeeze is much simpler. Now random variates with density proportional to the difference between hat and squeeze $(h - s)$ can easily be generated, since then for an interval (b_{i-1}, b_i) we find $h(x) - s(x) = h(x) - \nu_i h(x) = (1 - \nu_i) h(x)$, i.e. its density is proportional to the hat $h(x)$. Thus we can decompose the hat h into the density below the squeeze and the density between squeeze and hat, where both densities are proportional to the hat function in every interval (b_{i-1}, b_i). We then utilize

Algorithm 4.3 TDR-Immediate-Acceptance

Require: T-concave quasi-density $f(x)$ with domain (b_l, b_r);
transformation $T(x)$, $\tilde{f}(x) = T(f(x))$, construction points $p_1 < \ldots < p_N$.
Output: Random variate X with density prop. to f.
/* Setup: Construct hat $h(x)$ and squeeze $s(x)$. */
1: Compute $\tilde{f}(p_i) = T(f(p_i))$ and $\tilde{f}'(p_i)$, for $i = 1, \ldots, N$.
2: Compute b_i as the intersection point of the tangents of $\tilde{f}(x)$ in the points p_i and p_{i+1}, for $i = 1, \ldots, N-1$.
 Set $b_0 \leftarrow b_l$ and $b_N \leftarrow b_r$.
3: Compute $\nu_i = \min\left(f(b_{i-1})/h(b_{i-1}), f(b_i)/h(b_i)\right)$, for $i = 1, \ldots, N$.
 Set $\nu_1 \leftarrow 0$ if $b_l = -\infty$, and $\nu_N \leftarrow 0$ if $b_r = \infty$.
4: Compute areas A_i below hat $h(x)$ for each subinterval (b_{i-1}, b_i) using (4.2), for $i = 1, \ldots, N$.
5: Set $A \leftarrow \sum_{i=1}^{N} A_i$.
/* Generator */
6: **loop**
7: Generate $V \sim U(0, A)$.
  /* Generate J with probability vector proportional to $(A_1, \ldots, A_N)$. */
8: $J \leftarrow \min\{J : A_1 + \cdots + A_J \geq V\}$. /* use *Indexed Search* (Sect. 3.1.2) */
  /* Recycle uniform random number */
9: $V \leftarrow V - (A_1 + \cdots + A_{J-1})$. /* $V \sim U(0, A_J)$. */
  /* Generate X with density proportional to h_J. */
10: **if** $V \leq \nu_J A_J$ **then** /* below squeeze, immediate acceptance */
11:   Compute $X \leftarrow H_J^{-1}(V/\nu_J)$ using (4.4).
12:   **return** X.
13: **else** /* accept or reject from $h_J - s_J$. */
14:   Compute $X \leftarrow H_J^{-1}((V - \nu_J A_J)/(1 - \nu_J))$ using (4.4).
15:   Compute $h(X) \leftarrow T^{-1}(\tilde{f}(p_J) + \tilde{f}'(p_J)(X - p_J))$.
16:   Generate $Y \sim U(\nu_J h(X), h(X))$.
17:   **if** $Y \leq f(X)$ **then** /* evaluate density */
18:    **return** X.

the composition method; the area below the squeeze can be used as "region of immediate acceptance", i.e. the generated random variate X can be immediately returned without the requirement of a second uniform random number (see Sect. 2.3.3). So we can easily compile Algorithm 4.3 (TDR-Immediate-Acceptance). Notice that the setup step of this new algorithm is identical to that of Algorithm 4.2 (TDR-Proportional-Squeeze). Using the results of Thm. 2.17 we observe that $E(\#f) = \frac{A_{h-s}}{A_f} \leq \varrho_{hs} - 1$ is the same as for Algorithm 4.2 whereas $E(\#U) = \frac{2A_{h-s}}{A_f} \leq 2\varrho_{hs} - 1/\varrho_{hs}$ is clearly smaller and close to one as long as N is not too small.

4.5.4 Three Design Points

We want to demonstrate here how the results from the previous sections can be used to design universal algorithms. One advantage of transformed density

4.5 Algorithms and Variants of Transformed Density Rejection

rejection is the freedom to choose the number of design points N. Large values of N (e.g. around 100) lead to very fast algorithms with a slow setup and large tables, whereas small values of N can be used to design comparatively simple universal algorithms with small tables and moderate setups. For very short universal algorithms practically without setup see Chap. 6 below.

We start here with the case $N = 3$ (Hörmann, 1995) which is more convenient than $N = 2$ if the mode m of the distribution is known. In addition with one construction point in the mode there is no danger that values of the transformed hat function do not belong to the domain of T^{-1}. We have discussed above that we can use an arbitrary value of c, $-1 < c \leq 0$. Due to the simplicity of the necessary transforms T, T^{-1} and F_T^{-1} it is most convenient to use $c = -1/2$ or $c = 0$. The case $c = -1/2$ leads to an automatic algorithm that works for a larger class than $c = 0$. Moreover, it leads to faster implementations. The necessary changes for other values of c can be easily obtained.

For the following we set $c = -1/2$. We take $p_1 < m$, $p_2 = m$ and $p_3 > m$ as construction points, where m denotes the mode of the distribution. For the choice of p_1 and p_3 we can use the results of Thm. 4.14; the setup is fastest if we use formula (4.19). It turns out that this choice leads to points p_1 and p_3 that are too far away from the mode for most bell-shaped densities. So we suggest to start with the formula

$$p_{3,1} = m \pm 0.664/f(m) \qquad (4.24)$$

which minimizes the rejection constant $\alpha = A_h/A_f$ for the normal distribution. We tested this choice of the construction points for gamma, beta and t-distribution with many different parameters greater or equal to one. α was always below 1.6, for most of the nearly symmetric distributions it was close to 1.32. But it also may happen that $\alpha > 4$ for this choice. In such a case we restart the setup using formula (4.19) (i.e. we replace 0.664 by 2 in (4.24)). Then we know from Thm. 4.14 that α is bounded by 4. The algorithm is a special case of Algorithm 4.1 (TDR). To make it more accessible we have collected all details as Algorithm 4.4 (TDR-3-Points).

This algorithm also uses secants instead of tangents (see Sect. 4.5.1) to remove the burden of coding the derivative of the given density. The formula for Δ of Sect. 4.5.1 is used in the algorithm and guarantees that no serious round-off errors can occur in computing the slope of the secants. Another detail of the algorithm we have not explained yet refers to the case when p_1 lies outside the support of the distribution. If this is the case the hat is constructed with only two points of contact and the corresponding area A_1 is set to zero. But this also implies that we have no squeeze left of the mode. As this slows down the algorithm considerably we decided to define the point p_1 in such a case just for computing the squeeze as: $p_1 = m - 0.6(m - b_0)$, where b_0 denotes the left border of the domain.

Algorithm 4.4 TDR-3-Points (for $c = -1/2$)

Require: $T_{-1/2}$-concave quasi-density $f(x)$ with A_f close to 1; location of mode m; b_0 (lower) and b_3 (upper) bound of the domain.
Output: Random variate X with density prop. to f.
/* Setup: */
1: Set $\tilde{f}(m) \leftarrow -1/\sqrt{f(m)}$, $g \leftarrow 0.664$.
2: **for** $i = 1, j = 1, k = -1$ and $i = 3, j = 2, k = +1$ **do**
3:     Set $p_i \leftarrow m + kg/f(m)$.
4:     **if** $kp_i < kb_{j+k}$ **then**
5:        Set $\tilde{y}_i \leftarrow -1/\sqrt{f(p_i)}$, $\tilde{s}'_i \leftarrow (\tilde{f}(m) - \tilde{y}_i)/(m - p_i)$.
6:        Set $\Delta \leftarrow \max(|p_i|, |\tilde{y}_i/\tilde{s}'_i|) \cdot 10^{-5}$
7:        $y_i \leftarrow -1/\sqrt{f(p_i - k\Delta)}$, $a_i \leftarrow -k(y_i - \tilde{y}_i)/\Delta$.
8:        Set $b_j \leftarrow p_i + (\tilde{f}(m) - y_i)/a_i$, $d_i \leftarrow y_i - a_i p_i$, $A_i \leftarrow k/(a_i \tilde{f}(m))$.
9:        **if** b_{j+k} finite **then**
10:           Set $A_i \leftarrow A_i - k/(a_i (a_i b_{j+k} + d_i))$.
11:        **else**
12:           Set $A_i \leftarrow 0$.
13:     **if** $kb_{j+k} > km$ **then** /* We set p_i for the squeeze only */
14:        Set $p_i \leftarrow m + (b_{j+k} - m)\,0.6$, and $\tilde{s}'_i \leftarrow (\tilde{f}(m) + 1/\sqrt{f(p_i)})/(m - p_i)$.
15: Set $A_2 \leftarrow (b_2 - b_1) f(m)$, and $A \leftarrow A_1 + A_2 + A_3$.
16: **if** $A \geq 4$ **then**
17:     **if** $g = 0.664$ **then**
18:        Set $g \leftarrow 2$, and start with the for-loop (Step 2) again.
19:     **else**
20:        **Exit** with error message.
/* Generator */
21: **loop**
22:     Generate $U \sim U(0, A)$.
23:     **if** $U \leq A_1$ **then**
24:        set $X \leftarrow -d_1/a_1 + 1/(a_1^2 (U + 1/(a_1 \tilde{f}(m))))$,
          and $l_X \leftarrow (a_1 (U + 1/(a_1 \tilde{f}(m))))^2$.
25:     **else**
26:        **if** $U \leq A_1 + A_2$ **then**
27:           Set $X \leftarrow (U - A_1)(b_2 - b_1)/A_2 + b_1$, and $l_X \leftarrow f(m)$.
28:        **else**
29:           Set $l_X \leftarrow U - A_1 - A_2 - 1/(a_3 \tilde{f}(m))$,
          $X \leftarrow -d_3/a_3 - 1/(a_3^2 l_X)$, and $l_X \leftarrow (a_3 l_X)^2$.
30:     Generate $V \sim U(0, l_X)$.
31:     **if** $X < m$ **then**
32:        **if** $X \geq p_1$ and $V (\tilde{f}(m) - (m - X)\tilde{s}'_1)^2 \leq 1$ **then**
33:           **return** X.
34:     **else**
35:        **if** $X \leq p_3$ and $V (\tilde{f}(m) - (m - X)\tilde{s}'_3)^2 \leq 1$ **then**
36:           **return** X.
37:     **if** $V \leq f(X)$ **then**
38:        **return** X.

4.6 Other Transformations

We have seen in Thm. 4.1, that the family T_c includes all scale invariant transformations (i.e. all transformations where a change of the multiplicative factor of the density does not change the shape of the hat). That's the reason why the family T_c plays a key-role for transformed density rejection. Nevertheless, it could be worthwhile to think about other possible transforms.

It seems possible to improve the T_c family when noticing that many of the most important distributions are bell-shaped. So it is not optimal to use $c \leq 0$ between the inflection points of such a density as f is concave in that region anyway. Thus $c = 1$ or even higher values of c would lead to better fitting hats. Of course we could also try to reach this aim by using different transformations in different regions of the domain of the densities (cf. Sect. 4.7.3) but in the moment we are concentrating on the choice of the transformation.

We experimented quite a lot but did not find anything that was really worth to replace T_c in the design of automatic algorithms. Nevertheless, there are two suggestions that we want to present here (for other examples see Exercises 4.8, 4.9, and 4.10).

The first example is based on an – as far as we know – unpublished idea of M. Evans. Use the inverse of a density as transformation T. Then transformed density rejection could be used to produce a piecewise hat that follows the local characteristics of the density. This idea can be also used to get the class T_c from above. If we start from the density of the exponential distribution we arrive at the transformation $T_0 = \log$, (for other values of c see Exercise 4.11).

For practically all other densities we have the problem that we would need a closed form representation for the inverse of the density, the integral of the density and its inverse. This is a reason that, for example, the normal distribution and many other standard densities cannot be used fruitfully for that approach. The only distribution that gives practically applicable results that are not included in the class T_c seems to be the Cauchy-distribution. But for the Cauchy density $f(x) = 1/(1 + x^2)$ and the induced transformation $T_{\text{Cauchy}}(x) = -\sqrt{1/x - 1}$ the local behavior changes considerably with x. Therefore we do not have the nice property that the shape of the constructed hat is not influenced by the scaling constant of the density, as shown for the class T_c in Thm. 4.1. Another problem is that T_{Cauchy} is not defined for values $x \geq 1$. Therefore we have to know the mode m of the density and have to scale the density such that $f(m) \leq 1$. But even then the term T_{Cauchy}-concave is not well defined as it can happen, that for one scaling factor the density is T_{Cauchy}-concave but for another scaling factor it is not (see Exercise 4.12 for an example). Figure 4.14 shows the use of T_{Cauchy} for the positive part of the normal distribution. The figure shows the original scale for the normal distribution with $f(0) = 1$ (l.h.s.) and $f(0) = 0.1$ (r.h.s.). Notice the difference between the two constructed hats.

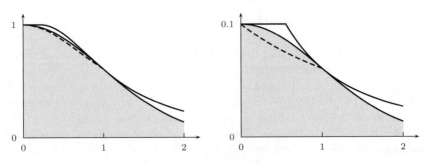

Fig. 4.14. Hat and squeeze for the half-normal distribution with different scaling factor using T_{Cauchy} and points of contact 0 and 1

A second class of new transformations that could be especially suitable to generate from bell-shaped densities can be found by sticking two members of the T_c-family together, such that the resulting transformation remains differentiable. This is of no practical importance as we can obtain the same effect much easier using different values of c for different subintervals as explained in Sect. 4.7.3 below. Nevertheless, we think that a short discussion here is in place, mainly to demonstrate that we can find usable transforms that are not in the class T_c.

The transformation $T_{(0|1)}(x) = \log(x)$ for $x \leq e$ and $T_{(0|1)}(x) = x/e$ for $x > e$ seems to be the simplest version regarding the simplicity of the functions $F_T(x)$ and $F_T^{-1}(x)$, see Exercise 4.13 for a discussion of that fact. (In Exercise 4.10 another transformation $T_{(-1/2|1)}$ is introduced.) Although this new transformation is changing its local behavior like T_{Cauchy} above, we now have the advantage that we know where this change occurs. Thus we can use transformation $T_{(0|1)}$ for all symmetric or monotone log-concave densities f when we choose a constant a such that $a\,f(x) = e$ holds for the inflection point(s) of the density.

Comparisons of the transformations discussed here with the class T_c will be based on a comparison of the area A_h below the hat for a fixed number of design points N. In Exercise 4.16 we see that we can obtain moderate gains for the acceptance probability of Algorithm 4.1 (TDR) by using these two special transformations instead of T_0 and $T_{-1/2}$. But this gain becomes negligible if we increase the number of design-points. For demonstrating the generality of the mathematical theory with examples we think that the introduction of the special transformations here is justified. For the design of good universal algorithms they are of no practical relevance, even more as the same hat as for transform $T_{(0|1)}$ can be easily obtained by using different transformations from the T_c-class for different intervals of the density (see Sect. 4.7.3).

4.7 Generalizations of Transformed Density Rejection

It is not difficult to generalize TDR. But the main benefits of a good universal random variate generator are its simplicity and its generality, i.e. its applicability to a large class of distributions. So we think that the enhanced complexity of the method due to the generalization is only justified if the class of applicable distributions is really increased. As the original version of TDR is restricted to T-concave distributions we are concentrating here on simple generalizations that can be used to sample from non-T_c-concave distributions including distributions with unbounded densities.

4.7.1 T-Convex Distributions

If we know that a density f is T-convex, i.e. $T(f(x))$ is convex on its domain, we can reverse the role of tangent and secants in standard TDR to construct hat and squeeze. Now the hat consists of secants connecting neighboring construction points whereas the squeeze is the maximum of the tangents touching the density in the construction points. The idea for this generalization is due to Evans and Swartz (1998) (see also Evans and Swartz (2000)[1]). This idea is of practical interest especially for unbounded densities as they cannot be treated by standard TDR. Notice that the boundary points of the domain (b_l, b_r) of the density f have to be used as construction points for secants to get a hat function. Figure 4.15 shows hat and squeeze for a beta distribution with two poles and construction points 0, 0.1, 0.7 and 1. Since 0 and 1 are poles of the density they cannot be used for construction points of tangents and so the squeeze is only the minimum of two tangents in this example.

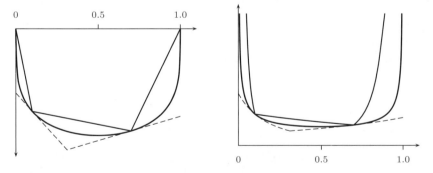

Fig. 4.15. Hat and Squeeze for the beta distribution with parameters $a = b = 1/2$. Construction points at 0, 0.1, 0.7 and 1. Transformed scale for $c = -1/2$ (l.h.s.) and original scale (r.h.s.).

---

[1] They also allow decreasing transformations which can swap the meaning of T-convex and T-concave, depending on the transformation!

To compile the details of such an algorithm for a T-convex density f with domain (b_l, b_r) we consider construction points $b_l = p_1 < p_2 < \ldots < p_N = b_r$. Notice that (b_l, b_r) must be bounded as f is T-convex. Again we write $\tilde{f}(x) = T(f(x))$. Then we construct a hat function by means of secants by

$$h_i(x) = T^{-1}\left(\tilde{f}(p_i) + \tilde{h}'(p_i)(x - p_i)\right) \quad \text{for} \quad p_i \leq x < p_{i+1}, \quad i = 1, \ldots, N-1$$

where

$$\tilde{h}'(p_i) = \frac{\tilde{f}(p_{i+1}) - \tilde{f}(p_i)}{p_{i+1} - p_i} \ .$$

Problems arise when b_l or b_r is a pole of the density f. Then we have to set

$$\tilde{f}(p_0) = \lim_{x \searrow b_l} \tilde{f}(x) \quad \text{and} \quad \tilde{f}(p_N) = \lim_{x \nearrow b_r} \tilde{f}(x) \ .$$

Of course this only works if these limits exist and are finite. That is, the following additional condition must hold.

(T4) $\lim_{x \searrow b_l} \tilde{f}(x) < \infty$ and $\lim_{x \nearrow b_r} \tilde{f}(x) < \infty$.

Notice that (T4) is satisfied for all transformations T_c with $c < 0$ but always wrong for $c \geq 0$.

In close analogy to Sect. 4.1 we obtain the following formulas for the areas A_i below the hat function and for the inverse cdf H^{-1} of the hat function in the subintervals:

$$A_i = \frac{1}{\tilde{h}'(p_i)}\left(F_T(\tilde{f}(p_i) + \tilde{h}'(p_i)(p_{i+1} - p_i)) - F_T(\tilde{f}(p_i))\right), \quad (4.25)$$

$$H_i^{-1}(u) = p_i + \frac{1}{\tilde{h}'(p_i)}\left(F_T^{-1}\left(\tilde{h}'(p_i)\,u + F_T(\tilde{f}(p_i))\right) - \tilde{f}(p_i)\right), \quad (4.26)$$
$$\text{for } 0 \leq u \leq A_i.$$

The squeeze in the interval (p_i, p_{i+1}) is simply the minimum of the tangents in p_i and p_{i+1} transformed back into the original scaled:

$$s_i(x) = \min\left(T^{-1}(\tilde{f}(p_i) + \tilde{f}'(p_i)(x - p_i)), T^{-1}(\tilde{f}(p_{i+1}) + \tilde{f}'(p_{i+1})(x - p_{i+1}))\right) \ .$$

Again we have to take care about poles at the boundary of the domain (b_l, b_r). Then we use tangents at the points that are parallel to the y-axis, i.e. we set $\tilde{f}'(b_l) = -\infty$ and $\tilde{f}'(b_r) = \infty$, respectively.

The details for a generator for T-convex distributions are compiled in Algorithm 4.5 (TDR-T-convex). However, the treatment of the special case where the density f has poles is omitted.

For finding appropriate construction points we can use any of the methods for standard TDR as described in Sect. 4.4. Of course other variants like squeeze proportional to the hat (Sect. 4.5.2) can be adapted for T-convex densities (then we have a hat proportional to the squeeze).

4.7 Generalizations of Transformed Density Rejection

Algorithm 4.5 TDR-T-convex

Require: T-convex quasi-density $f(x)$ with bounded domain (b_l, b_r); transformation $T(x)$, construction points $b_l = p_1 < \ldots < p_N = b_r$.
Output: Random variate X with density prop. to f.
/* Setup: Construct hat $h(x)$ and squeeze $s(x)$. */
1: Compute $\tilde{f}(p_i) = T(f(p_i))$ and $\tilde{f}'(p_i)$, for $i = 1, 2, \ldots, N$.
2: Compute $\tilde{h}'(p_i) = (\tilde{f}(p_{i+1}) - \tilde{f}(p_i))/(p_{i+1} - p_i)$, for $i = 1, 2, \ldots, N-1$.
3: Compute b_i as the intersection point of the tangents of $\tilde{f}(x)$ in the points p_i and p_{i+1}, for $i = 1, \ldots, N-1$.
4: Compute areas A_i below hat $h(x)$ for each subinterval (p_i, p_{i+1}) using (4.25) for $i = 1, \ldots, N-1$.
5: Set $A \leftarrow \sum_{i=1}^{N-1} A_i$.
/* Generator */
6: **loop**
7:    Generate $V \sim U(0, A)$.
/* Generate J with probability vector proportional to $(A_1, \ldots, A_{N-1})$. */
8:    $J \leftarrow \min\{J: A_1 + \cdots + A_J \geq V\}$. /* use *Indexed Search* (Sect. 3.1.2) */
/* Recycle uniform random number */
9:    $V \leftarrow V - (A_1 + \cdots + A_{J-1})$. /* $V \sim U(0, A_J)$. */
/* Generate X with density proportional to h in the interval (p_J, p_{J+1}). */
10:    Compute $X \leftarrow H_J^{-1}(V)$ using (4.26).
/* Evaluate hat and squeeze */
11:    Compute $h(X) \leftarrow T^{-1}(\tilde{f}(p_J) + \tilde{h}'(p_J)(X - p_J))$.
12:    **if** $X < b_J$ **then**
13:      Compute $s(X) \leftarrow T^{-1}(\tilde{f}(p_J) + \tilde{f}'(p_J)(X - p_J))$.
14:    **else** /* $X \geq b_J$ */
15:      Compute $s(X) \leftarrow T^{-1}(\tilde{f}(p_{J+1}) + \tilde{f}'(p_{J+1})(X - p_{J+1}))$.
/* Accept or reject */
16:    Generate $U \sim U(0, 1)$.
17:    **if** $U h(X) \leq s(X)$ **then** /* evaluate squeeze */
18:      **return** X.
19:    **if** $U h(X) \leq f(X)$ **then** /* evaluate density */
20:      **return** X.

4.7.2 Non-T-Concave Distributions

With Algorithms 4.1 (TDR) and 4.5 (TDR-T-convex) we have methods available for sampling from T-concave and T-convex distributions, respectively. If the inflection points of the transformed density are known we can combine these two algorithms. Using these inflection points we split the domain of the density f into subdomains where f is either T-concave or T-convex and apply the appropriate algorithm in each of these. We then arrive at Algorithm 4.6 (TDR-Mixed) that works for all densities including unbounded ones, provided that the inflection points of the transformed density are known. Unfortunately this is only practically applicable if the number of inflection points of the transformed density is bounded (and small). This condition also implies that

Algorithm 4.6 TDR-Mixed

Require: quasi-density $f(x)$ with domain (b_l, b_r);
 transformation $T(x)$, $\tilde{f}(x) = T(f(x))$;
 list $b_l = w_1 < w_2 < \ldots < w_N = b_r$ of inflection points of $\tilde{f}$ and poles of f;
 a function to decide whether $\tilde{f}$ is convex or concave on an interval (w_i, w_{i+1}).
Output: Random variate X with density prop. to f.
 /* Setup: */.
1: **for** $i = 1, \ldots, N-1$ **do**
2: **if** $\tilde{f}|_{(w_i, w_{i+1})}$ is concave **then**
3:   Construct hat for T-concave density (setup of Alg. 4.1, TDR).
4: **else** /* $\tilde{f}|_{(w_i, w_{i+1})}$ is convex */
5:   Construct hat for T-convex density (setup of Alg. 4.5, TDR-T-convex).
6: Compute areas A_i below the hat for each subinterval (w_i, w_{i+1}).
 /* Generator */
7: **loop**
8: Generate J with probability vector proportional to $(A_1, \ldots, A_{N-1})$.
  /* use *Indexed Search* (Sect. 3.1.2) */
9: Generate random variate X from hat distribution for subdomain (w_J, w_{J+1}) (one trial in resp. Alg. 4.1, TDR, or 4.5, TDR-T-convex).
10: **if** X has been accepted at first trial **then**
11: **return** X.
12: **else**
13: Redo loop.

---

f becomes inevitably T_c-concave in the tails for a $c > -1$. For an example of a density where this condition is violated see Exercise 4.17.

One practical disadvantage of Algorithm 4.6 (TDR-Mixed) is the fact that the user has to know the inflection points of the transformed density. We can also imagine that the setup becomes rather slow when coded for arbitrary densities with possibly many inflection points. Nevertheless, Algorithm 4.6 (TDR-Mixed) can be used to sample from distributions when the density f and its derivative f' are given. If it is not T-concave, then we can determine the inflection points as the local maxima and minima of f' (cf. Exercise 4.18). For bounded intervals Algorithm 4.5 (TDR-T-convex) can also be used without transformation, i.e. the case $c = 1$ in our notation. Then the hat consists of piecewise tangents in the concave regions of the density and of piecewise secants in the convex regions. For the squeeze it is exactly the other way round.

4.7.3 Varying Values of c

Another variant of TDR also suggested by Evans and Swartz (1998) is the use of different values of c for different regions of the density. Of course this makes the algorithm slightly more complicated and less elegant but it can lead to better fitting hat functions. If we think of densities with unbounded domains that are T_c-concave only for a $c < -1$ we cannot use ordinary TDR with fixed

4.7 Generalizations of Transformed Density Rejection

c as this would lead to a hat that is not a quasi-density. But in most cases such a density is in the tails T_c-concave for a larger c. Thus different values of c in different regions can make Algorithm TDR applicable to such distributions.

If we are interested in a very good fit of the hat, perhaps because we have a very complicated density that is very difficult to evaluate, we can try to use an optimal c that is constant in an interval around every construction point. As we can see from Fig. 4.3 on p. 64 such an optimal c is the largest possible value for c. A good candidate for such a c for a region around p_0 is the local concavity $\mathrm{lc}_f(p_0)$ which we have defined in Sect. 4.3, since it transforms the density into an approximately linear function around the point. This guarantees that the linear hat in the transformed scale has an optimal fit. If we want to use the TDR principle we have to take the minimal value of lc_f over the interval thus guaranteeing the T_c-concavity over the interval. In general it is certainly not worth the time to find the minimum over lc_f for an interval, but for many of the most important standard distributions (including the normal, gamma, beta and t-distributions) it turns out that $\mathrm{lc}_f(x)$ is unimodal with the same mode as the density itself. In this case we have no problem to find the minimum of lc_f over an interval as it is always in an endpoint. To construct a simple hat using this observation we decompose the domain of the distribution into intervals. Let us assume as an example that $\mathrm{lc}_f(x)$ is (like the density) monotonically decreasing in the interval (p_1, p_2). Then we take the infimum over the interval:

$$c = \inf_{p_1 < x < p_2} \mathrm{lc}_f(x) = \mathrm{lc}_f(p_2)$$

as c for that interval. Thus for the interval (p_1, p_2) the density is $T_{\mathrm{lc}_f(p_2)}$-concave. The second derivative of the transformed density is 0 in p_2 and the tangent touching in p_2 is a very close fitting hat for the transformed density.

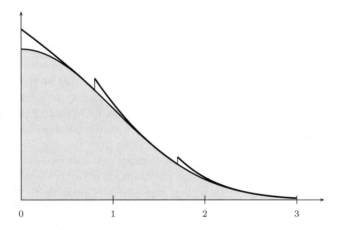

Fig. 4.16. Hat for the half-normal distribution with design points 0.8, 1.7, and 3. The c-values are 1.56, 0.35, and 0.11 in the respective subintervals.

Algorithm 4.7 TDR-c-Optimal

Require: T-concave monotone density $f(x)$ and its monotone local concavity $\mathrm{lc}_f(x) = 1 - \frac{f''(x)\,f(x)}{f'(x)^2}$; transformation $T_c(x) = \mathrm{sign}(c)\,x^c$, construction points $-\infty < p_0, \ldots, p_N < \infty$ with p_0 (lower) and p_N (upper) bound of the domain.
Output: Random variate X with density prop. to f.
  /* Setup: Construct hat $h(x)$ and squeeze $s(x)$. */
1: Compute and store $c_i \leftarrow \mathrm{lc}_f(p_i)$ for $i = 2, 3 \ldots, N$.
2: Compute and store $\tilde{f}(p_i) = T_{c_i}(f(p_i))$ and $\tilde{f}'(p_i) = T_{c_i}(f(p_i))'$ for $i = 1, 2, \ldots, N$.
3: Store $\tilde{s}'(p_i) = \frac{T_{c_i}(f(p_i)) - T_{c_i}(f(p_{i-1}))}{p_i - p_{i-1}}$ for $i = 2, \ldots, N$.
4: Store $A_i \leftarrow (F_{T c_i}(\tilde{f}(p_i)) - F_{T c_i}(\tilde{f}(p_i) + \tilde{f}'(p_i)\,(p_{i-1} - p_i)))/\tilde{f}'(p_i)$ for $i = 2, \ldots N$.
  /* $A_i = \int_{p_{i-1}}^{p_i} h_i(x)\,dx$ for $h_i(x) = T_{c_i}^{-1}\left(\tilde{f}(p_i) + \tilde{f}'(p_i)(x - p_i)\right)$. */
5: Set $A \leftarrow \sum_{i=2}^{N} A_i$.
  /* Generator */
6: **loop**
7:   Generate J with probability vector proportional to $(A_2, \ldots, A_N)$.
8:   Generate $V \sim U(0, A_J)$. /* Recycle uniform random number */
9:   Set $X \leftarrow p_J + (F_{T c_J}^{-1}(\tilde{f}'(p_J) V + F_{T c_J}(\tilde{f}(p_J) + \tilde{f}'(p_J)(p_{J-1} - p_J))) - \tilde{f}(p_J))/\tilde{f}'(p_J)$. /* X has now density proportional to h_J. */
10: Compute $h(X) \leftarrow T^{-1}(\tilde{f}(p_J) + \tilde{f}'(p_J)(X - p_J))$.
11: Generate $U \sim U(0,1)$.
12: Set $s(X) \leftarrow T^{-1}(\tilde{f}(p_J) + \tilde{s}'(p_J)(X - p_J))$.
13: **if** $U\,h(X) \leq s(X)$ **then** /* evaluate squeeze */
14:    **return** X.
15: **if** $U\,h(X) \leq f(X)$ **then** /* evaluate density */
16:    **return** X.

If we do the same for all intervals the back-transformed hat is no longer continuous but it has a very good fit as in all construction points the second derivative of the hat and the density coincide. Figure 4.16 shows this idea for the half-normal distribution on $[0, 3]$ and design points at 0.8, 1.7, and 3.

It is possible to show that for this variant of TDR the area between density and hat is no longer $O(1/N^2)$ but $O(1/N^3)$ even for equidistant design points.

A critical reader could raise the question, why the way to construct a hat described here with using the "optimal c" should be better than using the Taylor series of the density in several points or a spline polynomial instead. It is better as we want to generate random variates from the density. Therefore we need an approximation (or hat) of the density that has an easily invertible antiderivative which is not the case for polynomials of order higher than two.

It is also a natural question to ask whether it is possible to change the above construction such that the c is not changing in discrete jumps but continuously. We could use the optimal c (i.e. lc_f) in every point. The idea is good but it implies that every density is transformed into a straight line. So the transformed hat is the same as the transformed density. Transformed back again the hat is the same as the density. That means that TDR results in ordinary inversion and only works if the standard inversion algorithm works as well.

4.7.4 Generalized Transformed Density Rejection

Devroye (2001) has drawn our attention to the fact that several standard distributions (like e.g. the gamma distribution) are getting log-concave when we transform the random variate with the logarithm. We could define the family of *log-log-concave* distributions: i.e. distributions X where the random variate $Y = \log X$ with density $f_Y(y) = f(\exp(y))\exp(y)$ is log-concave. For positive random variates the family of log-log-concave distributions is including all monotone log-concave distributions and also unbounded densities with pole at 0. Also the gamma distributions with $\alpha > 0$ and the positive part of all t-distributions with $\nu \geq 1$ is contained in this class. Using Algorithm 4.1 (TDR) and the density $f_Y(y)$ we can easily generate random variates $Y = \log X$ that are transformed to $X = \exp(Y)$. So at the expense of one additional exponentiation we can enlargen the class of possible distributions. It is possible to generalize this idea to c-log-concave, log-c-concave, and even to c-c-concave distributions. Details about the properties of these new classes require further investigations.

4.8 Automatic Ratio-of-Uniforms Method

The ratio-of-uniforms method introduced by Kinderman and Monahan (1977) can be seen as a variant of the rejection method (Sect. 2.4). If we want to sample from a distribution with quasi-density f with domain (b_l, b_r) then we have to generate random points (U, V) uniformly in the set

$$\mathcal{A}_f = \{(u,v): 0 < v \leq \sqrt{f(u/v)},\ b_l < u/v < b_r\}$$

and return the ratio U/V. For sampling these random points rejection from a convenient enveloping region for $\mathcal{A}_f$ is used, usually the *minimal bounding rectangle*. This method is very flexible and can be adjusted to a large variety of distributions. It has become a popular transformation method to generate non-uniform random variates, since it results in exact, efficient, fast and easy to implement algorithms. Typically these algorithms have only a few lines of code (Barabesi (1993) gives a survey and examples of FORTRAN codes for several standard distributions). However, the parameters of these bounding rectangles are computed analytically. Moreover, the rejection constants are quite large, e.g. we find $\alpha = 4/\sqrt{e\pi} \approx 1.37$ for the normal distribution.

In this section we will show that the ratio-of-uniforms method is also well suited for building universal methods. To do so we have to find a method for finding a good fitting envelope automatically.

4.8.1 The Relation to Transformed Density Rejection

Stadlober (1989b) and Dieter (1989) have clarified the relationship of the ratio-of-uniforms method to the ordinary acceptance/rejection method (see also Sect. 2.4). But there is also a deeper connection to transformed density rejection. The regions $\mathcal{A}_f$ of many standard distributions are convex, especially those of log-concave distributions (see e.g. Figs. 2.8, 4.17, or 6.2). Indeed we find the following useful characterization for densities with convex region $\mathcal{A}_f$ (Leydold, 2000a).

Theorem 4.17. $\mathcal{A}_f$ *is convex if and only if* $f(x)$ *is* T_c-*concave for* $c = -1/2$, *i.e. for transformation* $T(x) = -1/\sqrt{x}$.

Proof. Again consider the transformation (2.1)

$$\mathbb{R} \times (0,\infty) \to \mathbb{R} \times (0,\infty), \quad (U,V) \mapsto (X,Y) = (U/V, V^2)$$

from Sect. 2.4. Since $T(x) = -1/\sqrt{x}$ is strictly monotonically increasing, the transformation $(X,Y) \mapsto (X, T(Y))$ maps $\mathcal{G}_f = \{(x,y): 0 < y \le f(x),\ b_l < x < b_r\}$ one-to-one onto $\mathcal{T}_f = \{(x,y): y \le T(f(x)),\ b_l < x < b_r\}$, i.e. the region below the transformed density. Hence by $T(v^2) = -1/v$,

$$\mathbb{R} \times (0,\infty) \to \mathbb{R} \times (-\infty, 0), \quad (U,V) \mapsto (X,Y) = (U/V, -1/V) \qquad (4.27)$$

maps $\mathcal{A}_f$ one-to-one onto $\mathcal{T}_f$. Notice that f is T-concave if and only if $\mathcal{T}_f$ is convex. Thus it remains to show that $\mathcal{A}_f$ is convex if and only if $\mathcal{T}_f$ is convex, and consequently that straight lines remain straight lines under transformation (4.27). Now let $a\,x + b\,y = d$ be a straight line in $\mathcal{T}_f$. Then $a\,(u/v) - b/v = d$ or, equivalently, $a\,u - d\,v = b$, is a straight line in $\mathcal{A}_f$. Analogously we find for a straight line $a\,u + b\,v = d$ in $\mathcal{A}_f$ the line $a\,x + d\,y = -b$ in $\mathcal{T}_f$. □

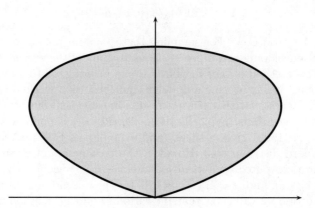

Fig. 4.17. The region $\mathcal{A}_f$ is convex for the normal distribution.

4.8.2 Enveloping Polygons

Leydold (2000a) suggests the use of polygonal envelops and squeezes. In our treatment we follow this presentation. (There is also a remark without any details by Dagpunar (1988).) To simplify the description of this method we first assume unbounded support for the (quasi-) density f. For such a distribution it is easy to make an enveloping polygon: Select a couple of points $\mathbf{p}_i$, $i = 0,\ldots,N$, on the boundary of $\mathcal{A}_f$ and use the tangents at these points as edges of the enclosing polygon $\mathcal{P}^e$ (see Fig. 4.18). We denote the vertices of $\mathcal{P}^e$ by $\mathbf{m}_i$. These are simply the intersection points of the tangents. The squeezing region is the inside of the polygon $\mathcal{P}^s$ with vertices $\mathbf{p}_i$.

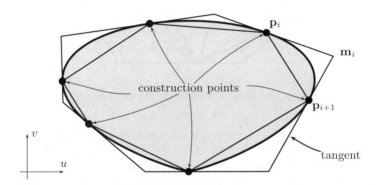

Fig. 4.18. Polygonal envelope $\mathcal{P}^e$ and squeeze $\mathcal{P}^s$ for the convex set A_f.

Notice that the origin $(0,0)$ is always contained in the polygon $\mathcal{P}^e$. Moreover every straight line through the origin corresponds to an $x = u/v$ and thus its intersection with $\mathcal{A}_f$ is always connected. Therefore we use $\mathbf{p}_0 = (0,0)$ for the first construction point and the u-axis as its tangent.

For a construction point $\mathbf{p}$ at a point $x = u/v$ we need (p_u, p_v) on the "outer boundary" of $\mathcal{A}_f$ and the tangent line of $\mathcal{A}_f$ at this point. These are given by the positive root of $v^2 = f(x)$ and the total differential of $v^2 - f(u/v)$, hence

$$\text{boundary: } p_v = \sqrt{f(x)}\,, \quad p_u = x\, p_v\,;$$
$$\text{tangent: } a_u\, u + a_v\, v = a_c = a_u\, p_u + a_v\, p_v\,, \qquad (4.28)$$
$$\text{where } a_v = 2\, p_v + f'(x)\, x/p_v \quad \text{and} \quad a_u = -f'(x)/p_v\,.$$

4.8.3 The Algorithm

To sample uniformly from the enclosing polygon we use the triangulation of $\mathcal{P}^e$ and $\mathcal{P}^s$ given by segments S_i, $i = 0,\ldots,N$, at vertex $\mathbf{p}_0$, see Fig. 4.19. Segment S_i has the vertices $\mathbf{p}_0, \mathbf{p}_i, \mathbf{m}_i$ and $\mathbf{p}_{i+1}$, where $\mathbf{p}_{N+1} = \mathbf{p}_0$ for the last segment. Each segment is divided into the triangle S_i^s inside the squeeze (dark shaded) and a triangle S_i^o outside (light shaded). Notice that the segments S_0 and S_N have only three vertices and that there exist no squeezing triangles S_0^s and S_N^s.

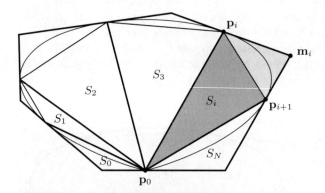

Fig. 4.19. Triangulation of enveloping polygon.

Now we can generate a point (U,V) uniformly from the polygon $\mathcal{P}^e$ using the composition method (Sect. 2.3) together with an algorithm for sampling from the triangles S_i^s and S_i^o (see Exercise 11.3). Then accept the ratio $X = U/V$ whenever $V^2 \leq f(X)$. By Thm. 4.17 this algorithm is in some sense equivalent to Algorithm TDR. It can be seen as a different method to generate points uniformly distributed in the region below the hat function.

However, this algorithm can be much improved by the concept of immediate acceptance (Sect. 2.3.3). Since the triangle S_i^s has vertex $\mathbf{p}_0 = (0,0)$, we can generate a uniformly distributed point in this triangle by $\mathbf{y} = (R_1 - R_2)\,\mathbf{p}_i + R_2\,\mathbf{p}_{i+1}$, where R_1 and R_2 are independent $U(0,1)$ uniform random numbers with $0 \leq R_2 \leq R_1 \leq 1$ (for a detailed explanation see Sect. 11.2.2). Notice that each generated point in a triangle S_i^s inside the squeeze can immediately be accepted without evaluating the probability density function and thus we are only interested in the ratio of the components. Therefore we arrive at

$$x = \frac{u}{v} = \frac{(R_1 - R_2)\,p_{i,1} + R_2\,p_{i+1,1}}{(R_1 - R_2)\,p_{i,2} + R_2\,p_{i+1,2}} = \frac{p_{i,1} + R\,(p_{i+1,1} - p_{i,1})}{p_{i,2} + R\,(p_{i+1,2} - p_{i,2})} \qquad (4.29)$$

where $p_{i,j}$ is the j-th component of vertex $\mathbf{p}_i$, and $R = R_2/R_1$ again is a $U(0,1)$ uniform random number by the ratio-of-uniforms Thm. 2.20, since $0 \leq R_2 \leq R_1 \leq 1$ (Kinderman and Monahan, 1977). Notice that formula (4.29) is equivalent to generating a point uniformly on the line from $\mathbf{p}_i$ to $\mathbf{p}_{i+1}$ and return the ratio of its coordinates.

To complete the list of ingredients we have to consider the case where the domain (b_l, b_r) of the density is bounded. The simplest way is to introduce

4.8 Automatic Ratio-of-Uniforms Method

Algorithm 4.8 Adaptive-RoU (Adaptive Ratio-of-Uniforms)

Require: T-concave quasi-density $f(x)$ with domain (b_l, b_r);
construction points $p_1 < \ldots < p_N$.
Output: Random variate X with density prop. to f.
  /* Setup: Construct envelope $\mathcal{P}^e$ and squeeze $\mathcal{P}^s$. */
1: Compute $\mathbf{p}_i$ and $\mathbf{a}_i$ for each construction point p_i using (4.28).
2: Set $\mathbf{p}_0 \leftarrow (0,0)$, $\mathbf{p}_{N+1} \leftarrow (0,0)$, and compute $\mathbf{a}_0$ and $\mathbf{a}_{N+1}$ using (4.30).
3: Compute $\mathbf{m}_i$ as the intersection point of the tangents in $\mathbf{p}_i$ and $\mathbf{p}_{i+1}$,
   for $i = 0, \ldots, N$.
4: Compute areas $|S_i^s|$, $|S_i^o|$, and $|S_i| = |S_i^s| + |S_i^o|$, for each segment S_i,
   $i = 0, \ldots, N$.
5: Set $A \leftarrow \sum_{i=0}^{N} |S_i|$.
  /* Generator */
6: **loop**
7:   Generate $R \sim U(0, A)$.
     /* Generate J with probability vector proportional to $(|S_0|, \ldots, |S_N|)$. */
8:   $J \leftarrow \min\{J: |S_1| + \cdots + |S_J| \geq R\}$. /* use *Indexed Search* (Sect. 3.1.2) */
     /* Recycle uniform random number */
9:   $R \leftarrow R - (|S_1| + \cdots + |S_{J-1}|)$. /* $R \sim U(0, |S_J|)$ */
10: **if** $R \leq |S_J^s|$ **then** /* inside squeeze, S_J^s */
11:    **return** $(|S_J^s| p_{J,1} + R(p_{J+1,1} - p_{J,1})) / (|S_J^s| p_{J,2} + R(p_{J+1,2} - p_{J,2}))$.
       /* (4.29) */
12: **else** /* outside squeeze, S_J^o */
13:    $R_1 \leftarrow (R - |S_J^s|)/|S_J^o|$. /* $R_1 \sim U(0,1)$ */
14:    Generate $R_2 \sim U(0,1)$.
15:    **if** $R_1 > R_2$, swap R_1, R_2. /* $R_1 \leq R_2$ */
16:    $R_3 \leftarrow 1 - R_2$, $R_2 \leftarrow R_2 - R_1$.
17:    $X \leftarrow (p_{J,1} R_1 + p_{J+1,1} R_2 + m_{J,1} R_3) / (p_{J,2} R_1 + p_{J+1,2} R_2 + m_{J,2} R_3)$.
18:    **if** $V^2 \leq f(X)$ **then**
19:      **return** X.

two different tangents to the vertex $\mathbf{p}_0 = (0,0)$. If $b_l > -\infty$ then we restrict the triangular segment S_0 by using the "left" tangent line $u - b_l v = 0$, i.e. we set

$$a_u u + a_v v = 0,$$
where $a_u = \cos(\arctan(b_l))$, and $a_v = -\sin(\arctan(b_l))$. (4.30)

We set $\arctan(b_l) = -\pi/2$ if $b_l = -\infty$. Analogously for $b_r < \infty$ we use the "right" tangent $u - b_r v = 0$. It is convenient to use $\mathbf{p}_{N+1} = (0,0)$ to realize these two tangents. Algorithm 4.8 (Adaptive-RoU) summarizes the *automatic ratio-of-uniforms* method. A more detailed description can be found in Leydold (2000a).

4.8.4 Construction Points and Performance

As for Algorithm 4.1 (TDR) the performance of Algorithm 4.8 (Adaptive-RoU) can be determined by the ratio $\varrho_{hs} = |\mathcal{P}^e|/|\mathcal{P}^s|$ between the area inside the envelope and the area inside the squeeze. For the expected number of uniform random numbers $\#U$ and the number of evaluations of the density $\#f$ we have the bounds: $\mathrm{E}(\#U) \leq 2\,\varrho_{hs} - 1/\varrho_{hs}$ and $\mathrm{E}(\#f) \leq \varrho_{hs} - 1$. Notice that ϱ_{hs} is the same for Algorithm 4.1 (TDR) with $c = -\frac{1}{2}$ and Algorithm 4.8 (Adaptive-RoU) for a given set of construction points by Thm. 4.17.

Another consequence of this observation is that we can use the same construction points as described in Sect. 4.4. Especially (derandomized) adaptive rejection sampling can be applied completely analogously. Furthermore, the idea for the equiangular rule of thumb (4.20) in Sect. 4.4.4 is now natural, since it results in segments $|S_i|$ with equal angles at the origin.

4.8.5 Non-Convex Region

Algorithm Adaptive-RoU can be modified such that it works with non-convex regions $\mathcal{A}_f$ too. Adapting the idea we have described in Sect. 4.7.1 we have to partition $\mathcal{A}_f$ into segments using the inflection points of the transformed density with transformation $T(x) = -1/\sqrt{x}$. In each segment of $\mathcal{A}_f$ where $T(f(x))$ is not concave but convex, we have to use secants for the boundary of the enveloping polygon P^e and tangents for the squeeze P^s (see Fig. 4.20). Notice that the squeeze region in such a segment is a quadrangle $\mathbf{p}_0\mathbf{p}_i\mathbf{m}_i\mathbf{p}_{i+1}$ and has to be split into two triangles. The necessary changes of Algorithm 4.8 (Adaptive-RoU) are straightforward.

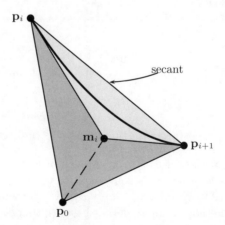

Fig. 4.20. Non-convex set $\mathcal{A}_f$. The squeeze polygon (dark shaded area) has to be divided into two triangles.

4.9 Exercises

The Class T_c of Transformations

Exercise 4.1. Show that for $T(x) = \log(x)$ and $T(x) = -1/\sqrt{x}$ the hat constructed by TDR is not influenced by the scale parameter of the density.
Hint: Construct the local hat in p_0 for an arbitrary density $f(x)$ and for $\gamma f(x)$ (where $\gamma > 0$ is constant). Then check that the two hats are multiples of each other.

T_c-Concave Distributions

Exercise 4.2. Proof the following necessary conditions for $T_{-1/2}$-concave distributions:
f is continuous on its domain and unimodal, the tails of f go to zero as least as fast as x^{-2} (i.e. f has sub-quadratic tails).
Hint: For continuity and unimodality first show that $T(f(x))$ is continuous and unimodal. For the subquadratic tails consider the case with $T(f(x)) = -1 - |x|$. Show that this $f(x)$ has quadratic tails and that no $T_{-1/2}$-concave distribution can have tails higher than a multiple of that f.

Exercise 4.3. Show that the density $f(x) = \exp(-\sqrt{|x|})$ cannot be made T_c-concave for any c.

Construction Points

Exercise 4.4. For one design point: Compute the optimal design point and the acceptance probabilities for the positive part of the normal distribution and the transformations: $T_{-9/10}$, $T_{-1/2}$, $T_{-1/4}$, T_0. Compare the acceptance probabilities.

Exercise 4.5. Do the same as in Exercise 4.4 and for two design points, one in the mode, the second optimally placed.

Algorithms and Variants of Transformed Density Rejection

Exercise 4.6. Compute the acceptance probability for the half-normal distribution, $T(x) = \log x$ and a single, optimal point of contact.
How is the acceptance probability changed if you replace the tangent by a "shifted secant"? Try several values of Δ between 10^{-5} and 0.5.

Exercise 4.7. For Algorithm 4.4 (TDR-3-Points), the normal, the Cauchy and the Gumbel distribution (with parameter $k = 1$) compute the acceptance probability and the expected number of uniforms $E(\#U)$ and the expected number of density evaluations $E(\#f)$ when – as suggested in the algorithm description – the simple rule $p_{3,1} = m \pm 0.664/f(m)$ is used to obtain the design points.
How do these values change, when these design points are replaced by the optimal design points?

Other Transformations

Exercise 4.8. Investigate how the inverse of the positive part of the normal distribution could be used as transformation T_{normal} for TDR if we assume that we have a good numerical approximation for the cdf of the normal distribution. How can we circumvent the use of F^{-1} which is numerically difficult?

Exercise 4.9. T_{normal} of Exercise 4.8 is only of limited use for a black-box algorithm. Show that the exponential distribution is not T_{normal}-concave no matter which scaling constant we use.

Exercise 4.10. Find a practically useful definition for $T_{(-1/2|1)}$, a transform that behaves partly like $T_{-1/2}$ and partly like a linear function. Find F_T and F_T^{-1}.

Exercise 4.11. Find densities whose inverse leads to the class T_c of distributions.
Hint: Compute T_c^{-1} first. Notice that you have to restrict the domain to obtain a (quasi-) density.

Exercise 4.12. Show that the exponential distribution is T_{Cauchy}-concave for some scalings and not for some others.
Hint: Use $f(x) = a \exp(-x)$ and try different values for a between 0 and 1.

Exercise 4.13. Try different change-points for the definition of the transformation $T_{(0|1)}$. Which one results in the simplest form for F_T^{-1}?

Exercise 4.14. If we choose the scaling constant a such that $a\,f(x) = e$ in the inflection point of the density, show that transformation $T_{(0|1)}$ can be used for all symmetric or monotone log-concave densities, that have an inflection point.

Exercise 4.15. Find for $T_{(-1/2|1)}$ as you have defined it in Exercise 4.10, a rule as in Exercise 4.14. Show that with this rule transformation $T_{(-1/2|1)}$ can be used for all symmetric or monotone $T_{-1/2}$-concave densities, that have an inflection point.

Exercise 4.16. For one design point: Compute the optimal design point and the acceptance probabilities for the positive part of the normal distribution and the transformations: $T_{(-1/2|1)}$, $T_{(0|1)}$ (using the rules of Exercise 4.14 and 4.15). Compare the acceptance probabilities with the results of Exercise 4.4

Generalizations of Transformed Density Rejection

Exercise 4.17. Find an example of a density that is not T_c-concave for any interval (a, ∞) and arbitrary transformation T_c.
Hint: Take a density where f' is oscillating in the tail.

Exercise 4.18. Compute all information necessary that Algorithm 4.5 (TDR-T-convex) can be applied to the exponential-power distribution $f(x) = e^{-|x|^a}$ for $a = 0.8$ and $a = 0.5$. (Use $c = -0.5$.)

Automatic Ratio-of-Uniforms Method

Exercise 4.19. Find the inverse of the mapping of (4.27) in the proof of Thm. 4.17. Calculate how this inverse mapping transforms the lines $x = 3$, $y = -1 - x$ and $y = -x$.

5

Strip Methods

Transformed density rejection as developed in Chap. 4 is a very flexible and efficient universal method for generating non-uniform random variates at the price of some computational effort and the necessity of some mathematical theory. In this chapter we follow a different strategy: Make the generation method as simple as possible. Thus we try to cover the region $\mathcal{A}_f$ between a density f and the x-axis by a union of rectangles. When making this enveloping region fit better and better these rectangles become more and more skinny, i.e. *strips*. The resulting algorithms are quite simple and (very) fast. It should be noted here, however, that this method obviously only works for bounded densities with bounded domains.

This idea of decomposing into many rectangular pieces has been used for a long time to create very fast generators for standard distributions (see e.g. Marsaglia et al., 1964; Marsaglia and Tsang, 1984; Ahrens and Dieter, 1989). We are, as always, mainly interested in automatic algorithms.

There are two possibilities how to construct such strips. The first approach uses *vertical strips*. It is equivalent to rejection from a piecewise constant hat on (small) subintervals of the domain. Important contributions to this method are due to Ahrens (1993, 1995). Therefore we have been calling this method *Ahrens Method* in our internal discussions for several years. We have the freedom to use this name here as well, in honor of Jo Ahrens, one of the great pioneers in the field of random variate generation. The Ahrens method is developed in detail in Sect. 5.1.

It is also possible to use *horizontal strips*. This method has been proposed by Marsaglia and Tsang (1984). We give a short description of this approach in Sect. 5.2.

Devroye (1986a, Sect. VIII.2 and VIII.3) presents not only the strip method but also a variant that he calls *grid method*. There he uses a grid of small rectangles with equal area to cover a given region. Although it also works for sampling from univariate distributions, it is mainly designed for sampling uniformly from any region in higher dimensions. Since it is less efficient for the task here it will be treated in Sect. 11.2.3 as method for generating random points uniformly in a polytope.

It is obvious that the methods of this chapter have many things in common with transformed density rejection with a large number of design points. As the use of rectangles simplifies the whole approach we think that it is justified to present some details of these rectangle methods here. TDR is more complicated but superior as it can generate from infinite tails in an exact way and provides a better (local) fit for the constructed hats.

5.1 Staircase-Shaped Hat Functions ("Ahrens Method")

If we are given a bounded density f over a bounded domain $[b_l, b_r]$ it is easy to sample variates by rejection from a constant hat. It is enough to have an upper bound for the density, see Algorithm 2.2 (Rejection-from-Uniform-Hat). If in addition the density is known to be monotonically decreasing then this upper bound is easily obtained by the value of f at the boundary b_l. We also can use a lower bound of the density as constant squeeze. The performance of this algorithm, however, is usually bad. It can be improved by splitting the domain of f into two subintervals, as it has been done in Examples 2.12 and 2.13. We even can do better when we continue with splitting the domain to get more and more subintervals. We will end up with a piecewise constant hat and squeeze, i.e., a staircase shaped function, see Fig. 5.1. We denote the splitting points by $b_l = b_0 < b_1 < b_2 < \ldots < b_N = b_r$. Then for a monotonically decreasing quasi-density f, hat and squeeze are given by

$$h(x) = \begin{cases} f(b_{i-1}) & \text{for } x \in [b_{i-1}, b_i), \ i = 1, \ldots, N, \\ 0 & \text{otherwise.} \end{cases}$$

$$s(x) = \begin{cases} f(b_i) & \text{for } x \in [b_{i-1}, b_i), \ i = 1, \ldots, N, \\ 0 & \text{otherwise.} \end{cases}$$

We should notice here that using staircase-shaped hat functions can be seen as applying transformed density rejection with transformation T_c at the limiting case $c \to \infty$ where the construction points in TDR become the splitting points (see also Fig. 4.3 on p. 64 and the discussion at the beginning of Sect. 4.3). Thus it is not astonishing that we will find similarities between these two methods.

5.1.1 The Algorithm

We restrict our attention to bounded monotone densities on a bounded interval. If the local extrema of a density are known we can, of course, use the described algorithms for all monotone slopes separately and put them together by means of the composition method (Sect. 2.3). For densities with

5.1 Staircase-Shaped Hat Functions ("Ahrens Method")

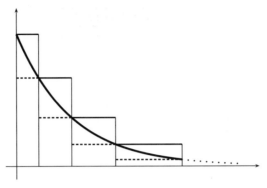

Fig. 5.1. Staircase-shaped functions a hat and squeeze. Notice that we have to chop off the tail in case of unbounded domain

infinite tails staircase shaped hat functions are useless, but from a practical point of view it is possible to use "save cut-off" rules as we have to work with finite precision uniform random numbers (see Ahrens, 1993, 1995). As these rules depend, especially for heavy tailed distributions, on the floating point arithmetic and the uniform generator used, and as we have TDR available for infinite tails anyway, we do not include the discussion of cut-off rules here.

When we use the piecewise constant hat and the squeeze given above for a rejection algorithm we arrive at Algorithm 5.1 (Ahrens). It is straightforward to modify it for monotonically increasing quasi-densities.

Algorithm 5.1 Ahrens (Basic method)

Require: monotonically decreasing quasi-density $f(x)$ on $[b_l, b_r]$;
design points $b_l = b_0 < b_1 < b_2 < \ldots < b_N = b_r$.
Output: Random variate X with density prop. to f.
/* Setup */
1: Compute and store all values $f(b_i)$ for $i = 0, \ldots, N$.
2: $A_i \leftarrow f(b_{i-1})(b_i - b_{i-1})$ for $i = 1, \ldots, N$. /* area below hat in (b_{i-1}, b_i) */
3: Set $A \leftarrow \sum_{i=1}^{N} A_i$.
/* Generator */
4: **loop**
5:     Generate $V \sim U(0, A)$.
    /* Generate J with probability vector proportional to $(A_1, \ldots, A_N)$. */
6:     $J \leftarrow \min\{J: A_1 + \cdots + A_J \geq V\}$. /* use *Indexed Search* (Sect. 3.1.2) */
7:     $V \leftarrow \frac{1}{A_J}(V - (A_1 + \cdots + A_{J-1}))$. /* $V \sim U(0,1)$. Recycle uniform r.n. */
8:     $X \leftarrow b_{J-1} + V \cdot (b_J - b_{J-1})$. /* $X \sim U(b_{J-1}, b_J)$. */
9:     Generate $Y \sim U(0, f(b_{J-1}))$.
10:    **if** $Y \leq f(b_J)$ **then** /* evaluate squeeze */
11:       **return** X.
12:    **if** $Y \leq f(X)$ **then** /* evaluate density */
13:       **return** X.

Algorithm 5.2 Ahrens-Immediate-Acceptance

Require: monotonically decreasing quasi-density $f(x)$ on $[b_l, b_r]$;
design points $b_l = b_0 < b_1 < b_2 < \ldots < b_N = b_r$.
Output: Random variate X with density prop. to f.
/* Setup */
1: Compute and store all values $f(b_i)$ for $i = 0, \ldots, N$.
2: Compute $\nu_i = f(b_i)/f(b_{i-1})$ for $i = 1, \ldots, N$.
3: $A_i \leftarrow f(b_{i-1})(b_i - b_{i-1})$ for $i = 1, \ldots, N$. /* area below hat in (b_{i-1}, b_i) */
4: Set $A \leftarrow \sum_{i=1}^{N} A_i$.
/* Generator */
5: **loop**
6:    Generate $V \sim U(0, A)$.
/* Generate J with probability vector proportional to $(A_1, \ldots, A_N)$. */
7:    $J \leftarrow \min\{J \colon A_1 + \cdots + A_J \geq V\}$. /* use Indexed Search (Sect. 3.1.2) */
8:    $V \leftarrow \frac{1}{A_J}(V - (A_1 + \cdots + A_{J-1}))$. /* $V \sim U(0,1)$. Recycle uniform r.n. */
9:    **if** $V \leq \nu_J$ **then** /* below squeeze, immediate acceptance */
10:       $X \leftarrow b_{J-1} + (V/\nu_J) \cdot (b_J - b_{J-1})$. /* $X \sim U(b_{J-1}, b_J)$. */
11:       **return** X.
12:    **else** /* accept or reject from $h_J - s_J$. */
13:       $X \leftarrow b_{J-1} + (V - \nu_J)/(1 - \nu_J) \cdot (b_J - b_{J-1})$. /* $X \sim U(b_{J-1}, b_J)$. */
14:       Generate $Y \sim U(f(b_J), f(b_{J-1}))$.
15:       **if** $Y \leq f(X)$ **then** /* evaluate density */
16:       **return** X.

Since the region below the squeeze and the region between hat and squeeze are so simple in each subinterval $[b_{i-1}, b_i)$ it is easy to apply the concept of *immediate acceptance* to reduce the required number of uniform random variates (Sect. 2.3.3). The details are compiled in Algorithm 5.2 (Ahrens-Immediate-Acceptance). For large N it requires slightly more than one uniform random number to generate one random variate X.

As for transformed density rejection the choice of appropriate design points b_i for the subintervals is crucial for the performance of the algorithm. For increasing number N of points the ratio ϱ_{hs} of the area below the hat and the area below the squeeze tends to 1. When this happens Algorithms 5.1 and 5.2 are very close to the inversion method.

5.1.2 Equidistant Design Points

The easiest choice of design points are equidistant points, i.e. we use the following points for a given interval $[b_l, b_r]$

$$b_i = b_l + (b_r - b_l) i/N \quad \text{for } i = 0, \ldots, N.$$

In this case we have the following simple theorem partly given in a different formulation by Devroye (1986a, p. 362). Notice that the expected number of iterations of Algorithms 5.1 (Ahrens) and 5.2 (Ahrens-Immediate-Acceptance) is the same provided that the same design points are used.

5.1 Staircase-Shaped Hat Functions ("Ahrens Method")   117

Theorem 5.1. *For a monotonically decreasing quasi-density $f(x)$ and equidistant design points the area A_{h-s} between hat and squeeze is given by*

$$A_{h-s} = \frac{(b_r - b_l)(f(b_l) - f(b_r))}{N}.$$

Furthermore the expected number of iterations for Algorithm 5.1 (Ahrens) with equidistant design points is bounded by

$$\mathrm{E}(I) \leq 1 + \frac{(b_r - b_l)(f(b_l) - f(b_r))}{A_f N}$$

where $A_f = \int_{b_l}^{b_r} f(x)\,\mathrm{d}x$. The expected number of evaluations of f is bounded by

$$\mathrm{E}(\#f) = \frac{(b_r - b_l)(f(b_l) - f(b_r))}{A_f N}.$$

Proof. By decomposing (b_l, b_r) into N intervals of equal length we can easily compute

$$A_{h-s} = \left(\sum_{i=1}^{N} f(b_l + (i-1)/N\,(b_r - b_l)) - f(b_l + i/N\,(b_r - b_l))\right)\frac{b_r - b_l}{N} =$$
$$= \frac{(f(b_l) - f(b_r))(b_r - b_l)}{N}.$$

The bound for the expected number of iterations is obvious, if we remember that $\mathrm{E}(I) = A_h/A_f = (A_f + A_{h-s})/A_f$ and use the trivial bound $A_s \leq A_f$ (A_h is the area below the hat function). The bound for the expected number of iterations follows directly from Thm. 2.9 and the formula for A_{h-s}. □

It is important to note that this theorem does not provide uniform bounds on the performance of Algorithm 5.1, since the inequalities depend on the area A_f. For a fixed domain $[b_l, b_r]$ and fixed values $f(b_l)$ and $f(b_r)$, the area A_f can be for example $(b_r - b_l)(f(b_l) - f(b_r))/2$ if f is linear or close to $(b_r - b_l)f(b_r)$ if f is strongly convex.

5.1.3 Optimal Design Points

As a criterion for the optimality of the design points it is reasonable to minimize the expected number of iterations $\mathrm{E}(I)$ or the expected number $\mathrm{E}(\#f)$ of evaluations of f. By Thm. 2.9 we have to minimize either the area A_h below the hat or the area A_{h-s} between hat and squeeze. First we consider the case where an interval $[b_0, b_2]$ is decomposed into two intervals by the point b_1. In Exercise 5.1 we prove that both criteria lead to the same result for linear densities. Therefore we expect that for short intervals the difference between minimizing A_h and A_{h-s} is small. As A_h is a simpler objective function than A_{h-s} it is not astonishing that minimizing A_h leads to simpler results. Ahrens (1993) demonstrates that it is even possible to compute the design points that exactly minimize A_h by a simple recursion. For the choice of b_1 that minimizes A_h for a fixed domain $[b_0, b_2]$ we can use simple calculus (as done by Ahrens, 1993) to verify the necessary condition $0 = A'_h(b_1) = f(b_0) - f(b_1) + f'(b_1)(b_2 - b_1)$, or

$$b_2 = b_1 + \frac{f(b_0) - f(b_1)}{-f'(b_1)} . \tag{5.1}$$

Here $A_h(b_1)$ denotes the area below the hat as a function of the new design point b_1.

Inspecting the second derivative of $A_h(b_1)$ it is easy to see that the condition is sufficient for convex densities (Exercise 5.2). For concave densities it is possible that $A_h(b_1)$ has several local extrema. This can happen for monotone densities which have an inflection point in a region where f is very flat, or for densities that have a rapidly changing first derivative. $f(x) = \min(1 - x/10, 9 - 9x)$ on $[0, 1]$ and $f(x) = -x^3/6 + x^2/4 - 0.15x + 0.5$ on $[0, 1.6]$ are examples of such densities (see Exercise 5.3). So for large values of N where we have short subintervals the problem of two local minima only becomes a practical problem for strangely shaped monotone densities. However, it is possible to construct examples (e.g. a staircase-shaped function where the flat parts have slope -0.1 and the steep parts have slope -9) where the problem of more than one local minimum can occur in many subintervals.

It is useful to notice that (5.1) can be used to compute optimal design points recursively. We only need a first interval length $b_1 - b_0$ and the possibility to evaluate f' which could also be approximated numerically. In practice we have to solve two problems: How do we decide about the first interval length? This could be done by trial and error but then the setup becomes very slow. The second problem is that the border b_N of the last constructed interval need not be equal to the right boundary b_r of the given domain of f. Of course we can simply set $b_N \leftarrow b_r$ but this implies that b_{N-1} is not the optimal division of the interval $[b_{N-2}, b_N]$. Nevertheless, we know that the first $N - 1$ design points are optimal on $[b_0, b_{N-1}]$ and thus it is clear that the difference between the area A_h below the optimal hat and the area below the hat we have achieved by applying this recursion cannot be greater than $(f(b_{N-1}) - f(b_r))(b_r - b_{N-1})$.

Ahrens (1993) also explains a heuristic procedure to obtain nearly optimal design points. As it is more complicated than the exact procedure above we are not giving the details. More important for applications is in our opinion the equal area approach explained below.

5.1.4 (Derandomized) Adaptive Rejection Sampling (D|ARS)

The idea of (derandomized) adaptive rejection sampling as described in the respective Sects. 4.4.5 and 4.4.6 for transformed density rejection can analogously be applied to the Ahrens method. (Zaman (1996) describes such an algorithm.)

Notice that in adaptive rejection sampling the expected point in an interval (b_{i-1}, b_i) is just the arithmetic mean of the boundary points b_{i-1} and b_i. However, using this splitting point for DARS is not recommended in the tails of a distribution and leads to poor design points for some densities. Better performance for derandomized adaptive rejection sampling gives the "arc-mean" (see (4.23) on p. 85) of the boundary point.

This consideration remains valid for adaptive rejection sampling as well. Simply using a rejected point as new construction point is not recommended for densities with heavy tails. The setup is faster for a desired rejection constant when the "arc-mean" of the boundary points is used, even at the price of one additional evaluation of the density.

The DARS principle can be used for minimizing A_h or A_{h-s}. It is also possible to use the asymptotic theory explained in Sect. 4.4.3 to compute N "asymptotically optimal" design points. As we can only consider bounded domains here, these points are the same for minimizing A_{h-s} or minimizing A_h. The resulting algorithm for computing the design points is quite fast but probably more complicated than using the DARS principle.

5.1.5 The Equal Area Approach

The equal area approach is in some aspects comparable with the grid methods (Sect. 11.2.3; see also Devroye, 1986b) as speed and simplicity are gained by decomposing the area below the hat into pieces of equal area. For the grid method these pieces are squares or small rectangles whereas strips of equal area are used by Ahrens (1995). He suggests to construct the design points such that the area below the hat is the same for all subintervals. What we gain is simplicity and speed as we no longer need a guide table for fast generation of the interval index J in Algorithm 5.1. Instead we sample from the discrete uniform distribution by simply computing the integer part of a continuous uniform random variate.

The necessary design points b_i can be computed by a simple recursion. To have subintervals where the area below the hat is equal to some constant γ the design points must satisfy $(b_i - b_{i-1})f(b_{i-1}) = \gamma$. Thus we arrive at

$$b_i = b_{i-1} + \gamma/f(b_{i-1}) \quad \text{for } i = 1, \ldots, N \ . \tag{5.2}$$

Ahrens (1995) shows that for $f(x) = 1/x$ on a fixed interval the design points obtained by this recursion minimize the area A_h below the hat (see Exercise 5.4). He also shows that for arbitrary monotone f and $N \to \infty$, A_h converges to the area A_f below the density when using the equal area approach.

In practice the control over the performance of the algorithm is simple if it is possible to compute γ such that the desired value for the rejection constant

α is obtained; for coding the algorithm it is also useful to have an upper bound for N for given γ. If f is monotone and the area below the density A_f is known, we have simple solutions for both tasks utilizing results of Hörmann (2002).

Theorem 5.2. *We are given a monotonically decreasing quasi-density $f(x)$ on $[b_l, b_r]$. Using the recursion (5.2). we construct a piecewise constant hat such that the area below the hat is γ in every subinterval. Then we have the following bounds for the area between hat and squeeze A_{h-s} and for the maximal number of intervals N:*

$$A_{h-s} \leq (\log f(b_l) - \log f(b_r) + 1)\gamma , \qquad (5.3)$$

$$N \leq \frac{A_f}{\gamma} + \log f(b_l) - \log f(b_r) + 1 . \qquad (5.4)$$

If $f(b_r)$ is close or equal to 0 the bounds below (that are also generally correct) will yield better results:

$$A_{h-s} \leq (\log f(b_l) + \log(b_r - b_l) - \log \gamma + 2)\gamma , \qquad (5.5)$$

$$N \leq \frac{A_f}{\gamma} + \log f(b_l) + \log(b_r - b_l) - \log \gamma + 2 . \qquad (5.6)$$

Proof. See Hörmann (2002). □

The choice of γ to obtain at least the desired rejection constant α_0 is not difficult. Using the bound (5.3) we obtain

$$\alpha = \frac{A_h}{A_f} \leq 1 + \frac{A_{h-s}}{A_f} \leq 1 + \frac{(\log f(b_l) - \log f(b_r) + 1)\gamma}{A_f} .$$

So for fixed α_0 we can compute

$$\gamma_0 = \frac{(\alpha_0 - 1)A_f}{\log f(b_l) - \log f(b_r) + 1}$$

which guarantees a rejection constant smaller or equal to α_0. For the case that $f(b_r)$ is very small or 0 we can get a similar result using formula (5.5):

$$\gamma_0 = \frac{(\alpha_0 - 1)A_f}{\log f(b_l) + \log(b_r - b_l) - \log \gamma_0 + 2} .$$

To approximately solve this equation we have to resort to numerical methods. Using bisection or regula falsi it is no problem to find a solution for γ_0. After computing γ it is no problem to obtain an upper bound for N by just taking the minimum of the two bounds (5.4) and (5.6).

If A_f is unknown we can estimate A_f by a rough numeric integration or use trial and error to find γ.

5.1 Staircase-Shaped Hat Functions ("Ahrens Method") 121

We also have to deal with the fact that, by applying recursion (5.2), the last design point may be left of the boundary point b_r, i.e., $b_N > b_r$. But we must not change this design point b_N as this would spoil the equal area property for the last interval $[b_{N-1}, b_N]$. Instead we have to provide code that the quasi-density f returns 0 for all $x > b_r$. Thus all generated points with $X > b_r$ are rejected.

To get a fast algorithm we again use the idea of immediate acceptance (see Algorithm 5.2, Ahrens-Immediate-Acceptance). Notice that by recursion (5.2) we have

$$f(b_{i-1}) = \gamma/(b_i - b_{i-1}) \,. \tag{5.7}$$

Thus we find for the ratios $\nu_i = f(b_i)/f(b_{i-1}) = (b_i - b_{i-1})/(b_{i+1} - b_i)$. Consequently we have the case of immediate acceptance if for a uniform random number $V \sim U(0, 1)$ we find $V(b_{i+1} - b_i) \leq (b_i - b_{i-1})$. Moreover, in this case (i.e., when $V \sim U(0, \nu_i)$), $X = b_{i-1} + (V/\nu_i) \cdot (b_i - b_{i-1}) = b_{i-1} + V \cdot (b_{i+1} - b_i)$ is uniformly distributed in (b_{i-1}, b_i). Using this trick we only have to store the design points b_i and can avoid to store the values $f(b_i)$. However, we have to consider the last interval, i.e., when $i = N$ which implies that we would need b_{N+1} to decide about immediate acceptance. The simplest solution is to treat this interval as a special case and not to use immediate acceptance at all in this interval. Algorithm 5.3 (Ahrens-Equal-Area) gives all details for a monotonically decreasing quasi-density.

Remark. Ahrens (1995) presented a more complicated version of this algorithm. He uses the auxiliary point $b_{N+1} = b_N + \gamma/f(b_r)$ to store $f(b_N)$ which is needed for immediate acceptance in the terminal interval. An algorithm that utilizes this idea has to deal with the special case $f(b_r) = 0$ where $b_{N+1} = \infty$. Moreover, we have to reject a generated point X whenever $X > b_r$.

It is possible to improve this algorithm further by chopping the terminal interval $[b_{N-1}, b_N]$ to $[b_{N-1}, b_r]$. Then the probability of selecting the terminal sector, i.e., the probability for $J = N$ has to be modified by replacing Step 10 $J \leftarrow 1 + \lfloor VN \rfloor$ by $J \leftarrow 1 + \lfloor V \cdot (N - 1 + f(b_{N-1}) \frac{b_r - b_{N-1}}{b_N - b_{N-1}}) \rfloor$.

5.1.6 Comparison of the Different Variants

To assess the performance of the different variants of the Ahrens method introduced above we can use Thm. 5.1. So we know that, for the case of equidistant design points, the expected number of iterations $E(I) = 1 + O(1/N)$ and the expected number of evaluations of f, $E(\#f) = O(1/N)$. Optimizing the design points only changes the multiplicative constant of the O term. We can expect the best performance for optimal design points which require the most complicated and slowest setup. But even for these optimal points we can see that for a fixed number N of design points the performance of the Ahrens method is not uniformly bounded in the class of all bounded monotone densities on a bounded interval. The same is true for all other known algorithms.

Algorithm 5.3 Ahrens-Equal-Area

Require: monotonically decreasing quasi-density $f(x)$ on $[b_l, b_r]$;
an estimate for A_f, approximate desired number of intervals $\tilde{N}$.
Output: Random variate X with density prop. to f.

/* Setup */
1: $\gamma \leftarrow (A_f/\tilde{N}) + 2\log \tilde{N}/\tilde{N}^2$.
2: $i \leftarrow 0, b_0 \leftarrow b_l$.
3: **repeat**
4: $\quad i \leftarrow i + 1, b_i \leftarrow b_{i-1} + \gamma/f(b_{i-1})$.
5: **until** $b_i \geq b_r$ or $f(b_i) = 0$
6: $N \leftarrow i$. /* number of intervals */
7: $f_r \leftarrow f(b_r)$. /* $= f(b_N)$ */
/* Generator */
8: **loop**
9: $\quad$ Generate $V \sim U(0, 1)$.
10: $\quad J \leftarrow 1 + \lfloor VN \rfloor$. /* $J \sim U(\{1, 2, \ldots, N\})$ */
11: $\quad V \leftarrow 1 + VN - J$. /* $V \sim U(0,1)$. Recycle uniform random number. */
12: $\quad$ **if** $J < N$ **then** /* main part */
13: $\quad\quad Q \leftarrow V \cdot (b_{J+1} - b_J)$.
14: $\quad\quad$ **if** $Q \leq (b_J - b_{J-1})$ **then** /* below squeeze, immediate acceptance */
15: $\quad\quad\quad X \leftarrow b_{J-1} + Q$. /* $X \sim U(b_{J-1}, b_J)$. */
16: $\quad\quad\quad$ **return** X.
17: $\quad\quad$ **else** /* accept or reject from $h_J - s_J$. */
18: $\quad\quad\quad Q \leftarrow (Q - (b_J - b_{J-1})) \cdot (b_J - b_{J-1})/(b_{J+1} - 2b_J + b_{J-1})$.
19: $\quad\quad\quad X \leftarrow b_{J-1} + Q$. /* $X \sim U(b_{J-1}, b_J)$. */
20: $\quad\quad\quad f_{J-1} \leftarrow \gamma/(b_J - b_{J-1}), f_J \leftarrow \gamma/(b_{J+1} - b_J)$.
21: $\quad\quad\quad$ Generate $Y \sim U(f_J, f_{J-1})$.
22: $\quad\quad\quad$ **if** $Y \leq f(X)$ **then** /* evaluate density */
23: $\quad\quad\quad\quad$ **return** X.
24: $\quad$ **else** /* $i = N$, terminal interval */
25: $\quad\quad X \leftarrow b_{J-1} + V \cdot (b_J - b_{J-1})$. /* $X \sim U(b_{J-1}, b_J)$. */
26: $\quad\quad$ Generate $Y \sim U(0, \gamma/(b_J - b_{J-1}))$. /* $Y \sim U(0, f(b_{J-1}))$. */
27: $\quad\quad$ **if** $Y \leq f(X)$ **then** /* evaluate density */
28: $\quad\quad\quad$ **return** X.

For the equal area rule it is also possible to describe its asymptotic performance using the following theorem.

Theorem 5.3. *We are given a monotonically decreasing quasi-density on the interval $[b_l, b_r]$. Using the recursion (5.2) we construct a piecewise constant hat such that the area below the hat is γ in every interval and $p_N = b$. Then we have the following two bounds for the area between hat and squeeze A_{h-s}.*

$$A_{h-s} \leq \frac{A_f}{\frac{N}{k_1} - 1} = A_f \sum_{i=1}^{\infty} \left(\frac{k_1}{N}\right)^i \tag{5.8}$$

for $N > k_1$, with $k_1 = \log f(b_l) - \log f(b_r)$, and

$$A_{h-s} \le \frac{A_f}{\frac{N}{k_2} - 1} = A_f \sum_{i=1}^{\infty} \left(\frac{k_2}{N}\right)^i \qquad (5.9)$$

for $N > k_2$, with $k_2 = \log f(b_l) + \log(b_r - b_l) + \log N - \log A_f + 1$.

Proof. See Hörmann (2002). □

The results of this theorem imply that for $f(b_r) > 0$ the equal area rule leads to the same asymptotic behavior as for optimal design points and for equidistant design points. For $f(b_r) = 0$ we get $\mathrm{E}(I) = 1 + O(\log N/N)$ which indicates that for $f(b_r) = 0$ and large values of N the equal area rule is not good. This can be explained by the observation that the equal area rule does not place enough design points close to b_r in that case. We should not forget that the above theorem is an asymptotic result; our experiments show that for moderate values of N the equal area rule is better than equidistant design points for several (especially convex) densities, even with $f(b_r) = 0$, whereas equidistant design points are better for linear and concave densities.

In generation practice the main advantage of Algorithm 5.3 (Ahrens-Equal-Area) is the simplicity of its setup. Due to the fact that the values $f(b_i)$ need not be stored it has also less memory requirements than Algorithm 5.1. So for most applications it is a matter of taste which of the two algorithms is preferred.

5.2 Horizontal Strips

Marsaglia and Tsang (1984) propose the *ziggurat method*, which is based on horizontal strips rather than on vertical bars. They also use an equal area approach for these strips (see Sect. 5.1.5). Since all these horizontal strips have the same area the tail and the region near the mode, called *cap*, of the density f are not covered. For each strip the cap is rotated and rescaled such that it fits into the region of rejection. If a generated point falls into such a region, it is transformed back into a point in the cap. Thus together with the idea of immediate acceptance (Sect. 2.3.3) this results in a very fast algorithm that requires hardly more than one uniform random number for each generated nun-uniform random variate. The (small) tail region must be generated by a different method. Due to the construction of the strips the tail-probabilities cannot be arbitrarily small. Therefore they cannot be simply chopped-off.

Although the ziggurat method can be used to create very fast algorithms we think that it is not well suited for designing an automatic algorithm. Compared to the Ahrens method an automatic algorithm based on the ziggurat method is more complicated and generation from the tail region requires a different automatic algorithm. In addition it is never close to inversion. Thus we do not provide the details of such an algorithm. The interested reader is referred to the original paper.

5.3 Exercises

Exercise 5.1. Show that for the Ahrens method with a linear density and $N = 2$, A_{h-s} (the area between hat and squeeze) and A_h (the area below the hat) is minimized for $x_1 = (x_0 + x_2)/2$.
Hint: Compute a formula for A_{h-s} for general value of x_1. Then minimize it with respect to x_1.

Exercise 5.2. Proof that (5.1) is correct. Inspecting the second derivative $A_h''(x_1)$ find a sufficient condition that A_h has only one local minimum. Using it show that the condition of display (5.1) is sufficient for convex densities.

Exercise 5.3. We are given the following densities:
$f_0(x) = \min(1 - x/10,\ 9 - 9x)$ on $(0, 1)$;
$f_1(x) = (1 - 0.19x)(1 - x^{10})$ on $(0, 1)$; and
$f_2(x) = -x^3/6 + x^2/4 - 0.15x + 0.5$ on $(0, 1.6)$.
Plot $A_h(x_1)$ for these three densities to see that they really have two local minima. Looking at $A_h''(x_1)$ of Exercise 5.2 try to find other simple examples of densities where $A_h(x_1)$ has more than one local minimum.

Exercise 5.4. Show that for $f(x) = 1/x$ and arbitrary interval (b_l, b_r) with $0 < b_l < b_r < \infty$ the equal area rule leads to design points that minimize A_h.
Hint: Compute the formula of A_h for arbitrary design point x_0 in (b_l, b_r). Then find the x_0 that minimizes A_h and check that for this x_0 the are below the hat in (b_l, x_0) and in (x_0, b_r) are equal.

Exercise 5.5. Rewrite Algorithm 5.3 (Ahrens-Equal-Area) for monotonically increasing quasi-densities.

6

Methods Based on General Inequalities

The methods developed in Chaps. 4 and 5 result in fast algorithms at the price of a complex and expensive setup. In this chapter we do it the other way round and try to find methods with hardly any setup by means of universal inequalities. The resulting algorithms are short and more robust against round-off errors since no complicated hat function has to be computed.

We start with a simple example that illustrates the type of algorithms we are developing. We are given the density (as a black-box) together with some additional information about properties that can be used to construct a global hat function. The important point is that we can find a hat that is valid for all densities that share these properties.

Example 6.1. We are given a bounded quasi-density f with bounded domain $[b_l, b_r]$. Then it easy to find a bounding rectangle for the region $\mathcal{G}_f$ below the quasi-density. We have used this idea as the simplest example for the rejection method, see Algorithm 2.2 (Rejection-from-Uniform-Hat) on p. 17 for the details.

If the mode m of the quasi-density is known we can obviously use the (best possible) bound $M = f(m)$. Then the setup section of this algorithm consists only of computing $f(m)$.

The main characteristic of this algorithm is the expected number of iterations $\mathrm{E}(I)$ which is equal to the rejection constant α. We have

$$\mathrm{E}(I) = M\,(b_r - b_l)/A_f$$

by Thm. 2.9. Here and in the sequel A_f denotes the area below the density, i.e. $A_f = \int_{b_l}^{b_r} f(x)\,\mathrm{d}x$.

Obviously $\mathrm{E}(I)$ can be very large if the density f has high spikes, or long tails, or if no good global bound M is available. To improve the performance of this simple algorithm we need more information about the distribution. We start with distributions with monotone densities in Sect. 6.1. In Sect. 6.2 we shortly consider Lipschitz densities. In Sects. 6.3 and 6.4 we develop short and simple algorithms for T_c-concave distributions.

6.1 Monotone Densities

The "spirit" of this section and much of the material is taken from Devroye (1986a, Sect. VII.3). We also have added new algorithms. For the sake of simplicity we first assume that the left border of the domain is at 0.

The first help that we can get from the monotonicity is that we can easily compute the global bound $M = f(0)$. But we cannot find a better bound for f unless we have some extra information about the area below the density or about moments. We can formulate the following two general inequalities for monotone densities that will help us to construct hat functions.

Theorem 6.2. *Let f be a monotone density on $[0, \infty)$. If $\int x^r f(x)\,\mathrm{d}x \leq \mu_r < \infty$ for an $r \geq 0$ then*
$$f(x) \leq \frac{(r+1)\mu_r}{x^{r+1}}.$$
In addition we have for all a with $a > 0$
$$f(x) \leq \frac{\left(\int_0^\infty f^a(t)\,\mathrm{d}t\right)^{1/a}}{x^{1/a}}.$$
Notice that the tails of both upper bounds are integrable only if $r > 0$ and $a < 1$, respectively.

Proof. To show the first inequality we use the definition of the r-th moment together with the monotonicity. Then we get
$$\mu_r = \int_0^\infty t^r f(t)\,\mathrm{d}t \geq \int_0^x t^r f(t)\,\mathrm{d}t \geq f(x)\int_0^x t^r\,\mathrm{d}t = f(x)\frac{x^{r+1}}{r+1}$$
and the result follows. For the second bound we use monotonicity to see
$$\int_0^\infty f^a(t)\,\mathrm{d}t \geq \int_0^x f^a(t)\,\mathrm{d}t \geq \int_0^x f^a(x)\,\mathrm{d}t = x\,f^a(x). \quad \square$$

6.1.1 Monotone Densities with Bounded Domains

If for a monotone density f the integral $\int_0^\infty f(x)\,\mathrm{d}x$ is known we can use the inequalities of Thm. 6.2 above to construct upper bounds for f.

Theorem 6.3 (Devroye 1986a). *For all monotone densities f on $[0, \infty)$*
$$f(x) \leq \frac{A_f}{x}.$$
The bound is optimal in the sense that for every x there exists a monotone density for which the bound is attained.

6.1 Monotone Densities

Proof. This is a special case of both inequalities of Thm. 6.2. Take either $r = 0$ in the first inequality or $a = 1$ in the second.
To see that this inequality is optimal take $f(t) = f(x)$ for $0 \leq t \leq x$ and 0 otherwise. □

Combining the result of Thm. 6.3 with the trivial constant bound results in the following general dominating function for monotone densities

$$f(x) = h(x) \leq \min\left(f(0), \frac{\check{A}_f}{x}\right)$$

where $\check{A}_f \geq A_f = \int f(x)\,dx$ denotes an upper bound for the area below the density. Figure 6.1 (l.h.s.) on p. 129 shows an example of this hat function.

Clearly this bound is not integrable on the real line but can be used for monotone distributions with bounded domains. We start with monotone densities on $[0, 1]$. The "cdf" of the hat (we are not norming the area to 1) is

$$H(x) = \begin{cases} f(0)\,x & \text{for } 0 \leq x \leq \frac{\check{A}_f}{f(0)}, \\ \check{A}_f\left(1 + \log\left(\frac{f(0)}{\check{A}_f}x\right)\right) & \text{for } \frac{\check{A}_f}{f(0)} \leq x \leq 1. \end{cases}$$

The area below the hat is $H(1) = \check{A}_f\left(1 + \log(f(0)/\check{A}_f)\right)$ and the expected number of iterations is therefore

$$\mathrm{E}(I) = (\check{A}_f/A_f)\left(1 + \log(f(0)/\check{A}_f)\right) \geq 1 + \log(f(0)/A_f).$$

where the second equality holds when $\check{A}_f$ is the exact area below the density, i.e. $\check{A}_f = A_f$. It is no problem to invert this "cdf" $H(x)$ of the hat function and thus we arrive at Algorithm 6.1 (Monotone-01).

We can use this algorithm for an arbitrary monotone density f over a bounded domain $[b_l, b_r]$, since it can be transformed into a monotone density $\tilde{f}$ with domain $[0, 1]$ by $\tilde{f}(x) = (b_r - b_l)f((b_r - b_l)x + b_l)$. Algorithm 6.2 (Monotone-Bounded) contains the details.

To find the expected number of iterations for Algorithm 6.2 (Monotone-Bounded) we use $\tilde{f}(0) = (b_r - b_l)f(b_l)$ and the result for Algorithm 6.1. We arrive at

$$\mathrm{E}(I) = (\check{A}_f/A_f)\left(1 + \log((b_r - b_l)f(b_l)/\check{A}_f)\right) \geq 1 + \log((b_r - b_l)f(b_l)/A_f),$$

where again the second equality holds when $\check{A}_f = A_f$. This shows – not surprisingly – that the algorithm is optimal for the uniform distribution, since then $A_f = (b_r - b_l)f(b_l)$. Comparing with the performance of Algorithm 2.2 (Rejection-from-Uniform-Hat) we see the considerable improvement that comes from the new information about the monotonicity of the density and the upper bound for A_f together with the inequality of Thm. 6.3. Unfortunately $\mathrm{E}(I)$ is not uniformly bounded for all monotone densities with

Algorithm 6.1 Monotone-01 (Simple rejection for monotone density on $[0, 1]$)

Require: bounded monontone quasi-density $f(x)$ on $[0, 1]$ with mode 0;
upper bound $\check{A}_f \geq \int_0^1 f(x)\,dx$.
Output: Random variate X with density prop. to f.
  /∗ Setup ∗/
1: $f_0 \leftarrow f(0)$.
2: $A_h \leftarrow \check{A}_f \left(1 + \log(f_0/\check{A}_f)\right)$. /∗ area below hat ∗/
  /∗ Generator ∗/
3: **loop**
4:   Generate $U \sim U(0, A_h)$ and $V \sim U(0, 1)$.
5:   **if** $U \leq \check{A}_f$ **then**
6:     Set $X \leftarrow U/f_0$ and $h \leftarrow f_0$.
7:   **else**
8:     Set $X \leftarrow (\check{A}_f/f_0) \exp(U/\check{A}_f - 1)$ and $h \leftarrow \check{A}_f/X$.
9:   **if** $V h \leq f(X)$ **then**
10:     **return** X.

Algorithm 6.2 Monotone-Bounded

Require: bounded monontone quasi-density $f(x)$ on $[b_l, b_r]$ with mode b_l;
upper bound $\check{A}_f \geq \int_{b_l}^{b_r} f(x)\,dx$.
Output: Random variate X with density prop. to f.
  /∗ Setup ∗/
1: $f_0 \leftarrow (b_r - b_l) f(b_l)$.
2: $A_h \leftarrow \check{A}_f \left(1 + \log(f_0/\check{A}_f)\right)$. /∗ area below hat ∗/
  /∗ Generator ∗/
3: **loop**
4:   Generate $U \sim U(0, A_h)$ and $V \sim U(0, 1)$.
5:   **if** $U \leq \check{A}_f$ **then**
6:     Set $\tilde{X} \leftarrow U/f_0$ and $h \leftarrow f_0$.
7:   **else**
8:     Set $\tilde{X} \leftarrow (\check{A}_f/f_0) \exp(U/\check{A}_f - 1)$ and $h \leftarrow \check{A}_f/\tilde{X}$.
9:   $X \leftarrow (b_r - b_l)\tilde{X} + b_l$.
10: **if** $V h \leq (b_r - b_l) f(X)$ **then**
11:     **return** X.

bounded domains. For densities with a high peak and/or long tails the algorithm still works but becomes arbitrarily slow. If we look at the performance characteristics of the algorithms of Sect. 5.1 we can conclude that using large tables leads to a fast performance for arbitrary monotone densities; but the size of the tables must increase together with the peak of the distribution. For fixed table size there is no universal algorithm available that is uniformly fast in the class of all monotone distributions.

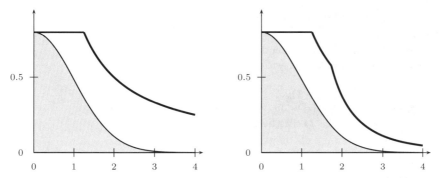

Fig. 6.1. Truncated normal distribution on $[0,4]$ with the hat of Algorithm 6.2 (Monotone-Bounded) (l.h.s.) and the (truncated) hat of Algorithm 6.3 (Monotone-Unbounded) with $r = 2$ (r.h.s.). (The area below the hat A_h is 2.16 for the l.h.s. and 1.73 for the truncated hat on the r.h.s.)

6.1.2 Monotone Densities with Unbounded Domain

The hat function used for Algorithm 6.1 (Monotone-01) in the previous section is not integrable for unbounded domains. Therefore we need further information on the given distribution in this case. In this section we assume that we have upper bounds for the moments of the desired distribution available. Then we can use the inequalities from Thm. 6.2 to construct integrable hat functions. To simplify the notation we assume throughout this section that the density f has integral one (i.e. $A_f = 1$). The changes for the general case with quasi-densities are straightforward and are discussed in Exercise 6.1. Among the many possibilities to construct universal algorithms for monotone densities with one known moment we will present the details of a new algorithm that uses the information about the area below the density (implicitly given by the normed density). We think that it is a sensible compromise between simplicity and nearly optimal performance. It assumes that we know either for one $r > 0$ a bound for the r-th moment μ_r or for one a, $0 < a < 1$, a bound for $(\int f^a)^{1/a}$. Then using Thm. 6.2 we get bounds for the density that can be written as k/x^{r+1} and $k/x^{1/a}$, respectively. We will now concentrate on the first case with given moments where we have

$$k = (r+1)\mu_r .$$

For the details necessary for the second variant of the algorithm see Exercise 6.5. We arrive at the hat function

$$f(x) \leq h(x) = \min\left(f(0), 1/x, \frac{k}{x^{r+1}}\right) = \begin{cases} f(0) & \text{for } x \leq 1/f(0), \\ 1/x & \text{for } 1/f(0) < x \leq k^{1/r}, \\ k/x^{r+1} & \text{for } x > k^{1/r}. \end{cases}$$

Note that the definition of the three different regions for this hat function cannot result in a problem as $1/f(0) \leq ((r+1)\mu_r)^{1/r} = k^{1/r}$ holds for all

Algorithm 6.3 Monotone-Unbounded

Require: bounded monotone density $f(x)$ on $[0, \infty)$ (with $A_f = 1$);
 upper bound μ_r for the r-th moment of the density for some $r > 0$.
Output: Random variate X with density f.
 /* Setup */
1: $f_0 \leftarrow f(0)$, $\quad k \leftarrow (r+1)\mu_r$.
2: $A_2 \leftarrow \log(f_0) + \log((r+1)\mu_r)/r$, $\quad A_3 \leftarrow 1/r$.
3: $A_h \leftarrow 1 + A_2 + A_3$. /* area below hat */
 /* Generator */
4: **loop**
5: $\quad$ Generate $U \sim U(0, A_h)$ and $V \sim U(0, 1)$.
6: $\quad$ **if** $U \leq 1$ **then**
7: $\quad\quad$ Set $X \leftarrow U/f_0$ and $h \leftarrow f_0$.
8: $\quad$ **else**
9: $\quad\quad$ **if** $U \leq 1 + A_2$ **then**
10: $\quad\quad\quad$ Set $X \leftarrow \exp(U - 1 - \log(f_0))$ and $h \leftarrow 1/X$.
11: $\quad\quad$ **else**
12: $\quad\quad\quad$ Set $X \leftarrow (k/(1 - r(U - 1 - A_2)))^{1/r}$ and $h \leftarrow k/X^{r+1}$.
13: $\quad$ **if** $Vh \leq f(X)$ **then**
14: $\quad\quad$ **return** X.

---

monotone densities. To see this we can easily check that equality is true for arbitrary $r > 0$ for the uniform distribution. Moreover, for a fixed value of $f(0)$, all monotone distributions have larger moments μ_r than the uniform distribution. (We may say that among all monotone distributions with fixed $f(0)$ the uniform distribution is "stochastically smallest".) Figure 6.1 (r.h.s.) shows as an example the hat for the normal distribution restricted to $[0, 4]$.

To generate variates from the above hat function we can use the inversion method. Therefore we need the unnormalized "cdf" of this hat function which is given by

$$H(x) = \begin{cases} f(0)x & \text{for } x \leq 1/f(0), \\ 1 + \log(f(0)x) & \text{for } 1/f(0) < x \leq k^{1/r}, \\ 1 + \log(f(0)) + \frac{\log(k)}{r} + \frac{1}{r}\left(1 - \frac{k}{x^r}\right) & \text{for } x > k^{1/r}. \end{cases}$$

The areas below the three parts of the hat are $A_1 = 1$, $A_2 = \log(f(0)) + \log((r+1)\mu_r)/r$, and $A_3 = 1/r$, respectively. The details for a generation method that utilizes this hat are given in Algorithm 6.3.

The expected number of iterations of this algorithm is given by

$$E(I) = 1 + \log(f(0)) + \frac{\log((r+1)\mu_r) + 1}{r} = 1 + \frac{1 + \log(r+1) + \log(f(0)^r \mu_r)}{r}. \tag{6.1}$$

Hence $E(I)$ is bounded for any monotone density with fixed upper bound μ_r. If $f(0)$ or μ_r is tending to infinity, $E(I)$ is not uniformly bounded, but, at least the increase for growing $f(0)$ or growing μ_r is very slow. In general we cannot hope to design a uniformly fast universal algorithm for monotone densities as this class contains members with arbitrarily sharp peaks and arbitrarily long tails.

Think, for example, of the mixture of two uniform distributions with density $f(x) = 10^6$ for $0 \le x \le 10^{-12}$ and $f(x) = 10^{-6}$ for $10^{-12} < x \le 10^6 - 1 + 10^{-12}$. Using the above formula we then find for, e.g., $r = 2$, $E(I) = 29.13$. It is not astonishing that for distributions with such a shape the above Algorithm 6.3 (Monotone-Unbounded) is not effective.

On the other hand a closer look at (6.1) above gives us a hint for which distribution families $E(I)$ will be uniformly bounded. The summands 1 and $(1+\log(r+1))/r$ are no problem if $r > 0$. So we can see that for a certain family of distributions $E(I)$ is uniformly bounded if for a fixed $r > 0$ there exists an upper bound for $f(0)^r \mu_r$ for all densities from this family. We can interpret $f(0)^r \mu_r$ as a measure for the length of the tail together with the height of the spike of a monotone distribution. Thus it can be seen as a measure of the difficulty of generating from a certain distribution. We could use this as a definition for families of monotone densities as well, for example the family of all monotone densities over $[0, \infty)$ with $f(0)\mu_1 \le C$ or with $f(0)^2 \mu_2 \le C$. These are the "natural" distribution families associated with Algorithm 6.3. The expected number of iterations is uniformly bounded for them.

From these considerations we can also conclude why it was possible for us to construct black-box algorithms with uniformly bounded performance characteristics for T_c-concave distributions. The T_c-concave families do not include members with arbitrarily long tails and arbitrarily high spikes. Mathematically we can express this property by a bound for the moments of these distributions.

Theorem 6.4. *Let f be a monotonically decreasing density on $[0, \infty)$. If in addition f is log-concave, then*

$$\mu_r \le \frac{\Gamma(r+1)}{f(0)^r}$$

where $\Gamma(.)$ denotes the gamma function.
If f is T_c-concave with $-1 < c < 0$, then we find

$$\mu_r \le \frac{\Gamma(r+1)\, \tilde{c}^r\, \Gamma(\tilde{c}-r)/\Gamma(\tilde{c})}{f(0)^r} \quad \text{for } r < \tilde{c}$$

where $\tilde{c} = -(1 + \frac{1}{c})$.

Proof. If we assume that $f(0) = 1$ fixed and that f is T_c-concave, it is possible to understand that the largest possible value for μ_r occurs for the density $f_{\max}$, that has a linear transformed density. Any other T_c-concave density with $f(0) = 1$ has – compared with the density $f_{\max}$ – some mass moved to the left. (We therefore call the random variate $X_{\max}$ with density $f_{\max}$ the stochastically largest T_c-concave random variable X with $f(0) = 1$.)

For the log-concave case ($c = 0$) we can find $f_{\max}$ with $f_{\max}(0) = 1$ by considering all functions e^{-ax}. Clearly we have to take $a = 1$ and thus $f_{\max} = e^{-ax}$ to obtain a density. Computing the integral $\int_0^\infty x^r e^{-x}\, dx$ shows that $\mu_r = \Gamma(r+1)$ for the exponential distribution.
For $-1 < c < 0$ we can easily see that $f(0) = 1$ implies $T(f_{\max}(0)) = -1$. So the transformed density must be of the form $T(f_{\max}(x)) = -1 - ax$. If we choose the unique a that leads to $A_f = 1$ we obtain $f_{\max}(x) = (x/\tilde{c} + 1)^{1/c}$ and can compute: $\mu_r = \int_0^\infty x^r f_{\max}(x)\, dx = \Gamma(r+1)\tilde{c}^r \frac{\Gamma(\tilde{c}-r)}{\Gamma(\tilde{c})}$ for $f_{\max}$. The inequalities of the theorem are thus valid for $f(0) = 1$.

To see that the inequalities are correct for arbitrary values of $f(0)$ we consider bX instead of X. Clearly $\mu_r(bX) = b^r \mu_r(X)$. In addition the random variate bX has the density $f_{bX}(x) = \frac{1}{b} f\left(\frac{x}{b}\right)$. So $f_{bX}(0) = \frac{1}{b}$. So changing the scale parameter b adds a multiplicative factor of b^r on either side of our inequalities and is therefore not influencing their validity. □

Note that again (as expected) the result for $c = 0$ is the limiting case of the result for $c < 0$, which can be seen by a straightforward computation using Stirling's approximation of the gamma function.

We can use Thm. 6.4 to obtain global bounds for the moments of T_c-concave distributions. As the formulas are a bit unfriendly we spell out the most important special cases. They allow us to use Algorithm 6.3 (Monotone-Unbounded) for any T_c-concave distribution without needing any moment. Instead it is enough to know a c that makes the density T_c-concave and to evaluate $f(0)$.

Corollary 6.5. *For $c > -1/2$ we get*

$$\mu_1 \le \frac{\frac{1+c}{1+2c}}{f(0)}.$$

For $c > -1/3$ we get

$$\mu_2 \le \frac{\frac{2(1+c)^2}{(3c+1)(2c+1)}}{f(0)^2}.$$

For $c = -1/2$

$$\mu_{1/2} \le \frac{\pi/2}{f(0)^{1/2}}.$$

Proof. The results are all direct consequences of Thm. 6.4 together with basic properties of the gamma function. The first equality can be obtained using $r = 1$, the second using $r = 2$ and the last using $r = 1/2$ and $c = -1/2$. □

We can also use Thm. 6.4 together with (6.1) to compute upper bounds for $E(I)$ for Algorithm 6.3 (Monotone-Unbounded) when applied to T_c-concave distributions. The performance is uniformly bounded for any fixed c but not uniformly bounded if we consider the case $c \to -1$ (but this is also not true for the TDR algorithms suggested in Sect. 4.4.2). For fixed c we have the freedom to choose r such that $E(I)$ is minimized.

Table 6.1 gives these upper bounds for different values of c and some of the best choices of r. It shows that for log-concave densities the difference compared to the black-box algorithms developed in following Sect. 6.3 is not very large. For $c = -1/2$ the fit of the hat is – due to the heavy tails – much worse.

Table 6.1. Upper bounds for $E(I)$ for Algorithm 6.3 (Monotone-Unbounded) for T_c-concave distributions with different c

c	r	bound for $f(0)^r \mu_r$	$E(I)$
0	1	1	2.693
0	2	1	2.395
0	2.5	1.329	2.381
0	3	2	2.392
-1/5	1	1.333	2.980
-1/5	2	5.333	2.886
-1/3	1	2	3.386
-1/2	0.5	1.571	4.714

6.1.3 Other Hat Functions for Monotone Densities

In Exercise 6.2 we develop the details of another algorithm for monotone distributions as suggested by Devroye (1986a, p.317). It is shorter than Algorithm 6.3 (Monotone-Unbounded) above as it does not use the middle-part of our hat with shape $1/x$. On the other hand it is not using all available information. If we assume that we know that the area below the density is one why should we "waste" this information? The expected number of iterations for Devroye's Algorithm is given by (see Exercise 6.3)

$$E(I) = \frac{r+1}{r}((r+1)\mu_r)^{1/(r+1)} f(0)^{r/(r+1)}$$

which is considerably larger than (6.1), the bound for Algorithm 6.3, especially if $f(0)^r \mu_r$ is large. For log-concave distributions the difference of the $E(I)$ is not so big, only about 15 percent.

If we have estimates for several moments available (as it is the case, e.g., for log-concave distributions) we could use the minimum of several of the bounds of Thm. 6.2 as hat function. But our experiments with log-concave distributions showed that compared with the enhanced complexity the gain in terms of $E(I)$ is very modest. For example for log-concave distributions a hat based on $r = 0, 1$, and 2 results in $E(I) \leq 2.35$, whereas using $r = 0, 1$,

2, 3, 4, and 5 gives $\mathrm{E}(I) \le 2.277$. Theoretically it would be possible to use the pointwise minimum of the hats for all values of $r > 0$, but even then the expected number of iterations remains above 2.26. If we compare this with our results for Algorithm 6.3 (Monotone-Unbounded) and the log-concave case as shown in Table 6.1 the gain is not more than about 5 percent which is – in our opinion – not enough for further pursuing this idea.

6.2 Lipschitz Densities

We say that a function is *Lipschitz(C)* when

$$\sup_{x \ne y} \frac{|f(x) - f(y)|}{|x - y|} \le C \;.$$

It is possible to derive an upper bound for such Lipschitz densities.

Theorem 6.6. *When the density f is Lipschitz for some $C > 0$ and the cdf F is available then we have*

$$f(x) \le \sqrt{2C \min(F(x), 1 - F(x))} \;.$$

Proof. See Devroye (1986a, p. 320). □

If we combine the above theorem with bounds for tail-probabilities we can obtain global hat functions for subclasses of Lipschitz densities. If we know, for example, for some $r > 0$ a moment μ_r of the density we can use Chebyshev's inequality and get

$$1 - F(x) \le \frac{\mu_r}{|x|^r} \;.$$

Using Thm. 6.6 results in the inequality

$$f(x) \le \sqrt{2C} \min(1, \sqrt{\mu_r} |x|^{-r/2}) \;.$$

The details of an Algorithm for Lipschitz densities are posed in Exercise 6.7. In Exercise 6.8 the reader is asked to show that $\mathrm{E}(I) = \sqrt{8C} \mu_r^{1/r} r/(r-2)$ which is only finite for $r > 2$. But even for $r > 2$, $\mathrm{E}(I)$ is very big unless C is really small. This is not astonishing as Lipschitz densities can have a lot of possible shapes and need not be unimodal; also Chebyshev's inequality that we have used here is a very general (and therefore weak) inequality. On the other hand we should not forget that this is the only universal algorithm we present in this book that works for multi-modal distributions if the locations of the modes are not known! For more on algorithms for Lipschitz densities see Devroye (1986a, Sect. VII.3.3).

6.3 Generators for $T_{-1/2}$-Concave Densities

In Sect. 6.1 we have seen that general inequalities that use upper bounds for moments result in rejection algorithms with suitable rejection constants when the given density is log-concave. Using directly the log-concavity of such distributions, however, results in even more efficient algorithms. The following theorem holds for every log-concave density.

Theorem 6.7 (Devroye 1986a). *If f is a log-concave density with mode $m = 0$ and $f(0) = 1$, then writing q for $F(0)$, where F denotes the cdf of the distribution, we have*

$$f(x) \leq \begin{cases} \min(1, e^{1-x/(1-q)}) & \text{for } x \geq 0, \\ \min(1, e^{1+x/q}) & \text{for } x < 0. \end{cases}$$

The area under the bounding curve is 2.

This theorem has to be modified accordingly if f is monotone, i.e., if $q = 0$ or $q = 1$, because then the respective tail of the hat vanishes.

Proof. We only show the case where $q = 0$. We obviously have $f(x) \leq 1$ by assumption. Now suppose $f(x) > \exp(1-x)$ for some $x > 1$. Then by the log-concavity of f we must have $f(t) > \exp(-kt)$ for all $t \in [0, x]$ where $k = 1 - 1/x$. Hence $1 \geq \int_0^x f(t)\,dt > \int_0^x e^{-(1+1/x)t}\,dt = \frac{x}{x-1}(1 - e^{1-x}) > 1$, a contradiction. Thus $f(x) \leq \exp(1-x)$ holds for all $x > 1$. □

Notice that the area below this bounding curve is just the upper bound for the area of the optimal hat that we have constructed using transformed density rejection and two or three points of contact, respectively (see Sect. 4.4.2). However, in opposition to the method developed in Chap. 4 this theorem also uses the fact that the area A_f below the density is known and equal to 1. Using this inequality to build an algorithm for log-concave distributions is straightforward and left to the reader (Exercise 6.9).

If $F(m)$ is not known, a modified universal hat exists with area 4 (see Devroye, 1986a, Sect. VII.2.3). In both cases these universal hats are not optimal. Devroye (1984c) derives the properties of the optimal hat and provides a (rather expensive) generator for the corresponding density. The areas below the optimal bounding curves are $\pi^2/6$ and $\pi^2/3$, respectively, i.e. about 18% better.

In this section we use analogous global inequalities to derive algorithms that are simpler, faster, and applicable to the larger families of $T_{-1/2}$-concave distributions.

6.3.1 Generators Based on the Ratio-of-Uniforms Method

The universal hat given in Thm. 6.7 is very simple. In this section, however we will derive a universal hat function for the larger class of $T_{-1/2}$-concave

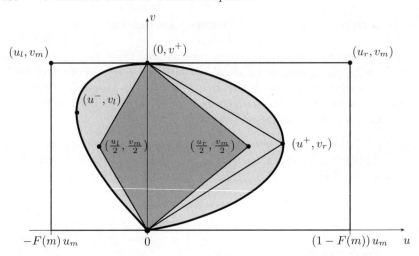

Fig. 6.2. $\mathcal{A}_f$ and universal bounding rectangle $\mathcal{R}_f$ and squeeze $\mathcal{S}_f$ for gamma(3) distribution

distributions. Moreover, the resulting algorithm is simpler, faster, and has the same rejection constant 2 (or 4). Our treatment of the material follows Leydold (2001).

In Thm. 4.17 we have seen that for a given quasi-density f the region

$$\mathcal{A}_f = \{(u,v) \colon 0 < v \leq \sqrt{f(u/v+m)},\ b_l < u/v + m < b_r\}$$

is convex if and only if f is $T_{-1/2}$-concave. This can be used to find a bounding rectangle for $\mathcal{A}_f$. Let m be the mode of f and denote coordinates of the minimal bounding rectangle by v^+, u^- and u^+, respectively, see (2.2). Then $v^+ = \sqrt{f(m)}$ and $\mathcal{A}$ has the extremal points $(0, v^+)$, (u^-, v_l) and (u^+, v_r), for respective v_l and v_r. Define $\mathcal{A}^+ = \{(u,v) \in \mathcal{A}_f \colon u > 0\}$ and analogously $\mathcal{A}^-$. By the convexity of $\mathcal{A}$, the triangle with vertices at $(0,0)$, $(0, v^+)$ and (u^+, v_r) is contained in $\mathcal{A}^+$ and thus has a smaller area (see Fig. 6.2). Consequently

$$\tfrac{1}{2} v^+ u^+ \leq |\mathcal{A}^+|$$

where $|\mathcal{A}^+|$ denotes the area of $\mathcal{A}^+$. From the fact that the Jacobian of the transformation (2.1) is 2 (see proof of Thm. 2.20, p. 34) it follows immediately that

$$|\mathcal{A}^+| = \tfrac{1}{2} \int_m^{b_r} f(x)\,\mathrm{d}x = \tfrac{1}{2}(1 - F(m))\,A_f\ .$$

Hence if the cdf $F(x)$ at the mode m is known, we find

$$u^+ \leq u_r = (1 - F(m))\,A_f / \sqrt{f(m)}\ .$$

Analogously we have

$$u^- \geq u_l = -F(m)\, A_f/\sqrt{f(m)}\,.$$

There also exists a region that is always contained in $\mathcal{A}_f$ and thus can be used as universal squeeze. It is the intersection of the triangle with vertices at $(u_l, 0)$, $(0, v^+)$ and $(u_r, 0)$ and the triangle with vertices at (u_l, v^+), $(0,0)$ and (u_l, v^+), see Fig. 6.2. We can summarize our result in the following theorem.

Theorem 6.8. *Suppose f is a $T_{-1/2}$-concave quasi-density with mode m and cdf F. Let*

$$\begin{aligned}
\mathcal{R}_f &= \{(u,v)\colon u_l \leq u \leq u_r, 0 \leq v \leq v_m\}\,,\\
\mathcal{Q}_f &= \{(u,v)\colon -u_m \leq u \leq u_m, 0 \leq v \leq v_m\}\,,\\
\mathcal{S}_f &= \{(u,v)\colon \tfrac{u_l}{v_m} \leq \tfrac{u}{v} \leq \tfrac{u_r}{v_m} \text{ and } \tfrac{u_l}{v_m} \leq \tfrac{u}{v_m-v} \leq \tfrac{u_r}{v_m}\}
\end{aligned} \qquad (6.2)$$

where

$$\begin{aligned}
v_m &= \sqrt{f(m)}\,, & u_l &= -F(m)\, u_m\,,\\
u_m &= A_f/\sqrt{f(m)}\,, & u_r &= (1-F(m))\, u_m\,.
\end{aligned}$$

Then

$$\mathcal{S}_f \subset \mathcal{A}_f \subset \mathcal{R}_f \subset \mathcal{Q}_f$$

and

$$2\,|\mathcal{S}_f| = |\mathcal{A}_f| = \tfrac{1}{2}|\mathcal{R}_f| = \tfrac{1}{4}|\mathcal{Q}_f|\,. \qquad (6.3)$$

Proof. The enveloping regions immediately follow from the above considerations. The ratios (6.3) between the respective regions follow from the fact that we are dealing with triangles inscribed into a rectangle. The universal squeeze must be contained in all triangles with vertices $(0,0)$ and $(0, v^+)$ and an edge with u-coordinate u_l or u_r. $\mathcal{S}_f$ is just the intersection of these triangles. To show that this region is contained in every convex $\mathcal{A}_f$ it is sufficient to show that $(\tfrac{u_r}{2}, \tfrac{v_m}{2})$ is in this region. If this were not the case we could split the rectangle given by the vertices $(0,0)$ and (u_r, v_m) by a straight line in a way that the left hand part contains $\mathcal{A}^+$ and the right hand part the point $(\tfrac{u_r}{2}, \tfrac{v_m}{2})$. But then the left hand part must be smaller than the right hand part, i.e., smaller than half of $2\,|\mathcal{A}^+|$, a contradiction. For further details see Leydold (2001). □

Notice that any quadrangle with vertices at $(0, v^+)$ and $(0,0)$, one arbitrary vertex at the left edge of $\mathcal{R}_f$ and one on the right edge has area $|\mathcal{R}_f|/2 = |\mathcal{A}_f|$. Moreover every such quadrangle corresponds to a $T_{-1/2}$-concave quasi-density. Thus $\mathcal{R}_f$ is optimal for the class of all $T_{-1/2}$-concave distributions, i.e. any other universal enveloping region must contain $\mathcal{R}_f$. Analogously this holds for $\mathcal{Q}_f$ when $F(m)$ is not known. Moreover, any other universal squeeze is contained in $\mathcal{S}_f$.

We can apply Thm. 6.8 to construct a rejection algorithm based on the ratio-of-uniforms method (Sect. 2.4). The details are compiled in Algorithm 6.4 (Simple-RoU). The rejection constant α and thus the expected number of iterations $\mathrm{E}(I)$ is 2 when $F(m)$ is known and 4 otherwise.

Algorithm 6.4 Simple-RoU (Simple Ratio-of-Uniforms method)

Require: $T_{-1/2}$-concave quasi-density $f(x)$, area A_f, mode m;
  cdf at mode $F(m)$ (optional).
Output: Random variate X with density prop. to f.

/* Setup */
1: $v_m \leftarrow \sqrt{f(m)}, \quad u_m \leftarrow A_f/v_m$.
2: **if** $F(m)$ is provided **then**
3: $\quad u_l \leftarrow -F(m)\, u_m, \quad u_r \leftarrow u_m + u_l$.
4: **else**
5: $\quad u_l \leftarrow -u_m, \quad u_r \leftarrow u_m$.
/* Generator */
6: **repeat**
7: $\quad$ Generate $U \sim U(u_l, u_r)$ and $V \sim U(0, v_m)$.
8: $\quad X \leftarrow U/V + m$.
9: **until** $V^2 \leq f(X)$.
10: **return** X.

Algorithm Simple-RoU does not make use of the universal squeeze, but it could easily be included. Indeed, it is not very useful unless the density function is very expensive to evaluate.

It should be noted here that this algorithm is numerically more stable than Algorithm 4.1 (TDR) in the sense that given the T_c-concavity of the density we always have a reliable hat function for the rejection algorithm. For Algorithm TDR, however, it might happen that for "strange" densities the computation of the hat fails due to round-off errors in ill-conditioned equations.

Remark. If only upper and lower bounds for A_f, $F(m)$, $f(m)$ or m are available, an accordingly modified version of Algorithm Simple-RoU still works.

6.3.2 The Mirror Principle

Devroye (1984c) suggests the usage of a hat function for $f(x) + f(-x)$ when $F(m)$ is not known to reduce the expected number of uniform random numbers. To apply this idea for Algorithm 6.4 (Simple-RoU) we need the following result.

Lemma 6.9. *Let $f_1(x)$ and $f_2(x)$ be two nonnegative functions with respective bounding rectangles $\mathcal{R}_1$ and $\mathcal{R}_2$ for $\mathcal{A}_{f_1}$ and $\mathcal{A}_{f_2}$ with common left lower vertex $(0,0)$ and the respective right upper vertices (u_1, v_1) and (u_2, v_2). Then*

$$\mathcal{A}_{f_1+f_2} \subseteq \left\{ (u,v) \colon 0 \leq u \leq \sqrt{u_1^2 + u_2^2},\ 0 < v \leq \sqrt{v_1^2 + v_2^2} \right\}.$$

Proof. Let $(u,v) \in \mathcal{A}_{g_1+g_2}$. Obviously $u \geq 0$. By (2.2) $v^2 \leq \sup(f_1(x) + f_2(x)) \leq \sup f_1(x) + \sup f_2(x) \leq v_1^2 + v_2^2$ and $u^2 \leq \sup(x-m)^2 \, (f_1(x)+f_2(x)) \leq \sup(x-m)^2 f_1(x) + \sup(x-m)^2 f_2(x) \leq u_1^2 + u_2^2$ as proposed. □

Algorithm 6.5 Simple-RoU-Mirror-Principle

Require: $T_{-1/2}$-concave quasi-density $f(x)$, area A_f, mode m.
Output: Random variate X with density prop. to f.

/* Setup */
1: $v_m \leftarrow \sqrt{f(m)}, \quad u_m \leftarrow A_f/v_m$.
/* Generator */
2: **loop**
3:     Generate $U \sim U(-u_m, u_m)$ and $V \sim U(0, \sqrt{2}\, v_m)$.
4:     $X \leftarrow U/V$.
5:     **if** $V^2 \leq f(X+m)$ **then**
6:         **return** $X + m$.
7:     **if** $V^2 \leq f(X+m) + f(-X+m)$ **then**
8:         **return** $-X + m$.

Theorem 6.10. *For any $T_{-1/2}$-concave density f with mode m let*

$$\widehat{\mathcal{R}}_f = \{(u,v): -u_m \leq u \leq u_m,\ 0 < v \leq \sqrt{2}\, v_m\} \tag{6.4}$$

where u_m and v_m are as defined in Thm. 6.8. Then

$$\mathcal{A}_{f(x)+f(-x)} \subset \widehat{\mathcal{R}}_f \quad \text{and} \quad |\widehat{\mathcal{R}}_f| = 4\sqrt{2}\,|\mathcal{A}_f|. \tag{6.5}$$

Proof. Let $\mathcal{R}(p) = \{(u,v): -p\,u_m \leq u \leq (1-p)\,u_m,\ 0 < v \leq v_m\}$. By Thm. 6.8 we find $\mathcal{A}_{f(x)} \subset \mathcal{R}(F(m))$ and $\mathcal{A}_{f(-x)} \subset \mathcal{R}(1-F(m))$. Hence by Lemma 6.9, $\mathcal{A}_{f(x)+f(-x)} \subset \bigcup_{p \in [0,1]} \{(u,v): -u(p) \leq u \leq u(p),\ 0 < v \leq \sqrt{2}\,v_m\} = \widehat{\mathcal{R}}_f$, where $u(p) = \sqrt{p^2\,u_m^2 + (1-p)^2\,u_m^2}$. Equation (6.5) follows immediately from (6.3). □

Using this theorem we can compile Algorithm 6.5 (Simple-RoU-Mirror-Principle). It reduces the expected number of uniform random numbers at the expense of more evaluations of $f(x)$. By (6.5) the rejection constant of this algorithm is $2\sqrt{2}$ in opposition to 4 in Algorithm 6.4 (Simple-RoU) when $F(m)$ is not known.

Remark. $\widehat{\mathcal{R}}_f$ is not optimal. However, the optimal envelope $\widehat{\mathcal{R}}_{\text{opt}}$ is not rectangular and contains the rectangles $\{(u,v): -u_m/2 \leq u \leq u_m/2,\ 0 < v \leq \sqrt{2}\,v_m\}$ and $\{(u,v): -u_m \leq u \leq u_m,\ 0 < v \leq v_m\}$. Thus we can find the estimate $|\widehat{\mathcal{R}}_{\text{opt}}| \geq (1 - \frac{1}{2}(\sqrt{2}-1))\,|\widehat{\mathcal{R}}_f| > 0.79\,|\widehat{\mathcal{R}}_f|$.

6.3.3 Universal Hats for $T_{-1/2}$-Concave Densities

For the proof of Thm. 4.17 on p. 104 we have used transformation (2.1)

$$\mathbb{R} \times (0, \infty) \to \mathbb{R} \times (0, \infty), \quad (U, V) \mapsto (X, Y) = (U/V, V^2)$$

together with the transformation $T_{-1/2}$ to verify that each straight line in $\mathcal{A}_f$ is transformed into a straight line in the region $\mathcal{T}_f$ below the transformed

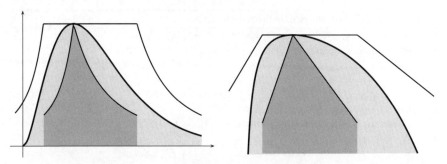

Fig. 6.3. Universal hat and squeeze for gamma(3) distribution. Original scale (l.h.s.) and transformed scale (r.h.s.)

density $\tilde{f}$. This can be further used to transform bounding rectangle and squeeze from Thm. 6.8 into hat function and squeeze for the original density f. Figure 6.3 gives an example. We arrive at the following theorem.

Theorem 6.11. *For any $T_{-1/2}$-concave density f with mode m let v_m, u_m, u_l and u_r be defined as in Thm. 6.8 and let $x_m = u_m/v_m$, $x_l = u_l/v_m$ and $x_r = u_r/v_m$. Define*

$$\check{h}(x) = \begin{cases} f(m) & \text{for } -x_m \leq x - m \leq x_m, \\ \frac{u_m^2}{(x-m)^2} & \text{otherwise.} \end{cases}$$

$$h(x) = \begin{cases} \frac{u_l^2}{(x-m)^2} & \text{for } x - m < x_l, \\ f(m) & \text{for } x_l \leq x - m \leq x_r, \\ \frac{u_r^2}{(x-m)^2} & \text{for } x - m > x_r. \end{cases}$$

$$s(x) = \begin{cases} \left(\frac{u_l\, v_m}{u_l + v_m\,(x-m)}\right)^2 & \text{for } x_l \leq x - m < 0, \\ \left(\frac{u_r\, v_m}{u_r + v_m\,(x-m)}\right)^2 & \text{for } 0 \leq x - m < x_r, \\ 0 & \text{otherwise.} \end{cases}$$

$$\check{s}(x) = \begin{cases} f(m)/4 & \text{for } x_l \leq x - m \leq x_r, \\ 0 & \text{otherwise.} \end{cases}$$

Then
$$\check{h}(x) \geq h(x) \geq f(x) \geq s(x) \geq \check{s}(x) \quad \text{for all } x.$$

Furthermore we find
$$\frac{1}{4}\int \check{h}(x)\,dx = \frac{1}{2}\int h(x)\,dx = \int f(x)\,dx = 2\int s(x)\,dx = 4\int \check{s}(x)\,dx. \tag{6.6}$$

6.3 Generators for $T_{-1/2}$-Concave Densities

Algorithm 6.6 Simple-Setup-Rejection

Require: $T_{-1/2}$-concave quasi-density $f(x)$, area A_f, mode m;
  cdf at mode $F(m)$ (optional).
Output: Random variate X with density prop. to f.
  /* Setup */
1: $f_m \leftarrow f(m)$, $v_m \leftarrow \sqrt{f_m}$, $u_m \leftarrow A_f/v_m$.
2: **if** $F(m)$ is provided **then**
3:   $A \leftarrow 2\,A_f$. /* Area below hat */
4:   $a_l \leftarrow F(m)\,A/2$, $a_r \leftarrow A/2 + a_l$.
5:   $u_l \leftarrow -F(m)\,u_m$, $u_r \leftarrow u_m + u_l$.
6: **else**
7:   $A \leftarrow 4\,A_f$.
8:   $a_l \leftarrow A/4$, $a_r \leftarrow 3A/4$.
9:   $u_l \leftarrow -u_m$, $u_r \leftarrow u_m$.
  /* Generator */
10: **repeat**
11:   Generate $V \sim U(0, A)$.
12:   **if** $V < a_l$ **then**
13:     $X \leftarrow -u_l^2/V + m$. /* Compute X by inversion */
14:     $Y \leftarrow V^2/u_l^2$. /* Compute $h(X)$. */
15:   **else if** $V \leq a_r$ **then**
16:     $X \leftarrow u_l/v_m + (V - a_l)/v_m^2 + m$.
17:     $Y \leftarrow f_m$.
18:   **else** /* $V > a_r$ */
19:     $X \leftarrow u_r^2/(v_m\,u_r - (V - a_r)) + m$.
20:     $Y \leftarrow (A - V)^2/u_r^2$.
21:   Generate $U \sim U(0,1)$.
22: **until** $U\,Y \leq f(X)$.
23: **return** X.

Moreover
$$\int_m^{m+x_r} h(x)\,\mathrm{d}x = \int_{m+x_r}^{\infty} h(x)\,\mathrm{d}x\ . \tag{6.7}$$

Completely analogous results hold for the left hand tail of $h(x)$ and for $\check{h}(x)$.

Proof. We get this upper and lower bound for the density immediately when we apply transformation (2.1) to the bounding curves for the enveloping region and the squeeze in Thm. 6.8, see Leydold (2001, Thm. 7). □

We can now use this theorem to compile a universal generator for $T_{-1/2}$-concave distributions based on the acceptance/rejection technique. Algorithm 6.6 (Simple-Setup-Rejection) generates a random variate with density proportional to the hat function by inversion. The squeeze is omitted.

Obviously Algorithm 6.6 (Simple-Setup-Rejection) is more complex (and slower) than Algorithm 6.4 (Simple-RoU), but it has the advantage that the rejection constant can be easily decreased when the domain of density is an interval $(b_l, b_r) \subset \mathbb{R}$. Just replace $(0, A)$ in Step 11 by (A_l, A_r), where $A_l = \int_{-\infty}^{b_l} h(x)\,\mathrm{d}x$ and $A_r = \int_{-\infty}^{b_r} h(x)\,\mathrm{d}x$.

6.4 Generators for T_c-Concave Densities

6.4.1 Generalized Ratio-of-Uniforms Method

The results of Sect. 6.3.1 can be further generalized. Wakefield et al. (1991) and Stefănescu and Văduva (1987) have suggested the following generalization of Thm. 2.20.

Theorem 6.12 (Wakefield, Gelfand, and Smith 1991). *Let $f(x)$ be a quasi-density with domain (b_l, b_r) not necessarily finite. Let θ be a strictly increasing differentiable function on $[0, \infty)$ such that $\theta(0) = 0$, and let k and m be constants. Suppose the pair of variables (U, V) is uniformly distributed over the region*

$$\mathcal{A}_f^{(\theta)} = \{(u,v): 0 < v \leq \theta^{-1}[k\,f(u/\theta'(v) + m)],\ b_l < u/\theta'(v) + m < b_r\}, \quad (6.8)$$

then $X = U/\theta'(V) + m$ has probability density function $f(x)/A_f$.

Wakefield, Gelfand, and Smith (1991) have suggested power functions $\theta_r(v) = v^{r+1}/(r+1)$, $r > 0$ and $k = 1/(r+1)$. Then (6.8) becomes

$$\mathcal{A}_f^{(r)} = \{(u,v): 0 < v \leq \sqrt[r+1]{f(u/v^r + m)},\ b_l < u/v^r + m < b_r\}. \quad (6.9)$$

The *minimal bounding rectangle* for $\mathcal{A}_f^{(r)}$ is given by

$$\mathcal{R}_{\mathrm{mbr}} = \{(v, u): v^- \leq v \leq v^+, 0 \leq u \leq u_m\}$$

where

$$\begin{aligned}
v^+ &= \sup_{b_l < x < b_r} \sqrt[r+1]{f(x)}, \\
u^- &= \inf_{b_l < x < b_r} (x-m)[f(x)]^{r/(r+1)}, \\
u^+ &= \sup_{b_l < x < b_r} (x-m)[f(x)]^{r/(r+1)}.
\end{aligned} \quad (6.10)$$

There also exists a generalization of Thm. 4.17. Consider the following regions

$$\begin{aligned}
\mathcal{B}_f^{(r)} &= \{(u,w): 0 < w \leq [f(u/w + m)]^{r/(r+1)},\ b_l < u/w + m < b_r\}, \\
\mathcal{G}_f &= \{(x,y): 0 < y \leq f(x),\ b_l < x < b_r\}, \\
\mathcal{T}_f^{(c)} &= \{(x,y): 0 < y \leq -(f(x))^c,\ b_l < x < b_r\},
\end{aligned} \quad (6.11)$$

and transformations

6.4 Generators for T_c-Concave Densities

$$\begin{aligned}
&\Phi_{\mathcal{AB}}\colon \mathbb{R}\times\mathbb{R}^+ \to \mathbb{R}\times\mathbb{R}^+,\ (U,V)\mapsto (U,W)=(U,V^r)\,,\\
&\Phi_{\mathcal{BG}}\colon \mathbb{R}\times\mathbb{R}^+ \to \mathbb{R}\times\mathbb{R}^+,\ (U,W)\mapsto (X,Y)=(U/W+m,W^{(r+1)/r})\,,\\
&\Phi_{\mathcal{GT}}\colon \mathbb{R}\times\mathbb{R}^+ \to \mathbb{R}\times\mathbb{R}^-,\ (X,Y)\mapsto (X,Z)=(X,-Y^c)\,.
\end{aligned} \qquad (6.12)$$

These transformations between the above regions play a crucial role for the further development of the theory for this type of generation methods. Notice that $\Phi_{\mathcal{AB}}$ maps $\mathcal{A}_f^{(r)}$ one-to-one onto $\mathcal{B}_f^{(r)}$. Analogously for $\Phi_{\mathcal{BG}}$ and $\Phi_{\mathcal{GT}}$. Figure 6.4 shows these regions for the normal distribution. Notice that the transformation $\Phi_{\mathcal{AG}} = \Phi_{\mathcal{BG}} \circ \Phi_{\mathcal{AB}}\colon (U,V)\mapsto (X,Y) = (U/V^r+m, V^{r+1})$ has constant Jacobian $r+1$. Thus Thm. 6.12 follows for the power function $\theta(v) = v^{r+1}/(r+1)$ analogously to the proof of Thm. 2.20. (It is not difficult to show that the Jacobian is constant for any such function $\theta(v)$.) Moreover

$$|\mathcal{A}_f^{(r)}| = A_f/(r+1)\,. \qquad (6.13)$$

Theorem 6.13. $\mathcal{B}_f^{(r)}$ *is convex if and only if* $f(x)$ *is* T_c*-concave with* $c = -\frac{r}{r+1}$.

Proof. See Leydold (2003, Thm. 6). □

Now we use this theorem and transformations (6.12) to build universal hat functions for T_c-concave densities. The treatment of this section follows Leydold (2003). However, details are omitted and the interested reader is referred to the original paper.

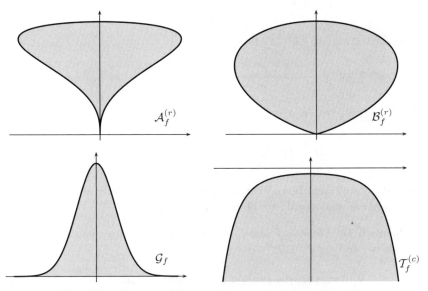

Fig. 6.4. $\mathcal{A}_f^{(r)}$, $\mathcal{B}_f^{(r)}$, $\mathcal{G}_f$, and $\mathcal{T}_f^{(c)}$ for the normal distribution with $r = 3$ and $c = -3/4$. Notice that $\mathcal{B}_f^{(r)}$ is convex if and only if $\mathcal{T}_f^{(c)}$ is convex (Thm. 6.13)

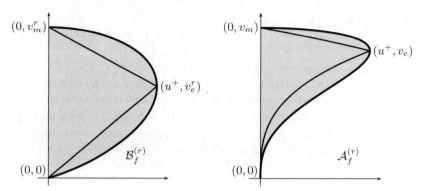

Fig. 6.5. The region $\mathcal{B}_f^{(r)}$ with enclosed triangle with vertices at $(0,0)$, $(0, v_m^r)$ and extremal point (u^+, v_e^r) (l.h.s.). $\mathcal{A}_f^{(r)}$ and the corresponding enclosed region on the r.h.s.

Let $f(x)$ be a T_c-concave density with $c = -\frac{r}{r+1}$ and mode m. Define $\mathcal{A}^+ = \{(u,v) \in \mathcal{A}_f^{(r)} : u \geq 0\}$, $\mathcal{A}^- = \{(u,v) \in \mathcal{A}_f^{(r)} : u \leq 0\}$, and analogously $\mathcal{B}^+$ and $\mathcal{B}^-$. By Thm. 6.13, $\mathcal{B}^+$ contains the triangle with vertices at $(0,0)$, $(0, v_m^r)$ and the extremal point with coordinates (u^+, v_e^r), where $v_m = v^+$ and $v_e \geq 0$ (see Fig. 6.5). Then transformation Φ_{AB}^{-1} maps this triangle into a three-side region in $\mathcal{A}_f^{(r)}$ of area

$$\int_0^{u^+} \left([v_m^r + u(v_e^r - v_m^r)/u^+]^{1/r} - [u\, v_e^r/u^+]^{1/r} \right) du = u^+ v_m^r \frac{r}{r+1} \frac{v_e - v_m}{v_e^r - v_m^r}$$

which cannot exceed $|\mathcal{A}^+|$. Consequently

$$u^+ \leq |\mathcal{A}^+| \frac{1}{v_m^r} \frac{r+1}{r} \frac{v_e^r - v_m^r}{v_e - v_m} = |\mathcal{A}^+| \frac{1}{v_m} \frac{r+1}{r} \frac{(v_e/v_m)^r - 1}{(v_e/v_m) - 1}.$$

Using (6.13) and the fact that $|\mathcal{A}^+| = (1 - F(m))\, |\mathcal{A}_f^{(r)}|$ we arrive at

$$u^+ \leq (1 - F(m))\, A_f \frac{1}{r\, v_m} \frac{(v_e/v_m)^r - 1}{(v_e/v_m) - 1}. \tag{6.14}$$

An analogous (lower) bound can be derived for u^-. There also exists a universal squeeze analogously to Thm. 6.8.

Theorem 6.14. *Suppose f is a T_c-concave quasi-density with $c = -\frac{r}{r+1}$, $r > 0$, mode m, and cdf F. Let*

$$\mathcal{R}_f^{(r)} = \{(u,v) : u_l(v) \leq u \leq u_r(v), 0 \leq v \leq v_m\},$$
$$\mathcal{Q}_f^{(r)} = \{(u,v) : -u_m(v) \leq u \leq u_m(v), 0 \leq v \leq v_m\},$$
$$\mathcal{S}_f^{(r)} = \{(u,v) : \tfrac{\bar{u}_l}{v_m^r} \leq \tfrac{u}{v^r} \leq \tfrac{\bar{u}_r}{v_m^r} \text{ and } \tfrac{\bar{u}_l}{v_m^r} \leq \tfrac{r\, u}{v_m^r - v^r} \leq \tfrac{\bar{u}_r}{v_m^r}\}.$$

6.4 Generators for T_c-Concave Densities

where
$$v_m = \sqrt[r+1]{f(m)}, \qquad u_m(v) = A_f/(r\, v_m)\, \frac{(v/v_m)^r - 1}{(v/v_m) - 1},$$
$$u_l(v) = -F(m)\, u_m(v), \quad u_r(v) = (1 - F(m))\, u_m(v),$$
$$\bar{u}_l = -F(m)\, A_f/v_m, \quad \bar{u}_r = (1 - F(m))\, A_f/v_m.$$

Then
$$\mathcal{S}_f^{(r)} \subset \mathcal{A}_f^{(r)} \subset \mathcal{R}_f^{(r)} \subset \mathcal{Q}_f^{(r)}$$

and
$$|\mathcal{R}_f^{(r)}| = \tfrac{r+1}{r}(\gamma + \psi(r+1))\, |\mathcal{A}_f^{(r)}|, \quad |\mathcal{Q}_f^{(r)}| = 2\, |\mathcal{R}_f^{(r)}|, \quad \text{and}$$
$$|\mathcal{S}_f^{(r)}| = (1 - 1/(r+1)^{1/r})\, |\mathcal{A}_f^{(r)}|,$$

where $\gamma = 0.577216\ldots$ denotes Euler's (gamma) constant and $\psi(z) = \Gamma'(z)/\Gamma(z)$ the Digamma function (also known as Euler's psi function). The envelopes $\mathcal{R}_f^{(r)}$ and $\mathcal{Q}_f^{(r)}$ and the universal squeeze $\mathcal{S}_f^{(r)}$ are optimal.

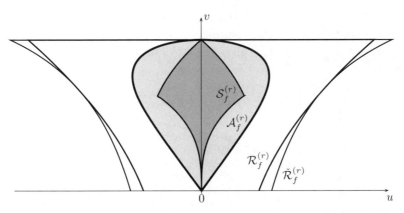

Fig. 6.6. $\mathcal{A}_f^{(r)}$, universal bounding envelopes $\mathcal{R}_f^{(r)}$ and $\check{\mathcal{R}}_f^{(r)}$, and universal squeeze $\mathcal{S}_f^{(r)}$ for the Cauchy distribution. ($r = 3$)

Figure 6.6 shows $\mathcal{A}_f^{(r)}$, universal bounding envelopes $\mathcal{R}_f^{(r)}$, and universal squeeze $\mathcal{S}_f^{(r)}$ for the Cauchy distribution. Sampling points uniformly on $\mathcal{R}_f^{(r)}$ is difficult and requires an appropriate enveloping region itself. This can be done by means of transformed density rejection (see Chap. 4) by using the fact that $\frac{x-1}{x^r - 1}$ is a convex function on $x \in [0, 1)$ for $r \geq 1$. A possible envelope is given by the following theorem.

Theorem 6.15. *Let f be a T_c-concave quasi-density with $c = -\frac{r}{r+1}$, $r > 1$, mode m, and cdf F. Let*
$$x_r = 1 - 2.187/(r + 5 - 1.28/r)^{0.9460}$$

and denote $a_r = \tau(x_r) - x_r\,\tau'(x_r)$ and $b_r = \tau'(x_r)$, where $\tau(x) = -(x-1)/(x^r - 1)$. Let

$$\check{\mathcal{R}}_f^{(r)} = \{(u,v)\colon \check{u}_l(v) \leq u \leq \check{u}_r(v), 0 \leq v \leq v_m\}\,,$$
$$\check{\mathcal{Q}}_f^{(r)} = \{(u,v)\colon -\check{u}_m(v) \leq u \leq \check{u}_m(v), 0 \leq v \leq v_m\}$$

where

$$v_m = \sqrt[r+1]{f(m)}\,,\qquad \check{u}_m(v) = A_f/(r\,v_m)\,\frac{-1}{a_r + b_r\,(v/v_m)}\,,$$
$$\check{u}_l(v) = -F(m)\,\check{u}_m(v)\,,\quad \check{u}_r(v) = (1 - F(m))\,\check{u}_m(v)\,.$$

Then $\mathcal{A}_f^{(r)} \subset \check{\mathcal{R}}_f^{(r)} \subset \check{\mathcal{Q}}_f^{(r)}$ and

$$|\check{\mathcal{R}}_f^{(r)}| = \frac{r+1}{r}\,\frac{1}{b_r}\,\log\left(\frac{a_r}{a_r + b_r}\right)\,|\mathcal{A}_f^{(r)}| \qquad\text{and}\qquad |\check{\mathcal{Q}}_f^{(r)}| = 2\,|\check{\mathcal{R}}_f^{(r)}|\,.$$

Proof. See Leydold (2003). □

Figure 6.7 shows the respective areas of the various enveloping and squeezing regions defined in the above theorems. These can also be used to get an idea of the performance of the resulting rejection algorithm. In particular the ratio between $|\check{\mathcal{R}}_f^{(r)}|$ and $|\mathcal{A}_f^{(r)}|$ gives the rejection constant.

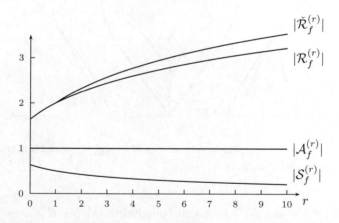

Fig. 6.7. Respective ratios of $|\check{\mathcal{R}}_f^{(r)}|$, $|\mathcal{R}_f^{(r)}|$ and $|\mathcal{S}_f^{(r)}|$ to $|\mathcal{A}_f^{(r)}|$ as functions of r. (Notice that $\check{\mathcal{R}}_f^{(r)}$ is not defined for $r \leq 1$)

To generate points uniformly in $\check{\mathcal{R}}_f^{(r)}$ we have to generate random variates with quasi-density $h(x) = -1/(a + b\,x)$ on $[0,1]$. The cdf and its inverse are given by

$$H(x) = \frac{\log(a/(a+b\,x))}{\log(a/(a+b))} \quad\text{and}\quad H^{-1}(u) = \tfrac{a}{b}(\exp(-u\,\log(a/(a+b))) - 1)\,.$$

6.4 Generators for T_c-Concave Densities

Algorithm 6.7 Generalized-Simple-RoU

Require: T_c-concave quasi-density $f(x)$, area A_f, mode m;
  parameter r ($r \geq -c/(1+c)$ and $r > 1$);
  cdf at mode $F(m)$ (optional).
Output: Random variate X with density prop. to f.
  /* Setup */
1: $p \leftarrow 1 - 2.187/(r + 5 - 1.28/r)^{0.9460}$.
2: $b \leftarrow (1 - r\, p^{r-1} + (r-1)\, p^r)/(p^r - 1)^2$.
3: $a \leftarrow -(p-1)/(p^r - 1) - p\, b$.
4: $v_m \leftarrow \sqrt[r+1]{f(m)}$, $u_m \leftarrow A_f/(r\, v_m)$.
5: **if** $F(m)$ is provided **then**
6: $\quad u_l \leftarrow -F(m)\, u_m$, $u_r \leftarrow u_m + u_l$.
7: **else**
8: $\quad u_l \leftarrow -u_m$, $u_r \leftarrow u_m$.
  /* Generator */
9: **repeat**
10: $\quad$ Generate $W \sim U(0, \log(a/(a+b)))$.
11: $\quad$ Generate $Z \sim U(u_l, u_r)$.
12: $\quad V \leftarrow (\exp(-W) - 1)\, a/b$.
13: $\quad U \leftarrow -Z/(a + b\, V)$, $V \leftarrow V\, v_m$.
14: $\quad X \leftarrow U/V^r + m$.
15: **until** $V^{r+1} \leq f(X)$.
16: **return** X.

The details of a rejection algorithm that is based on these derived inequalities is given in Algorithm 6.7 (Generalized-Simple-RoU). There is no need for an algorithm for $r < 1$ since then the much simpler and faster Algorithm 6.4 (Simple-RoU) can be used.

Analogously to Thm. 6.11 it is possible to compute universal hats for T_c-concave distributions. However, the resulting hat functions are very complicated and cannot be used for rejection algorithms (see Leydold, 2003).

Remark. By our construction of the envelope in Thm. 6.14 we have for the rejection constant α

$$\alpha = \alpha_{\text{univ}}(r) = \frac{r+1}{r}(\gamma + \psi(r+1)) \ .$$

(To be more precise: this is the rejections constant when we use $F(m)$ for constructing the hat.) It is interesting to compare this with transformed density rejection with three optimal points of contact to construct a table-mountain shaped hat. Here Thm. 4.14 gives sharp upper bounds for the rejection constant α for all T_c-concave distributions. By setting $c = -r/(r+1)$ (Thm. 6.13) we can write them as

$$\alpha \leq \bar{\alpha}_{\text{tdr3}}(r) = \frac{1}{1 - (1/(r+1))^{1/r}} \ .$$

148   6 Methods Based on General Inequalities

It is interesting to note that $\alpha_{\text{univ}}(1) = \bar{\alpha}_{\text{tdr3}}(1)$, $\alpha_{\text{univ}} > \bar{\alpha}_{\text{tdr3}}$ for $r < 1$, and $\alpha_{\text{univ}} < \bar{\alpha}_{\text{tdr3}}$ for $r > 1$, see Fig. 6.8. Thus the performance of the simple generator that is based on Thm. 6.14 is comparable to the worst-case performance of a more specialized generator based on transformed density rejection with three points of contact. However, it must be noted that the latter method does neither require the area A_f nor $F(m)$. If $F(m)$ is known the general upper bound for $\bar{\alpha}_{\text{tdr3}}$ is halved (see the remark after Thm. 4.14). Moreover, we should not forget that $\bar{\alpha}_{\text{tdr3}}$ is an upper bound for the real rejection constant which usually is smaller.

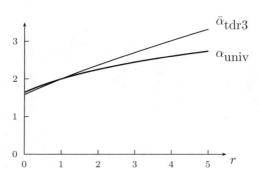

Fig. 6.8. α_{univ} and $\bar{\alpha}_{\text{tdr3}}$

6.4.2 Log-Concave Distributions

Devroye (1984c) gives a universal hat for log-concave densities. Let f be a log-concave density on $[0, \infty)$ with mode $m = 0$, $f(0) = 1$, and $A_f = 1$. Then

$$f(x) \leq \begin{cases} 1, & \text{for } 0 \leq x \leq 1, \\ \text{the unique solution } y < 1 \text{ of } y = \exp(x(y-1)), & \text{for } x > 1. \end{cases}$$
(6.15)

The area below this optimal bounding curve is $\pi^2/6$. In Thm. 6.14 we have constructed an optimal universal enveloping region for $\mathcal{A}_f^{(r)}$ for T_c-concave densities f where $c = -r/(r+1)$. Now for $r \to 0$ we have $c \to 0$ and $T_c \to \log$, see Sect. 4.2. It is not surprising that the limit for the rejection constant for an algorithm that utilizes this envelope is $\lim_{r \to 0} |\mathcal{R}_f^{(r)}|/|\mathcal{A}_f^{(r)}| = \lim_{r \to 0} \frac{r+1}{r}(\gamma + \psi(r+1)) = \pi^2/6$. If we have a closer look at $\mathcal{A}_f^{(r)}$ we find for $r \to 0$, $\mathcal{A}_f^{(0)} = \{(u,v): 0 < v \leq f(u), b_l < u < b_r\} = \mathcal{G}_f$, i.e. the region between the graph of the density and x-axis. (We assume that f has the same properties as the log-concave density above.) For the bounding curve from Thm. 6.14 we find $\mathcal{R}_f^{(0)} = \{(u,v): 0 \leq u \leq u_m(v), 0 \leq v \leq 1\}$ with $u_m(v) = \lim_{r \to 0} \frac{1}{r} \frac{v^r-1}{v-1} =$

$\frac{\log v}{v-1}$. Notice that $x = \frac{\log y}{y-1}$ is equivalent to (6.15). Thus Devroye's bounding curve is a special case of Thm. 6.14.

6.4.3 Heavy-Tailed T_c-Concave Distributions

We call a distribution a *heavy-tailed T_c-concave distribution*, for a $c < 0$, if its density $f(x)$ with mode m is T_c-concave and $(x-m)f(x)^{-c}$ is monotonically increasing on $\mathbb{R}$. Notice that for densities with support not equal to $\mathbb{R}$ this only can hold if the density is monotone with support $(-\infty, m]$ or $[m, \infty)$.

This heavy-tailed property is rather restrictive. If f is the density of a heavy-tailed T_c-concave distribution for a particular c, then f cannot be heavy-tailed T_c-concave for any other $c' \neq c$. Indeed $(x-m)f(x)^{-c}$ is increasing and bounded only if f behaves asymptotically as $x^{1/c}$ for $x \to \infty$. Thus for any $c' \neq c$, $(x-m)f(x)^{-c}$ either converges to 0 or is unbounded. For a monotonically decreasing T_c-concave density we find that f is heavy-tailed if and only if $c = \liminf_{x \to \infty} \mathrm{lc}_f(x) > -1$, where $\mathrm{lc}_f(x)$ denote the local concavity of f at x, see (4.9). A well known example for a heavy-tailed T_c-concave distribution for $c = -1/2$ is the Cauchy distribution. The t-distribution with a degrees of freedom is heavy-tailed T_c-concave with $c = -1/(1+a)$ (see Exercise 6.10).

A consequence of this property is that the extremal points of the region $\mathcal{A}_f^{(r)}$, for $r = -c/(c+1)$, are $(u^-, 0)$ and/or $(u^+, 0)$, see Fig. 6.9. Then, if we proceed as in Sect. 6.4.1, we can find a quadrangle contained in the set $\mathcal{B}^+$. This can be used to construct universal hats and squeezes for the class of heavy-tailed T_c-concave distributions. (Again the presentation follows Leydold, 2003.)

Theorem 6.16. *Let $f(x)$ be a heavy-tailed T_c-concave density with $c = -\frac{r}{r+1}$, $r > 0$, mode m, and cdf F. Let*

$$\mathcal{R}_f^{(r)} = \{(u,v) \colon u_l(v) \leq u \leq u_r(v), 0 \leq v \leq v_m\},$$
$$\mathcal{Q}_f^{(r)} = \{(u,v) \colon -u_m(v) \leq u \leq u_m(v), 0 \leq v \leq v_m\},$$
$$\mathcal{S}_f^{(r)} = \{(u,v) \colon \bar{u}_l \leq u(r+1) \leq \bar{u}_r \text{ and } \bar{u}_l \leq \frac{r\,u}{1-(v/v_m)^r} \leq \bar{u}_r\},$$

where

$$v_m = \sqrt[r+1]{f(m)}, \qquad u_m(v) = A_f/(r\,v_m)\,\frac{(v/v_m)^r - 1}{(v/v_m)^{r+1} - 1},$$
$$u_l(v) = -F(m)\,u_m(v), \quad u_r(v) = (1 - F(m))\,u_m(v),$$
$$\bar{u}_l = -F(m)\,A_f/v_m, \quad \bar{u}_r = (1 - F(m))\,A_f/v_m.$$

Then

$$\mathcal{S}_f^{(r)} \subset \mathcal{A}_f^{(r)} \subset \mathcal{R}_f^{(r)} \subset \mathcal{Q}_f^{(r)}$$

and

6 Methods Based on General Inequalities

$$|\mathcal{R}_f^{(r)}| = -\frac{1}{r}\left(\gamma + \psi\left(\frac{1}{r+1}\right)\right)|\mathcal{A}_f^{(r)}|, \quad |\mathcal{Q}_f^{(r)}| = 2|\mathcal{R}_f^{(r)}|, \quad \text{and}$$

$$|\mathcal{S}_f^{(r)}| = \left(1 - 1/(r+1)^{(r+1)/r}\right)|\mathcal{A}_f^{(r)}|$$

where $\gamma = 0.577216\ldots$ denotes Euler's (gamma) constant and $\psi(z) = \Gamma'(z)/\Gamma(z)$ the Digamma function (also known as Euler's psi function).

Proof. Define $\mathcal{A}^+$ and $\mathcal{B}^+$ as in Sect. 6.4.1. Since f is heavy-tailed T_c-concave, $\mathcal{B}^+$ is convex and contains the triangle with vertices at $(0,0)$, $(0, v_m^r)$ and $(u^+, 0)$. Moreover, if $\mathcal{B}^+$ contains a boundary point (u_b, v_b^r), $u_b > 0$, then $\mathcal{B}^+$ also contains the quadrangle with vertices at $(0,0)$, $(0, v_m^r)$, (u_b, v_b^r) and $(u_b, 0)$. By transformation Φ_{AB}^{-1} (see (6.12)) this is mapped into a four-sided region in $\mathcal{A}^+$ of area

$$\int_0^{u_b} (v_m^r + u(v_b^r - v_m^r)/u_b)^{1/r}\, du = u_b \frac{r}{r+1} \frac{v_b^{r+1} - v_m^{r+1}}{v_b^r - v_m^r}$$

which cannot exceed $|\mathcal{A}^+|$. Thus we find analogously to Sect. 6.4.1

$$u_b \leq (1 - F(m))\, A_f \frac{1}{r\, v_m} \frac{(v_b/v_m)^r - 1}{(v_b/v_m)^{r+1} - 1}.$$

An analogous (lower) bound can be deduced for a boundary point (u_b, v_b^r) with $u_b < 0$. From this the enveloping regions $\mathcal{R}_f^{(r)}$ and $\mathcal{Q}_f^{(r)}$ can be derived. Computing the respective areas is straightforward. For further details and for the proof of the universal squeeze see Leydold (2003). □

As in Sect. 6.4.1 we need a bounding region for the universal hat of Thm. 6.16. It is simplest to use rectangles, see Fig. 6.9. They are given by

$$\check{\mathcal{R}}_f^{(r)} = \{(u,v): \check{u}_l \leq u \leq \check{u}_r, 0 \leq v \leq v_m\},$$
$$\check{\mathcal{Q}}_f^{(r)} = \{(u,v): -\check{u}_m \leq u \leq \check{u}_m, 0 \leq v \leq v_m\}$$

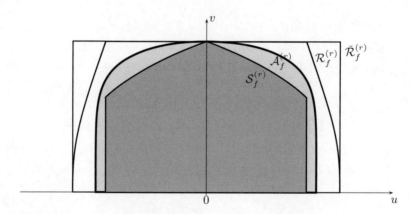

Fig. 6.9. $\mathcal{A}_f^{(r)}$ and universal bounding envelopes $\mathcal{R}_f^{(r)}$ and $\check{\mathcal{R}}_f^{(r)}$, and universal squeeze $\mathcal{S}_f^{(r)}$ for Student's distribution with $\nu = 1/3$ degree of freedom. ($r = 3$)

6.4 Generators for T_c-Concave Densities

Algorithm 6.8 Heavy-Tailed-Simple-RoU

Require: heavy-tailed T_c-concave quasi-density $f(x)$, $c \leq -1/2$, area A_f, mode m;
  parameter $r = -c/(1+c)$;
  cdf at mode $F(m)$ (optional).
Output: Random variate X with density prop. to f.
  /* Setup */
1: $v_m \leftarrow \sqrt[r+1]{f(m)}$, $u_m \leftarrow A_f/(r\,v_m)$.
2: **if** $F(m)$ is provided **then**
3:   $u_l \leftarrow -F(m)\,u_m$, $u_r \leftarrow u_m + u_l$.
4: **else**
5:   $u_l \leftarrow -u_m$, $u_r \leftarrow u_m$.
  /* Generator */
6: **repeat**
7:   Generate $U \sim U(u_l, u_r)$ and $V \sim U(0, v_m)$.
8:   $X \leftarrow U/V^r + m$.
9: **until** $V^{r+1} \leq f(X)$.
10: **return** X.

where $v_m = \sqrt[r+1]{f(m)}$, $\check{u}_m = A_f/(r\,v_m)$, $\check{u}_l = -F(m)\,\check{u}_m$, and $\check{u}_r = (1 - F(m))\,\check{u}_m$. For the resulting rejection constant we find $|\check{\mathcal{R}}_f^{(r)}| = \frac{r+1}{r}|\mathcal{A}_f^{(r)}|$ and $|\check{\mathcal{Q}}_f^{(r)}| = 2\,|\check{\mathcal{R}}_f^{(r)}|$.

Figure 6.10 shows the respective areas of the various enveloping and squeezing regions defined in the above theorems. The details for a rejection algorithm are compiled in Algorithm 6.8 (Heavy-Tailed-Simple-RoU).

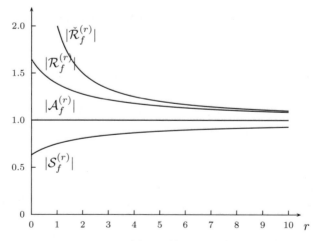

Fig. 6.10. Respective ratios of $|\check{\mathcal{R}}_f^{(r)}|$, $|\mathcal{R}_f^{(r)}|$ and $|\mathcal{S}_f^{(r)}|$ to $|\mathcal{A}_f^{(r)}|$ as functions of r for heavy-tailed distributions

152   6 Methods Based on General Inequalities

Although the envelope $\check{\mathcal{R}}_f^{(r)}$ is not optimal, we find that $|\check{\mathcal{R}}_f^{(r)}|$ converges to $|\mathcal{R}_f^{(r)}|$ for $r \to \infty$. In particular we have $|\check{\mathcal{R}}_f^{(r)}| < 1.1\,|\mathcal{R}_f^{(r)}|$ for $r \geq 4$. So it is only worthwhile to find sophisticated generators for the distribution with density proportional to $\frac{z^r-1}{z^{r+1}-1}$ if r is small. For $r = 1$, $\frac{z^r-1}{z^{r+1}-1}$ reduces to $1/(z+1)$.

Remark. To know the parameter c for a heavy-tailed T_c-concave distribution is quite a bit of information. But then – especially for c close to -1 – this information leads to very well fitting hat functions (see Fig. 6.10) and small rejection constants. Algorithm 6.8 (Heavy-Tailed-Simple-RoU) is a universal algorithm but nevertheless a fast and simple algorithm to sample for example from the t-distribution with shape-parameter smaller than 1.

6.5 Exercises

Exercise 6.1. How can Algorithm 6.3 (Monotone-Unbounded) be formulated for densities that have known integral unequal to one? Which is the "new" hat?
Hint: Use the fact that an arbitrary density f can be rescaled to have integral one by taking $f_1(x) = f(x)/A_f$.

Exercise 6.2. Describe the details of a rejection-algorithm for monotone densities on $(0, \infty)$ with given moment μ_r that uses as hat function the minimum of $f(0)$ and $(r+1)\mu_r/x^{r+1}$.
(This algorithm is described in Devroye, 1986a, p.317.)

Exercise 6.3. Compute the expected number of iterations for the Algorithm of Exercise 6.2.

Exercise 6.4. Show how the Algorithm of Exercise 6.2 can be applied to log-concave distributions. Which r Should be taken to minimize the area below the hat?

Exercise 6.5. Develop the details of an Algorithm for monotone densities with integral 1 and a known bound for $(\int f^a)^{1/a}$ for some $0 < a < 1$.
Hint: Use Thm. 6.2 and develop an Algorithm similar to Algorithm 6.3 (Monotone-Unbounded); as right-most part of the hat use $k/x^{1/a}$ with $k = (\int f^a)^{1/a}$.

Exercise 6.6. Compute the expected number of iterations for the Algorithm of Exercise 6.5.

Exercise 6.7. Work out the details of an Algorithm for Lipschitz densities with given Lipschitz C and a given moment.

Exercise 6.8. Compute the expected number of iterations of the algorithm developed in Exercise 6.7.

Exercise 6.9. Use the inequality from Thm. 6.7 to create an algorithm for arbitrary distributions. Use quasi-density f, area A_f below the density, mode m, and cdf at m as its input.

Exercise 6.10. Show that the t-distribution with a degrees of freedom is heavy tailed T_c-concave with $c = -1/(1+a)$.
Hint: Show that the t-distribution is T_c-concave for the given c and that $x\,f(x)^{-c}$ is monotonically increasing.

7
Numerical Inversion

The inversion method (see Sect. 2.1) is one of the most important key-stones of random variate generation. In this chapter we try to demonstrate how the inversion principle can be applied to obtain universal algorithms. If the inverse of the cumulative distribution function F^{-1} is computable, we can generate variates of the desired distribution without any additional information. So we could speak of a *"true"* black-box algorithm in this case, as we simply can return $F^{-1}(U)$. Unfortunately most important distributions have a cdf that cannot be expressed in terms of elementary functions and the inverse cdf is even more difficult. Hence the assumption that F^{-1} is available as a black-box is often of little practical relevance.

Nevertheless, by means of numerical integration or approximation techniques it is possible to provide (sometimes very) slow black-boxes for evaluating the cdf of most standard distributions. In this chapter we will assume that the cdf is available and discuss how numerical methods can be used to invert the cdf which is necessary to generate random variates.

In Sect. 7.1 we discuss variants with a short setup and without using auxiliary tables. The resulting algorithm is quite simple but slow for most distributions as the average number of required evaluations of the cdf is high. The number of evaluations of the cdf can be reduced (almost) to 0 if we use large tables at the price of a slow setup and if we sacrifice a bit of the desired precision. The details are explained in Sect. 7.2.

We have already remarked in Sect. 2.1 that the inversion method has theoretic advantages that make it attractive for simulation purposes. It preserves the structural properties of the underlying uniform pseudo-random number generator and can thus be used for variance reduction techniques, for sampling from truncated distributions, from marginal distributions, and from order statistics. It is also important in the framework of quasi-Monte Carlo computing, since there no other methods for importance sampling have been developed, yet. Moreover, the quality of the generated random variables depends only on the underlying uniform (pseudo-) random number generator.

These are reasons why many researchers in the field of simulation consider inversion to be the best method for generating non-uniform random variates, see e.g. Bratley et al. (1983).

On the other hand some authors do not like the idea of numerical inversion as it is slow and only approximate for many distributions. In the idealized arbitrary precision arithmetic model assumed by Devroye (1986a) inversion is not possible in finite time unless the inverse of the cdf is known explicitly. Therefore Devroye considers a combination of inversion and rejection that he calls *inversion-rejection* for the case that the cdf and the density are computable. There the real line is partitioned into intervals (that need not be stored) by simple recursions, and inversion by sequential search is used to determine the correct interval. Then rejection from a constant hat function is used to generate the exact variate since inversion is assumed to be impossible. Devroye discusses the trade off between short and long intervals using bounds on the expected number of iterations. Short intervals are good for the rejection constant but require a slower search. Whereas longer intervals lead to a faster search but result in a slower rejection algorithm. Details are provided especially for unimodal and for Lipschitz densities, see Devroye (1984b) and Devroye (1986a, Sect. VII.4.). The discussion of sequential search procedures for finding the correct interval given there can be instructive for numeric inversion algorithms as well.

We do not want to contribute to the old discussion if inversion or rejection should be preferred. We have seen in the last three chapters that rejection algorithms can provide us with very fast or with simple but slower universal algorithms. In this chapter we demonstrate that the inversion principle can be used for obtaining universal algorithms as well. They may be slower or less accurate than the rejection algorithms and they may need a longer setup and larger tables. But due to their theoretic advantages they should be included in every good random variate generation library as well.

7.1 Search Algorithms Without Tables

For inversion algorithms we have to solve the equation

$$F(X) = U$$

for fixed U. We assume that the cdf is available and for continuous distributions we know that it is continuous and monotonically increasing. So we may try to solve the above equation numerically. As it is known that the Newton method and the *regula falsi* converge quickly for smooth functions if we can provide good starting values, we can realize numerical inversion close to machine precision in many cases. For bad starting values or strangely shaped functions regula falsi is known to be more stable than the Newton algorithm. But even for regula falsi we can encounter very slow convergence if the cdf changes its slope rapidly (i.e. for densities with several modes and especially with high spikes). For these cases the last resort is the bisection method which is slow but known to converge for any cdf even when it is not continuous. So we can design a hybrid algorithm using regula falsi to start with and switch to the bisection method if regula falsi converges slowly. In our experiments we made good experiences when switching to bisection for the current step if the second successive step of the regula falsi resulted in updating the same side of the interval. After 50 iterations we change to bisection for all further iterations.

Doing this we can guarantee that the convergence is at least as fast as that for the bisection rule where the length of the interval after n recursion steps is $2^{-n}\Delta$ when Δ denotes the length of the starting interval. Thus to obtain an error in x-direction not bigger than ϵ_x we have to use at most

$$n = \log_2 \frac{\Delta}{\epsilon_x}$$

steps which is around 50 for common values for Δ and ϵ_x. This is of course quite a lot but it is only necessary in a region where a cdf is very steep or even not continuous. Otherwise the regula falsi will lead to much faster convergence. It is important to understand here that the bisection method guarantees a convergence of the x-values but cannot give an error-guarantee for the cdf values (u-direction) as we want the method to work for not continuous cdfs as well.

As a convergence criterion we suggest stopping if the relative error is smaller than ϵ_x and to use the absolute error ϵ_x^2 if the interval is very close to 0. But for very flat regions of the cdf in the tails the stopping criterion based on the x-error only may lead to a lot of unnecessary work. So we suggest using the absolute error in u-direction as second stopping criterion. As any uniform random number generator generates discrete numbers, and as the distance of these numbers is often different from machine precision we define ϵ_U as the maximal allowed absolute error in u-direction. It is certainly not necessary to choose ϵ_U much smaller than the minimal possible positive distance between two uniform random numbers (typically $2^{-32} \approx 10^{-9}$). Summarizing we have: The iterations of numerical inversion are stopped whenever the absolute u-error is smaller than ϵ_U, or the relative x-error is smaller than ϵ_x, or the absolute x-error is smaller than ϵ_x^2.

What is left is the search for a starting interval. In extensive experimentation we tried to optimize average performance for the distributions of Table 7.1 with many different parameter values. Therefore we suggest to approximately compute the upper and lower quartile in the setup. Then use these approximate quartiles as starting interval for the regula falsi. If the generated u-value is outside of the quartiles we use a search procedure starting with a step length which is a fixed fraction (e.g. 40%) of the inter-quartile range. This steps size is doubled until the desired interval is found. Then the regula falsi combined with bisection can start.

158   7 Numerical Inversion

Table 7.1 shows the average number of evaluations of the cdf we observed in our experiments. The main strength of inversion without table is its flexibility. It works for discrete and continuous distributions and also for mixtures. It is really a black-box algorithm as we need no information about discrete or continuous, location of modes or poles of the density, or anything else. So it is the method of this book that works for the largest class of distributions. The disadvantage is the high number of evaluations of the cdf required and the fact that the cdf is quite expensive for most standard distributions. Of course it is a good idea to store more than one interval and its cdf-values in the setup. The details of such an algorithm will be discussed in Sect. 7.2 below.

Table 7.1. Average number of evaluations of the cdf for numerical inversion

Distribution	cdf evaluations $\epsilon_x = \epsilon_U = 10^{-12}$	cdf evaluations $\epsilon_x = \epsilon_U = 10^{-8}$
Normal	6.0	5.3
Gamma(10)	6.1	5.4
Exponential	7.0	6.2
Gamma(0.5)	7.8	6.9
Gamma(0.01)	32.1	19.9
Cauchy	6.9	6.0
Discrete Uniform (0,1,2,3,...,9)	40.8	28.5

The results of Table 7.1 show that we can expect an average of between 5 and 10 evaluations of the cdf if the distribution has a smooth density. In these cases the search algorithm is practically only using regula falsi and converges fast. So a change of the error bound ϵ has little influence on the performance (about one additional evaluation of the cdf for increasing the precision by 10^4). For the case of the gamma(0.01) and the discrete uniform distribution the situation is different. The search algorithm is mainly using bisection steps as the cdf is very flat for most values of x. Thus increasing the precision by a factor of 10^4 requires approximately 13 additional steps (as $2^{13} = 8192$).

7.2 Fast Numerical Inversion

The evaluation of the cdf is known to be slow for many popular distributions. It is therefore interesting to see how big tables can speed up numerical inversion if we assume that we have a possibility to evaluate the cdf $F(x)$, possibly using a numerical integration package to compute the area below a given density. As we are discussing an approximate method we have to specify the desired accuracy ϵ_U, i.e. the maximal absolute error $|U - F(X)|$, where U denotes the generated uniform random number and X the non-uniform variate computed by approximately inverting the cdf. For many of todays computer environments and random number sources we can call our numerical inversion algorithm "exact" if we set $\epsilon = 10^{-10}$. But there remains the general drawback of numerical approximation algorithms that for a different computing platform or a random number source with higher precision the value of ϵ_U has to be adjusted. Then the table sizes for the fast algorithms explained in this section may explode.

For all algorithms described here we have to cut off the tails of the distribution. As the cdf is computable this is no big practical problem; we can use e.g. $F^{-1}(\epsilon_U/2)$ and $F^{-1}(1 - \epsilon_U/2)$ as "save" cut off points.

7.2.1 Approximation with Linear Interpolation

It is possible to store a large table with pairs $(x, u = F(x))$ or with pairs $(u, x = F^{-1}(U))$ in a slow setup step. Then we can use indexed search (see Sect. 3.1.2) to select the (short) interval in which the generated X will fall. Using equidistributed values for u makes this step even simpler. The problems remaining are how to generate X within the short interval and how to choose the u or x-values of the stored pairs.

Linear interpolation of the cdf without any refinements is suggested in some simulation books. This is of course fast and simple but not accurate, which is apparent if we remember that a linear cdf corresponds to a uniform density. So using linear interpolation of the cdf means that we are approximating the desired density by a step-function shaped density. So we should use rejection (as done in the Sect. 5.1) to get exact results. Otherwise we need too many intervals for practical purposes, e.g. about 2500 for the half-normal distribution and $\epsilon_U = 10^{-7}$. As the error of linear interpolation is $O(1/N^2)$, obtaining one additional decimal digit of precision requires about $\sqrt{10}$ times the number of intervals. Thus linear interpolation is fast and simple but it is producing only an approximate inversion. So we do not give any further details.

7.2.2 Iterated Linear Interpolation

To make the above algorithm exact we can use numerical inversion (as described in Sect. 7.1 above) within the short intervals to find the exact X. This means that we can use regula falsi with the interval borders as starting points. Notice that if we just return the result of the first iteration of the regula falsi as the final X, without checking the size of the error, this is exactly the same as the linear interpolation of the cdf and does not need any evaluations of F during the sampling procedure. Iterating until the convergence criterium $F(X_i) - F(X_{i+1}) < \epsilon_U$ is fulfilled is slow, as we have to evaluate F at least once. On the other hand we easily get very accurate results as the regula falsi is known to converge very fast if we have a smooth F. For the Newton method is even known to converge quadratically. This means that adding one iteration in the numerical procedure squares the error. So e.g. changing ϵ_U form 10^{-8} to 10^{-16} requires on average only one additional iteration of the search step. So the theory of the convergence of the numerical methods shows that numerical inversion with the Newton algorithm leads to very accurate results. Regula falsi is known to converge at almost the same speed; and in opposition to the Newton method we need not evaluate the derivative $F'(x) = f(x)$. This observation implies – and was also supported by our empirical experiments – that increasing the size of the table with starting values reduces the necessary number of evaluations of F only very slowly. So we suggest to use a table of moderate size (e.g. $N = 100$) for the starting values in this case. If we use the combination of regula falsi and bisection described in Sect. 7.1 we also have the advantage that the resulting algorithm is very robust and practically working for every distribution, even mixtures of discrete and continuous distributions without any additional knowledge.

For the setup it is necessary to decide which pairs $(x_i, F(x_i))$ should be stored. If we use the pairs

$$(F^{-1}(i/N), i/N) \quad \text{for} \quad i = 1, \ldots, N-1$$

all intervals have the same probability and we can generate the index of the interval fast and without a table. This is therefore simple but, as Ahrens and Kohrt (1981) point out, this interval design leads to a higher number of necessary iterations in the tails for most popular standard distributions.

7.2.3 Approximation with High-Order Polynomials

Ahrens and Kohrt (1981) developed details of a numerical inversion method especially designed for a fast assembler sampling code. They suggest a fixed decomposition into 99 subintervals which have a constant probability $1/64$ in the center. Towards the tails these probabilities are halved so that the right most and left most intervals have probability 2^{-24}. Inside these intervals the rescaled inverse cdf is approximated by Lagrange interpolation on nine points which is then converted into Chebyshev polynomials. If the precision is higher than required these polynomials are truncated and the resulting Chebyshev expansion is reconverted into a common polynomial which is used as approximate inverse cdf. Ahrens and Kohrt (1981) report good marginal speed but a very slow setup for this method. For the details we refer to the original paper. Unfortunately the setup of the algorithm is too complicated for routine use. Therefore we present a new algorithm based on similar ideas which is easier to implement.

7.2.4 Approximation with Cubic Hermite Interpolation

When we consider the methods of the last two subsections we can see that the problem of linear interpolation is that we need too many intervals to obtain exact results. On the other hand Ahrens and Kohrt (1981) use few intervals but high order polynomials which makes their method quite complicated. The main point of our new algorithm (presented in more detail in Hörmann and Leydold, 2003) is that we use a moderate order (three) for the polynomials that approximate the inverse cdf. The number of intervals is determined such that the required accuracy is reached.

For all algorithms of this chapter it is necessary to evaluate $F(b_i)$ and $F(b_{i+1})$ for all small subintervals $[b_i, b_{i+1}]$ as we need the probabilities of the intervals for the decomposition part of the algorithm. For most distributions it is comparatively cheap to evaluate the density; so we decided to use $f(b_i)$ and $f(b_{i+1})$ together with $F(b_i)$ and $F(b_{i+1})$ to approximate F^{-1} by a polynomial. (This method is known as cubic Hermite interpolation in the literature.) In order to increase the numerical stability we rescale the inverse function for a subinterval $[b_0, b_1]$ such that it maps $[0, 1]$ to $[b_0, b_1]$, i.e. its argument $\tilde{u}$ is between 0 and 1. We denote this function by $\tilde{F}^{-1}(\tilde{u}) = F^{-1}((1 - \tilde{u}) F(b_0) + \tilde{u} F(b_1))$. Then we find for the boundaries of the interval

$$\tilde{F}^{-1}(0) = b_0, \ (\tilde{F}^{-1})'(0) = (F(b_1) - F(b_0))/f(b_0) ,$$
$$\tilde{F}^{-1}(1) = b_1, \ (\tilde{F}^{-1})'(1) = (F(b_1) - F(b_0))/f(b_1) .$$

To approximate $\tilde{F}^{-1}$ by cubic Hermite interpolation we have to choose the coefficients of $p(\tilde{u}) = \alpha + \beta \tilde{u} + \gamma \tilde{u}^2 + \delta \tilde{u}^3$ such that p and $\tilde{F}^{-1}$ and their derivatives coincide at 0 and 1. Thus we find

$$\alpha = b_0, \quad \beta = \frac{F(b_1) - F(b_0)}{f(b_0)} ,$$
$$\gamma = 3 (b_1 - b_0) - (F(b_1) - F(b_0)) \left(\frac{2}{f(b_0)} + \frac{1}{f(b_1)}\right) , \quad \text{and}$$
$$\delta = 2 (b_0 - b_1) + (F(b_1) - F(b_0)) \left(\frac{1}{f(b_0)} + \frac{1}{f(b_1)}\right) .$$

So by this method we are approximating the inverse cdf by piecewise cubic polynomials. In all design points $b_i = F^{-1}(u_i)$ the polynomial approximation has the exact value for the inverse cdf and its derivative. If we consider the density $g(x)$ of the distribution we have constructed by approximating F^{-1} we clearly have $f(b_i) = g(b_i)$ for all b_i and $\int_{b_i}^{b_{i+1}} f(x)\,dx = \int_{b_i}^{b_{i+1}} g(x)\,dx$ for all intervals. Figure 7.1 illustrates the situation for the normal distribution on the subinterval $[0, 1]$. If $f(b_i) = 0$ then the above approximation does not work but it can be easily replaced by linear interpolation in that interval.

Notice that our approximation method leads to exact results if the inverse cdf is a polynomial of at most order three.

From the construction of the approximating polynomial $p(\tilde{u})$, explained above, we expect that the approximation error will be largest for $\tilde{u}$ close to $1/2$. Our intensive numerical experiments support this conjecture. So we get a good idea of the possible maximal error (in the u-direction) if we compute the error for $\tilde{u} = 0.5$ which is given by

$$|F(\tilde{p}(0.5)) - (F(b_i) + F(b_{i+1}))/2| .$$

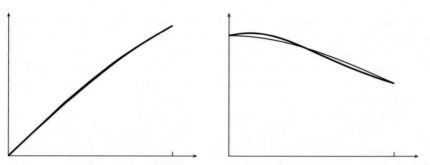

Fig. 7.1. Approximation of the cdf (l.h.s.) and of the density (l.h.s.) for normal distribution on subinterval $[0, 1]$

It follows from well known properties of Hermite interpolation that for a four times continuously differentiable density f, the above error is $O((b_{i+1} - b_i)^4)$ for $b_{i+1} \to b_i$. So if we use many subintervals we can expect small errors for smooth densities.

We can obtain the construction points in the following way: First we divide the domain of the distribution into a small number (e.g. 10) of intervals of equal length; then we half these intervals recursively and check the error in each subinterval at $\tilde{u} = 0.5$ until it is smaller than ϵ_U. Table 7.2 gives the resulting number of intervals for several standard distributions. All of our experiments showed that after the construction of the intervals the true error is clearly smaller than the desired ϵ_U. However, we have no proof that the error is guaranteed to be smaller as we have only checked it for $\tilde{u} = 0.5$. It could be possible to construct distributions with rapidly oscillating densities where either the number of intervals explodes or the error is under-estimated. The only guaranteed error bound (in the u-direction) that applies to arbitrary distributions is the probability of the subinterval. If we assume that the desired distribution has a continuous density and all local extrema of this density are known and used as design points we know that the cdf is convex or concave in all subintervals. Then we could use the tangents and the secant as simple upper and lower bounds for the cdf. This means that we use the linear interpolation of the cdf as an upper bound for the error which is quite pessimistic for smooth densities.

Table 7.2. Necessary number of intervals to obtain required precision

distribution	interval	$\epsilon_U = 10^{-8}$	$\epsilon_U = 10^{-10}$	$\epsilon_U = 10^{-12}$
Exponential	$(0, 25)$	212	640	2029
Half-normal	$(0, 7)$	182	623	1861
Half-Cauchy	$(0, 10^{-10})$	297	912	2825
Gamma(2)	$(10^{-5}, 30)$	288	929	2879
Gamma($\frac{1}{2}$)	$(10^{-20}, 25)$	229	709	2232

The results of Table 7.2 show that for the (well-behaved) distributions we have selected (including one with a very heavy tail and one with an unbounded density at 0) $\epsilon_U = 10^{-10}$ can be reached with less than 1000 intervals. This ϵ_U is sufficient for calling our algorithm "exact" for most of todays computing environments and random number sources. But we can also see that – depending on the distribution – two additional decimal places of precision require about three times the number of subintervals, a nice illustration of the $O(1/N^4)$ law for the maximal error. So we expect up to 10^5 intervals to get $\epsilon_U = 10^{-16}$ which would be necessary to obtain an "exact" method for 64-bit integer random number sources. Algorithm 7.1 (Fast-Table-Inversion) collects the necessary details of our fast numerical inversion algorithm using cubic Hermite interpolation. For a version also allowing for quintic Hermite interpolation see Hörmann and Leydold (2003).

Algorithm 7.1 Fast-Table-Inversion

Require: cdf $F(x)$ and density $f(x)$;
$\quad \epsilon_U$ the maximal allowed error, i.e. $|U - F(X)|$ (e.g. $\epsilon_U = 10^{-10}$);
$\quad$ number n_s of starting intervals, (10 or 20 are recommended).
Output: Random variate X with (approximate) cdf F.
$\quad$ /* Setup */
1: Compute b_l and b_r such that $0.1\,\epsilon_U < F(b_l) < \epsilon_U$ and $1 - \epsilon_U < F(b) < 1 - 0.1\,\epsilon_U$.
2: **for** $i = 0$ to n_s **do**
3: $\quad$ Set $b_i \leftarrow b_l + i(b_r - b_l)/n_s$.
4: $\quad$ Compute $F_i \leftarrow F(b_i)$ and $f_i \leftarrow f(b_i)$.
5: Set $i \leftarrow 0$
6: **repeat**
7: $\quad$ Compute the coefficients of the approximating polynomial $p(\tilde u)$:
$\quad\quad \alpha_i \leftarrow b_i,\ \beta_i \leftarrow (F_{i+1} - F_i)/f_i,$
$\quad\quad \gamma_i \leftarrow 3\,(b_{i+1} - b_i) - (F_{i+1} - F_i)\,(2/f_i + 1/f_{i+1}),$
$\quad\quad \delta_i \leftarrow 2\,(x_i - x_{i+1}) + (F_{i+1} - F_i)\,(1/f_i + 1/f_{i+1}).$
8: $\quad$ **if** $|F(b_i + 0.5(\beta_i + 0.5\,(\gamma_i + 0.5\,\delta_i))) - 0.5\,(F_i + F_{i+1})| < \epsilon_U$ **then**
9: $\quad\quad$ Set $i \leftarrow i + 1$. /* error below bound, continue with next interval */
10: $\quad$ **else**
11: $\quad\quad$ Split subinterval (b_i, b_{i+1}) at $b'_{i+1} = (b_i + b_{i+1})/2$ and insert in position $i+1$ into list (increment indices of all elements after position i by one).
12: $\quad\quad$ Compute $F_{i+1} \leftarrow F(b'_{i+1})$ and $f_{i+1} \leftarrow f(b'_{i+1})$.
13: **until** $b_i = b_r$
14: Set $N \leftarrow i$.
15: Compute guide table for Algorithm 3.2 (Indexed-Search) to generate discrete distribution with cumulative probabilities F_i for $i = 1, 2, \ldots, N$.
$\quad$ /* Generator */
16: Generate $U \sim U(0,1)$.
17: $J \leftarrow \max\{J \colon F_J < U\}$. /* use *Indexed Search* (Sect. 3.1.2) */
18: $\tilde U \leftarrow (U - F_J)/(F_{J+1} - F_J)$.
19: Set $X \leftarrow b_J + \tilde U(\beta_i + \tilde U(\gamma_i + \tilde U \delta_i))$.
20: **return** X.

7.3 Exercises

Exercise 7.1. Compute a formula for the maximal x-error ϵ_x for numerical inversion using the bisection method. Assume that $b_l = 0$ and $b_r = 1$. What is the maximal number of iterations necessary to guarantee $\epsilon_x < 10^{-10}$.

Exercise 7.2. Consider the distributions on (0,1) with $F(x) = \sqrt{x}$ and $F(x) = \sqrt[3]{x}$. Using the formulas for the coefficients of cubic Hermite interpolation (given in Sect. sec:cubic-hermite) compute the approximate inverse cdf for the whole interval (0,1). Check that the approximation is exact for these two distributions.

8

Comparison and General Considerations

In the previous Chaps. 4, 5, 6, and 7 we have presented a lot of algorithms for sampling from univariate continuous distributions. The variety of methods is at a first glance confusing. On the other hand, it gives the reader the possibility to select the algorithm that is best suited for the given simulation problem. We have implemented most of the presented algorithms in a library called UNU.RAN (Universal Non-Uniform RANdom number generators) which can be downloaded from our web site as C source, see Sect. 8.1. It is not only a ready to use implementation of these algorithms, it is also a detailed reference of how these algorithms can be implemented in a programming language.

To give the reader an idea about the performance of the presented algorithms we used this library and ran simulations with different distributions (Sect. 8.2). We have mainly used common standard distributions for two reasons. First they are well known and the reader can easily classify them with respect to heavy tails, (un-)bounded domains, skewness, etc. Another reason for using standard distributions is that we want to demonstrate that these universal algorithms are also well suited to sample from these distributions. Indeed, these generators have advantages that make them preferable for standard distributions, even for the normal distribution. We also added a hyperbolic distribution as an example of a distribution that has recently become important in financial mathematics. This example shows that automatic methods provide generators of excellent speed and quality easily usable for non-experts in the field of random variate generation. Another special application is sampling from order statistics. For this task mainly numerical inversion is suggested in the simulation literature. Of course the resulting timing tables heavily depend on the used computer, operating system, and compiler. Nevertheless, they give an impression about the performance of the algorithms. To make these figures less dependent from the used computing environment we report generation times relative to the sampling of exponentially distributed random variates (within the same framework of our library). The interested reader can easily run these timings on her own computer using the programs provided with our UNU.RAN library to obtain the timing tables valid for her personal computing environment.

The quality of the generated pseudo-random sample is an important albeit hardly investigated problem for non-uniform random variate generation. In Sect. 8.3 we try to enlighten the situation by discussing which algorithms tend to preserve the structure (and thus also the quality) of the underlying uniform generator.

In Sect. 8.4 we demonstrate that some of the algorithms are well suited for sampling from truncated distributions, for sampling from order statistics and for correlation induction. We conclude with Sect. 8.5 where we give a short survey on the basic properties of the described algorithms, their requirements, advantages and disadvantages.

8.1 The UNU.RAN Library

We have implemented most of the important universal algorithms described in this monograph in a library called UNU.RAN (Universal Non-Uniform RANdom number generators) which can be downloaded from our web site (Leydold et al., 2002). It is a ready to use implementation of these algorithms in a robust and portable way using ISO C. Moreover, the source code of this library contains (of course) all the necessary coding details we have to some extent omitted in the book. The UNU.RAN source code can thus be seen as a demonstration how universal methods can be implemented in practice.

The main problem when implementing universal algorithms for generating random variates from arbitrary distributions in a software library is certainly the application programming interface (API). Considering generators for a fixed distribution everything is simple. Any programmer can easily guess that the following C statements are assigning a realization from a uniform, a standard normal, and a gamma(5) random variate to the respective variables.

```
x  = random();
xn = randnormal();
xg = randgamma(5.);
```

For universal methods the situation is clearly more complicated. First, the setup often is very expensive compared to the marginal generation time. Thus using an object-oriented design is more appropriate than simply using procedures as above. That is, we first make an instance of a *generator object* for the required distribution in a setup step which is then used for sampling. Second, we need a lot more parameters for building this generator object. All information about the given distribution (like e.g. the density and its mode) are implicitly available in a routine like `randgamma(5.)` but must be provided explicitly for an automatic algorithm. Some of these parameters are required, others are optional but can be adjusted to increase the performance of the generation algorithm. Additionally, the automatic algorithms themselves have design parameters. For all of them appropriate default values that work for almost all possible distributions should be set in the library. However, in some exceptional cases these default values must be modified to adjust the algorithm to the given distribution. This may be the case when optimal performance is desired, or if the given distribution has extreme properties. All these parameters would result in a very long argument list for the setup routine but there is no need for an average user to ever change them.

To solve these problems in UNU.RAN in a flexible way we have decided to use three types of objects: a *distribution object* for all data related to the given distribution, a *parameter object* for the chosen method and its parameters, and a *generator object* that is used to sample from the given distribution. This means that using one of the universal generators for a particular distribution requires several steps:

1. *Create a distribution object* that holds the necessary data for the required distribution. For easy use the library provides creators for many standard distributions that contain all the necessary functions and data.
   For example, creating an instance `distr` of a distribution object that holds a normal distribution with mean 2 and standard deviation 0.5 is done by
   ```
   fparams[] = {2., 0.5};
   distr = unur_distr_normal(fparams,2);
   ```
2. *Choose an algorithm.* Most universal algorithms have lots of parameters that can be used to adjust the algorithm for the given sampling problem (like e.g. a bound for ϱ_{hs} for TDR using ARS). This is realized by creating an instance of a *parameter object* that contains a marker for the chosen algorithm together with all necessary default values for these parameters. These defaults can be overwritten later.
   For example, if we want to sample from the above distribution by means of TDR with transformation $T_0(x) = \log(x)$ (i.e. with parameter $c = 0$, see Sect. 4.2) we have to create an instance `par` of the parameter object by
   ```
   par = unur_tdr_new(distr);
   unur_tdr_set_c(par, 0.);
   ```
   If the call to `unur_tdr_set_c` is omitted a default value for parameter c is used (in this example: $c = 0.5$).
3. *Initialize the generator object.* By executing the setup of the algorithm all necessary tables and constants are calculated and stored in the generator object. It can then be used for sampling from the given distribution.
   In our example, we initialize the generator object and get an instance `gen` by
   ```
   gen = unur_init(par); .
   ```
   Then we can sample one random variate x from the above distribution by
   ```
   x = unur_sample_cont(gen); .
   ```

More examples can be found in the UNU.RAN user manual (Leydold et al., 2002).

This procedure may look complicated, but according to our experience there is no simpler way to handle the complexity of such a library in a flexible way. This is caused by the fact that UNU.RAN is not a collection of special generators for particular distributions but a collection of widely applicable automatic generators. And this object-oriented design also has big advantages. The generator object is created once and can then be used during the whole simulation. It is also very easy to replace a generator object by another one. For example, first we run a simulation with, e.g., a gamma distribution. Then we want to replace the gamma distribution by some other distribution or even by an empirical distribution. To make this changes using UNU.RAN it is enough to change the distribution at a single place in the code.

To facilitate the automated usage of UNU.RAN we also have provided a *string API*. There all necessary information to build a generator object can be formulated as a single character string. The generator object **gen** from the above example can then be created using

```
gen = unur_str2gen("normal(2,0.5) & method=tdr; c=0.");
```

Of course the string API is itself utilizing the distribution object and the parameter object we have described above.

An Automatic Code Generator

Our experiences show that the implementation of universal algorithms in a flexible, reliable, and robust way results in rather large computer code. This is not only caused by setup and adaptive steps where all required constants for the generator have to be computed or updated. Even more effort is necessary to check the given data and the density functions provided by the user, since not every method works for every distribution. On the other hand the sampling routine itself is very simple and consists only of a few lines of code. Installing and using such a library might seem too tedious for "just a random number generator" at a first glance, especially when only a generator for one particular distribution is required.

As a solution to this problem we can use universal methods to run the setup and use the computed constants to produce a single piece of code that can then generate variates from the desired distribution. This concept of an *automatic code generator for random variate generation* has the advantage that it is also comparatively easy to generate code for different programming languages as only the sampling part of the algorithms has to be rewritten. Moreover, a graphical user interface (GUI) can be used to simplify the task of obtaining a generator for the desired distribution for a practitioner or researcher with little programming knowledge and/or little background in random variate generation. This user interface is also saving time for those who just need a generator for a single distribution and do not want to install a library and read the manual for that task.

We have implemented a proof of concept study using a web based interface. It can be found at http://statistik.wu-wien.ac.at/anuran/. Currently, program code in C, Fortran, and Java can be generated and downloaded. For more details we refer to Leydold et al. (2003).

8.2 Timing Results

The generation speed is an important indicator for the performance of the described algorithms. Usually such timing results presented in the literature distinguish between setup time and marginal generation time. Although this might be interesting to assess the performance of the algorithm, the total (or average) generation time for all generated random variates is of more interest in practice. Thus we have decided to give average generation times for various sample sizes (i.e. average generation time per generated random variate including setup).

In our timing experiments we have experienced a great variability of the results. They depended not only on the computing environment, the compiler, and the uniform pseudo-random number generator but also on coding details like stack variables versus heap for storing the constants, etc. To decrease this variability a little bit we decided to define the *relative generation time* of an algorithm as the generation time divided by the generation time for the exponential distribution using inversion which is done by $X \leftarrow -\log(1-\text{random}())$. Of course this time has to be taken in exactly the same programming environment, using the same type of function call, etc. The relative generation time is still influenced by many factors and we should not consider differences of less than 25 %. Nevertheless, it can give us a crude idea about the speed of a certain random variate generation method.

All our timing results are obtained using our UNU.RAN library. These experiments are included into the "source distribution" of the library and can be repeated by a user for her own computing environment. Table 8.1 lists the distributions that have been used for our experiment. It includes some common standard distribution since their respective density functions, distribution functions, domains, and other properties are well known. Additionally we have added a hyperbolic distribution as these types of distributions have been recently suggested to model stock returns in financial simulation (Sect. 15.1). We have inserted its density via UNU.RAN's string API. Thus the evaluation of its density is slow compared to, e.g., the normal distribution. We also have included a truncated normal distribution (see Sect. 8.4.1) and, as an example of an order statistic (see Sect. 8.4.3), the median of a sample of 199 iid. power-exponential distributed variates (with quasi-density $\exp(-|x|^3)$).

Table 8.2 lists the algorithms and variants we used for our timing experiments. We selected them to demonstrate the performance of algorithms based on transformed density rejection and on Ahrens method as well as on the inversion method. The results for different sample sizes are given in Ta-

Table 8.1. Distributions used for timing tests

Symbol	Distribution
N	standard normal
C	Cauchy
E	standard exponential
Ga	gamma(3)
B_1	beta(3,4)
B_2	beta(30,40)
H	hyperbolic with quasi-density $\exp(-\frac{1}{2}\sqrt{1+x^2})$
TN	truncated normal distribution with domain $(0, 0.5)$
OS	median of a sample of 199 r. v. with quasi-density $\exp(-|x|^3)$

Table 8.2. Algorithms used for timing tests

Symbol	Algorithm
GW0	4.1 (TDR), $c = 0$, DARS, $\varrho_{hs} \approx 1.01$
GWS	4.1 (TDR), $c = -1/2$, DARS, $\varrho_{hs} \approx 1.01$
IA	4.3 (TDR-Immediate-Acceptance), $c = -1/2$, DARS, $\varrho_{hs} \approx 1.01$
PSh	4.2 (TDR-Proportional-Squeeze), $c = -1/2$, DARS, $\varrho_{hs} \approx 1.01$
PSm	4.2 (TDR-Proportional-Squeeze), $c = -1/2$, DARS, $\varrho_{hs} \approx 1.10$
PSl	4.2 (TDR-Proportional-Squeeze), $c = -1/2$, DARS, $\varrho_{hs} \approx 2$
PSA	4.2 (TDR-Proportional-Squeeze), $c = -1/2$, ARS, $\varrho_{hs} \geq 1.01$
TDR3	4.4 (TDR-3-Points)
AROU	4.8 (Adaptive-RoU), DARS, $\varrho_{hs} \approx 1.01$
AHRh	5.2 (Ahrens-Immediate-Acceptance), DARS, $\varrho_{hs} \approx 1.01$
AHRl	5.2 (Ahrens-Immediate-Acceptance), DARS, $\varrho_{hs} \approx 1.10$
SROU	6.4 (Simple-RoU)
SROUm	6.4 (Simple-RoU) with given cdf at mode
GSROU	6.7 (Generalized-Simple-RoU) with $r = 2$
SSR	6.6 (Simple-Setup-Rejection)
SSRm	6.6 (Simple-Setup-Rejection) with given cdf at mode
NINV	numerical inversion with regula falsi
NINVT	inversion with regula falsi and table of size 100
HINV	7.1 (Fast-Table-Inversion), $\epsilon_U = 10^{-9}$

bles 8.3–8.6. A dash '–' indicates that the combination of distribution and algorithm does not work. For the Cauchy distribution this happens as it is not log-concave which makes the case $c = 0$ impossible. We did not include the very difficult cdf of the hyperbolic distribution and hence it cannot work with any algorithm based on the inversion method. The reason why the log-concave density for the order statistic (OS) did not work for IA and PSh is more sophisticated. The UNU.RAN algorithms try to check whether the conditions for the chosen transformation method are satisfied. Due to numerical

problems the test for T-concavity fails in a region where the density is very close to 0. This shows the problem of implementing black-box algorithms in a secure way. The corresponding routines in the library could be adjusted such that this distribution works; however, the price we have to pay is that then the routine might not detect non-T-concave distributions and could produce incorrect random variates.

Tables 8.3–8.6 demonstrate that we always have the trade off between setup time and marginal generation time. If only a few random variates have to be generated algorithms with small or no tables (like SROU, GSROU, SSR and TDR3) are fastest. But as we can see in Table 8.5 algorithms with moderate tables (like AROU, AHR1 and PSm) are faster for sample size 10^4. Algorithms with extremely large tables (like PSh, AHRh and HINV) are competitive only for sample sizes of at least 10^6. Even then cache-effects may slow down the speed of the real-world implementations of such algorithms. Otherwise AHRh should be faster than AROU. For this reason we have not included linear approximation of the cdf (Sect. 7.2.1) in our list as it generates huge tables in an extremely expensive setup. Due to these huge tables slow memory look-ups result only in a moderately fast marginal generation time.

Table 8.3. Relative average generation times for sample size $n = 1$ (changing parameter case)

	N	C	E	Ga	B_1	B_2	H	TN	OS
GW0	641	–	66	405	567	507	483	57	1110
GWS	667	560	229	619	577	740	991	53	1381
IA	1019	824	452	781	852	1148	1474	64	–
PSh	1015	822	452	781	853	1148	1474	64	–
PSm	259	202	128	278	331	422	431	64	1338
PSl	84	76	53	129	98	178	222	64	708
PSA	48	41	52	65	73	76	132	60	421
TDR3	19	16	16	21	28	28	57	19	66
AROU	263	253	116	299	299	360	584	38	1010
AHRh	2047	4462	1895	2193	1940	2500	4185	53	6064
AHRl	197	508	159	202	235	298	461	26	820
SROU	16	12	14	17	17	19	50	17	50
SROUm	14	10	12	14	16	16	47	16	29
GSROU	24	22	22	26	26	28	62	24	64
SSR	16	14	12	16	17	19	50	16	48
SSRm	14	12	16	16	16	17	48	16	29
NINV	69	57	59	153	157	529	–	36	912
NINVT	950	970	962	3073	3385	10740	–	710	14230
HINV	5512	5583	2796	8910	7732	25750	–	241	41300

Table 8.4. Relative average generation times for sample size $n = 100$

	N	C	E	Ga	B_1	B_2	H	TN	OS
GW0	8.91	–	3.48	6.61	8.18	7.62	7.45	3.00	13.63
GWS	7.99	6.95	3.63	7.51	7.10	8.72	11.21	1.87	15.14
IA	10.97	9.05	5.38	8.64	9.35	12.25	15.50	1.52	–
PSh	11.16	9.25	5.59	8.86	9.56	12.47	15.71	1.74	–
PSm	3.80	3.12	2.45	3.94	4.46	5.38	5.62	1.74	14.84
PSl	2.47	2.10	2.04	2.98	3.05	3.71	4.46	1.74	11.19
PSA	3.94	2.36	2.57	4.04	4.02	4.77	6.40	1.73	14.15
TDR3	1.99	1.40	2.45	2.81	2.73	2.78	3.87	1.45	8.07
AROU	3.42	3.31	1.94	3.80	3.80	4.39	6.69	1.14	10.99
AHRh	21.45	45.61	19.90	22.87	20.37	25.95	42.65	1.32	61.72
AHRl	2.95	5.98	2.53	3.05	3.40	4.04	5.81	1.07	9.85
SROU	6.61	4.31	4.97	6.97	7.58	10.12	15.78	4.29	43.37
SROUm	3.37	2.27	3.08	3.94	4.77	5.43	8.13	3.13	20.36
GSROU	14.21	11.47	12.30	14.46	14.58	17.31	24.80	11.31	57.50
SSR	6.80	4.62	3.32	5.93	5.89	9.83	16.10	1.84	43.34
SSRm	3.54	2.42	5.76	3.66	4.43	5.50	8.28	2.42	20.56
NINV	11.77	10.80	11.45	34.68	35.76	118.60	–	7.16	157.60
NINVT	17.10	16.98	16.39	52.33	57.22	181.60	–	12.30	240.10
HINV	55.84	55.25	29.25	90.07	77.38	255.20	–	3.34	409.90

Table 8.5. Relative average generation times for sample size $n = 10^4$

	N	C	E	Ga	B_1	B_2	H	TN	OS
GW0	2.69	–	2.91	2.71	2.66	2.70	2.80	2.51	2.81
GWS	1.48	1.49	1.46	1.49	1.49	1.50	1.54	1.39	1.64
IA	1.03	1.01	0.97	1.02	1.03	1.06	1.10	0.93	–
PSh	1.24	1.22	1.18	1.23	1.24	1.27	1.31	1.14	–
PSm	1.28	1.20	1.23	1.29	1.28	1.33	1.44	1.14	1.75
PSl	1.68	1.39	1.57	1.73	2.13	2.02	2.33	1.14	4.34
PSA	1.25	1.23	1.18	1.26	1.28	1.29	1.37	1.14	1.60
TDR3	1.84	1.27	2.38	3.16	2.51	2.54	3.41	1.28	7.61
AROU	0.82	0.81	0.80	0.82	0.83	0.83	0.87	0.78	0.97
AHRh	1.05	1.32	1.02	1.07	1.05	1.12	1.30	0.80	1.53
AHRl	0.98	0.94	0.94	1.03	1.05	1.08	1.30	0.82	1.73
SROU	6.61	4.24	4.94	6.85	7.53	10.20	15.78	4.25	43.85
SROUm	3.34	2.21	3.03	3.83	4.74	5.43	7.86	3.39	20.60
GSROU	14.29	11.61	12.30	14.54	14.71	17.56	24.95	11.24	58.98
SSR	6.78	4.49	3.29	5.80	5.94	9.89	16.07	1.73	44.41
SSRm	3.42	2.30	5.67	3.60	4.39	5.41	8.04	2.30	20.84
NINV	11.34	10.72	11.60	33.74	35.51	119.00	–	6.98	155.00
NINVT	7.97	8.03	7.74	22.82	24.83	81.00	–	5.85	105.00
HINV	1.57	1.57	1.24	1.89	1.76	3.59	–	0.98	5.10

Table 8.6. Relative average generation times for sample size $n = 10^6$

	N	C	E	Ga	B_1	B_2	H	TN	OS
GW0	2.65	–	2.93	2.69	2.63	2.68	2.77	2.53	2.80
GWS	1.45	1.46	1.46	1.45	1.45	1.45	1.47	1.42	1.52
IA	0.95	0.95	0.95	0.96	0.97	0.97	0.97	0.95	–
PSh	1.17	1.16	1.16	1.18	1.18	1.18	1.18	1.16	–
PSm	1.28	1.20	1.24	1.28	1.27	1.31	1.42	1.16	1.64
PSl	1.69	1.40	1.59	1.75	2.16	2.03	2.34	1.16	4.31
PSA	1.17	1.17	1.17	1.17	1.18	1.18	1.19	1.15	1.24
TDR3	1.89	1.29	2.39	2.68	2.57	2.61	3.44	1.31	7.67
AROU	0.81	0.80	0.79	0.81	0.83	0.82	0.83	0.81	0.89
AHRh	0.87	0.86	0.86	0.87	0.87	0.88	0.91	0.82	0.95
AHRl	0.99	0.91	0.95	1.04	1.05	1.07	1.25	0.84	1.67
SROU	6.67	4.36	4.98	7.07	7.61	10.27	15.94	4.24	43.85
SROUm	3.37	2.23	3.06	3.93	4.78	5.46	7.94	3.08	20.57
GSROU	14.49	11.67	12.50	14.64	14.76	18.33	24.97	11.39	58.01
SSR	6.97	4.61	3.30	5.98	5.94	9.96	16.07	1.76	43.69
SSRm	3.50	2.34	5.75	3.56	4.41	5.43	8.10	2.31	21.01
NINV	11.45	10.56	11.17	34.46	35.15	116.80	–	7.03	152.70
NINVT	7.94	7.66	7.39	22.52	24.40	78.30	–	5.70	101.50
HINV	0.98	0.96	0.98	0.99	0.99	1.00	–	0.97	1.00

Remark. Till the eighties of the last century memory requirements were considered an important criterion in the random variate generation literature. However, memory requirements are hardly a restriction in modern computing environments used for numerical computations (except maybe in some embedded systems). Thus we do not include tables for memory requirements. It is obvious that with the exception of TDR3, SROU and SSR and its variants all the automatic methods of Table 8.2 need a lot of memory, much more than specialized algorithms, like the Box-Muller method.

8.3 Quality of Generated Samples

In the literature on uniform (pseudo-) random number generation speed is only one aspect when designing a pseudo-random number generator (PRNG). A large part of the literature deals with the quality of the generated samples (also called streams) of (pseudo-) random numbers, i.e., whether the generated point set looks like a set of "real" random numbers. This is done by both theoretical investigations of the structural properties and by empirical studies. For the theoretical part period length, equidistribution properties and lattice structure are of interest. For empirical studies small stochastic models are simulated using streams of uniform (pseudo-) random numbers. The distribution of the output variable is then compared with the theoretical distribution using various statistical tests. For all these investigations, theoretical and empirical, the distributions of points in higher dimensions are of interest. We refer the interested reader to the literature on uniform pseudo-random number generation, in particular to Knuth (1998), Niederreiter (1992), Marsaglia (1985), L'Ecuyer (1998), or to the pLab website maintained by Hellekalek (2002). Collections of empirical tests have been compiled, e.g., by Marsaglia (1996) in his DIEHARD test battery, by L'Ecuyer and Simard (2002) in their TestU01 package, and by the National Institute of Standards and Technology (NIST).

8.3.1 Streams of Non-Uniform Random Variates

In the area of non-uniform random variate generation the situation is different. Speed, memory requirements, and consumption of uniform random numbers are often the only described characteristics of newly proposed algorithms in the literature. Concerning quality it is only stated that the methods are exact, that means that perfect iid. uniform random numbers (which are *never* available) would be transformed into independent random numbers of the correct distribution.

To stress the importance of both theoretical and empirical tests for non-uniform random variate generators, we want to start with a generation method for normal random variates proposed by Kinderman and Ramage (1976). It is sometimes called the "fastest normal variate generator" (which is not true anymore) and it is used in some numerical computing libraries, like the IMSL Fortran Library (routine RNNOA, see IMSL, 2003). However, recently it has been discovered that there is a small error in this algorithm. A point set generated by this algorithm has slightly too many numbers near 0 (see Tirler, Dalgaard, Hörmann, and Leydold, 2003). This effect is rather small but could have serious consequences in a stochastic model. This example shows that even "obvious" errors in the description of an algorithm may not be detected for years. To find faulty results that are caused by interactions between the structure of the PRNG and the transformation method heavy testing of the generated streams of non-uniform random numbers is required.

However, contributions to the problem of the quality of the generated point sets are rare. Devroye (1982), for example, derives some measures for the error that is committed when the exact density f is approximated by a density g and gives some bounds. Monahan (1985) discusses the problem of accuracy that is caused by approximations and discretization errors on a digital computer, but the randomness of the pseudo-random number generator is not an issue. Deng and Chhikara (1992) propose a new criterion of robustness to compare the effects of imperfect uniform PRNGs on different transformation methods and give some examples. None of these papers considers the combination of a particular transformation method with uniform generators used in practice.

The first transformation method whose quality was considered in the literature is the Box-Muller method (Box and Muller, 1958). Neave (1973) found defects when it is used in conjunction with linear congruential generators (LCG) with bad lattice structure and discourages the use of this method (like several other authors did in the 1970ies). However, such bad PRNGs should not be used anyway[1]. (Indeed, Tezuka (1995) uses this "Neave effect" as a test for the quality of the underlying uniform PRNG.) Later Afflerbach and Wenzel (1988) have shown that the Box-Muller method (as a special case of the *polar method*, see Devroye, 1986a) can be viewed as a two-dimensional inversion method. The two-dimensional structure of the uniform generator is not preserved but transformed into a system of intersecting spirals, see Fig. 8.1. Thus the structure of the normal variates is different to the structure of the uniform random numbers but the quality of the uniform generator is preserved.

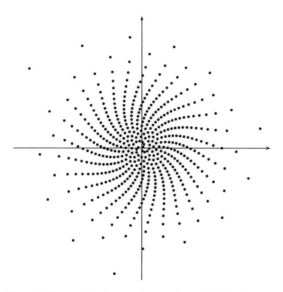

Fig. 8.1. The Box-Muller method transforms the grid of a linear congruential generator into spirals, illustrated by means of a "baby" PRNG

Two of us have studied the quality of non-uniform random variates between 1990 and 1993. The results are restricted to one-dimensional distributions and to LCGs only (Afflerbach and Hörmann, 1992; Hörmann and Derflinger, 1993; Hörmann, 1994a,b). For the ratio-of-uniforms method it turned out that the combination with an LCG always results in a defect. Due to the lattice structure of random pairs generated by an LCG there is always a hole without a point with probability of the order $1/\sqrt{M}$, where M is the

---

[1] Uniform PRNGs shipped out with compiler libraries or software are very often not state-of-the-art. Such built-in PRNGs must *never* be used for serious stochastic computations. As a rule of thumb do not use the internal generator of any language, if the simulation is critical.

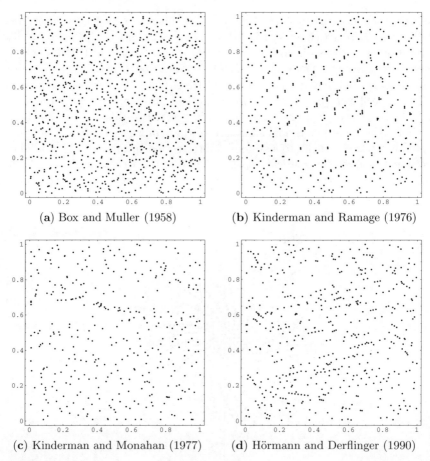

Fig. 8.2. Scatter plots of all overlapping tuples $(x_0, x_1), (x_1, x_2), (x_2, x_3), \ldots$ of a sequence of random numbers $x_1, x_2, x_3, \ldots$ generated by some normal variate generators. The points were transformed back into uniform pseudo-random numbers by means of the cdf; see Fig. 8.3(a) for the underlying uniform "baby" PRNG

modulus of the LCG; see Fig. 8.2(c), where the ratio-of-uniforms method with rejection from the minimal bounding rectangle has been used (see Sect. 2.4, Algorithm 2.9). However, as with the defects reported for the Box-Muller method, this observation is not crucial for state-of-the-art generators which should be used anyway in serious computations.

There also exist a few empirical studies. Chen and Burford (1981) have selected five sets of statistical tests to study the combination between 7 particular uniform PRNGs and three transformation methods for generating normal random variates. They also take a closer look to the implicit assumption that any "good" generator can be used in conjunction with any "good" transformation method. It is supposed that the resulting random variates will have good properties of randomness and the desired distribution fit. However, the results of this empirical research show that this assumption is false. Hence, they conclude, a simulation practitioner must be cautious in the selection of both the uniform PRNG and the transformation method. Furthermore, research is needed to identify which uniform generators are good for which kind of transformation method.

Leydold, Leeb, and Hörmann (2000) have made similar tests but have considered higher dimensional properties of the generated point sets. They also conclude that the combination of some linear uniform PRNG and a (non-linear) transformation method often results in unpredictable quality of the generated sequence of normal (pseudo-) random variates. They suggest that transformation methods which preserve the structural properties of the underlying uniform PRNG should be used.

Empirical tests are very important for finding deficiencies. However, it should be noted here that the RANDU uniform PRNG behaves very well in the research of Chen and Burford (1981). This is remarkable since nowadays it is infamous for its extremely bad lattice structure in dimension three, where all points fall into not more than 15 planes (see Park and Miller, 1988). Also the above faulty results with the generator by Kinderman and Ramage (1976) have not been detected by Leydold et al. (2000), since the least sample size that is necessary to detect this demerit is about 10^7. These observations clearly show that methods for generating non-uniform random variates also need strong theoretical foundations.

8.3.2 Inversion Method and Transformed Density Rejection

As a first résumé we can state that it is necessary to test each combination of a uniform PRNG and a particular transformation method for sampling from every distribution. However, such tests are very expensive and the number of necessary tests explodes with the number of distributions and uniform PRNGs that are available in a numerical library. When a new generator is added it is not only necessary to test this PRNG but also all combinations with each of the transformation methods. Moreover, the increase in computing power also increases the requirements for random number generators. Thus "good" generators may turn into generators with bad statistical properties when the sample size is increased due to new (faster) hardware, see the examples with RANDU or the generator by Kinderman and Ramage (1976) above. A solution to this problem might be to analyze the transformation methods, as it was done for the Box-Muller method. However, specialized algorithms, where speed and little memory requirements were the only design goals, often scramble the stream of uniform random numbers in a complicated way, as can be seen from Fig. 8.2(b,d).

Nevertheless, there is a possible way out of this problem. Let us start with the following assumption.

Postulate 8.1. A stream $\langle x_i \rangle$ of non-uniform pseudo-random variates that follow a distribution with cdf F is of *"good"* quality if the transformed stream $\langle u_i \rangle = \langle F(x_i) \rangle$ is a stream of $(0, 1)$-uniform pseudo-random numbers of *"good"* quality.

Here we do not want to repeat the discussion what a stream of uniform pseudo-random numbers of *"good"* quality is and refer the interested reader to the literature on uniform random number generation, see e.g. Knuth (1998).

Postulate 8.1 immediately implies that by using the inversion method a stream of *"good"* uniform pseudo-random numbers is always transformed into a stream of *"good"* non-uniform pseudo-random variates. At least this proposition holds if we have the inverse F^{-1} of the cdf F available (see also Sect. 8.3.4 for some critical remarks). As a consequence when applying the inversion method it is sufficient to use a state-of-the-art uniform PRNG (e.g. the Mersenne twister, see Matsumoto and Nishimura, 1998, or combined generators proposed by L'Ecuyer) to guarantee streams of *"good"* quality. Thus together with some other advantages the inversion method becomes more attractive even if faster methods are available. In practice, however, one often has to sample from distributions where only inexact or slow numerical methods are available for performing the inversion method. Thus there might be some risk of decreasing the quality of the transformed sequence.

The situation is not completely different for the rejection method when the hat distribution is generated by inversion. Here the hat distribution can be seen as a rough approximation of the required distribution. Thus if the rejection step is omitted we have an inexact method and the generated point set does not follow the required distribution (which cannot be tolerated, of course). Switching on the rejection step removes some of these points to correct this error. If we look at the scatter plot of all points that are generated by a uniform PRNG then "holes" appear, see Fig. 8.3(e). Using the cdf to transform these points we obtain a sequence of uniform pseudo-random numbers whose quality is worse in some sense than the quality of the underlying generator. At least we cannot argue that the quality remains the same under the transformation method as we have done with the inversion method. On the other hand it is clear that this effect becomes smaller for tighter hat functions, i.e. if the rejection constant and thus the ratio between area below the hat and the area below the density is small and close to one. This constant is bounded by the ratio ϱ_{hs} between the area below hat and squeeze.

For all variants of transformed density rejection the ratio ϱ_{hs} can be made as close to one as desired (Sect. 4.4). Figure 8.3 illustrates the situation for Algorithm 4.8 (Adaptive-RoU) and some values of ϱ_{hs}. It shows scatter plots of all overlapping tuples $(x_0, x_1), (x_1, x_2), (x_2, x_3), \ldots$ of a sequence of random numbers $x_1, x_2, x_3, \ldots$. For this experiment the "baby" generator $u_{n+1} = 869\, u_n + 1 \bmod 1024$ is used to generate normal random variates using different numbers of construction points. Plot (a) shows the scatter plot of the uniform generator for the whole period. Plots (b)–(f) show the tuples of the normal

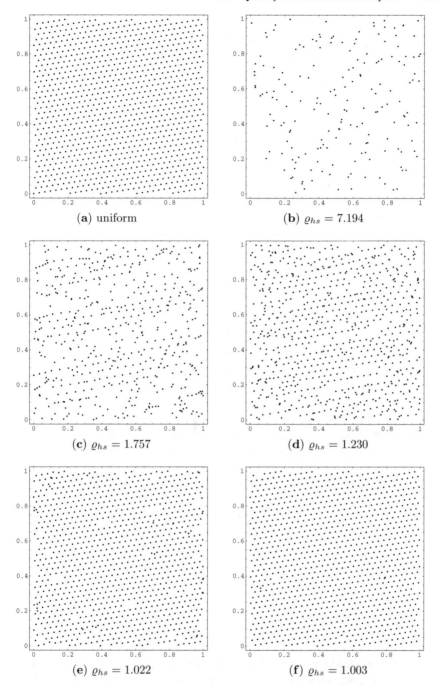

Fig. 8.3. Scatter plots of "baby" generator $u_{n+1} = 869\,u_n + 1 \bmod 1024$ (**a**) and of normal variates using Algorithm 4.8 (Adaptive-RoU) with 2, 4, 6, 29 and 79 equiangular construction points (**b–f**)

random variates transformed back to uniform random numbers using the cdf of the standard normal distribution. With decreasing ϱ_{hs} the structure of the underlying generator appears again. Thus for ϱ_{hs} close to one, the quality is close to the quality of the underlying uniform random number generator. So TDR may be preferred to inversion in particular when the cdf of the distribution is expensive, since then TDR is much faster. The same arguments also hold for Ahrens method (Sect. 5.1). However, this method requires much more construction points to gain the same ratio ϱ_{hs}.

8.3.3 Discrepancy

The above consideration are supported by the empirical study of Leydold et al. (2000). However, as alread mentioned above, we also need theoretical foundations since we cannot rely in our conclusions on just a few computing experiments.

The discrepancy as defined e.g. by Niederreiter (1992) – although hard to compute – has become an important measure for the uniformity of sets of numbers. For a sequence $\omega_u = (u_1, \ldots, u_n) \subset [0,1)^d$ it is given by

$$D(n, \omega_u) = \sup_R \left| \frac{\#(\omega_u \cap R)}{n} - \lambda(R) \right|$$

where the supremum is taken over all rectangles $R = [a_1, b_1) \times \ldots \times [a_d, b_d) \subset [0,1)^d$. $\#(M)$ denotes the cardinality of a set M and $\lambda(\cdot)$ the d-dimensional Lebesgue measure. The discrepancy is used to measure the uniformity of points generated by uniform pseudo-random number generators in Monte Carlo integration. For truly random sequences the discrepancy is of order $D(n, \omega_u) = O(\sqrt{\log \log n}/\sqrt{n})$ (Niederreiter, 1992). The discrepancy of points sets produced by a pseudo-random number generator should not deviate from that of truly random points; otherwise it produces pseudo-random numbers of bad quality.

For non-uniformly distributed sequences an analogous concept can be introduced. For a sequence $\omega = (x_1, \ldots, x_n) \subset \mathbb{R}^d$ we define its f-discrepancy by

$$D_f(n, \omega) = \sup_R \left| \frac{\#(\omega \cap R)}{n} - \lambda_f(R) \right|$$

where the supremum is taken over all rectangles $R = (a_1, b_1) \times \ldots \times (a_d, b_d) \subset \mathbb{R}^d$. However, now the measure $\lambda_f(M) = \int_M f(x)\,\mathrm{d}x$ is used. A sequence with f-discrepancy converging to zero can be interpreted as a set of points that follows a distribution with density f. Hlawka and Mück (1972a,b) call such a sequence f-equidistributed. There exist several slightly different definitions of discrepancy of non-uniformly distributed sets, see e.g. Hlawka and Mück (1972a); Fang and Wang (1994); Henderson, Chiera, and Cooke (2000).

In what follows we restrict our investigations to the case $d = 1$ for simplicity. The case $d \geq 2$ is similar. Let F denote the cdf of a distribution with density f, and let F^{-1} denote its inverse. Notice that for any set M, $\lambda_f(M) = \int_M \mathrm{d}F(x)$. Thus we find for a sequence $\omega = (F^{-1}(u_1), \ldots, F^{-1}(u_n)) = F^{-1}(\omega_u)$,

$$D_f(n, \omega) = D_f(n, F^{-1}(\omega_u)) = D(n, \omega_u) \;.$$

(For $d \geq 2$ see Hlawka and Mück (1972b) of a replacement for F and F^{-1}). Equivalently, for a sequence $\omega = (x_1, \ldots, x_n)$ we have

$$D_f(n, \omega) = D(n, F(\omega)) = D(n, \omega_u) \;, \tag{8.1}$$

where $F(\omega) = (F(x_1), \ldots, F(x_n))$. Thus we arrive at the well-known fact that the inversion method preserves the structure of the underlying uniform PRNG and hence the quality of the generated stream. Furthermore, the quality of the generated non-uniform pseudo-random variates is known and does not depend on the particular distribution.

The inverse of the cdf of a distribution is often not available in a simple form and thus the application of slow and/or inexact numerical methods is necessary. Thus algorithms that approximate the inverse cdf are proposed, see Leobacher and Pillichshammer (2002) for an example. Universal algorithms based on Newton's method or approximations by polynomials are described in detail in Chap. 7.

We want to take a closer look what happens when we use an inexact method. Let h and H denote the density and cdf, respectively, of the "real" distribution that is generated by the approximate inversion and let $\tilde{\omega} = (\tilde{x}_1, \ldots, \tilde{x}_n) = (H^{-1}(u_1), \ldots, H^{-1}(u_n))$ be such a sequence. Then relation (8.1) does not hold. Define a function Φ by

$$\Phi \colon (0,1) \to (0,1), u \mapsto \Phi(u) = F(H^{-1}(u)) \;.$$

In the case of a perfect transformation method we have $\Phi(u) = u$ and thus $|D_f(n, \tilde{\omega}) - D(n, \omega_u)| = 0$ (where again $\omega_u = F(\omega)$). Otherwise we find using (Niederreiter, 1992, Lemma 2.5),

$$|D_f(n, \tilde{\omega}) - D(n, \omega_u)| = |D_f(n, \tilde{\omega}) - D_f(n, \omega)| \leq 2 \max_{x \in \omega} |F(x) - H(x)| \;.$$

For the dimension $d \geq 2$ we have

$$|D_f(n, \tilde{\omega}) - D(n, \omega_u)| \leq 2d \max_{x \in \omega} |F(x) - H(x)| \;.$$

As a consequence of this formula we expect that an approximation of the cdf generates a sequence of similar discrepancy as the underlying uniform PRNG as long as the approximation error $\max_{x \in \mathbb{R}} |F(x) - H(x)|$ is small compared to this discrepancy. Notice that this error is just the parameter ϵ_U of Algorithm 7.1 (Fast-Table-Inversion) that can be used to adjust the algorithm to the particular simulation problem. However, we do not claim that this concept is fool-proved. The practitioner must be very carefully since the discrepancy $D(n, \omega_u)$ depends on the sample size n.

The largest drawback of this idea is that for ϵ_U not very close to 0 Algorithm 7.1 does not sample from the correct distribution, which should not be accepted for a stochastic simulation if an exact method is available. Thus we repeat these considerations for the rejection method (Sect. 2.2). Let $\alpha \geq 1$ be chosen such that αh is a hat function for the density f (α is then the rejection constant for the algorithm). If we switch off the rejection step and use this hat function as a rough approximation for the density f we find by a straightforward computation

$$|D_f(n,\tilde{\omega}) - D(n,\omega_u)| \leq 2 \max_{x \in \omega} |F(x) - H(x)| \leq 2(\alpha - 1). \qquad (8.2)$$

Using h as a rough approximation of the density introduces an error that is "repaired" when we accomplish the rejection step. However, by this step we remove some of the generated points and thus "introduce" some uncertainty in our estimation of the discrepancy. Nevertheless, this effect is small for small values of the rejection constant α. For transformed density rejection we can make α as small as desired. Thus the deviation in (8.2) as well as the effect of the rejection step become small for small values of α. Hence there are strong theoretical arguments for the claim that (with ϱ_{hs} very close to 1) TDR and its variants produce streams of non-uniform random variates of a quality that only depends on the quality of the underlying uniform PRNG but not on the particular distribution. Of course, there remain open questions that require additional research.

8.3.4 Floating Point Arithmetic

In the idealized model of arbitrary precision by Devroye (1986a) it is sufficient that a transformation method is exact. Real-world computers, however, work with so called floating point numbers, where there is only a fixed number of digits for the mantissa and the exponent of a number. Consequently we only have a discrete set of points on the real line on which we have to make our computations. Most modern computers are compliant with the IEEE 754-1985 standard for floating point arithmetic[2], see Overton (2001) for an introduction and Goldberg (1992) for further reading. In its *double* format there are about 16 decimal digits for the mantissa, and numbers can range from about 10^{-307} to 10^{308}. It is important to note that some rules for computing with real numbers do not hold for floating point numbers any more, e.g. the associative law.

Limited precision has of course severe consequences for random variate generation. Let us look at the following example: We want to generate beta distributed random variates with shape parameters $a = 1$ and $b = 0.1$, i.e. a distribution with density $f(x) = 0.1\,(1-x)^{-0.9}$ and cdf $F(x) = 1 - (1-x)^{0.1}$.

---

[2] In 1989, it received international endorsement as IEC 559, later designed IEC 60559.

Appling the inversion method we get the simple algorithm $X \leftarrow F^{-1}(U) = 1-(1-U)^{10}$ for sampling from this beta distribution, where $U \sim U(0,1)$. When we analyze this transformation we find that the interval $(0.975, 1)$ is mapped into the interval $(1 - 2^{-53}, 1) \approx (1 - 10^{-16}, 1)$. However, using the IEEE double format there are no floating points in the interval $(1 - 2 \times 10^{-16}, 1)$. Using *round to the nearest* (which is the default rounding rule) gives 1 for all numbers in this interval. This means that an interval of probability 0.025 is mapped into a single point. The situation is even worse if we use the shape parameter $b = 0.01$. Then it is the interval $(0.308, 1)$ that is mapped to 1 when using the double format of IEEE floating point arithmetic. When we compute the function $F(F^{-1}(u))$ by $x \leftarrow F^{-1}(u)$ and $y \leftarrow F(x)$ using this format and plot the results for $u \in (0,1)$ then we get the function in Fig. 8.4 instead of the identity function. Thus if one runs, e.g., a chi-square goodness of fit test with 100 bins all the points in 69 bins will collapse into the last bin and the test fails.

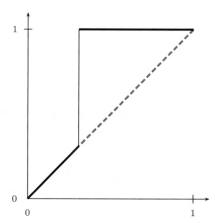

Fig. 8.4. Beta distribution with $a = 1$ and $b = 0.01$. Function $F(F^{-1}(u))$ computed via $x \leftarrow F^{-1}(u)$ and $y \leftarrow F(x)$ using the double format of IEEE 754 standard for floating point arithmetic. The correct function is $y \leftarrow u = F(F^{-1}(u))$ (dashed line)

The situation can become worse for specialized algorithms based on rejection and composition like those of Cheng (1978). There one even recognizes the same behavior near 0 when we use $a = 0.01$ and $b = 1$. We do not want to analyze this algorithm here but we note that such approximation and discretization error may also depend on small implementation details.

In this section we only can stress the importance of discretization and round-off errors that inevitably occur when exact methods are implemented in a real-world computer. They can cause serious biases in simulation results and a researcher who just uses a random variate generator might not be aware of the problem since there often are no warnings in the software manuals. There hardly exists a general solution to this problem (see Monahan, 1985, for one of the rare contributions) and there are a lot of open questions left. Nevertheless, we are convinced that universal methods are a good approach to tackle the problem of building a library of safe algorithms (at least in the sense that there is a warning when the generated sequence might introduce some bias).

8.4 Special Applications

In the previous sections of this chapter we have used standard distributions to demonstrate the properties of automatic methods. We also have collected arguments that show that such universal algorithms have some advantages compared to specialized algorithms even for sampling from common distributions including the Gaussian distribution. Nevertheless, automatic methods originally have been developed for non-standard simulation problems, where no ready to use algorithm could be found in simulation libraries. In this section we will discuss such problems and show that universal methods can be used almost without modifications.

8.4.1 Truncated Distributions

Sampling from truncated or tail distributions can be tedious for a practitioner who is unexperienced in the field of random variate generation. Looking into a software library one often finds only generators for the Gaussian tail distribution. For other distributions rejection from the untrucated distribution must be used. If the domain of interest has low probability then the rejection constant is (very) high and this method is (very) slow.

On the other hand, the inversion method (Sect. 2.1) can be easily used to sample from truncated distributions. Algorithm 8.1 (Inversion-Truncated) shows the necessary modifications to the basic algorithm on page 14.

Algorithm 8.1 Inversion-Truncated

Require: Inverse of the cumulative distribution function $F^{-1}(u)$, domain (b_l, b_r).
Output: Random variate X with cdf F on domain (b_l, b_r).
/∗ Setup ∗/
1: $u_{\min} \leftarrow F(b_l)$, $u_{\max} \leftarrow F(b_r)$.
/∗ Generator ∗/
2: Generate $U \sim U(u_{\min}, u_{\max})$.
3: Set $X \leftarrow F^{-1}(U)$.
4: **return** X.

For universal algorithms based on the rejection method the situation is similar. All these algorithms work with quasi-densities on an arbitrary domain. If, moreover, the hat distribution is generated by inversion, then it is easy to truncate the domain of the original distribution after the setup. For example, in Algorithm 4.1 (TDR) on page 60 we can add an adjustment step which is executed when the domain is truncated to a subinterval $(b'_l, b'_r) \subseteq (b_l, b_r)$. There we set $A_{\min} \leftarrow \int_{b_l}^{b'_l} h(x)\,\mathrm{d}x$ and $A_{\max} \leftarrow \int_{b_l}^{b'_r} h(x)\,\mathrm{d}x$. Then we have to replace Step 7 by

7': Generate $V \sim U(A_{\min}, A_{\max})$.

Analogous modifications also work for Algorithms 4.2 (TDR-Proportional-Squeeze), 5.1 (Ahrens), and 6.6 (Simple-Setup-Rejection).

It is important to note that this modification only works if the algorithm is based on "pure" rejection. Algorithms using immediate acceptance like 4.3 (TDR-Immediate-Acceptance), 4.8 (Adaptive-RoU), or 5.2 (Ahrens-Immediate-Acceptance) cannot be modified in this simple way.

8.4.2 Correlation Induction

Common random numbers and antithetic variates are two of the best known variance reduction methods for simulation experiments (see Bratley et al. (1983); for an application example of common random numbers see Example 15.2). Both methods require the generation of correlated random variates. Using the inversion method it is no problem to induce the strongest possible positive or negative correlation when generating two random variate streams (even with different distributions). For positive correlation (*common random numbers*) we simply use the same uniform numbers for both streams, for negative correlation (*antithetic variates*) we take U_i for the generation of the first stream and $1 - U_i$ for the second stream. This is one of the main reasons why Bratley et al. (1983) call inversion the method of choice for generating random variates.

As we have already mentioned inversion is for many distributions extremely slow or even impossible. Therefore Schmeiser and Kachitvichyanukul (1990) have suggested rules to realize correlation induction for rejection algorithms. The results of their paper and of Hörmann and Derflinger (1994b) show that this idea only works well if the rejection constant is close to one. For transformed density rejection the rejection constant can be made as small as desired. Thus universal algorithms based on TDR are well suited for correlation induction when the following (slightly modified) rules by Schmeiser and Kachitvichyanukul (1990) are used (see also Leydold et al., 2002):

(1) Conceptually, two independent uniform random number streams are used. If not needed the second stream is ignored.
(2) For each random variate generated, the same number n_1 of random numbers from the first stream are used.
(3) The first of these random numbers is used to sample from the hat distribution by inversion.

(4) The second stream is used if all of the random numbers from the first stream are exhausted by rejection. Thus the number of random numbers from the second stream is random, the expected number should be small.

Rule (2) is necessary to guarantee that the streams of uniform random numbers for generating correlated random variates always run synchronously, whereas rule (3) tries to induce correlation between corresponding random variables, when the first number is accepted for both generators. (This only works well if inversion is used to sample from the hat distribution in the rejection algorithms.)

The pair of generated random variates is uncorrelated only when rejection occurs for at least one of the two variate generators. However, this is rare for small values of the respective rejection constants α_1 and α_2 and its frequency is $1 - 1/(\alpha_1 \alpha_2)$, see Thm. 2.9. This probability itself is bounded by $2(\bar{\varrho}_{hs} - 1)$ where $\bar{\varrho}_{hs}$ is the minimum of the respective ratios between hat and squeeze for the two algorithms.

The choice of n_1 (the fixed number of uniform random numbers from the first stream used for generating one variate) depends mainly on the rejection algorithms used. Using $n_1 = 4$ (as suggested by Schmeiser and Kachitvichyanukul (1990)) is a waste of uniform random numbers for ϱ_{hs} (and hence α) close to one. Leydold et al. (2002) suggest to choose n_1 such that it is the number of uniform random numbers required for one acceptance/rejection loop in the corresponding algorithm, i.e., $n_1 = 2$ for Algorithms 4.1 (TDR), 4.2 (TDR-Proportional-Squeeze), and 5.1 (Ahrens), and $n_1 = 1$ for Algorithms 4.3 (TDR-Immediate-Acceptance), 4.8 (Adaptive-RoU), and 5.2 (Ahrens-Immediate-Acceptance).

Table 8.7 shows the respective observed induced correlation of common and antithetic random variates for some distributions using the inversion method (where the induced correlation is maximal) and Algorithms TDR-Proportional-Squeeze and TDR-Immediate-Acceptance. It shows that both algorithms work quite well even if the ratio between hat and squeeze, ϱ_{hs}, is close to 1.11 (tables on r.h.s.). If $\varrho_{hs} \approx 1.01$ then there is almost no difference to the inversion method (tables on l.h.s.). However, in Sect. 8.2 we have seen that TDR is usually faster than numerical inversion.

8.4.3 Order Statistics

If $X_1, X_2, \ldots, X_n$ are iid. random variables, then the order statistics for this sample are denoted by $X_{(1)}, X_{(2)}, \ldots, X_{(n)}$ where $X_{(1)} \leq X_{(2)} \leq \ldots \leq X_{(n)}$. Order statistics are an important notion of statistics and it is of practical importance in many applications to have a simple possibility to sample from order statistics. Hörmann and Derflinger (2002) discuss the problem of generating independent replications of a single order statistic (in particular of the minimum, the maximum or the median of a sample). Among other methods they use the well known fact that for an arbitrary continuous distribution

8.4 Special Applications

Table 8.7. Observed induced correlations for normal (N), exponential (E), gamma(2) (G), beta(1,2) (B_1), and beta(10,20) (B_2) distributed random variates, using inversion (INV) and Algorithms 4.2 (PS) and 4.3 (IA) with parameters $\varrho_{hs} = 1.01$ and 1.11

Common random numbers (positive correlation)

$\varrho_{hs} = 1.01$

		N	E	G	B_1	B_2	U
	PS	0.99					
N	IA	0.99					
	INV	1.00					
	PS	0.89	0.99				
E	IA	0.89	0.98				
	INV	0.90	1.00				
	PS	0.94	0.98	0.99			
G	IA	0.93	0.97	0.98			
	INV	0.94	0.99	1.00			
	PS	0.96	0.93	0.96	0.99		
B_1	IA	0.96	0.92	0.96	0.99		
	INV	0.97	0.94	0.97	1.00		
	PS	0.99	0.91	0.95	0.97	0.99	
B_2	IA	0.98	0.91	0.95	0.97	0.99	
	INV	0.99	0.92	0.96	0.98	1.00	
	PS	0.97	0.85	0.91	0.97	0.97	1
U	IA	0.97	0.86	0.91	0.97	0.97	1
	INV	0.97	0.86	0.91	0.97	0.97	1

$\varrho_{hs} = 1.11$

		N	E	G	B_1	B_2	U
	PS	0.96					
N	IA	0.91					
	INV	1.00					
	PS	0.80	0.95				
E	IA	0.77	0.85				
	INV	0.90	1.00				
	PS	0.88	0.84	0.96			
G	IA	0.84	0.83	0.89			
	INV	0.94	0.99	1.00			
	PS	0.90	0.87	0.90	0.97		
B_1	IA	0.87	0.84	0.87	0.94		
	INV	0.97	0.94	0.97	1.00		
	PS	0.94	0.82	0.89	0.91	0.96	
B_2	IA	0.89	0.80	0.86	0.89	0.91	
	INV	0.99	0.92	0.96	0.98	1.00	
	PS	0.92	0.82	0.86	0.94	0.92	1
U	IA	0.91	0.80	0.87	0.93	0.92	1
	INV	0.97	0.86	0.91	0.97	0.97	1

Antithetic random variates (negative correlation)

		N	E	G	B_1	B_2	U
	PS	-0.98					
N	IA	-0.98					
	INV	-1.00					
	PS	-0.88	-0.63				
E	IA	-0.89	-0.63				
	INV	-0.90	-0.64				
	PS	-0.93	-0.71	-0.78			
G	IA	-0.93	-0.71	-0.78			
	INV	-0.94	-0.72	-0.79			
	PS	-0.96	-0.78	-0.85	-0.92		
B_1	IA	-0.96	-0.78	-0.85	-0.92		
	INV	-0.97	-0.79	-0.86	-0.93		
	PS	-0.98	-0.86	-0.91	-0.95	-0.98	
B_2	IA	-0.98	-0.86	-0.91	-0.95	-0.98	
	INV	-0.99	-0.87	-0.92	-0.96	-0.99	
	PS	-0.97	-0.86	-0.91	-0.97	-0.97	-1
U	IA	-0.97	-0.86	-0.91	-0.97	-0.97	-1
	INV	-0.97	-0.86	-0.91	-0.97	-0.97	-1

		N	E	G	B_1	B_2	U
	PS	-0.87					
N	IA	-0.88					
	INV	-1.00					
	PS	-0.77	-0.60				
E	IA	-0.76	-0.55				
	INV	-0.90	-0.64				
	PS	-0.82	-0.65	-0.70			
G	IA	-0.82	-0.63	-0.70			
	INV	-0.94	-0.72	-0.79			
	PS	-0.88	-0.73	-0.79	-0.88		
B_1	IA	-0.86	-0.71	-0.77	-0.86		
	INV	-0.97	-0.79	-0.86	-0.93		
	PS	-0.87	-0.76	-0.80	-0.87	-0.87	
B_2	IA	-0.88	-0.73	-0.81	-0.86	-0.88	
	INV	-0.99	-0.87	-0.92	-0.96	-0.99	
	PS	-0.91	-0.81	-0.86	-0.95	-0.92	-1
U	IA	-0.91	-0.80	-0.86	-0.94	-0.92	-1
	INV	-0.97	-0.86	-0.91	-0.97	-0.97	-1

with density f and cdf F the i-th order statistic from a sample of size n has density
$$f_{X_{(i)}}(x) = k\, f(x) F(x)^{i-1} (1 - F(x))^{n-i}\ .$$
In Sect. 4.3 we have shown that densities of order statistics of log-concave or $T_{-1/2}$-concave distributions are again log-concave and $T_{-1/2}$-concave, respectively. Hence transformed density rejection is well suited for sampling from a particular order statistics. For the performance of the particular algorithms see Tables 8.3–8.6, distribution OS (Table 8.1). It is possible to use Ahrens algorithm as well but it is not easy to find the mode and save cut off points for the tails. In addition the set-up is much slower as much more design points are necessary.

A totally different simulation problem is the simulation of all order statistics, in other words the generation of an ordered sample. Devroye (1986a, Chap. V) contains a detailed presentation of different methods to accomplish this task. We will use uniform spacings (i.e. the differences between uniform order statistics) in Sect. 11.2.2.

8.5 Summary

As we have collected quite a number of different automatic algorithms for continuous distributions it is probably in place to give a short overview of these methods. So we list here the most important algorithm groups and their requirements, advantages and disadvantages. The performance of some of the methods can be adjusted by a simple *control parameter*. *Setup* and *Speed* give the time required for the setup and the marginal generation time, respectively. *Tables* and *Code* give the size for necessary tables and the length of the computer program that implements the algorithm in question.

- **Transformed density rejection**
  Algorithm 4.1 (TDR) and its variants 4.2 (TDR-Proportional-Squeeze) and 4.3 (TDR-Immediate-Acceptance), Algorithm 4.8 (Adaptive-RoU)

  Applicable to: T_c-concave distributions.
  Required: quasi-density f and its derivative; *optional:* mode m.
  Control parameter: ϱ_{hs} (Thm. 2.9) – a value of 1.1 leads to moderate setup and generation times, whereas a value of 1.01 or smaller leads to a longer setup and a (very) fast sampling procedure.
  Setup: moderate to slow depending on ϱ_{hs}.
  Speed: fast to very fast depending on the variant and on ϱ_{hs}; Algorithm 4.8 (Adaptive-RoU) is fastest.
  Tables: moderate to large depending on ϱ_{hs}.
  Code: long.

Special advantages:
> Quality for ϱ_{hs} close to 1 comparable to inversion method;
> well suited for correlation induction;
> possibility to sample from truncated distributions;
> well suited to generate order statistics.

- **Transformed density rejection with 3 design points**
  Algorithm 4.4 (TDR-3-Points)

  Applicable to: T_c-concave distributions.
  Required: quasi-density f, its mode m, and approximate area A_f.
  Setup: fast.
  Speed: moderate.
  Tables: not required.
  Code: moderate.
  Special advantages: possibility to sample from truncated distributions.

- **Simple ratio-of-uniforms**
  Algorithm 6.4 (Simple-RoU) and its variants 6.5 (Simple-RoU-Mirror-Principle), 6.6 (Simple-Setup-Rejection) and 6.7 (Generalized-Simple-RoU).

  Applicable to: T_c-concave distributions.
  Required: Quasi-density f, area A_f and mode m;
  > *optional:* cdf at mode $F(m)$.

  Setup: very fast.
  Speed: slow.
  Tables: not required.
  Code: short.

- **Ahrens method**
  Algorithm 5.1 (Ahrens) and its variants 5.2 (Ahrens-Immediate-Acceptance) and 5.3 (Ahrens-Equal-Area)

  Applicable to: Unimodal distributions with bounded domain.
  Required: quasi-density f and its mode m / extremal points.
  Control parameter: ϱ_{hs} (Thm. 2.9) – a value of 1.1 leads to moderate setup and generation times, whereas a value of 1.01 or smaller leads to a very long setup and a very fast sampling procedure.
  Setup: moderate to very slow depending on ϱ_{hs}.
  Speed: fast to very fast depending on the variant and on ϱ_{hs}.
  Tables: large to very large depending on ϱ_{hs}.
  Code: moderate to long.
  Special advantages:
  > Quality for ϱ_{hs} close to 1 comparable to inversion method;
  > well suited for correlation induction;
  > possibility to sample from truncated distributions.

  Disadvantages:
  > Tails have to be truncated.

- **Rejection from universal hat**
  Algorithm 6.2 (Monotone-Bounded), Algorithm 6.3 (Monotone-Unbounded)

  Applicable to: Distributions with monotone densities.
  Required: Quasi-density f and upper bound for area A_f;
        optional: upper bound μ_r for the r-th moment of the density for some $r > 0$ (*required* for unbounded domain).
  Setup: very fast.
  Speed: slow.
  Tables: not required.
  Code: short.

- **Numerical inversion with root finding algorithms**
  Algorithm explained in Sect. 7.2.2

  Applicable to: Continuous distributions.
  Required: cdf $F(x)$; *optional:* density $f(x)$.
  Control parameter: ϵ_U, the maximal allowed error in u-direction; size of auxiliary table.
  Setup: fast to slow depending whether auxiliary tables are used.
  Speed: slow to very slow depending on size of auxiliary tables.
  Tables: any size.
  Code: moderate when external root finding algorithm is used.
  Special advantages:
        Easy to sample from truncated distributions and order statistics; simple correlation induction;
        works almost everywhere.
  Disadvantage:
        Not exact, but approximation error can be controlled and can be close to machine precision.

- **Fast inversion**
  Algorithm 7.1 (Fast-Table-Inversion)

  Applicable to: Continuous distributions.
  Required: cdf $F(x)$ and density $f(x)$; *optional:* location of mode m and of discontinuities of the density.
  Control parameter: ϵ_U, the maximal allowed error in u-direction.
  Setup: very slow.
  Speed: fast.
  Tables: large to very large depending on ϵ_U.
  Code: long.
  Special advantages:
        Known quality of generated points;
        easy to sample from truncated distributions and order statistics; simple correlation induction;
        Does not require any function calls for cdf or density after the setup.

Disadvantage:
>Tails have to be truncated;
>not exact, but approximation error can be controlled and can be close to machine precision.

It is not possible to draw any general conclusions here. There is no best automatic method. Instead the choice depends on the respective distribution and simulation problem. Due to the quality considerations presented above we recommend to use Algorithm 7.1 (Fast-Table-Inversion) or one of the Algorithms of the TDR family with ϱ_{hs} close to 1 if large samples of a fixed distribution are required. For example Algorithm 7.1 (Fast-Table-Inversion) is a good choice if the cdf of the desired distrbution is available and large samples of a fixed distribution are required as it is fast for this situation and inversion is known to produce good quality variates. If the cdf is not available Algorithm 7.1 is simply not applicable and if the required distribution in the simulation changes in every step, Algorithm 7.1 should not be used as it is much too slow (see Table 8.3).

As a second example let us assume that we need a sample of size 10^8 of the t-distribution with parameter $\nu = 2.3$. The cdf is very unfriendly and probably not easily available for us. So we will decide to use one of the Algorithms of the TDR family with ϱ_{hs} close to 1. Algorithm 5.1 (Ahrens) is not well suited as the heavy tails of the required distribution would lead to very large tables.

9

Distributions Where the Density Is Not Known Explicitly

There are situations in the practice of statistical research where continuous distributions are not characterized by their density or cumulative distribution function. Instead some other functions like the hazard rate (Sect. 9.1), the characteristic function (Sect. 9.4), or a sequence of Fourier coefficients (Sect. 9.3) are known. In such cases it can be very useful to have the possibility to sample directly from these distributions without computing the density, which is often very difficult and cumbersome in such situations. Luc Devroye (1981; 1984a; 1986c; 1986d; 1989) has introduced most of these methods. We give an overview and present the details of those algorithms that seem to be most useful in practice; among them a new automatic algorithm for distributions with increasing hazard rate.

In this chapter one might also expect algorithms for the situation where only moments of the required distribution are known. In this case, however, the distribution is not uniquely determined and we have to make additional assumptions. Designing (and choosing) generation methods is therefore linked with modeling a stochastic distribution. Thus we have decided to deal with this situation in Chap. 12 (Combination of Generation and Modeling) to make this fact more transparent (see Sect. 12.4).

9.1 Known Hazard-Rate

The hazard rate (or failure rate) is a mathematical way of describing aging. It thus plays a crucial role in reliability studies and all applications involving lifetime distributions. If the lifetime X is a random variable with density $f(x)$ and cdf $F(x)$ the *hazard rate* $h(x)$ and the *cumulative hazard rate* $H(x)$ are defined as

$$h(x) = \frac{f(x)}{1 - F(x)} \quad \text{and} \quad H(x) = \int_0^x h(t)\,dt = -\log(1 - F(x))\ .$$

In other words, $h(x)$ represents the (conditional) rate of failure of a unit that has survived up to time x with probability $1 - F(x)$. It is easy to see that

$$f(x) = h(x)\, e^{-H(x)} \quad \text{and} \quad F(x) = 1 - e^{-H(x)} \qquad (9.1)$$

which also shows that the hazard rate uniquely determines a distribution. Note that $\int_0^\infty h(x)\,dx$ must be unbounded, i.e. $\lim_{x \to b_r} H(x) = \infty$ for a distribution with (bounded or unbounded) support (b_l, b_r).

The key distribution is the standard exponential distribution as it has constant hazard rate $h(x) = 1$. Hazard rates tending to infinity describe distributions with sub-exponential tails whereas distributions with hazard rates tending to zero have heavier tails than the exponential distribution.

For hazard rates $h(x)$ that are not easy to integrate we often cannot find a closed form for the cdf or for the density. For these situations we recommend the algorithms of this section. We first demonstrate that the inversion, the decomposition, and the rejection methods for generating random variates with given density or cdf have their analogues for hazard rates. We also explain the connection between the hazard rate and non-homogeneous Poisson processes. The pioneers in the field of sampling from a non-homogeneous Poisson process are Lewis and Shedler (1979). Devroye (1986a, Chap. VI) summarizes their ideas and adds a black-box algorithm. We mainly follow these presentations and add a new algorithm for increasing hazard rates.

9.1.1 Connection to the Poisson Process

The hazard rate can be interpreted as the rate function of a non-homogeneous Poisson process on $[0, \infty)$. Therefore generating a random variate with a given hazard rate $h(x)$, $x > 0$, is equivalent to generating the first observation T_1 of a non-homogeneous Poisson process. For the second observation T_2 we can use the same methods with the same hazard rate $h(x)$, but now for $x > T_1$ instead of $x > 0$. Continuing in this way we can generate the i-th observation T_i recursively by generating variates with hazard rate $h(x)$ for $x > T_{i-1}$. Therefore we can use all algorithms of this section to generate from non-homogeneous Poisson-processes as well. However, the given performance characteristics of the algorithms for a fixed hazard rate may change considerably, as the shape of the hazard rate far away from 0 has little or no influence on the generation of T_1 but can become important when many observations of the Poisson process are generated.

9.1.2 The Inversion Method

To derive the inversion method (Algorithm 9.1) we start from $F(X) = U$ as for the standard inversion method in Sect. 2.1. Using the cumulative hazard rate $H(x)$ and equation (9.1) we obtain $1 - \exp(-H(X)) = U$ which is easily transformed to

$$X = H^{-1}(-\log(1-U)) = H^{-1}(E),$$

where E denotes an exponential random variate. The inversion method is of limited practical value for direct generation as a distribution with simple cumulative hazard rate also has a simple cdf and can thus be generated by standard inversion. Nevertheless, inversion of the cumulative hazard rate is important as building block for the next two methods. It is also useful to generate variates from distributions with step-function shaped hazard rate (see Exercise 9.1).

Algorithm 9.1 Hazard-Rate-Inversion

Require: Cumulative hazard rate $H(x)$.
Output: Random variate X with cumulative hazard rate $H(x)$.
1: Generate exponentially distributed random variate E (e.g. by $E \leftarrow -\log(1-U)$).
2: Set $X \leftarrow H^{-1}(E)$ (e.g. by solving $E = H(X)$ numerically).
3: **return** X.

9.1.3 The Composition Method

In analogy to the composition method for densities it is possible to formulate the composition principle for hazard rates as well. We start with the formal statement of the basic theorem.

Theorem 9.1. *Assume $h(x) = h_1(x) + h_2(x)$ can be written as a sum of two hazard rates. Then we can generate a random variate X with hazard rate h, by generating X_1 with hazard rate h_1 and X_2 with hazard rate h_2 and taking $X = \min(X_1, X_2)$.*

Proof. Using the cumulative hazard rate $H(x) = H_1(x) + H_2(x)$ and equation (9.1) we get $\text{Prob}(\min(X_1, X_2) \geq x) = (1 - F_1(x))(1 - F_2(x)) = e^{-H_1(x)} e^{-H_2(x)} = e^{-H(x)} = \text{Prob}(X \geq x)$. □

Of course it is trivial to generalize this method to cases with $n > 2$ summands, see Algorithm 9.2 (Hazard-Rate-Composition). Note that it is possible for decompositions of h that $\int_0^\infty h_i(x)\,dx < \infty$ for some (but not for all) i. If this is the case h_i is no hazard rate and it can happen that during generation of X_i no random variate is generated. Then we set $X_i \leftarrow \infty$ which is no problem for the minimum operation that determines X.

We can see from Algorithm 9.2 that the composition method for hazard rates is clearly more time consuming than that for densities, since we have to generate from all pieces of the decomposition to obtain the minimum.

Algorithm 9.2 Hazard-Rate-Composition

Require: Hazard rate that can be written as a sum $h(x) = \sum_{i=1}^{n} h_i(x)$.
Algorithms to generate from hazard rates h_i.
Output: Random variate X with hazard rate $h(x)$.
1: **for** $i = 1$ to n **do**
2:     Generate X_i with hazard rate h_i.
3:     If no event was generated set $X_i \leftarrow \infty$.
4: **return** $\min_{1 \leq i \leq n} X_i$.

9.1.4 The Thinning Method

This is the analog to the rejection method. If we have a dominating hazard rate $g(x) \geq h(x)$ available, we generate observations X_i from the Poisson process with rate $g(x)$. Using a uniform random number V and the acceptance condition $V g(X) \leq h(X)$ we decide if we can accept X or have to continue sampling from the Poisson process with rate $g(x)$. To do this we start at the point X_i we have just generated but rejected and therefore the X_i are in increasing order. Note that the acceptance condition is exactly the same as for standard rejection (Sect. 2.2) but the names of the functions have changed, as the hazard rate $h(x)$ is now taking the place of the density f and the dominating hazard rate $g(x)$ that of the hat h.

Algorithm 9.3 Hazard-Rate-Thinning

Require: Hazard rate $h(x)$, dominating hazard rate $g(x) \geq h(x)$.
Output: Random variate X with hazard rate $h(x)$.
1: Set $X_0 \leftarrow 0$ and $i \leftarrow 1$.
2: **loop**
3:     Generate X_i with hazard rate $g(x)$ on (X_{i-1}, ∞).
4:     Generate $V \sim U(0, g(X_i))$.
5:     **if** $V \leq h(X_i)$ **then**
6:         **return** X_i.
7:     **else** /* continue */
8:         Set $i \leftarrow i + 1$.

The proof that the thinning method generates from a non-homogeneous Poisson process is very similar to the proof for ordinary rejection (see Sect. 2.2). The key idea is that the accepted pairs (X_i, V) are the occurrences of a homogeneous Poisson process on the set $\{(x, y) | x > 0 \text{ and } 0 < y < f(x)\}$, for details see Devroye (1986a, p.254). The analysis of the performance of the thinning method is much more difficult than that of ordinary rejection.

Theorem 9.2. *Let f and F be the density and cdf corresponding to hazard rate $h(x)$. Let $g(x) \geq h(x)$ be the dominating hazard rate having cumulative hazard rate $G(x)$. Then the expected number of iterations in Algorithm 9.3*

(Hazard-Rate-Thinning) is given by

$$\mathrm{E}(I) = \int_0^\infty g(x)\,(1 - F(x))\,\mathrm{d}x = \int_0^\infty f(x)\,G(x)\,\mathrm{d}x\ .$$

Proof. See Devroye (1986c, Sect. VI.2.4). □

For the important special case of $g(x) = M$ (spelled out as Algorithm 9.4, Hazard-Rate-Constant-Thinning) we immediately get the following observation.

Corollary 9.3. *For dominating hazard rate $g(x) = M$ the expected number of iterations is given by*
$$\mathrm{E}(I) = M\,\mathrm{E}(X)\ .$$

Proof. As $G(x) = cx$ this follows directly from Thm. 9.2 and the definition of the expectation. □

Corollary 9.3 implies that thinning with a constant dominating hazard rate can become very slow for heavy-tailed distributions.

Algorithm 9.4 Hazard-Rate-Constant-Thinning

Require: Bounded hazard rate $h(x) \leq M$.
Output: Random variate X with hazard rate $h(x)$.
1: Set $X \leftarrow 0$ and $\lambda \leftarrow M$.
2: **loop**
3:   Generate a $U \sim U(0,1)$.
4:   Set $E \leftarrow -\log(1-U)/\lambda$. /* random variate with hazard rate λ. */
5:   Set $X \leftarrow X + E$.
6:   Generate $V \sim U(0,\lambda)$.
7:   **if** $V \leq h(X)$ **then**
8:     **return** X.

9.1.5 Algorithms for Decreasing Hazard Rate

The family of distributions with decreasing hazard rate forms a subclass of all monotone densities. This subclass is closed under convex combinations and all densities have higher tails than the exponential distribution. For densities with decreasing hazard rate $h(x)$ we can obviously use the thinning Algorithm 9.4 with constant dominating hazard rate $g(x) = h(0) \geq h(x)$. But we have seen that this algorithm becomes very slow if $\mathrm{E}(X)$ is large. So it is worth-while to consider the following improvement due to Devroye (1986c): whenever a variate X_i is rejected, the dominating hazard rate is changed to $h(X_i)$. As $h(X_i) \geq h(x)$ for $x \geq X_i$ we again get a dominating hazard rate and the algorithm works correctly. It is obvious that decreasing the dominating hazard rate within the rejection loop results in a better performance than just using the constant hazard rate. The expected number of evaluations for Algorithm 9.5 (Decreasing-Hazard-Rate-Dynamic-Thinning) depends linearly on the first logarithmic moment; for Algorithm 9.4 we have seen that it depends linearly on the first moment of the distribution. For many heavy-tailed distributions dynamic thinning is thus much faster than the thinning algorithm with constant hat. (See Devroye (1986c) for details.)

Algorithm 9.5 Decreasing-Hazard-Rate-Dynamic-Thinning

Require: Bounded, decreasing hazard rate $h(x)$.
Output: Random variate X with hazard rate $h(x)$.
1: Set $X \leftarrow 0$.
2: **loop**
3:    Set $\lambda \leftarrow h(X)$.
4:    Generate a $U \sim U(0,1)$.
5:    Set $E \leftarrow -\log(1-U)/\lambda$. /* random variate with hazard rate λ. */
6:    Set $X \leftarrow X + E$.
7:    Generate $V \sim U(0, \lambda)$.
8:    **if** $V \leq h(X)$ **then**
9:       **return** X.

9.1.6 Algorithms for Increasing Hazard Rate

For distributions with increasing hazard rate it is possible to use a staircase-shaped dominating function $g(x)$ with $g(x) = h(p_i)$ for $p_{i-1} < x \leq p_i$ for design points $0 = p_0 < p_1 < p_2 < \ldots$. If the distribution has bounded support on $(0, b_r)$, the sequence of the p_i must converge to b_r. Using inversion to generate from the dominating hazard rate $g(x)$ thinning can then be used to generate an observation with hazard rate $h(x)$. However, the sequence of the p_i is infinite and thus this table and the dominating hazard rate cannot be fixed in the setup. In opposition to the case of decreasing hazard rate above there is no "natural" way to choose these points. We have to update this list if it happens that due to many rejection steps a large value for X has to be generated. Hence the implementation of such an algorithm may be quite troublesome (see Özgül, 2002). The reader may try to derive the details in Exercise 9.3.

To get a simpler algorithm we develop a method called *dynamic composition* that combines the composition and the thinning method and does not need any tables. We decompose the hazard rate into two summands $h(x) = h_1(x) + h_2(x)$ using a design point p_0, see Fig. 9.1.

$$h_1(x) = \min(h(x), h(p_0)), \qquad h_2(x) = \begin{cases} 0 & \text{for } x < p_0, \\ h(x) - h(p_0) & \text{for } x \geq p_0. \end{cases}$$

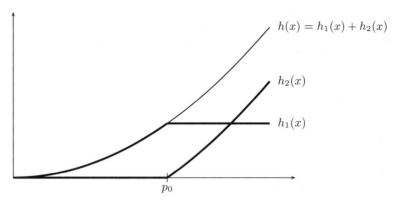

Fig. 9.1. Decomposition of hazard rate $h(x) = 3\,x^2$ of the Weibull distribution with shape parameter 3 into $h_1(x) + h_2(x)$

From both parts we are going to generate using thinning from a constant dominating hazard rate. For h_1 we obviously can take $g(x) = h(p_0)$. If the first variate X from hazard rate h_1 that is accepted by the thinning method satisfies $X \leq p_0$ we can immediately return X. There is no need to generate from h_2 as these variates are always greater or equal to p_0. Otherwise, if $X > p_0$, we set $p_1 \leftarrow X$. The main idea in this case is that it is not necessary to sample from the whole h_2. As the decomposition method for hazard rates requires that we return the minimum of the two generated variates, it is enough to generate from h_2 if the generated value is smaller than p_1. So we can use the constant hazard rate $h(p_1) - h(p_0)$ to sample a variate X from h_2 by means of the thinning algorithm. We return X if $X < p_1$. Otherwise we return p_1. Algorithm 9.6 is dynamic as the dominating hazard rate for the second loop depends on the outcome of the first loop. The performance characteristics of the method depend on the design parameter p_0 and were not established yet. Our first experiments indicate that setting p_0 to a value close to $\mathrm{E}(X)$ works quite well. For more empirical investigations see Özgül (2002).

9.1.7 Computational Experience

We will soon describe the experiences we made with our implementations of Algorithms 9.4 (Hazard-Rate-Constant-Thinning), 9.5 (Decreasing-Hazard-Rate-Dynamic-Thinning) and 9.6 (Increasing-Hazard-Rate-Dynamic-Composition). All these algorithms are using thinning so we can conclude from Thm. 9.2 that their performance strongly depends on the distribution we want to generate from.

We started our little timing experiment with the Pareto distribution with shape parameter $a > 0$ which has decreasing hazard rate $h(x) = \frac{a}{x+1}$, density $f(x) = \frac{a}{(1+x)^{a+1}}$ and expectation $\mathrm{E}(X) = \frac{1}{a-1}$ that only exists for $a > 1$. Table 9.1 shows our relative timing results, that clearly indicate that the

Algorithm 9.6 Increasing-Hazard-Rate-Dynamic-Composition

Require: Increasing hazard rate $h(x)$ for distribution with unbounded domain; appropriate value for parameter $p_0 > 0$.
Output: Random variate X with hazard rate $h(x)$.
  /∗ Generator (first loop) ∗/
1: Set $X \leftarrow 0$.
2: Set $\lambda_0 \leftarrow h(p_0)$.
3: **repeat**
4:    Generate $U \sim U(0,1)$.
5:    Set $E \leftarrow -\log(1-U)/\lambda_0$. /∗ random variate with hazard rate λ_0. ∗/
6:    Set $X \leftarrow X + E$.
7:    Generate $V \sim U(0, \lambda_0)$.
8: **until** $V \leq h(X)$
9: **if** $X \leq p_0$ **then**
10:    **return** X.
  /∗ Generator (second loop) ∗/
11: Set $p_1 \leftarrow X$.
12: Set $X \leftarrow p_0$.
13: Set $\lambda_1 \leftarrow h(p_1) - h(p_0)$.
14: **repeat**
15:    Generate $U \sim U(0,1)$.
16:    Set $E \leftarrow -\log(1-U)/\lambda_1$. /∗ random variate with hazard rate λ_1. ∗/
17:    Set $X \leftarrow X + E$.
18:    Generate $V \sim U(\lambda_0, \lambda_0 + \lambda_1)$.
19: **until** $V \leq h(X)$
20: **if** $X \leq p_1$ **then**
21:    **return** X.
22: **else**
23:    **return** p_1.

dynamic Algorithm 9.5 (Decreasing-Hazard-Rate-Dynamic-Thinning) is superior to Algorithm 9.4 (Hazard-Rate-Constant-Thinning) for small values of a. The results also support what we have already seen from Corollary 9.3: Algorithm 9.4 should not be used for heavy tailed distributions (e.g. the Pareto distribution with shape parameter $a < 1$).

Table 9.1. Relative generation times of Algorithms 9.4 (Hazard-Rate-Constant-Thinning) and 9.5 (Decreasing-Hazard-Rate-Dynamic-Thinning) for Pareto distribution with parameter a

Algorithm	$a = 0.1$	$a = 0.5$	$a = 1$	$a = 5$	$a = 10$
9.4	10^6	3500	22	2.3	2
9.5	9	4	3	2.3	2

To test Algorithm 9.6 (Increasing-Hazard-Rate-Dynamic-Composition) we used the Weibull distribution with shape parameter $b > 1$ which has hazard rate $h(x) = bx^{b-1}$ and expectation $E(X)$ between 0.885 and 1. When experimenting with the choice of p_0 we have observed that it has strongly influenced the performance of the algorithm for larger values of b but was less important for smaller values of b. This seems to be linked with the fact that the variance of the Weibull distribution decreases with increasing b. The value $p_0 = 1$ leads to a close to optimal performance for all the values of b that we tried. The timing results have shown that for $b = 1.5$ Algorithm 9.6 has been about 8 times slower than generating the exponential distribution. The relative generation time increases with increasing b and was 25 for $b = 10$.

9.2 The Series Method

In this section we introduce a method for generating random variates from densities without evaluating the density f exactly. Instead a convergent series representation of f is needed. In Sects. 9.3 and 9.4 we will see that this idea is very useful if the density is not given explicitly. It can also be applied if the density is very expensive to evaluate.

The series method is based on the rejection method. It utilizes the series representation of the density to construct a sequence of upper and lower bounds (squeezes) for the density. This sequence is used to decide whether the acceptance condition $V h(X) \leq f(X)$ is satisfied or not without computing the value $f(X)$. The series method is due to Devroye (1981).

9.2.1 The Convergent Series Method

We assume that the density f can be written as a convergent series $(s_i(x))$ and that we have an error bound R_n available, i.e.

$$f(x) = \sum_{i=1}^{\infty} s_i(x) \quad \text{and} \quad \left| \sum_{i=n+1}^{\infty} s_i(x) \right| \leq R_{n+1}(x) \ .$$

If we have a hat function $h(x) \geq f(x)$ available we can state Algorithm 9.7 (Convergent-Series).

Compared with the standard rejection algorithm we are in a new situation now, due to the evaluations of $s_i(X)$ in the inner loop. It should be clear that if the error bound converges to 0 the algorithm halts with probability 1. Devroye (1986a, p.156) even shows for the expected number of evaluations of the coefficients s_i

$$E(\#s_i) \leq 2 \int \left(\sum_{n=1}^{\infty} R_n(x) \right) \mathrm{d}x \ . \tag{9.2}$$

202   9 Distributions Where the Density Is Not Known Explicitly

Algorithm 9.7 Convergent-Series

Require: Convergent series $(s_i(x))$ for density f with error bound $R_n(x)$, hat function $h(x)$.
Output: Random variate X with density f.
1: **loop**
2:    Generate X with density proportional to $h(x)$.
3:    Generate $V \sim U(0, h(X))$.
4:    Set $S \leftarrow 0$ and $n \leftarrow 0$.
5:    **repeat**
6:      Set $n \leftarrow n+1$,
7:      Set $S \leftarrow S + s_i(X)$.
8:    **until** $|S - V| > R_{n+1}(X)$
9:    **if** $V \leq S$ **then**
10:     **return** X.

It seems clear from the above bound that not any convergent series is guaranteed to lead to an algorithm that has a bounded expected number of evaluations of s_i, although it halts with probability 1. So we have to be careful when constructing such algorithms. As simple illustration we present the following example.

Example 9.4. We consider $f(x) = \exp(x)$ for $0 \leq x \leq 1$. As series we take the Taylor series around 0,

$$f(x) = 1 + \sum_{i=1}^{\infty} \frac{x^i}{i!}, \qquad R_n(x) \leq \frac{x^n e}{n!},$$

and as hat function the constant e. Then it is no problem to verify the details of the below algorithm.

Output: Random variate X with density $f(x) = \exp(x)$ for $0 \leq x \leq 1$.
1: **loop**
2:    Generate $X \sim U(0, 1)$.
3:    Generate $V \sim U(0, e)$.
4:    Set $s \leftarrow 1$, $S \leftarrow 1$, $R \leftarrow X e$, and $n \leftarrow 0$.
5:    **repeat**
6:      Set $n \leftarrow n+1$, $s \leftarrow sX/n$, $R \leftarrow RX/(n+1)$, and $S \leftarrow S + s$.
7:    **until** $|S - V| > R$
8:    **if** $V \leq S$ **then**
9:      **return** X.

This is a simple example but nevertheless it demonstrates the virtues of the series method. We can generate random variates from the density $\exp(x)$ without ever calling the exponential function or the logarithm. On the other hand we should not forget that the algorithm is only guaranteed to work if we assume that we can handle and store real numbers with arbitrary precision. We should not forget here that – due to possible extinction – sums are known to be numerically unstable. In the example above this is not a problem as the partial sums are all between 1 and e and the sum is converging at exponential rate. Using (9.2) we can easily compute that E($\#s_i$) is bounded by $\sum_{i=1}^{\infty} 2e/(i+1)! \approx 3.905$. Our empirical experiments have shown that the exact value is around 2.25.

9.2.2 The Alternating Series Method

Many series representations $f(x) = \sum_{i=0}^{\infty} s_i(x)$ are alternating in the sense that for odd i the $s_i(x)$ are negative and for even i the $s_i(x)$ are positive. If the series also fulfills the inequalities

$$\sum_{i=0}^{k} s_i(x) \leq f(x) \leq \sum_{i=0}^{k+1} s_i(x), \quad \text{for all odd } k,$$

we can use the partial sums given above directly as upper and lower bounds for a rejection algorithm. Algorithm 9.8 (Alternating-Series) contains the details. For a performance analysis of the algorithm see Devroye (1981). It is well known that the evaluation of alternating series is – due to extinction – numerically very unstable. So it is often necessary in practice to use high precision arithmetic or rational arithmetic for the evaluation of these sums.

Algorithm 9.8 Alternating-Series

Require: Convergent alternating series $(s_i(x))$ for density f, hat function $h(x)$.
Output: Random variate X with density f.
1: **loop**
2:   Generate X with density proportional to $h(x)$.
3:   Generate $V \sim U(0, h(X))$.
4:   Set $S \leftarrow s_0(X)$ and $n \leftarrow 0$.
5:   **repeat**
6:     Set $n \leftarrow n+1$, $S \leftarrow S + s_i(X)$. /* S is now a lower bound for $f(X)$. */
7:     **if** $V \leq S$ **then** /* accept */
8:       **return** X.
9:     Set $n \leftarrow n+1$, $S \leftarrow S + s_i(X)$. /* S is now an upper bound for $f(X)$. */
10: **until** $V > S$ /* reject */

The outer loop of Algorithm 9.8 (Alternating-Series) seems to be an infinite loop but the algorithm returns X and exits from the inner loop. If the series is alternating and convergent the algorithm halts with probability 1. As illustration we present the following example.

Example 9.5. To generate from the exponential distribution without using subroutines for the logarithm or the exponential function we can use the Taylor series $f(x) = \exp(-x) = \sum_{i=0}^{\infty}(-x)^i/i!$. Its odd and even partial sums are upper and lower bounds for the density f and can thus be used for Algorithm 9.8 (Alternating-Series). As the hat function we take

$$h(x) = \frac{3^3}{e(x+2)^3}$$

which allows for simple and fast inversion and results in a rejection constant of 1.24159. Note that this hat is the optimal hat of transformed density rejection with one design point and $c = -1/3$ (see Sect. 4.4.1). The details are given in the algorithm below.

It is important not to overlook the fact that we cannot expect satisfactory results if we do not have arbitrary precision arithmetic. Assume a double precision floating point arithmetic which has about 16 significant digits in decimal notation, and consider the series for $x = 60$. The probability that a sample X from the hat distribution is greater than 60 is about 0.1 percent and thus this case will certainly occur when generating moderate sized samples. The largest positive summand in the series for $x = 60$ is then $60^{60}/60! > 10^{24}$ and the largest negative summand is in absolute value bigger than 10^{24} as well but the sum should converge to $\exp(-60)$ which is very close to 0. So using standard floating point arithmetic the sum will due to extinction not converge to the correct result. This also has happened with our implementation of the below algorithm. For example for $x = 49$ the alternating series "converges numerically" to 3607.19. This implies that the value $X = 49$ is always accepted independent of the value of V and the algorithm does not generate variates form the correct distribution. It produces by far too many variates in the far tails. We could try tricks to overcome these numerical problems but they are not easy to use without changing the entire method. The only simple way out is to use exact rational arithmetic for the evaluation of the sum but of course this slows down the algorithm considerably.

Require: Arbitrary precision arithmetic.
Output: Standard exponential distributed random variate X.
1: **loop**
2:     Generate $U \sim U(0,1)$.
3:     Set $X \leftarrow 2/\sqrt{1-U} - 2$. /* X has density prop. to hat */
4:     Set $h(X) \leftarrow 3^3/(e(X+2)^3)$.
5:     Generate $V \sim U(0, h(X))$.
6:     Set $s \leftarrow 1$, $S \leftarrow 1$, and $n \leftarrow 0$.
7:     **repeat**
8:         Set $n \leftarrow n+1$, $s \leftarrow s\,x/n$, and $S \leftarrow S - s$. /* now $S \leq f(X)$. */
9:         **if** $V \leq S$ **then**
10:            **return** X.
11:        Set $n \leftarrow n+1$, $s \leftarrow sx/n$ and $S \leftarrow S + s$. /* now $S \geq f(X)$. */
12:    **until** $V > S$

9.3 Known Fourier Coefficients

Given a density f on the interval $(-\pi, \pi)$, its Fourier coefficients are defined for $n = 0, 1, 2, \ldots$ by

9.3 Known Fourier Coefficients

$$a_n = \frac{1}{\pi}\int_{-\pi}^{\pi} f(x)\cos(nx)\,dx\;, \quad b_n = \frac{1}{\pi}\int_{-\pi}^{\pi} f(x)\sin(nx)\,dx\;.$$

Note that $a_0 = A_f/\pi$ and $b_0 = 0$. The Fourier coefficients uniquely determine f. If the Fourier coefficients are absolutely summable, i.e.

$$\sum_{k=1}^{\infty} (|a_k| + |b_k|) < \infty$$

then the trigonometric series is absolutely and uniformly convergent to f and we can write

$$f(x) = \frac{a_0}{2} + \sum_{k=1}^{\infty}(a_k \cos(kx) + b_k \sin(kx))\;.$$

Devroye (1989) suggests to use this trigonometric series to generate variates for the case that the density f is not known explicitly but Fourier coefficients of f are available. Writing $S_n(x, f)$ for the partial sum of the trigonometric series up to n we can obtain the trivial error bound

$$|S_n(x,f) - f(x)| \le \sum_{k=n+1}^{\infty}(a_k\cos(kx)+b_k\sin(kx)) \le \sum_{k=n+1}^{\infty}\sqrt{a_k^2+b_k^2} = R_{n+1}\;.$$

If we assume that the Fourier coefficients a_k and b_k, and R_1 (or upper bounds for R_n) are available we can use the convergent series method (see Algorithm 9.7) to generate random variates with density f. Note that if R_1 is known the R_k can be calculated easily by subtracting from R_1. The constant $h(x) = a_0/2 + R_1$ is used as hat-function. For easier use we collect the details in Algorithm 9.9 (Fourier-Coefficients). This algorithm halts with probability one if and only if the Fourier coefficients are absolutely summable. Devroye (1989) shows that the expected number of (a_i, b_i) pairs that have to be evaluated is bounded by

$$E(\#(a_i,b_i)) \le 1 + 4\pi\sum_{n=0}^{\infty}\sum_{k=n+1}^{\infty}\sqrt{a_k^2+b_k^2} \le 1 + 4\pi\sum_{k=1}^{\infty} k(|a_k|+|b_k|)\;.$$

To make this expectation finite it suffices to have Fourier coefficients that are of order $O(n^{-2-\epsilon})$ for some $\epsilon > 0$. However, as we already have discussed for the series method, Algorithm 9.9 has numerical problems, especially if the a_k or b_k have large values for small k or are converging slowly.

For the case of a cosine series with convex coefficients (i.e. the a_k are a convex sequence and the b_k are all 0) Devroye (1989) explains a direct generation algorithm that is not based on rejection. Instead a representation theorem similar to Thm. 9.6 below for Polya characteristic functions is used that allows to generate the distribution as a mixture of Fejer densities. For the details we refer to the original paper.

Algorithm 9.9 Fourier-Coefficients

Require: Fourier coefficients a_k and b_k for density f on $(-\pi, \pi)$,
$R_1 = \sum_{k=1}^{\infty} \sqrt{a_k^2 + b_k^2}$ and $a_0 = A_f/\pi$.
Output: Random variate X with density f.
1: **loop**
2:   Generate $X \sim U(-\pi, \pi)$.
3:   Generate $V \sim U(0, a_0/2 + R_1)$.
4:   Set $S \leftarrow a_0/2$, $R \leftarrow R_1$, and $n \leftarrow 0$.
5:   **repeat**
6:     Set $n \leftarrow n+1$, $S \leftarrow S + a_n \cos(nx) + b_n \sin(nx)$, and $R \leftarrow R - \sqrt{a_n^2 + b_n^2}$.
7:   **until** $|S - V| > R$
8:   **if** $V \leq S$ **then**
9:     **return** X.

9.3.1 Computational Experience

We coded Algorithm 9.9 (Fourier-Coefficients) and tested it for different Fourier coefficients with convergence rates $O(1/n^2)$ and $O(1/n^3)$. We also computed the density and the cdf for several examples and made goodness of fit tests to see if the algorithm works correctly. As the algorithm is based on the series method we also checked if the series is converging to the correct density value. We could not observe any numerical problems in our experiments with the four examples we tried. Probably due to the fact that $|x| \leq \pi$, Algorithm 9.9 seems to have no numerical problems when used with standard double precision floating point numbers.

As suggested by the bound for the expected number of evaluations the algorithm was clearly slower for the $O(1/n^2)$ Fourier coefficients. For our examples the algorithm was between 10 and 100 times slower than generating an exponential variate by inversion.

9.4 Known Characteristic Function

The characteristic function of a random variable X defined as

$$\phi(t) = \mathrm{E}\left(e^{itX}\right)$$

is an important description of a random variable used in different fields of statistics. The characteristic function determines a random variable uniquely. It is especially useful if sums of independent random variates are considered. Although we have the Fourier inversion formula available it can happen that we know the characteristic function of a random variate but have great difficulties to compute its density. In such cases we can make use of special algorithms that can generate random variates if only the characteristic function is known. These algorithms are due to Devroye (1984a, 1986d), see also Devroye (1986a, Sects. IV.6.7 and XIV.3). We are presenting here only the two simplest algorithms.

9.4.1 Polya Characteristic Functions

Real-valued, even, continuous characteristic functions $\phi(t)$, convex on $\mathbb{R}^+$ with $\phi(0) = 1$ and $\lim_{t \to \infty} \phi(t) = 0$ are called *Polya characteristic functions*. The following representation theorem is useful.

Theorem 9.6. *Let $\phi(t)$ be a Polya characteristic function. If Y and Z are two independent random variables, and Y has the Fejer-de la Vallee Poussin (FVP) density*

$$\frac{1}{2\pi} \left(\frac{\sin(x/2)}{x/2} \right)^2,$$

and Z has distribution function F with

$$F(0) = 0, \quad \text{and} \quad F(t) = 1 - \phi(t) + t\,\phi'(t), \text{ for } t > 0$$

where ϕ' denotes the right-hand derivative (which exists everywhere), then the random variate $X = Y/Z$ has characteristic function $\phi(t)$.

Proof. See Devroye (1986a, p.187). □

Algorithm 9.10 Characteristic-Function-Polya

Require: Polya characteristic function $\phi(t)$, right derivative $\phi'(t)$.
Output: Random variate X with characteristic function ϕ.
    /* Setup */
1: Prepare function $F(t) = 1 - \phi(t) + t\,\phi'(t)$ for $t > 0$ and 0 for $t \leq 0$.
    /* Generator */
2: Generate Z with cdf $F(t)$ using numerical inversion. /* see Chap. 7 */
3: Generate Y with FVP density. /* Algorithm 9.11 (FVP-Density) */
4: **return** Y/Z.

Algorithm 9.10 (Characteristic-Function-Polya) utilizes this theorem. It is quite short but we need to sample from a distribution with the FVP density. To do so notice that $h(x) = \min(1, 1/x^2)$ is a hat for $f(x) = (\sin(x)/x)^2$ (FVP density with scale parameter 1) and sampling can be done by rejection. The acceptance probability is $\pi/4$ and we can easily generate from this hat by inversion. The details necessary for generation from the FVP density are given in Algorithm 9.11 (FVP-Density).

A drawback of Algorithm 9.10 (Characteristic-Function-Polya) is that it is using numerical inversion which is slow and/or not exact. Therefore Devroye (1986d) has developed a black-box algorithm for a large subclass of distributions with Polya characteristic functions that is not based on inversion and which not even requires the right-derivative $\phi'(t)$. For details we refer to the original paper.

Algorithm 9.11 FVP-Density (Fejer-de la Vallee Poussin density)

Output: Random variate Y with FVP density.
1: Generate $U \sim U(0,1)$ and $V \sim U(0,1)$.
2: **if** $U < 0.25$ **then**
3:   Set $X \leftarrow -0.25/U$.
4:   **if** $V \leq \sin^2(X)$ **then**
5:     **return** $2X$.
6: **else**
7:   **if** $U < 0.75$ **then**
8:     Set $X \leftarrow 4(U - 0.5)$.
9:     **if** $V \leq \sin^2(X)/X^2$ **then**
10:       **return** $2X$.
11:   **else**
12:     Set $X \leftarrow 0.25/(U - 0.75)$.
13:     **if** $V \leq \sin^2(X)$ **then**
14:       **return** $2X$.

9.4.2 Very Smooth Densities

Unfortunately Algorithm 9.10 (Characteristic-Function-Polya) works only for convex characteristic functions. Devroye (1986a, Sect. XIV.3) discusses possible more general black-box methods. Again we are presenting here only the simplest algorithm that is working for very smooth densities. It is based on the series method (Sect. 9.2) together with the following theorem.

Theorem 9.7. *Assume that a given distribution has two finite moments, and that the characteristic function ϕ has two absolutely integrable derivatives. Then the distribution has a density f where the following two inequalities hold*

$$f(x) \leq \frac{1}{2\pi} \int |\phi(t)|\,dt, \quad \text{and} \quad f(x) \leq \frac{1}{2\pi x^2} \int |\phi''(t)|\,dt\;.$$

The area below the minimum of the two bounding curves is $\frac{2}{\pi}\sqrt{\int |\phi| \int |\phi''|}$.

Proof. See Devroye (1986a, p.697). □

The integrability condition for ϕ implies that the density is bounded and continuous. The integrability condition for ϕ'' translates into a strong tail condition. This is the reason that we can find the global hat function given in Thm. 9.7. It is not obvious how to use this hat function for a rejection algorithm as we do not know the density f. If we use the alternating series method (Algorithm 9.8) it suffices to find an alternating converging series for f. For most of the algorithms presented in Devroye (1986a, Sect. XIV.3) numerical integration of the Fourier inversion formula together with error bounds is used to obtain the converging series for f. Unfortunately these algorithms become rather complicated and are not easy to use. Therefore we are not giving the details here.

The situation becomes much easier if we consider the Taylor series expansion of f. If ϕ is absolutely integrable, symmetric, and nonnegative, then we get the alternating series necessary for Algorithm 9.8 as $f(x)$ is sandwiched between consecutive partial sums in the Taylor series around 0,

$$f(0) - \frac{x^2}{2!} f'(0) + \frac{x^4}{4!} f''(0) - + \dots .$$

This can be seen from the Fourier inversion theorem for the real part of f (f is real as ϕ is symmetric.)

$$f(x) = \frac{1}{2\pi} \int \phi(t) \cos(tx)\,dt = \nu_0 - \frac{x^2}{2!}\nu_2 + \frac{x^4}{4!}\nu_4 - + \dots$$

where $\nu_{2n} = \frac{1}{2\pi}\int t^{2n}\phi(t)\,dt$. If $\int t^{2n}\phi(t)\,dt$ is finite then the $2n$-th derivative $f^{(2n)}$ exists and $f^{(2n)}(0) = \nu_{2n}$. The finity condition on the integral is a moment condition on ϕ and translates into a smoothness condition on f. For extremely smooth f all moments of ϕ can be finite which is the case e.g. for the normal and the Cauchy density and all symmetric stable densities with parameter at least equal to one. Also all characteristic functions with compact support are included. If the series $x^{2n}\nu_{2n}/(2n)!$ is summable for all $x>0$ we see that f is fully determined by all its derivatives at 0. But we have to know the moment sequence of ϕ, this is the sequence $f^{(2n)}(0) = \frac{1}{2\pi}\int t^{2n}\phi(t)\,dt$ and this seems to be possible only if we know a closed form for ϕ. So for Algorithm 9.12 (Characteristic-Function-Smooth) the knowledge of only the characteristic function ϕ is not enough, we have to know the moment sequence of ϕ as well. Nevertheless, it is an interesting automatic algorithm for sampling from random variates with a given simple characteristic function.

Algorithm 9.12 Characteristic-Function-Smooth

Require: Characteristic function $\phi(t)$, symmetric, real-valued nonnegative;
   $a = f(0) = \frac{1}{2\pi}\int |\phi(t)|\,dt$, $b = \frac{1}{2\pi}\int |\phi''(t)|\,dt$;
   $f^{(2n)}(0) = \frac{1}{2\pi}\int t^{2n}\phi(t)\,dt$ for $n = 1, 2, \dots$.
Output: Random variate X with characteristic function ϕ.
1: **loop**
2:   Generate $U \sim U(0,1)$.
3:   Generate X from the hat function $h(x) = \min(a, b/x^2)$.
4:   Set $T \leftarrow U\,h(x)$, $S \leftarrow a$, $n \leftarrow 0$, $Q \leftarrow 1$.
5:   **while** $T \le S$ **do**
6:     Set $n \leftarrow n+1$, $Q \leftarrow -Q\,X^2/(2n(2n-1))$.
7:     Set $S \leftarrow S + Q\,f^{(2n)}(0)$.
8:     **if** $T \le S$ **then**
9:       **return** X.
10:    Set $n \leftarrow n+1$, $Q \leftarrow -Q\,X^2/(2n(2n-1))$.
11:    Set $S \leftarrow S + Q\,f^{(2n)}(0)$.

A simple example where we can find all necessary information for Algorithm 9.12 is the normal distribution which has $\phi(t) = \exp(-t^2)$. Here we can calculate $\int |\phi''|$ and also the recursion for $f^{(2n)}(0)$. But of course this is not a recommended method to generate normal random variates. It is also possible to use this algorithm, e.g., for $\phi(t) = \exp(-t^4)$. For the stable distribution with $\phi(t) = \exp(-t^\alpha)$ and $1 \leq \alpha < 2$ we could use Algorithm 9.12 only if we can show that the bounds of Thm. 9.7 remains valid although the second moment of this distribution does not exist.

It is important to stress here that Algorithm 9.12 (Characteristic-Function-Smooth) should only be used with arbitrary precision arithmetic as it is based on the alternating Taylor series at 0 which is numerically very unstable for large values of x.

9.4.3 Computational Experience

We coded Algorithm 9.10 (Characteristic-Function-Polya) utilizing the numeric inversion algorithm we have implemented in UNU.RAN. Our results show that the algorithm works correctly and at acceptable speed. E.g., for $\phi(t) = \exp(-|t|)$ (Cauchy distribution) and $\phi(t) = \frac{1}{1+|t|}$ it is about ten times slower than generating an exponential random variate by inversion.

For Algorithm 9.12 (Characteristic-Function-Smooth) the situation is different. We coded and tested it for the normal distribution using standard double precision floating point arithmetic. The results have only shown that our warnings have a good reason. Due to numerical inaccuracies the algorithm is generating far too many variates in the far tails, as the alternating series is not converging to 0 for $x > 10$ but has – depending on x – arbitrary positive and negative values. So we can only repeat the warning here: Do not use any algorithm that is based on alternating sums, like Algorithm 9.12, with standard precision arithmetic!

9.5 Exercises

Exercise 9.1. Spell out the details of an algorithm that generates variates with a step-function shaped hazard rate.

Exercise 9.2. Generalize Algorithm 9.4 (Hazard-Rate-Constant-Thinning) and the algorithm of Exercise 9.1 and spell out the details of a thinning algorithm with a step-function shaped dominating hazard rate $g(x)$.

Exercise 9.3. Develop an algorithm to generate a random variate with unbounded domain and increasing hazard rate. Use a dynamic version of the algorithm of Exercise 9.2 and take $g(x) = h(p_i)$ for $p_{i-1} < x \leq p_i$ for design points $0 = p_0 < p_1 < p_2 \ldots$.

Exercise 9.4. Using the alternating series method and the Taylor-series expansion around 0 develop a generator for the density $f(x) = \sin x$ on $[0, \pi]$ that is not using a call to the sine or cosine function.

Check if the bounds for f computed by the alternating series algorithm become wrong for larger values of x due to numerical problems.

Part III

Discrete Univariate Distributions

10
Discrete Distributions

A discrete random variable is a random variable taking only values on integers. The distribution of a discrete random variable X is determined by its *probability mass function*, $p_k = \text{Prob}(X = k)$. It is also called *probability vector* if its support is bounded from below. In the latter case we assume without loss of generality that X is only taking values on the nonnegative integers. We then write $(p_0, p_1, \ldots)$ for the probability vector. Analogously to the continuous case we use the terms *quasi-probability mass function* for a nonnegative summable function that need not be normalized.

Automatic generators for discrete distributions are well known in the literature for a longer time than those for continuous distributions. Many text books on simulation explain, for example, the alias method first suggested by Walker (1974, 1977), see Sect. 3.2, the alias-urn method (Peterson and Kronmal, 1982), and inversion by guide table also called indexed search (Chen and Asau, 1974), see Sect. 3.1.2. These are well known table methods that can be used to generate from arbitrary discrete distributions with bounded domain. All of them need a time consuming setup and large tables that grow linearly with the size of the domain but they have very fast marginal generation times. There are other table-based methods suggested in the literature (e.g. Marsaglia, 1963). As they are either less general or much slower than the alias and the guide table methods, we do not explain them here. Instead we explain in Section 10.1 a new variant of the guide table method that also works for distributions with unbounded domains, as long as the tails of the distribution are not too heavy.

Reading Sects. 3.1, 3.2, and 10.1 it may look as if the alias method or indexed search are the final solution of the problem to sample from discrete distributions. However, this is not true. The main disadvantages of these algorithms are the slow setup, the size of the necessary tables and the problem with rounding errors in the setup. This can become prohibitive if the probability vector of the desired distribution is long (e.g. $L = 10^5$ or 10^6). Nevertheless, the alias method or guide table inversion are the best choices for most every-day discrete generation problems. But for some applications universal rejection algorithms for discrete distributions are of practical importance, especially when the tails of the distribution are very heavy. The faster setup times and the low memory requirements are also advantages of these algorithms.

We have discussed several discrete versions of the rejection method in Sect. 3.3. To design automatic rejection methods for discrete distributions it is necessary to construct the hat functions automatically. As for continuous distributions, we can utilize two different principles: an adaption of transformed density rejection that we will call *transformed probability rejection*, or applying general inequalities. The former needs a moderate setup but results in quite fast sampling algorithms, whereas the latter principle leads to simple algorithms with almost no setup but slow marginal execution times. We present both ideas in detail below (Sects. 10.2 and 10.3). Sect. 10.4 describes methods for sampling from discrete distributions where the probability vector is not given explicitly. Sect. 10.5 compares the performance of the presented algorithms whereas Sect. 10.6 summarizes its properties.

10.1 Guide Table Method for Unbounded Domains

Section 3.1.2 describes the guide table aided inversion method. It requires a setup but has a very fast marginal generation speed. It works for all discrete distribution with finite probability vector. However, it does not work for distributions with very large or unbounded domains. In this section we present a new variant of the guide table method without this limitation.

10.1.1 Indexed Search for Distributions with Right Tail

To obtain optimal marginal speed without cutting off the tails Dagpunar (1988) suggested hybrid generation algorithms for discrete standard distributions. He used the alias method for the main part of the distribution and rejection from an exponential hat function for the tails. This type of combined algorithm has certainly good properties and very good marginal speed if large tables are used but it combines two algorithms of very different nature which also means a higher programming effort. Therefore we think that it is better to use the guide table method (as presented in Algorithm 3.2) in combination with sequential search.

Algorithm 10.1 (Indexed-Search-Tail) compiles such a variant. It can be described as indexed search on $(0, L - 1)$ with a guide table of size $C - 1$, combined with sequential search for the tail. It thus remains a pure discrete inversion algorithm and generates exactly the same variates as Algorithm 3.1 (Sequential-Search). The algorithm has two design parameters: The length of the table of stored cumulative probabilities (called L) and the size of the guide table (called C). To simplify the algorithm and to fully integrate the sequential search into the indexed search, L is computed as $L - 1 = F^{-1}(1 - 1/C)$.

Algorithm 10.1 Indexed-Search-Tail

Require: Probability mass function p_k, $k \geq 0$; size C of the guide table.
Output: Random variate X with given probability function.
  /* Setup */
1: Set $k \leftarrow -1$.
2: **repeat**
3:    Set $k \leftarrow k + 1$.
4:    Set $P_k \leftarrow \sum_{j=0}^{k} p_j$. /* cumulative probabilities */
5: **until** $P_k \geq 1 - 1/C$.
6: Set $g_0 \leftarrow 0$, $k \leftarrow 0$.
7: **for** $j = 1$ to $C - 1$ **do** /* loop to build the guide table */
8:    **while** $j/C > P_k$ **do**
9:      Set $k \leftarrow k + 1$.
10:   Set $g_j \leftarrow k$.
  /* Generator */
11: Generate $U \sim U(0, 1)$.
12: Set $J \leftarrow \lfloor UC \rfloor$.
13: Set $X \leftarrow g_J$.
14: **if** $J < C - 1$ **then** /* main part */
15:   **while** $U > P_X$ **do**
16:     Set $X \leftarrow X + 1$.
17:   **return** X.
18: **else** /* tail, sequential search */
19:   Set $S \leftarrow P_X$.
20:   **while** $U > S$ **do**
21:     Set $X \leftarrow X + 1$.
22:     Set $S \leftarrow S + p_X$.
23:   **return** X.

Thus the guide table method (and therefore no evaluations of the p_i) occur with probability $1 - 1/C$. Sequential search for the tail is only used with probability $1/C$. Using the formula of Sect. 3.1.2 we find for the expected number of comparisons the upper bound $1 + L/(C-1)$ for the main part of the distribution. For the tail we can use the result for the sequential search method (see Sect. 3.1.1) to see that on average we need $1 + E(X|X \geq L) - L$ comparisons and $E(X|X \geq L) - L$ evaluations of the p_i.

These performance characteristics clearly show that for heavy tailed distributions, or distributions with a very large domain the guide table method will become slow and/or need huge tables and a very slow setup. For such distributions we should therefore use a rejection method for discrete distributions as described in Sect. 3.3. Nevertheless, for distributions with moderate sized domains or with exponential tails we can obtain very fast algorithms.

Example 10.1. We consider the geometric distribution with probability mass function
$$p_k = p(1-p)^{k-1}, \text{ for } k = 1, 2, 3, \dots.$$

For Algorithm 10.1 we also need the size C of the guide table. To assess the performance of the algorithm we find for fixed parameter p and size C the index of the guide table entry by

$$g_{C-1} = \left\lceil 1 - \frac{\log C}{\log(1-p)} \right\rceil .$$

It is well known that for the geometric distribution, as for the exponential distribution, $E(X|X \geq L) - L$ is not effected by L and is equal to $1/p - 1$. Then using the formulas from above with L replaced by $L - 1$ (as 0 is not in the domain) we obtain

$$E(\# \text{ of comparisons}) \leq \left(1 + \tfrac{L-1}{C-1}\right) \tfrac{C-1}{C} + \tfrac{1}{pC} ,$$
$$E(\# \text{ evaluations of } p_k) = \tfrac{1-p}{C} .$$

The results for different values of p and C are contained in Table 10.1. These show that for a geometric distribution with $p \geq 0.01$ we can obtain a very fast generator using a guide table of size $C = 1000$. For smaller values of p larger tables are necessary which can reduce the speed of the method due to cache-effects. (Note that we can obtain a quite fast and very simple generator for the geometric distribution by inversion if we return $\lceil \log U / \log(1-p) \rceil$ for $U \sim U(0,1)$.)

Table 10.1. Bounds for the performance characteristics for Algorithm 10.1 applied to the geometric distribution

p	C	L	$E(\#$ of comparisons$)$	$E(\#$ evaluations of $p_k)$
0.5	10^2	8	1.08	0.01
0.1	10^2	45	1.53	0.09
0.1	10^3	67	1.08	0.01
0.01	10^3	689	1.79	0.10
0.01	10^4	918	1.10	0.01
0.001	10^4	9207	2.02	0.10
0.001	10^5	11509	1.13	0.01

10.1.2 Indexed Search for Distributions with Two Tails

As we restrict our attention in this section to discrete distributions with a domain on the nonnegative integers we could use Algorithm 10.1 for any of these distributions. But computing the cumulative probabilities by starting at 0 may become very slow and could lead to unnecessarily huge tables if the main part of the distribution is far away from 0. If we can evaluate the cumulative probabilities without summing all p_k we could design a pure discrete inversion algorithm similar to Algorithm 10.1 but including a sequential search to the left to generate from the left tail. We pose the design of this algorithm as Exercise 10.1.

Unfortunately such an algorithm is of limited practical importance as the cumulative probabilities are not available for most discrete distributions. In such a situation we cannot stick to the pure discrete inversion algorithm as we cannot compute the probabilities of the right and the left tail separately but we have to treat them together and then decide between right and left tail in an extra comparison.

Such an algorithm should start with a search for the shortest possible interval (i_l, i_r) with $\sum_{i=i_l}^{i_r} p_i \geq 1 - 1/C$. This is not difficult for unimodal distributions: We can start with a point near the mode (which has to be provided by the user); then we increase the size of the interval always in that direction that has the higher probability till the sum of all probabilities is greater than $1 - 1/C$. After we have computed i_l and i_r we compute in a second pass the cumulative probabilities and the guide table for this interval. For the tail generation flip the left tail to the right and compute the joint tail probabilities

$$\tilde{p}_k = p_k + p_{i_l+i_r-k}, \text{ for } k > i_r .$$

Of course this can also be done "on-the-fly" when a probability mass function with unbounded domain is given. When we have sampled a $\tilde{J}$ from this tail distribution we have to decide between left and right tail. We return a J from the right tail (i.e. $J = \tilde{J}$) if $\tilde{P}_{J-1} < U < \tilde{P}_{J-1} + p_J$, otherwise, if $\tilde{P}_{J-1} + p_J < U < \tilde{P}_J$, we return $J = i_l + i_r - \tilde{J}$ from the left tail. (We set $\tilde{P}_{i_r} = \sum_{i=i_l}^{i_r} p_i$.) The details are contained in Algorithm 10.2 (Indexed-Search-2-Tails). Bounds for the performance characteristics of this algorithm can be obtained in close analogy to the discussion above for Algorithm 10.1.

Remark. The search for the shortest possible interval $[i_l, i_r]$ in the setup of Algorithm 10.2 (Indexed-Search-2-Tails) can simply start from the mode for unimodal distributions. For multi-modal distributions the search need not lead to the shortest possible interval but can be used as well. If the distribution has probabilities between two of its mode that are very small or even zero, the search procedure does not work. For such cases the user must provide an interval that contains all modes and the search algorithm starts from that interval.

10.2 Transformed Probability Rejection (TPR)

Transformed probability rejection (TPR) is the variant of transformed density rejection (see Chap. 4) designed for discrete distributions. A discrete distribution with probability function p_i, with support $S \subset \mathbb{Z}$ is called T-concave if

Algorithm 10.2 Indexed-Search-2-Tails

Require: Probability mass function p_k, $k \in \mathbb{Z}$;
(small) interval $[i_l, i_r]$ that contains all modes ($i_l = i_r$ possible);
size C of the guide table.
Output: Random variate X with given probability function.
/* Setup */
1: Compute and store the p_k for $i_l \leq k \leq i_r$ and $s \leftarrow \sum_{i=i_l}^{i_r} p_i$.
2: Set $p^- \leftarrow p_{i_l-1}$, $p^+ \leftarrow p_{i_r+1}$.
3: **while** $s < 1 - 1/C$ **do** /* loop to compute i_l and i_r */
4:   **if** $p^- < p^+$ **then**
5:     Set $s \leftarrow s + p^+$, $i_r \leftarrow i_r + 1$, $p_{i_r} \leftarrow p^+$.
6:     Compute $p^+ \leftarrow p_{i_r+1}$.
7:   **else**
8:     Set $s \leftarrow s + p^-$, $i_l \leftarrow i_l - 1$, $p_{i_l} \leftarrow p^-$.
9:     Compute $p^- \leftarrow p_{i_l-1}$.
10: Compute cumulative probabilities $P_k = \sum_{j=i_l}^{k} p_j$ for $i_l \leq k \leq i_r$.
11: Set $g_0 \leftarrow i_l$, $k \leftarrow i_l$.
12: **for** $j = 1$ to $C - 1$ **do** /* compute guide table; note that $g_{C-1} = i_r$ */
13:   **while** $j/C > P_k$ **do**
14:     Set $k \leftarrow k + 1$.
15:   Set $g_j \leftarrow k$.
/* Generator */
16: Generate $U \sim U(0, 1)$.
17: Set $J \leftarrow \lfloor U C \rfloor$.
18: Set $X \leftarrow g_J$.
19: **if** $J < C - 1$ **then** /* main part */
20:   **while** $U > P_X$ **do**
21:     Set $X \leftarrow X + 1$.
22:   **return** X.
23: **else** /* tails, sequential search */
24:   Set $S \leftarrow P_X$.
25:   **if** $U < S$ **then**
26:     **return** X.
27:   **loop**
28:     Set $X \leftarrow X + 1$.
29:     Set $S \leftarrow S + p_X$.
30:     **if** $U < S$ **then**
31:       **return** X. /* right tail */
32:     Set $S \leftarrow S + p_{i_l+i_r-X}$.
33:     **if** $U < S$ **then**
34:       **return** $i_l + i_r - X$. /* left tail */

$$T(p_i) \geq \tfrac{1}{2}(T(p_{i-1}) + T(p_{i+1})), \quad \text{for all } i \in S .$$

For log-concave distributions we have $T(x) = \log(x)$ and $p_i^2 \geq p_{i-1}\, p_{i+1}$.

In the literature there are three papers presenting algorithms based on TPR. A first paper by Hörmann (1994a) was written before the development of TDR. The presented algorithm utilizes rejection from discrete uniform-geometric hats and only works for log-concave distributions. More recent algorithms are based on *rejection-inversion* (Hörmann and Derflinger, 1996, 1997). The latter have faster generation times and are superior in almost all aspects (except that they have a slower setup). Thus we will discuss only these algorithms in detail. The interested reader is referred to the original paper of Hörmann (1994a) for the former algorithm.

10.2.1 The Principle of Rejection-Inversion

Rejection-inversion has been introduced by Hörmann and Derflinger (1996). It is not a new universal algorithm itself but a special version of rejection for discrete distributions. For other versions of discrete rejection see Sect. 3.3. The main point of rejection inversion is the observation that we can save the second uniform random number of ordinary rejection. For this purpose we sample a continuous random variate X from the hat distribution, round it to next integer to get a discrete variate K, and retain X for rejection in horizontal direction.

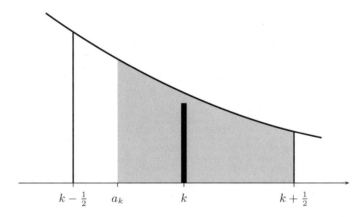

Fig. 10.1. The idea of rejection-inversion: the region of acceptance is shaded

We use a continuous hat function $h(x)$ as indicated in Fig. 10.1 (see also Fig. 3.5(c)), i.e. the area below the hat must be an upper bound for the probability p_k for k,

$$\int_{k-\frac{1}{2}}^{k+\frac{1}{2}} h(x)\, dx \geq p_k . \tag{10.1}$$

Assume for the moment a monotonically decreasing probability vector p_k, for $m \leq k \leq b_r \leq \infty$, and a monotonically decreasing hat function $h(x)$. Then each interval $(k-\frac{1}{2}, k+\frac{1}{2})$ is divided by a point a_k into the interval $(k-1/2, a_k)$ of rejection and interval $(a_k, k+\frac{1}{2})$ of acceptance, i.e. the acceptance condition now reads: accept X if $X \geq a_k$. In Fig. 10.1 the regions of acceptance (shaded) and rejection are shown. Of course a_k must have the property that

$$\int_{a_k}^{k+\frac{1}{2}} h(x)\,\mathrm{d}x = p_k \;.$$

Denote by $H(x) = \int h(x)\,\mathrm{d}x$ an antiderivative of the hat function h and by $H^{-1}(x)$ its inverse. Thus we find $p_k = \int_{a_k}^{k+\frac{1}{2}} h(x)\,\mathrm{d}x = H(k+\frac{1}{2}) - H(a_k)$, and hence $a_k = H^{-1}(H(k+\frac{1}{2}) - p_k)$. As we have constructed a_k such that acceptance occurs for $X \geq a_k$ we obtain the acceptance condition

$$X \geq H^{-1}(H(k+\tfrac{1}{2}) - p_k) \;.$$

As the use of the inversion method implies $X = H^{-1}(U)$, we can transform both sides of the acceptance condition with H to get the simplified acceptance condition

$$U \geq H(k+\tfrac{1}{2}) - p_k \;.$$

For the case $k = m$ we can avoid rejection by an extra trick: Start the hat function $h(x)$ for $x \geq a_m$ instead of $x \geq m - \frac{1}{2}$. This means that the uniform random number U used for inversion is not generated in the interval $(H(m-\frac{1}{2}), H(b_r+\frac{1}{2}))$ but in the interval $(H(a_m), H(b_r+\frac{1}{2}))$, where b_r denotes (the possibly infinite) right border of the domain.

Algorithm 10.3 (Rejection-Inversion) utilizes the idea of rejection-inversion to sample from discrete distribution by means of convex hat functions. In this case we can replace the hat condition (10.1) by the much simpler sufficient condition $p_k \leq h(k)$. Notice that it is not necessary for this algorithm that the given pmf is normalized.

Algorithm 10.3 Rejection-Inversion (Discrete)

Require: Non-increasing quasi-probability function p_k for $m \leq k \leq b_r \leq \infty$;
    convex hat function $h(x)$ with $h(k) \geq p_k$ ($H(x)$ denotes the antiderivative of h).
Output: Random variate X with given probability function.
    /* Setup */
1: Store $a_m \leftarrow H^{-1}(H(m+\frac{1}{2}) - p_m)$.
    /* Generator */
2: **loop**
3:     Generate $U \sim U(H(a_m), H(b_r+\frac{1}{2}))$.
4:     Set $X \leftarrow H^{-1}(U)$ and $K \leftarrow \lfloor X + \frac{1}{2} \rfloor$.
5:     **if** $U \geq H(K+\frac{1}{2}) - p_K$ **then**
6:         **return** K.

10.2.2 Rejection-Inversion and TPR

We have just seen that rejection inversion can be used to generate from discrete distributions if we have a convex monotonically decreasing hat function available. It is no problem to reformulate Algorithm 10.3 for monotonically increasing discrete distributions as well. We know from Chap. 4 that transformed density rejection (TDR) with the transformation class T_c is well suited to construct a convex hat function around a design-point p for a continuous distribution. We can use the same principle to construct hat functions for rejection inversion. For a design point k_0 we thus obtain the convex hat

$$h(x) = T_c^{-1}\left(\tilde{g}(k_0) + \tilde{g}'(k_0)(x - k_0)\right) \tag{10.2}$$

where $\tilde{g}(k) = T_c(p_k)$ denotes the transformed probabilities and $\tilde{g}'(k) = T_c(p_{k+1}) - T_c(p_k)$ the first difference of the transformed probabilities.

If we allow large tables and a slow setup, very fast automatic generation methods are available for discrete distributions. Therefore, we are only interested in the simplest general case of TDR, i.e. the case where we take the mode m as design-point together with one design-point x_l left and one point x_r right of the mode. Theorem 4.13 tells us how to place the design-points in that case for continuous distributions. To obtain a similar theorem for discrete distributions we need a continuation of the p_k's, i.e. a continuous function $f_p(x)$ with $f_p(k) = p_k$ for all k in the domain of the distribution. Generalizing the above definition, we define $\tilde{g}$ as the linear interpolation of the transformed probabilities

$$\tilde{g}(x) = T_c(p_{\lfloor x \rfloor})(1 - (x - \lfloor x \rfloor)) + T_c(p_{\lfloor x \rfloor + 1})(x - \lfloor x \rfloor) \;.$$

It is then obvious that the back-transformed version of $\tilde{g}$, given by

$$f_p(x) = T_c^{-1}(\tilde{g}(x)) \;,$$

is the pointwise minimal T_c-concave continuation for the discrete T_c-concave distribution p.

Theorem 10.2. *For a $c \leq 0$, let f_p be the the minimal T_c-concave continuation of a T_c-concave discrete distribution with probability function p_k with mode m, Then the area below the table-mountain shaped hat $h(x)$ constructed from the density f_p by using the TDR principle with construction points x_l, m, and x_r is minimized when x_r and x_l fulfill the condition*

$$f_p(x) = p_m \left(\frac{1}{c+1}\right)^{1/c} \text{ for } c < 0 \quad \text{and} \quad f_p(x) = \frac{p_m}{e} \text{ for } c = 0 \;.$$

The area below the hat function is equal to $p_m(x_r - x_l)$ and bounded for all T_c-concave discrete distributions by

$$t_o = \frac{1}{1 - (1/(1+c))^{1+1/c}} \text{ for } c < 0 \quad \text{and} \quad t_o = \frac{e}{e-1} = 1.582\ldots \text{ for } c = 0.$$

With the choice $x_l = m - t_o/p_m$ and $x_r = m + t_o/p_m$, the area below the hat function is less than or equal to $2\,t_o$ for arbitrary T_c-concave discrete distributions.

Proof. If the discrete distribution is T_c-concave the continuation f_p is T_c-concave as well. It is easy to see that (for $c \leq 1$) $A_{f_p} = \int f_p(x)\,dx \leq 1$. These two observations suffice to see that the given bounds are a direct consequence of Thm. 4.13. □

To construct the hat $h(x)$ according to Thm. 10.2 it is enough to determine the integer $\lfloor x_l \rfloor$ and to compute $\tilde{g}'(\lfloor x_l \rfloor) = \tilde{g}(x_l + 1) - \tilde{g}(x_l)$ and analogously for $\lfloor x_r \rfloor$. The hat is then computed using (10.2). Note that this rule is also correct for the case that x_l itself is an integer and $\tilde{g}'(x_l)$ does not exist. In this case it is possible to prove (see Hörmann and Derflinger, 1996) that we can replace the derivative by the value of the right or of the left derivative or a value in between.

We cannot use rejection-inversion for the whole table-mountain constructed according to the above theorem because it is not convex in the cutting points s_l and s_r of tail and center part and thus the condition $\int_{k-1/2}^{k+1/2} h(x)\,dx \geq p_k$ might not be satisfied. So we use the uniform head for $\tilde{s}_l \leq k \leq \tilde{s}_r$ where $\tilde{s}_l$ and $\tilde{s}_r$ are integers defined as $\tilde{s}_l = \lfloor s_l + 0.5 \rfloor$ and $\tilde{s}_r = \lfloor s_r + 0.5 \rfloor$. The area below the hat is therefore a bit larger than stated in Thm. 10.2 but we can use the "trick" used in Algorithm 10.3 (Rejection-Inversion) to avoid rejection for the border points $\tilde{s}_l, \tilde{s}_r, \tilde{s}_l - 1$ and $\tilde{s}_r + 1$. The last part of Thm. 10.2 remains correct and the area below the hat function is always smaller than $2\,t_o$ for that version.

One consideration that must be added to the description of rejection-inversion is that for $k \geq \tilde{s}_r + 1$ and for $\tilde{s}_l \leq k < m$ the area of acceptance is in the interval $(a_k, k + \frac{1}{2})$. For other values of k it is better to use $(k - \frac{1}{2}, a_k)$ as the region of acceptance, where the definition of a_k has to be changed appropriately. Figure 10.2 shows the three parts of the hat (thin lines) and the histogram (thick lines) of the desired distribution. The thick parts of the x-axis denote the regions of acceptance, the variable names ac (= a center) and at (= a tail), -1 (= left) and 1 (= right) are also used in the description of Algorithm 10.4 (Automatic-Rejection-Inversion).

For the choice of c for TDR we have discussed the trade-off between generality of the algorithm and good fitting hat for a fixed distribution in Sect. 4.2. For discrete distributions and TPR these considerations remain the same. Considering the simplicity of H and H^{-1} we conclude again that $c = -1/2$ is the best choice if we have to restrict ourselves to a single value of c. Of course the choice $c = 0$ is important as well as it implies a simple H and guarantees best-possible fit for all discrete log-concave distributions.

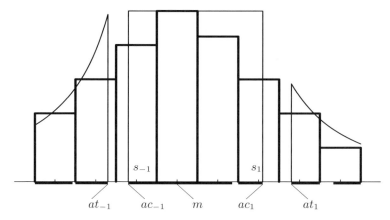

Fig. 10.2. The histogram, the hat-functions and the intervals of acceptance for Algorithm 10.4 (Automatic-Rejection-Inversion)

The best choice of the construction points depends on the information available for the desired distribution. If the location of the mode of the distribution is not known it is necessary to use a search procedure to find it. To compute the optimal values of x_l and x_r if the mode is known a search algorithm that includes many evaluations of the p_k is necessary. One advantage of this method is that it is enough to know the p_k's up to proportionality but the setup becomes really slow and depends on the size of the domain of the distribution. A second approach (if the sum of the probabilities is known) is to use the last part of Thm. 10.2 (the "minimax" approach) for the choice of x_l and x_r. It guarantees that the area below the hat is uniformly bounded for all T_c-concave distributions but is far away from optimal for many standard distributions. Therefore we think it is better to take a value close to the optimal value for the normal distribution and $c = -1/2$ (which is $x_r = m + 0.664/p_m$) as many of the classical discrete distributions have the normal distribution as limiting case. If the area below the hat becomes larger than $2\,t_o$ for this choice of x_r and x_l take the "minimax" approach as last resort that guarantees that the area below the hat is not larger than $2\,t_0$.

For the performance of random variate generation algorithms for discrete distributions the expected number of evaluations of the probabilities is of great importance as they are expensive to evaluate for nearly all discrete distributions. Hörmann and Derflinger (1996) have proven the following theorem that can be used to obtain a squeeze for TPR.

Theorem 10.3. *For a discrete T_c-concave distribution with probability vector p_k and mode m and a hat function $h(x)$ constructed using (10.2) with construction point k_0 we have*

$$k - a_k \text{ is nondecreasing for } m \le k \le k_0$$

with $a_k = H^{-1}(H(k + \tfrac{1}{2}) - p_k)$.

226   10 Discrete Distributions

Algorithm 10.4 Automatic-Rejection-Inversion (ARI)

Require: T_c-concave probability function p_k, mode m of the distribution.
Output: Random variate K with given probability function.
  /* Setup */
1: Prepare functions $T(x)$, $F(x)$, $F^{-1}(x)$ (taking a fixed value for c and the information given in Table 4.1).
2: Let b_{-1}, b_1 be smallest and largest integer of the domain which may be unbounded as well.
3: Set $d \leftarrow \max(2, \lfloor 0.664/p_m \rfloor)$.
4: **for** $i = \pm 1$ **do**
5:   Set $x_i \leftarrow m + i\,d$.
6:   **if** $(i\,x_i + 1 > i\,b_i)$ **then**
7:     Set $v_i \leftarrow 0$ and $s_i \leftarrow b_i$.
8:   **else**
9:     Set $y_i \leftarrow T(p_{x_i})$, $ys_i \leftarrow i\,(T(p_{x_i+i}) - y_i)$, $s_i \leftarrow \lfloor 0.5 + x_i + (T(p_m) - y_i)/ys_i \rfloor$.
10:    Set $Hat_i \leftarrow F(y_i + ys_i\,(s_i + i\,1.5 - x_i))/ys_i - i\,p_{s_i+i}$.
11:    Set $at_i \leftarrow x_i + (F^{-1}(ys_i\,Hat_i) - y_i)/ys_i$, $xsq_i \leftarrow i\,(at_i - (s_i + i))$.
12:    Set $v_i \leftarrow i\,(F(y_i + ys_i\,(b_i + i\,0.5 - x_i))/ys_i - F(y_i + ys_i\,(at_i - x_i))/ys_i)$.
13: Set $ac_i \leftarrow s_i + i\,(p_{s_i}/p_m - 0.5)$.
14: Set $v_c \leftarrow p_m\,(ac_1 - ac_{-1})$, $v_t \leftarrow v_c + v_{-1} + v_1$, $v_{cr} \leftarrow v_c + v_1$.
15: Set $t_o \leftarrow 1/(1 - (1/(1+c))^{(1+1/c)})$.
16: **if** $v_t > t_o$ **then**
17:   Set $d \leftarrow \lfloor t_o/p_m \rfloor$ and restart with this new value for d.
  /* Generator */
18: **loop**
19:   Generate $U \sim U(0, v_t)$.
20:   **if** $U \leq v_c$ **then**
21:     Set $X \leftarrow U\,(ac_1 - ac_{-1})/v_c + ac_{-1}$, $K \leftarrow \lfloor X + 0.5 \rfloor$.
22:     **if** $K < m$ **then** set $i \leftarrow -1$.
23:     **else** set $i \leftarrow 1$.
24:     Set $h \leftarrow 0.5 - p_K/p_m$.
25:     **if** $(h \leq i\,(K - X))$ **then**
26:       return K.
27:   **else**
28:     **if** $(U \leq v_{cr})$ **then**
29:       Set $i \leftarrow 1$, $U \leftarrow U - v_c$.
30:     **else**
31:       Set $i \leftarrow -1$, $U \leftarrow U - v_{cr}$.
32:     Set $U \leftarrow Hat_i + i\,U$, $X \leftarrow x_i + (F^{-1}(U\,ys_i) - y_i)/ys_i$ and $K \leftarrow \lfloor X + 0.5 \rfloor$.
33:     Set $h \leftarrow i\,F(y_i + ys_i\,(K + i\,0.5 - x_i))/ys_i - p_K$.
34:     **if** $(i\,U \geq h)$ **then**
35:       return K.

Proof. See Hörmann and Derflinger (1996, p.177). □

So we have $m - a_m$ as a lower bound for $k - a_k$ for k between m and the point of contact x_r. This bound can be used as a "squeeze" as k can be accepted without the evaluation of a probability if $m - a_m > k - x$.

A second way to reduce the necessary number of evaluations of the p_k is to add a table of arbitrary size that stores the right hand side of the acceptance condition for the values of k in an interval containing the mode. The important difference to the table-helped alias and inversion algorithms is that the table is not calculated in a setup. Instead each entry is computed during generation if a value is needed for the first time and stored for later use. Of course the speed-up of this variant depends on the size of the used help table and on the size of the main part of the distribution. For example, a table with 1000 floating point numbers and a table of 1000 logical variables that stores, whether a value was already computed, are taking only 9 Kb of memory on our machine and brings a real remarkable speed-up for most distributions of practical use. In the description of Algorithm 10.4 (Automatic-Rejection-Inversion) we have omitted the table and the squeeze for the sake of simplicity; both can be found in Hörmann and Derflinger (1997) and in the UNU.RAN source code.

10.3 Short Algorithms Based on General Inequalities

In the previous sections and Sects. 3.1.2 and 3.2 we have discussed (very) fast black-box algorithms for sampling from discrete distributions at the prize of some (expensive) setup. Now, analogously to the continuous case presented in Chap. 6, we are interested in short algorithms that have hardly any setup at the expense of slower marginal generation times. For this purpose we use general inequalities that hold for all probability functions of a given (large) class of discrete distributions.

10.3.1 Unimodal Probability Function with Known Second Moment

Assume that we are given the probability function p_k for a discrete unimodal distribution and the following three quantities:

1. m, the location of the mode.
2. M, an upper bound for the p_k, i.e. $M \geq p_m$.
3. s^2, an upper bound for the second moment about m. Note that if the variance σ^2 and the mean μ are known, then we have $s^2 = \sigma^2 + (m-\mu)^2$.

Then Devroye (1986a, p.494) shows the following general inequality.

Theorem 10.4. *For all discrete unimodal distributions,*

$$p_k \leq \min\left(M, \frac{3\,s^2}{|k-m|^3}\right).$$

In addition for all $x \in [k-\frac{1}{2}, k+\frac{1}{2})$,

$$p_k \leq g(x) = \min\left(M, \frac{3\,s^2}{(|x-m|-\frac{1}{2})_+^3}\right).$$

Furthermore,

$$\int_{-\infty}^{\infty} g(x)\,\mathrm{d}x = M + 3\,(3\,s^2)^{\frac{1}{3}}\,M^{\frac{2}{3}} = M + 3\,\theta\,.$$

Proof. For any $k > m$ we find

$$s^2 = \sum_{i=-\infty}^{\infty} (i-m)^2\,p_i \geq \sum_{k \geq i \geq m} (i-m)^2\,p_k \geq p_k \int_m^k (u-m)^2\,\mathrm{d}u = p_k\,(k-m)^3/3\,,$$

and the first inequality follows. For the second bound notice that for $k > m$, we have $x \geq k - \frac{1}{2}$ and $(k-m)^3 \geq (x-\frac{1}{2}-m)^3$ and thus $1/(k-m)^3 \leq 1/((x-m) - \frac{1}{2})^3$. $((y)_+ = \max(0, y)$ is used to make the denominator nonnegative). Finally, to compute the area below the bounding curve g define $\theta = (3\,s^2)^{\frac{1}{3}}\,M^{\frac{2}{3}}$. Notice that the term M is the minimum term on the interval $[m - \frac{1}{2} - \frac{\theta}{M}, m + \frac{1}{2} + \frac{\theta}{M}]$ and the area under this center part is thus $M + 2\,\theta$. Integrating the two tails of g gives θ. □

Algorithm 10.5 (Discrete-Unimodal-2ndMoment) is based on this theorem. It uses the upper bound $g(x)$ as continuous hat function for the histogram function $f(x) = p_{\mathrm{round}(x)}$. The generated (continuous) random variate X has then to be rounded to the next integer. To verify this algorithm notice that the right tail has domain $[m + \frac{1}{2} + \theta/M, \infty)$ and that the area below equals $\theta/2$. Moreover, we find $\int_{\frac{1}{2} + \theta/M}^{x} 3\,s^2/(x-\frac{1}{2})^3\,\mathrm{d}t = \frac{\theta}{2} - \frac{6\,s^2}{(2\,x-1)^2}$. Thus for a uniform random number $U \sim U(0, \theta/2)$ we get X via inversion by $X = \frac{1}{2} + \sqrt{\frac{3\,s^2}{\theta - 2\,U}}$ with simplifies to $X = \frac{1}{2} + \sqrt{3\,s/W}$ where $W = \theta - 2\,U \sim U(0, \theta)$.

The expected number of iteration in Algorithm 10.5 is given by $\mathrm{E}(I) = M + 3\,\theta$. Thus, the number of iterations is uniformly bounded over the class of unimodal discrete distributions with uniformly bounded $(1+s)\,M$.

10.3.2 *T*-Concave Distributions

For the class of log-concave distributions we can obtain simpler and better fitting upper bounds for the probabilities. The following theorem is due to Devroye (1987).

Algorithm 10.5 Discrete-Unimodal-2ndMoment

Require: Probability function p_k, mode m, upper bound M for p_m,
upper bound s^2 for second moment about m;
if expectation and variance are known use $s^2 = \sigma^2 + (m-\mu)^2$.
Output: Random variate X with given probability function.
  /* Setup */
1: Compute $\theta \leftarrow (3\,s^2)^{\frac{1}{3}} M^{\frac{2}{3}}$ and $u_0 \leftarrow (M+3\,\theta)/2$.
  /* Generator */
2: **loop**
3:   Generate $U \sim U(-u_0, u_0)$ and $V \sim U(0,1)$.
4:   **if** $|U| \leq u_0 - \theta/2$ **then** /* center part */
5:     $X \leftarrow m + U/M$.
6:     $Y \leftarrow V M$.
7:   **else** /* tails */
8:     $W \leftarrow 2(|U| - u_0) + \theta$. /* $W \sim U(0, \theta)$ */
9:     $X \leftarrow m + (\frac{1}{2} + \sqrt{3\,s^2/W})\,\text{sign}(U)$
10:    $Y \leftarrow V\,3\,s^2/(|X-m| - \frac{1}{2})^3$.
11:    $X \leftarrow \text{round}(X)$.
12:    **if** $Y \leq p_X$ **then** /* accept */
13:      **return** X.

Theorem 10.5. *For any discrete log-concave distribution with a mode at m and probabilities p_k, we have*

$$p_{m+k} \leq p_m \min(1, e^{1-p_m|k|}), \quad \text{for all } k.$$

It is easy to design a black-box algorithm that utilizes this theorem (Exercise 10.2). The expected number of iterations for such a generator is $4 + p_m$. Devroye (1987) also gives some hints how this number can be decreased to $2 + p_m$ in special cases.

Even shorter and faster algorithms are possible if we adopt the ratio-of-uniforms method to discrete distributions. This idea has been suggested by Stadlober (1989a). Leydold (2001) has shown that it is also well suited for black-box algorithms for $T_{-1/2}$-concave distributions using the approach of Sect. 6.3.1 for the histogram function $f_p(x) = p_{\lfloor x \rfloor}$. Then it is again possible to compute bounds for a bounding rectangle for the region of acceptance $\mathcal{A}_f$, see Fig. 10.3. However, the histogram function of a $T_{-1/2}$-concave discrete distribution is not $T_{-1/2}$-concave any more but has "spikes" and thus there remain some tedious technical problems. The interested reader is referred to the original paper for more details.

Theorem 10.6. *Let p_k be a $T_{-1/2}$-concave quasi-probability function of a discrete distribution with mode m. Let F denote the cdf of the distribution and let $\mathcal{R}_d = \mathcal{R}_d^- \cup \mathcal{R}_d^+$ with*

$$\mathcal{R}_d^- = \{(v,u)\colon -F(m-1)\sum p_i/\sqrt{p_{m-1}} \leq v \leq 0, 0 \leq u \leq \sqrt{p_{m-1}}\},$$
$$\mathcal{R}_d^+ = \{(v,u)\colon 0 \leq v \leq (1-F(m-1))\sum p_i/\sqrt{p_m}, 0 \leq u \leq \sqrt{p_m}\},$$

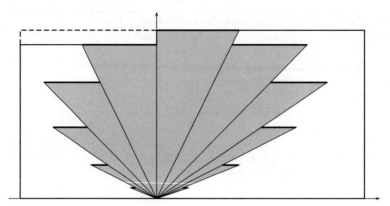

Fig. 10.3. Region of acceptance $\mathcal{A}_f$ and universal bounding rectangle for the histogram function of the Binomial(10,0.5) distribution. The dashed line indicates the enlargement of the bounding region for the simplified Algorithm 10.7 (Discrete-Simple-RoU-2)

and $\mathcal{Q}_d = \mathcal{Q}_d^- \cup \mathcal{Q}_d^+$ with

$$\mathcal{Q}_d^- = \{(v,u): -\sum p_i/\sqrt{p_{m-1}} \leq v \leq 0, 0 \leq u \leq \sqrt{p_{m-1}}\},$$
$$\mathcal{Q}_d^+ = \{(v,u): 0 \leq v \leq \sum p_i/\sqrt{p_m}, 0 \leq u \leq \sqrt{p_m}\}.$$

Then $\mathcal{A} \subset \mathcal{R}_d \subset \mathcal{Q}_d$ and

$$|\mathcal{R}_d| = 2|\mathcal{A}| \quad \text{and} \quad |\mathcal{Q}_d| = 4|\mathcal{A}|.$$

We set $\mathcal{R}_d^- = \mathcal{Q}_d^- = \emptyset$ whenever $p_{m-1} = 0$.

Proof. See Leydold (2001). □

Algorithm 10.6 (Discrete-Simple-RoU) is based on this theorem. It also works with any multiple of the probability function, i.e. $S = \sum p_k$ need not be one. The expected number of iterations is 2 when $F(m-1)$ is known and $2(2 - p_m/S)$ otherwise. Algorithm 10.7 (Discrete-Simple-RoU-2) shows a simplified variant of the first algorithm. It has a faster setup, since p_{m-1} has not to be computed, at the expense of slightly more iterations. It uses a larger bounding rectangle for the region of acceptance, see the dashed line in Fig. 10.3.

Remark. It is also possible to design an algorithm for T_c-concave distributions by means of the generalized ratio-of-uniforms method from Sect. 6.4.1 on p. 142. Details for such an algorithm can be found in Leydold (2003).

10.3 Short Algorithms Based on General Inequalities

Algorithm 10.6 Discrete-Simple-RoU (Discrete simple Ratio-of-Uniforms method)

Require: $T_{-1/2}$-concave quasi-probability function p_k, sum $S = \sum p_k$, mode m;
  optional: cdf at mode $F(m-1)$.
Output: Random variate X with given probability function.
  /∗ Setup ∗/
1: Set $v_l \leftarrow \sqrt{p_{m-1}}$, and $v_r \leftarrow \sqrt{p_m}$.
2: **if** $F(m-1)$ is provided **then**
3:    Set $a_l \leftarrow -F(m-1)\,S$, and $a_r \leftarrow (1-F(m-1))\,S$.
4: **else**
5:    Set $a_l \leftarrow -S - p_m$, and $a_r \leftarrow S$.
  /∗ Generator ∗/
6: **repeat**
7:    Generate $U \sim U(a_l, a_r)$.
8:    **if** $U < 0$ **then** /∗ left of mode ∗/
9:      Generate $V \sim U(0, v_l)$.
10:     Set $U \leftarrow U/v_l$.
11:    **else** /∗ right of mode ∗/
12:      Generate $V \sim U(0, v_r)$.
13:     Set $U \leftarrow U/v_r$.
14:    $X \leftarrow \lfloor U/V \rfloor + m$.
15: **until** $V^2 \le p_X$.
16: **return** X.

Algorithm 10.7 Discrete-Simple-RoU-2 (Discrete simple Ratio-of-Uniforms method)

Require: $T_{-1/2}$-concave quasi-probability function p_k, sum $S = \sum p_k$, mode m;
  optional: cdf at mode $F(m-1)$.
Output: Random variate X with given probability function.
  /∗ Setup ∗/
1: Store p_m and set $v_m \leftarrow \sqrt{p_m}$.
2: **if** $F(m-1)$ is provided **then**
3:    Set $u_l \leftarrow -F(m-1)\,S/v_m$, and $u_r \leftarrow (1-F(m-1))\,S/v_m$.
4: **else**
5:    Set $u_l \leftarrow -S/v_m$, and $u_r \leftarrow S/v_m$.
  /∗ Generator ∗/
6: **repeat**
7:    Generate $U \sim U(u_l, u_r)$ and $V \sim U(0, v_m)$.
8:    $X \leftarrow \lfloor U/V \rfloor + m$.
9: **until** $V^2 \le p_X$.
10: **return** X.

10.4 Distributions Where the Probabilities Are Not Known Explicitly

10.4.1 Known Probability Generating Function

Let X be a nonnegative integer-valued random variable with probability vector p_i. Then the *probability generating function* (or simply generating function) is defined as

$$k(s) = E(s^X) = \sum_{i=0}^{\infty} p_i s^i, \quad \text{for } 0 \leq s \leq 1 \;.$$

For most of the classical discrete distributions the formula for the generating function is less complicated than the formula for the probabilities p_i. Table 10.2 gives the most prominent examples.

Table 10.2. Generating Functions for some discrete distributions

Distribution	Generating Function	p_i	Domain
Binomial(n, p)	$(1 - p + ps)^n$	$\binom{n}{i} p^i (1-p)^{n-i}$	$0 \leq i \leq n$
Geometric(p)	$\frac{p}{1-(1-p)s}$	$p(1-p)^i$	$i \geq 0$
Neg. Binomial(n, p)	$\left(\frac{p}{1-(1-p)s}\right)^n$	$\binom{n+i-1}{i} p^n (1-p)^i$	$i \geq 0$
Poisson(λ)	$e^{-\lambda + \lambda s}$	$\frac{\lambda^i e^{-\lambda}}{i!}$	$i \geq 0$

To construct a generation algorithm (following Devroye, 1991) we will use a rejection method that is close to the series method for continuous random variables (see Sect. 9.2.2). So we have to find bounds for the probabilities p_i. The first step into the correct direction is the well known formula

$$k^{(r)}(0) = r! \, p_r \;.$$

But if we have k just given as a black-box we cannot compute all its derivatives. So we have to find approximations using differences. Define the *difference of order r* by

$$\Delta_t^r f(x) = \sum_{i=0}^{r} (-1)^{r-i} \binom{r}{i} f(x + i\,t) \;,$$

which can be computed in time proportional to r using the recursion for the binomial coefficients. Then the following bounds are proven by Devroye (1991),

$$\frac{\Delta_t^r k(0)}{r!\, t^r} \geq p_r \geq \frac{\Delta_t^r k(0)}{r!\, t^r} - \left((1-rt)^{-(r+1)} - 1\right) \sup_{i>r} p_i \;. \tag{10.3}$$

10.4 Distributions Where the Probabilities Are Not Known Explicitly

The evaluation times for these bounds are proportional to r. They are tighter for small t. On the other hand, however, for small values of t the evaluation of these differences is numerically unstable due to extinction even for moderate values of r. For example, for the Binomial distribution with $n = 100$ and $p = 0.5$, it is not possible to compute the bounds for p_{50} with $t = 10^{-4}$ when the working precision of the floating point arithmetic does not exceed 100 decimal digits. Therefore the algorithm presented in this section requires arbitrary precision rational arithmetic to evaluate the differences. Devroye (1991) discusses the choice of t in connection with the time complexity of the generation algorithm and suggests to take $t = \epsilon/(2r(r+1)^4)$, where $\epsilon > 0$ is some appropriate constant, and to half it if the precision is not enough to decide between acceptance and rejection. It is also possible to use auxiliary tables that store the bounds whenever they are computed during generation. This can speed up the generation of larger samples tremendously.

For a rejection algorithms we also need a dominating distribution. Directly from the definition of the generating function $k(s) = \sum_{i=0}^{\infty} p_i s^i$ and the fact that all summands of this series are positive we get the trivial bounds

$$k(s) \leq p_i s^i, \quad \text{and thus} \quad p_i \leq \frac{k(s)}{s^i}.$$

We can use these bounds to construct a dominating distribution if the probability function p_i has geometric or sub-geometric tails. Devroye (1991) suggests to use a single geometric dominating distribution and to optimize s such that the rejection constant α is minimized. Applying this suggestion to the Binomial distribution with $n = 100$ and $p = 0.5$ results in s close to 1.02 and $\alpha > 137$. This is disappointing even for a black-box algorithm. Thus we suggest to take the dominating distribution

$$q_i = \min\left(\frac{k(s_1)}{s_1^i}, 1, \frac{k(s_2)}{s_2^i}\right)$$

with $s_1 < 1$ and $s_2 > 1$. It is piecewise geometric and can be generated as a mixture of three geometric distributions. As it is log-concave as well it can also be generated efficiently with the methods of Sect. 10.2. Some experimentation with the distributions of Table 10.2 showed that with s_1 around 0.9 and $s_2 = 1/s_1$ we can expect rejection constants α between 10 and 20 which is still very big but ten times better than using just a single geometric hat.

We have collected the necessary details in Algorithm 10.8 (Generating-Function). Using arguments similar to those of Devroye (1991) it is possible to show that the expected execution time for the Algorithm is bounded. This paper also shows inequalities for deriving a hat function when the tails of the distribution are higher than exponential, as well as an inversion method that only uses the generating function.

10.4.2 Known Moment Sequence

The factorial moment sequence is defined as

Algorithm 10.8 Generating-Function

Require: Generating function $k(s)$ of a discrete distribution with exponential or sub-exponential tails.
The algorithm requires arbitrary precision rational arithmetic.
Output: Discrete distribution with given generating function $k(s)$.
/* setup */
1: Find constants $s_1 < 1$ and $s_2 > 1$ that minimize the area below the dominating distribution $q_i = \min\left(\frac{k(s_1)}{s_1^i}, 1, \frac{k(s_2)}{s_2^i}\right)$.
   Taking $s_1 = 0.9$ and $s_2 = 1.1$ is always possible.
2: Prepare table (M_i, t_i) for storing upper bound M_i for t_i in (10.3).
/* Generator */
3: **loop**
4:    Generate discrete variate X with probabilities prop. to q_i.
      Use Algorithm 10.4 (Automatic-Rejection-Inversion).
5:    Set $q_X \leftarrow \min\left(\frac{k(s_1)}{s_1^X}, 1, \frac{k(s_2)}{s_2^X}\right)$.
6:    Generate $V \sim U(0, q_X)$.
7:    **if** table entry at X present **then**
8:      Set $t \leftarrow t_X$.
9:      Set $M \leftarrow M_X$.
10:    **else**
11:      Set $t \leftarrow 0.1/(X+1)^5$.
12:      Set $M \leftarrow \Delta_t^X k(0) = \sum_{j=0}^{X}(-1)^{X-j}\binom{X}{j}k(jt)$.
13:      Update table $(M_X, t_X) \leftarrow (M, t)$.
14:    **repeat**
15:      **if** $V \leq S - (1 - Xt)^{-(X+1)} + 1$ **then**
16:        **return** X.
17:      Set $t \leftarrow t/2$.
18:      Set $M \leftarrow \Delta_t^X k(0) = \sum_{j=0}^{X}(-1)^{X-j}\binom{X}{j}k(jt)$.
19:      Update table $(M_X, t_X) \leftarrow (M, t)$.
20:    **until** $V > M$

$$M_i = \mathrm{E}(X(X-1)\ldots(X-i+1)) \qquad \text{for } i \geq 1\,.$$

As the factorial moment sequence M_i is linearly related to the moment sequence μ_i it is enough to give an algorithm for known factorial moment sequence. Based on the alternating series representation of the probabilities

$$p_j = \frac{1}{j!}\sum_{i=0}^{n}(-1)^i\frac{M_{j+i}}{i!}\,,$$

Devroye (1991) develops an algorithm for discrete distributions with known factorial moment sequence, which is in the spirit of the alternating series method for continuous distributions (see Sect. 9.2.2). As the (factorial) moment sequence does not always uniquely define the distribution this algorithm is interesting due to its relation to the famous moment problem but has limited importance for applications. For the details we therefore refer to the original paper.

10.4 Distributions Where the Probabilities Are Not Known Explicitly

10.4.3 Known Discrete Hazard Rate

The hazard rate for a discrete distribution with probability function p_k, $k = 0, 1, 2, \ldots$, is defined as

$$h_k = \frac{p_k}{\sum_{j=k}^{\infty} p_j} \, .$$

Using this definition we easily get

$$\sum_{j=k}^{\infty} p_j = \sum_{j=k-1}^{\infty} p_j - p_{k-1} = (1 - h_{k-1}) \sum_{j=k-1}^{\infty} p_j \, .$$

By induction this implies

$$p_k = h_k \sum_{j=k}^{\infty} p_j = h_k \prod_{j<k} (1 - h_j) \, .$$

So we can use the simple recursion

$$p_0 = h_0, \quad \text{and} \quad p_{k+1} = p_k \frac{h_{k+1}}{h_k} (1 - h_k)$$

to compute the probabilities if we know the discrete hazard rate. Thus inversion by sequential search starting from 0 can be used to generate from a discrete distribution with known hazard rate. Algorithm 10.9 (Discrete-Hazard-Rate-Inversion) contains the details. It is a real black-box algorithm as it works for arbitrary discrete distributions with known hazard rate. On the other hand we know from Sect. 3.1.1 that the expected number of iterations in the loop is $E(X)$ which can be very large or even infinity.

Algorithm 10.9 Discrete-Hazard-Rate-Inversion

Require: Hazard rate h_i of a discrete distribution for $i = 0, 1, 2, \ldots$.
Output: Discrete random variate X with hazard rate h_i.
1: Generate $U \sim U(0,1)$.
2: Set $X \leftarrow 0$, $p \leftarrow h_0$, $P \leftarrow p$.
3: **while** $U > P$ **do**
4:     Set $p \leftarrow p\, h_{X+1}(1 - h_X)/h_X$.
5:     Set $P \leftarrow P + p$ and $X \leftarrow X + 1$.
6: **return** X.

Shanthikumar (1985) (see also Devroye, 1986a, Sect. VI.3) suggested generating from a discrete hazard rate by sequentially testing if the random variate takes the values $0, 1, 2 \ldots$, see Algorithm 10.10 (Discrete-Hazard-Rate-Sequential). The expected number of iterations is $E(X)$ as for Algorithm 10.9 but it generates a uniform random variate inside the loop and is therefore not competitive. Shanthikumar (1985) suggests two improvements of Algorithm 10.10 that can be seen as discrete versions of the thinning and the dynamic thinning method (see Sect. 9.1 and Exercise 10.3). But as these improvements also use one uniform random number in every iteration of the loop and as they are less general than Algorithm 10.9, we do not include the details here.

Algorithm 10.10 Discrete-Hazard-Rate-Sequential

Require: Hazard rate h_i of a discrete distribution for $i = 0, 1, 2, \ldots$.
Output: Discrete random variate X with hazard rate h_i.
1: Set $X \leftarrow 0$.
2: **repeat**
3:     Generate $U \sim U(0, 1)$.
4:     Set $X \leftarrow X + 1$.
5: **until** $U < h_X$
6: **return** X.

10.5 Computational Experience

For simulations with large or very large sample sizes the fast generation of discrete standard distributions like e.g. the Binomial, Poisson and Hypergeometric distribution are of practical importance. If the distribution remains constant the setup time is not critical and thus the alias(-urn) method and the inversion with guide table are the first choice. But for discrete distributions with unbounded domain the use of the alias(-urn) method implies the necessity to cut off the tails. Assuming that the resolution of the uniform random number generator is known and called ϵ it cannot be denied that cutting off e.g. a tail with probability $\epsilon/100$ has no influence on the generated sample. The computation of a save cut-off point for a right tail is simple if the probability vector is normed to have sum one; however, we can encounter problems with rounding errors due to extinction, especially if we compute the cumulative probability function by simple summing probabilities. For most common standard distributions we can use the log-concavity property to obtain simple bounds for the tail probabilities. For the guide table method we can either cut off the tails or use the variants presented as Algorithms 10.1 or 10.2.

Our timing tests are conducted in the same manner as in Sect. 8.2 using our UNU.RAN library. Again the tables below show relative average generation times (i.e. the generation times divided by the time necessary to generate one exponential variate by inversion). Table 10.3 lists the distributions that have been used for our experiment. It includes distributions with random vectors as well as some standard distributions different domains. These distributions are given using their probability mass functions. The algorithms are listed

Table 10.3. Distributions used for timing tests

Symbol	Distribution
PV_1	random probability vector of length 100
PV_2	random probability vector of length 1000
Bin	binomial ($n = 50$, $p = 0.5$)
P_1	poisson ($\mu = 50$)
P_2	poisson ($\mu = 500$)
H_1	hypergeometric ($N = 5000$, $M = 500$, $n = 50$)
H_2	hypergeometric ($N = 500000$, $M = 50000$, $n = 5000$)
L_1	logarithmic ($\alpha = 0.01$)
L_2	logarithmic($\alpha = 0.99$)

Table 10.4. Algorithms used for timing tests

Symbol	Algorithm
SS	3.1 (Sequential-Search)
GTs	3.2 (Indexed-Search), table size $C = L/10$
GTe	3.2 (Indexed-Search), table size $C = L$
GTl	3.2 (Indexed-Search), table size $C = 4L$
AM	3.3 (Alias-Sample) (table size $C = L$)
AU	3.5 (Alias-Urn), table size $C = 4L$
ARIs	10.4 (Automatic-Rejection-Inversion), squeeze, no table
ARIt	10.4 (Automatic-Rejection-Inversion), no squeeze, table size = 100
ARIst	10.4 (Automatic-Rejection-Inversion), squeeze, table size = 100
ARIsl	10.4 (Automatic-Rejection-Inversion), squeeze, table size = 1000
DROU	10.6 (Discrete-Simple-RoU)
DROUm	10.6 (Discrete-Simple-RoU) with given cdf at mode

in Table 10.4. Notice, that the algorithms based on indexed search cut off the tails of distributions with unbounded domains. Results are displayed in Table 10.5.

Sequential search does not need a setup and thus the average generation time does not depend on the sample size. (The slower generation time for the changing parameter case is due to the fact that in practice memory has to be allocated and data have to be copied. This can be reduced by writing a sampling routine for this special case.) Its performance heavily depends on the shape of the distribution (i.e. on the expected value) and on the expense to evaluate their pmf. It is generally much slower than the other methods.

Table based methods (index search and alias method) require a setup. As one can clearly see from distributions PV_1 and PV_2 the required time depends on the length of the probability vector. For large sample sizes these methods are extremely fast, independent of the distribution (for sample size $n = 10^6$

Table 10.5. Relative average generation times

	PV_1	PV_2	Bin	P_1	P_2	H_1	H_2	L_1	L_2
\multicolumn{10}{c}{sample size $n = 1$ (changing parameter case)}
SS	10.25	27.30	140.00	132.00	1180.00	39.30	2480.00	10.30	13.70
GTs	18.80	94.00	265.00	251.00	1540.00	266.50	26700.00	17.10	1710.00
GTe	23.90	145.00	267.00	255.00	1560.00	268.20	27700.00	17.10	1990.00
GTl	37.60	270.00	273.00	263.00	1620.00	273.40	31500.00	18.80	2980.00
AM	32.50	232.00	272.00	261.00	1600.00	273.40	27000.00	18.80	1790.00
AU	66.60	576.00	292.00	292.00	1800.00	292.20	33000.00	18.80	2250.00
ARIs	–	–	63.20	39.30	39.30	63.22	66.60	18.80	17.10
ARIt	–	–	70.00	44.40	42.70	70.05	70.00	20.50	20.50
ARIst	–	–	64.90	42.70	42.70	64.93	64.90	22.20	20.50
ARIsl	–	–	64.90	63.20	61.50	64.93	88.80	41.00	41.00
DROU	–	–	35.90	27.30	27.30	34.17	39.30	13.70	13.70
DROUm	–	–	29.10	20.50	20.50	29.05	29.10	13.70	13.70
\multicolumn{10}{c}{sample size $n = 100$}
SS	1.50	9.33	132.00	123.00	1170.00	30.10	2470.00	1.15	20.13
GTs	0.92	1.67	3.38	3.23	16.10	3.37	267.00	0.80	18.15
GTe	0.91	2.10	3.31	3.19	16.20	3.31	277.00	0.80	20.55
GTl	1.01	3.33	3.37	3.26	16.90	3.37	319.00	0.80	30.45
AM	0.99	2.99	3.38	3.28	16.60	3.38	271.00	0.84	18.50
AU	1.30	6.39	3.55	3.55	18.60	3.54	330.00	0.82	23.14
ARIs	–	–	3.81	3.09	3.28	3.30	5.26	3.26	1.93
ARIt	–	–	3.16	2.68	3.59	2.53	5.69	1.84	1.71
ARIst	–	–	2.85	2.50	3.02	2.25	4.61	1.95	1.69
ARIsl	–	–	2.85	2.72	3.23	2.25	4.75	2.14	1.90
DROU	–	–	21.60	14.23	14.70	19.40	24.80	4.24	4.70
DROUm	–	–	12.30	7.67	7.64	12.30	12.70	4.27	4.68
\multicolumn{10}{c}{sample size $n = 10\,000$}
SS	1.43	9.18	134.00	125.00	1180.00	30.00	2530.00	1.08	19.80
GTs	0.74	0.75	0.76	0.76	0.91	0.70	3.34	0.64	0.84
GTe	0.67	0.69	0.68	0.69	0.82	0.67	3.42	0.64	0.84
GTl	0.65	0.68	0.67	0.67	0.80	0.66	3.83	0.64	0.94
AM	0.68	0.70	0.69	0.69	0.81	0.68	3.38	0.66	0.83
AU	0.65	0.70	0.67	0.67	0.82	0.67	4.02	0.64	0.86
ARIs	–	–	3.21	2.75	2.95	2.70	4.68	3.13	1.79
ARIt	–	–	1.15	1.28	1.66	1.07	2.01	1.65	1.17
ARIst	–	–	1.19	1.33	1.73	1.09	2.07	1.75	1.20
ARIsl	–	–	1.19	1.30	1.45	1.09	1.59	1.75	1.17
DROU	–	–	21.40	14.20	14.70	19.00	25.00	4.15	4.60
DROUm	–	–	12.20	7.55	7.53	12.30	12.70	4.16	4.60

there is no difference at all). Even the choice of the design parameters (size of the guide table, or size of the elongated probability vector for the alias-urn method) has very little influence on the marginal speed but increases the setup time. The speed of the indexed search and of the alias method are practically the same (but contrary to some statements in the literature GT seems to be slightly faster than AM). However, when such methods are used for distributions with (expensive) probability mass functions the setup is very slow. This is due to the fact that a finite probability vector has to be generated first which requires the evaluation of the pmf at every point. Using recursion formulas can reduce this price of course.

The speed of automatic rejection inversion (ARI) strongly depends on the generated distribution and is also influenced by the design parameters. If we use a large auxiliary table the marginal speed is not bad, not more than about twice the marginal speed of GT and AU. Only for the logarithmic distribution with $\alpha = 0.01$ ARI is significantly slower; for this distribution more than 99% of the mass is in 1 but the automatic choice of the design point right of the mode leads to a poor fit of the hat with rejection constant $\alpha = 1.99$, thus explaining the slow speed of ARI in this special case. For most distributions (especially those with large domains) the setup requires only a very small fraction necessary for the table methods. If the domain of the generated distribution is big this may lead to clear advantages for ARI. Even more important is the fact it can be used to generate from discrete distributions with heavy tails. In this case GT and AU cannot be used as the table sizes and the setup times explode.

The last method (DROU) is comparably fast in the changing parameter case since the setup is very short. However, its rejection constant is 2 and 4, respectively, and thus its marginal generation time strongly depends on the distribution.

10.6 Summary

As we have collected quite a number of different automatic algorithms for discrete distributions it is probably in place to give a short overview of these methods. So we list here the most important algorithm groups and their requirements, advantages and disadvantages.

- **Sequential search method**
  Algorithm 3.1 (Sequential-Search)

  Applicable to: Discrete distributions.
  Required: probability vector or probability mass function.
  Setup: none.
  Speed: slow, depending on $E(X)$.
  Tables: not required.
  Code: short.

- **Indexed search (guide table method)**
  Algorithms 3.2 (Indexed-Search), 10.1 (Indexed-Search-Tail) and 10.2 (Indexed-Search-2-Tails)

  Applicable to: Discrete distributions.
  Required: probability vector or probability mass function.
  Control parameter: size of guide table.
  Setup: (very) slow, depending on size of guide table.
  Speed: very fast.
  Tables: large, depending on the size of the domain of the distribution.
  Code: moderate.
  Special advantages: There are special versions available that work for unbounded domain.

- **Alias method**
  Algorithm 3.3 (Alias-Sample) and 3.4 (Alias-Setup); Algorithm 3.5 (Alias-Urn)

  Applicable to: Discrete distributions with bounded domain.
  Required: probability vector.
  Control parameter: size of alias table.
  Setup: (very) slow, depending on size of alias table.
  Speed: very fast.
  Tables: large, depending on the size of the domain of the distribution.
  Code: long.

- **Automatic rejection inversion**
  Algorithm 10.4 (Automatic-Rejection-Inversion)

  Applicable to: T-concave discrete distributions.
  Required: quasi-probability mass function, mode and approximate sum of the probabilities.
  Control parameter: size of auxiliary table.
  Setup: moderate.
  Speed: moderate.
  Tables: optional.
  Code: long.
  Special advantages: Setup time not influenced by the size of the domain.

- **Rejection from general hat**
  Algorithm 10.5 (Discrete-Unimodal-2ndMoment)

  Applicable to: Unimodal discrete distributions with bounded moments.
  Required: probability mass function, mode and upper bound for a moment μ_r for arbitrary $r > 0$.
  Setup: short.
  Speed: slow.
  Tables: not required.
  Code: short.

- **Discrete simple ratio-of-uniforms**
  Algorithms 10.6 (Discrete-Simple-RoU) and 10.7 (Discrete-Simple-RoU-2)

  Applicable to: T-concave discrete distributions
  Required: probability mass function and mode;
          optional: value of the cdf at the mode.
  Setup: short.
  Speed: slow.
  Tables: not required.
  Code: short.

As these methods work for different groups of distributions there is no single best method. If large samples of a fixed distribution with small or moderate domain are required the alias method or indexed search should be used. We prefer indexed search as it is an inversion method and also allows for variants that can cope with unbounded tails.

If the domain of a discrete distribution is very large, automatic rejection inversion is the best choice. In the case that the parameters of the distribution are constantly changing discrete simple ratio-of-uniforms or rejection from a general hat should be the best choice, automatic rejection inversion can be used as well.

10.7 Exercises

Exercise 10.1. Suppose we are given a discrete distribution with probabilities p_i, $i \geq 0$, which has two long tails; we have the possibility to evaluate the cumulative probabilities, $P_i = \sum_{j=0}^{i} p_j$, without computing all p_i's, and we know a value i not too far away from the mode.
Design an algorithm that, in analogy to Algorithm 10.2, uses the guide table method for the main part of the distribution, but use sequential search for both tails extra; to the left if $U < 1/C$ and to the right if $U > 1/C$.

Exercise 10.2. Write down the details for a black-box algorithm for discrete log-concave distributions that is based on Thm. 10.5. Try to improve your algorithm when the cdf at the mode m is known.
Hint: Take a look at Thm. 6.7 on p. 135.

Exercise 10.3. Assume that a discrete hazard rate h_i with $h_i < \theta$ is known. Discuss how to speed up Algorithm 10.10 by using a geometric(θ) random variate to jump ahead more than 1 (Shanthikumar, 1985). The resulting algorithm can be seen as a discrete version of the thinning Algorithm 9.3.

Part IV

Random Vectors

11

Multivariate Distributions

It is a commonplace for all fields of computational mathematics that in the multivariate situation everything becomes much more difficult. But on the other hand tools for handling multivariate distributions are of greatest importance as one of the most common modeling errors is to overlook dependencies between input variables of a stochastic system. This error can make simulation results totally useless. Thus it is important to have generation procedures for random vectors available. There are few commonly accepted standard distributions for random vectors; even for these distributions generation procedures are difficult to find. Nevertheless, there are quite a few papers discussing the design of multivariate families that are well suited for random vector generation. For many of these families the marginals are known as well. The monograph of Johnson (1987) is presenting this branch of multivariate simulation. You can also find many of these distributions in Devroye (1986a, Chap. XI).

We start this chapter with a section on general principles for random vector generation; we will need these methods for the universal generators presented below. We continue with Sect. 11.2 that discusses the generation of uniformly distributed random vectors over (simply shaped) bounded domains. These two sections give a first impression of the difficulties we have to overcome to generate random vectors. It seems to be due to these problems that the first automatic algorithms for multivariate distributions were developed only recently although their potential practical value is big.

Section 11.3 introduces such automatic random vector generators for log-concave distributions based on transformed density rejection as presented by Leydold and Hörmann (1998), Leydold (1998), or Hörmann (2000). Section 11.4 presents short universal methods for orthounimodal densities as suggested by Devroye (1997a).

It could be argued that the problem of generating random vectors with given marginal distributions and correlation structure would fit into this chapter as well. However, this problem is not well defined and thus has no unique solution. Therefore the generation of random vectors has to be combined with modeling considerations and we thus defer it to Sect. 12.5.

In Sect. 11.5 we shortly compare the performance of several of the algorithms we have presented before. Sect. 11.6 presents algorithms for multivariate discrete distributions.

11.1 General Principles for Generating Random Vectors

The inversion method cannot be generalized to higher-dimensions directly, but in some sense the *conditional distribution method* can be seen as its substitute for $d > 1$. Unfortunately we have to know all marginal distributions which makes the conditional distribution method not applicable for most multivariate distributions. The *rejection method* and the *ratio-of-uniforms method* can be directly generalized to higher dimensions. Their principles remain the same but we have to overcome many new difficulties when designing random vector generators, among them the famous "curse of dimensionality" that implies that the acceptance probability deteriorates quickly with rising dimension.

At the end of this section we also include some methods for sampling from the multinormal distribution. It is the most important multivariate distribution and thus we present these algorithms for their own sake. They serve also as examples how the general principles described in this section can be applied.

11.1.1 The Conditional Distribution Method

The conditional distribution method allows us to reduce the multivariate generation problem to d univariate generation problems. It can only be used if we know all $d-1$ *marginal distributions* $f_{1,2,\ldots,i}(x_1, x_2, \ldots, x_i)$ of the first i coordinates $(X_1, X_2, \ldots, X_i)$, for $i < d$. In this case we can write the joint density for the random vector $\mathbf{X} = (X_1, \ldots, X_d)$ as

$$f(x_1, \ldots, x_d) = f_1(x_1)\, f_2(x_2|x_1)\, f_3(x_3|x_1, x_2) \cdots f_d(x_d|x_1, \ldots, x_{d-1})$$

where the *conditional distributions* are given by

$$f_i(x_i|x_1, \ldots, x_{i-1}) = \frac{f_{1,\ldots,i}(x_1, x_2, \ldots, x_i)}{f_{1,2,\ldots,i-1}(x_1, x_2, \ldots, x_{i-1})} .$$

Remark. Notice that unless $i = d$, $f_i(x_i|x_1, \ldots, x_{i-1})$ is not a *full conditional distribution* (i.e. a conditional distribution of the multivariate distribution itself) but a conditional distribution of the marginal distribution of $(X_1, \ldots, X_i)$. We can compute this conditional distribution easily if we know the respective marginal distributions by using the formula from above. So the problem we encounter in practice is that we have to find all the marginal distributions.

In Sect. 14.1.2 we will discuss the Gibbs sampler, a very popular Markov Chain Monte Carlo method to sample from multivariate distributions. The

11.1 General Principles for Generating Random Vectors

main practical advantage of the Gibbs sampler lies in the fact that it only uses full conditional distributions and no marginal distributions are required. But for this reason we loose the independence of the generated vectors.

For a generation method based on this representation of the joint density f we start to generate a random variate X_1 from the marginal distribution f_1 to get the first component of the random vector $\mathbf{X}$. For the second component we have to sample from the conditional distribution of X_2 given X_1 with density $f_2(x_2|X_1)$; for the third component we have to sample from the conditional distribution of X_3 given (X_1, X_2) with density $f_3(x_3|X_1, X_2)$ and so on. Algorithm 11.1 (Conditional-Distribution-Method) compiles the whole generation method.

Algorithm 11.1 Conditional-Distribution-Method

Require: Dimension d, conditional distributions of the marginals, i.e. $f_1(x_1)$, $f_2(x_2|x_1)$, ..., $f_d(x_d|x_1, \ldots, x_{d-1})$.
(Or equivalently: all marginal distributions $f_{1,2,\ldots,i}(x_1, x_2, \ldots, x_i)$, for $i < d$).
Output: Random vector $\mathbf{X}$.
1: Generate X_1 with density f_1.
2: **for** $i = 2$ to d **do**
3:   Generate X_i with density $f_i(x_i|X_1, \ldots, X_{i-1})$.
4: **return** vector $\mathbf{X} = (X_1, \ldots, X_d)$.

When using the conditional distribution method for generating from a random variate with given joint distribution function a main problem is that we have to obtain all necessary marginal densities by integration. A second problem in practice is to find generators to sample from the conditional densities as they will vary depending on the random variates generated. The universal algorithms presented in Chap. 4 and 5 are of course good candidates. However, in general we need just one random variate from one density and thus have to rerun the setup every time.

We finish our explanation of the conditional distribution method by presenting two simple examples (see also Exercises 11.1 and 11.4).

Example 11.1 (Uniform distribution over a triangle). We consider the uniform distribution over the triangle with vertices at origin, $(2,0)$, and $(2,1)$. The marginal density is $f_1(x_1) = x_1/2$ for $0 \leq x_1 \leq 2$. So we find for the cdf $F_1(x_1) = x_1^2/4$ and for the inversion method $F_1^{-1}(u) = 2\sqrt{u}$. The conditional density $f_2(x_2|x_1)$ is uniform on $(0, x_1/2)$. So we arrive at:
Output: Random pairs uniformly over the triangle with vertices $(0,0)$, $(2,0)$,$(2,1)$.
1: Generate $U \sim U(0,1)$.
2: Set $X_1 \leftarrow 2\sqrt{U}$.
3: Generate $V \sim U(0,1)$.
4: Set $X_2 \leftarrow V X_1/2$.
5: **return** vector $\mathbf{X} = (X_1, X_2)$.

The next example will be used in Sect. 11.3.3 below to generate from the hat function in two-dimensional TDR.

Example 11.2 (A two-dimensional exponential-uniform distribution). Assume the density $f(x_1, x_2) = e^{-x_1}$ for $x_1 > 0$ and $-x_1 \leq x_2 \leq x_1$. The marginal density is then given by $f_1(x_1) = x_1 \exp(-x_1)$. This is the density of the gamma(2) distribution that can be easily generated as the sum of two independent exponential variates. The conditional density $f_2(x_2|x_1)$ is uniform on $(-x_1, x_1)$. So we arrive at the Algorithm:

Output: Random pairs following the density f given above.
1: Generate $U_1 \sim U(0,1)$ and $U_2 \sim U(0,1)$.
2: $X_1 \leftarrow -\log(U_1 U_2)$.
3: Generate $V \sim U(0,1)$.
4: $X_2 \leftarrow -X_1 + 2 V X_1$.
5: **return** vector $\mathbf{X} = (X_1, X_2)$.

11.1.2 The Multivariate Rejection Method

The idea of multivariate rejection is exactly the same as for dimension one (Sect. 2.2). First we have to find a hat function $h(\mathbf{x})$ with bounded integral that is an upper bound for the (quasi-) density $f(\mathbf{x})$. For the generation step a random vector $\mathbf{X}$ having a density proportional to the hat function $h(\mathbf{x})$ is generated. A random variate V uniformly distributed on $(0, h(\mathbf{X}))$ is then used to decide between acceptance and rejection. If $V \leq f(\mathbf{X})$ the random vector is accepted, otherwise it is rejected. Algorithm 11.2 (Multivariate-Rejection) gives the formal statement. Like its one-dimensional analogue it also works properly if f is any multiple of a density function, i.e. a quasi-density.

Algorithm 11.2 Multivariate-Rejection

Require: (Quasi-) density $f(\mathbf{x})$; hat function $h(\mathbf{x})$ with $f(\mathbf{x}) \leq h(\mathbf{x})$.
Output: Random vector $\mathbf{X}$ with density prop. to f.
1: **loop**
2:     Generate $\mathbf{X}$ with density proportional to $h(\mathbf{x})$.
3:     Generate $V \sim V(0,1)$.
4:     **if** $V h(\mathbf{X}) \leq f(\mathbf{X})$ **then**
5:         **return** vector $\mathbf{X}$.

As for the one-dimensional case the main task when designing a rejection algorithm is the construction of the hat function. Possible candidates include the multivariate normal or multivariate t-distribution with the same correlation matrix as the desired distribution, or the product of the marginal distributions. This means that we need some special information about the desired distribution like the correlation matrix or marginal distributions. Even then we have to solve a multidimensional minimization problem to find the minimal multiplicative factor β for a density function $g(\mathbf{x})$ such that $\beta g(\mathbf{x})$ is a hat function, i.e. $f(\mathbf{x}) \leq \beta g(\mathbf{x}) = h(\mathbf{x})$ for $\mathbf{x}$. This seems to be one important reason that rejection algorithms are rarely used to generate random vectors; the second is certainly the fact that with increasing dimension d the acceptance probability is rapidly falling to zero (cf. Table 11.1 on p. 254). Instead of using the rejection method special properties of distributions are often exploited. However, this approach is not useful for constructing automatic algorithms. In most cases it is not even suited to generate variates of the same distribution restricted to a certain domain.

11.1.3 The Composition Method

The composition method is not restricted to the one-dimensional case. Thus the basic idea of Sect. 2.3 holds for multivariate distributions and will again be a brick stone in the design of black-box algorithms. Especially, if our target density f can be written as a discrete finite mixture

$$f(\mathbf{x}) = \sum_{i=1}^{n} w_i\, f_i(\mathbf{x})$$

where the f_i's are given density functions and the w_i's form a probability vector, then we can use Algorithm 2.6 (Composition) on p. 27 "cum grano salis" to sample from the distribution.

11.1.4 The Ratio-of-Uniforms Method

Another possible technique for automatic random vector generation can be based on the multivariate version of the ratio-of-uniforms method (Stefănescu and Văduva, 1987; Wakefield et al., 1991). Using convexity properties of the region of acceptance an enveloping region analogously to the univariate case (see Sect. 4.8) can be build. Instead of enveloping polygons we now have polyhedra. However, there remains additional research to work out the details of such an algorithm. Notice that the multivariate analogue of the minimal bounding rectangle still exists (see Stefănescu and Văduva, 1987) but its rejection constant increases exponentially with dimension, similar to the situation shown in Table 11.1.

11.1.5 Vertical Density Representation

Vertical density representation was introduced by Troutt (1991, 1993) and was adapted for the generation of multivariate random vectors by Kotz, Fang, and Liang (1997). The latter authors also demonstrate its usability for spherically symmetric densities by considering a couple of practically important distributions (see also Pang, Yang, Hou, and Troutt, 2001; Fang, Yang, and Kotz, 2001, for more applications). For applying this idea to automatic algorithms there seem to remain a lot of open questions that require additional research. Thus we refer the interested reader to the cited papers.

11.1.6 The Multinormal Distribution

As the multinormal distribution is the most common multivariate distribution and a building block for some of the algorithms developed in later chapters we include its generation here. It is well known that the multinormal distribution is fully characterized by its mean vector $\boldsymbol{\mu}$ and its variance-covariance matrix $\boldsymbol{\Sigma}$. As it is no problem to add $\boldsymbol{\mu}$ after generation we will assume that $\boldsymbol{\mu}$ is the 0-vector in the sequel.

The best known standard method for generating vectors in $\mathbb{R}^d$ from the multinormal distribution with given covariance matrix $\boldsymbol{\Sigma}$ is based on the decomposition of $\boldsymbol{\Sigma}$ into $\boldsymbol{\Sigma} = \mathbf{L}\mathbf{L}'$ where $\mathbf{L}$ denotes a lower triangular matrix, which is usually obtained by Cholesky factorization. The details for this method are given in Algorithm 11.3 (Multinormal-Cholesky).

Algorithm 11.3 Multinormal-Cholesky

Require: Dimension d, variance-covariance matrix $\boldsymbol{\Sigma}$.
Output: Multinormal distributed vector $\mathbf{X}$ with mean $\mathbf{0}$ and covariance $\boldsymbol{\Sigma}$.
   /* Setup */
1: Compute Cholesky factor $\mathbf{L}$ of $\boldsymbol{\Sigma}$. /* use Algorithm 11.4 */
   /* Generator */
2: Generate vector $\boldsymbol{\varepsilon} = (\varepsilon_1, \varepsilon_2, \ldots, \varepsilon_d)'$ of d independent standard normal variates.
3: $\mathbf{X} \leftarrow \mathbf{L}\boldsymbol{\varepsilon}$.
4: **return** vector $\mathbf{X}$.

It is easy to prove that Algorithm 11.3 (Multinormal-Cholesky) works correctly as

$$E(\mathbf{X}\mathbf{X}') = \mathbf{L}E(\boldsymbol{\varepsilon}\boldsymbol{\varepsilon}')\mathbf{L}' = \mathbf{L}\mathbf{L}' = \boldsymbol{\Sigma}$$

shows that the generated vector has the desired variance-covariance matrix. That the generated vector is multinormal follows from the well known fact that a linear transform maps a multinormal vector into a multinormal vector. It is easy to see that after computing $\mathbf{L}$ in the setup the time-complexity is $O(d^2)$. A simple algorithm known as the square root method to compute the Cholesky factorization is presented as Algorithm 11.4 (Cholesky-Decomposition). Its time complexity, which is the setup time for generating from the multinormal distribution, is $O(d^3)$.

Remark. Notice that we can use a method similar to Algorithm 11.4 to sample from other distributions. Simply replace the normal variates in Step 2 by other random variates. The generated random vectors have the desired variance-covariance matrix $\boldsymbol{\Sigma}$. Notice, however, that such an algorithm is of limited practical use, because for all but the multinormal distribution the marginal distributions are changed considerably in Step 3. Section 12.5 deals with the problem of sampling from distributions where marginals and variance-covariance matrix are given.

11.1 General Principles for Generating Random Vectors

Algorithm 11.4 Cholesky-Decomposition

Require: Positive definite $d \times d$-matrix $\boldsymbol{\Sigma} = (\sigma_{ij})$.
Output: Lower triangular matrix $\mathbf{L} = (l_{ij})$ such that $\mathbf{LL}' = \boldsymbol{\Sigma}$.
1: **for** $i = 1$ to d **do**
2: $\quad l_{i1} \leftarrow \sigma_{i1}/\sqrt{\sigma_{11}}$.
3: **for** $i = 2$ to d **do**
4: $\quad$ **for** $j = 2$ to $i - 1$ **do**
5: $\quad\quad l_{ij} \leftarrow \left(\sigma_{ij} - \sum_{k=1}^{j-1} l_{ik} l_{jk}\right) / l_{jj}$.
6: $\quad x \leftarrow \sigma_{ii} - \sum_{j=1}^{i-1} l_{ij}^2$.
7: $\quad$ **if** $x \geq 0$ **then**
8: $\quad\quad l_{ii} \leftarrow \sqrt{x}$.
9: $\quad$ **else**
10: $\quad\quad$ **abort** ($\boldsymbol{\Sigma}$ not positive definite).
11: $l_{ij} \leftarrow 0$ for $1 \leq i < j \leq d$.
12: **return** matrix $\mathbf{L} = (l_{ij})$.

A second approach for sampling multinormal vectors uses the conditional distribution method. For the joint normal distribution all conditional distributions are normal and it is not difficult to compute their means and variances. We denote the variance-covariance matrix by $\boldsymbol{\Sigma} = (\sigma_{i,j})$ and use the notation

$$\boldsymbol{\Sigma}_{i+1} = \begin{bmatrix} \boldsymbol{\Sigma}_i & \boldsymbol{\Sigma}'_{1i} \\ \boldsymbol{\Sigma}_{1i} & \sigma_{i+1,i+1} \end{bmatrix} \quad \text{and} \quad \mathbf{X}_i = (X_1, \ldots, X_i)' \quad \text{for } i = 1, \ldots, n-1$$

where $\boldsymbol{\Sigma}_i$ is the upper $i \times i$ sub-matrix of $\boldsymbol{\Sigma}$ and $\boldsymbol{\Sigma}_{1i}$ is the last row-vector of $\boldsymbol{\Sigma}_{i+1}$ without its last element, i.e. $\boldsymbol{\Sigma}_{1i} = (\sigma_{i+1,1}, \sigma_{i+1,2}, \sigma_{i+1,3}, \ldots, \sigma_{i+1,i})$. It is not difficult to compute the conditional expectation and variance of X_{i+1} (see e.g. Ripley, 1987, p.99) and we can write:

$$\mathrm{E}(X_{i+1}|\mathbf{X}_i) = \boldsymbol{\Sigma}_{1i} \boldsymbol{\Sigma}_i^{-1} \mathbf{X}_i \;,$$

$$\mathrm{Var}(X_{i+1}|\mathbf{X}_i) = \sigma_{i+1,i+1} - \boldsymbol{\Sigma}_{1i} \boldsymbol{\Sigma}_i^{-1} \boldsymbol{\Sigma}'_{1i} \;.$$

The vectors $\boldsymbol{\Sigma}_{1i} \boldsymbol{\Sigma}_i^{-1}$ and the standard deviations can be stored in a setup step. For a general multinormal distribution this results in an algorithm with a setup which is slower than that of Algorithm 11.3 (Multinormal-Cholesky) whereas the marginal time to generate one vector is the same (see e.g. Ripley, 1987, p.99).

Although not evident, it is simple to prove that Algorithm 11.3 (Multinormal-Cholesky) and the conditional distribution method generate the same random vectors as long as the same vector of iid. Gaussian random variates is used. This simple fact was first observed by Hauser and Hörmann (1994). For the Cholesky factor $\mathbf{L}$ of $\boldsymbol{\Sigma}$ and for the vector $\boldsymbol{\varepsilon}$ we use a partition similar to that of $\boldsymbol{\Sigma}$ and write

$$\mathbf{L}_{i+1} = \begin{bmatrix} \mathbf{L}_i & 0 \\ \mathbf{L}_{1i} & L_{i+1,i+1} \end{bmatrix}, \quad \mathcal{E}_{i+1} = \begin{bmatrix} \mathcal{E}_i \\ \varepsilon_{i+1} \end{bmatrix}, \quad \text{for } i = 1, \ldots, n-1$$

Algorithm 11.5 Multi-t (Multivariate t-distribution)

Require: Dimension d, degree of freedom ν, variance-covariance matrix $\boldsymbol{\Sigma}$.
Output: Multi-t distributed random vector $\mathbf{X}$ with mean $\mathbf{0}$.
/∗ Setup ∗/
1: Compute Cholesky factor $\mathbf{L}$ of $\boldsymbol{\Sigma}$. /∗ use Algorithm 11.4 ∗/
/∗ Generator ∗/
2: Generate vector $\boldsymbol{\varepsilon} = (\varepsilon_1, \varepsilon_2, \ldots, \varepsilon_d)'$ of d independent standard normal variates.
3: Generate a chi-square variate χ with ν degrees of freedom.
4: $\boldsymbol{\varepsilon} \leftarrow \boldsymbol{\varepsilon}/\sqrt{\chi/\nu}$.
5: $\mathbf{X} \leftarrow \mathbf{L}\,\boldsymbol{\varepsilon}$.
6: **return** vector $\mathbf{X}$.

with $\mathcal{E}_i = (\varepsilon_1, \ldots, \varepsilon_i)'$ and $\mathcal{E}_{i+1} = (\varepsilon_1, \ldots, \varepsilon_{i+1})'$.

Theorem 11.3. *Multiplying with the Cholesky factor and the conditional distribution method generate identical vectors; i.e. we have for $i = 1, 2, \ldots, n-1$*

$$\mathbf{L}_{1i}\mathcal{E}_i + \mathbf{L}_{i+1,i+1}\varepsilon_{i+1} = \boldsymbol{\Sigma}_{1i}\boldsymbol{\Sigma}_i^{-1}\mathbf{X}_i + \sqrt{\sigma_{i+1,i+1} - \boldsymbol{\Sigma}_{1i}\boldsymbol{\Sigma}_i^{-1}\boldsymbol{\Sigma}'_{1i}}\,\varepsilon_{i+1}\,.$$

Proof. Using $\boldsymbol{\Sigma}_{i+1} = \mathbf{L}_{i+1}\mathbf{L}'_{i+1}$ in the partitioned representation we get representations of $\boldsymbol{\Sigma}_{i+1}$ and $\boldsymbol{\Sigma}_{i+1}^{-1}$ in terms of $\mathbf{L}$-matrices. Together with $\mathbf{X}_i = \mathbf{L}_i\mathcal{E}_i$ this yields $\boldsymbol{\Sigma}_{1i}\boldsymbol{\Sigma}_i^{-1}\mathbf{X}_i = \mathbf{L}_{1i}\mathcal{E}_i$, and $\sigma_{i+1,i+1} = \mathbf{L}_{1i}\mathbf{L}'_{1i} + L^2_{i+1,i+1} = \boldsymbol{\Sigma}_{1i}\boldsymbol{\Sigma}_i^{-1}\boldsymbol{\Sigma}'_{1i} + L^2_{i+1,i+1}$. □

11.1.7 The Multivariate t-Distribution

It is one of the oldest results of mathematical statistics that the t-distribution with ν degrees of freedom is the ratio of a standard normal variate and the square root of a chi-square variate with ν degrees of freedom divided by ν. The multivariate t-distribution with ν degrees of freedom, zero mean and scale matrix $\boldsymbol{\Sigma}$ is analogously the normal random vector with variance-covariance matrix $\boldsymbol{\Sigma}$ divided by the square root of a chi-square variate with ν degrees of freedom and divided by ν. The density of the resulting distribution is given by

$$f(\mathbf{x}) = \frac{\Gamma((\nu+d)/2)}{(\pi\nu)^{d/2}\Gamma(\nu/2)\sqrt{|\boldsymbol{\Sigma}|}}\left(1 + \frac{\mathbf{x}'\boldsymbol{\Sigma}^{-1}\mathbf{x}}{\nu}\right)^{-(\nu+d)/2}.$$

For the sake of completeness we collect all details for a generation method in Algorithm 11.5 (Multi-t).

11.2 Uniformly Distributed Random Vectors

The uniform distribution is a very simple distribution in dimension 1. In higher dimensions we have the freedom to choose very different sets as support. Thus the generation problem is not trivial any longer as we have already seen in Example 11.1 above. The algorithms presented here can generate uniform random vectors on several simple sets. They are useful for many applications and are building blocks for the universal algorithms developed below.

Note that an algorithm that can generate uniformly distributed random vectors over arbitrary sets could be used to sample from arbitrary distributions with density f on $\mathbb{R}^d$ as well. Just increase the dimension d by one and generate uniform vectors $\mathbf{X}$ on the set $\{\mathbf{x} \in \mathbb{R}^{d+1} \mid 0 \leq x_{d+1} \leq f(x_1, \ldots, x_d)\}$. It is a generalization of Thm. 2.4 that the first d components of $\mathbf{X}$ then follow the desired distribution.

11.2.1 The Unit Sphere

We write S^{d-1} and B^d for the unit sphere and unit ball in $\mathbb{R}^d$, respectively, i.e.
$$S^{d-1} = \{\mathbf{x} = (x_1, \ldots, x_d) \in \mathbb{R}^d : \|\mathbf{x}\| = 1\},$$
$$B^d = \{\mathbf{x} = (x_1, \ldots, x_d) \in \mathbb{R}^d : \|\mathbf{x}\| \leq 1\},$$
where $\|\mathbf{x}\| = \sqrt{x_1^2 + \cdots + x_d^2}$ denotes the usual Euclidean norm. We call a random vector $\mathbf{X} = (X_1, \ldots, X_d)$ in $\mathbb{R}^d$ *radially symmetric* if the distribution is invariant under rotations of the coordinate axes. $\mathbf{X}$ is called uniformly distributed on the sphere S^{d-1} if $\mathbf{X}$ is radially symmetric and $\|\mathbf{X}\| = 1$. $\mathbf{X}$ is called uniformly distributed in the ball B^d (or uniformly distributed in the sphere S^{d-1}) if $\mathbf{X}$ is radially symmetric and $\|\mathbf{X}\| \leq 1$.

Uniform on the Sphere

It is very easy to construct a generator for the uniform distribution on S^{d-1} if we remember that a vector of iid. normal random variates is radially symmetric. So it is enough to normalize a vector of independent normal variates to length one to obtain the uniform distribution on the sphere. The formal statement is given as Algorithm 11.6 (Uniform-Sphere).

Algorithm 11.6 Uniform-Sphere

Require: Dimension d.
Output: Random vector $\mathbf{X}$ uniformly distributed on sphere S^{d-1}.
1: Generate d iid. normal random variates $(N_1, \ldots, N_d)$.
2: $S \leftarrow \sqrt{N_1^2 + \ldots + N_d^2}$.
3: **return** vector $\mathbf{X} = (\frac{N_1}{S}, \ldots, \frac{N_d}{S})$.

There are other algorithms given in the literature as well, see Deák (1978), Rubinstein (1982), or Devroye (1986a). Using the normal distribution, however, is reported to be fastest for higher dimensions. Note also that the generation time increases only linearly with dimension d.

Algorithm 11.7 Uniform-Ball-Rejection

Require: Dimension d.
Output: Random vector $\mathbf{X}$ uniformly distributed in ball B^d.
1: **repeat**
2:    Generate d independent uniform random number $U_i \sim U(-1, 1)$.
3: **until** $\sum_{i=1}^{d} U_i^2 \leq 1$
4: **return** vector $\mathbf{X} = (U_1, \ldots, U_d)$.

Table 11.1. Acceptance probability for Algorithm 11.7 (Uniform-Ball-Rejection)

dimension	probability
2	0.785
3	0.524
4	0.308
5	0.164
6	0.081
10	2.5×10^{-3}
15	1.2×10^{-5}
50	1.5×10^{-28}

Remark. Notice that normalizing points uniformly distributed in the unit cube does not result in points uniformly distributed on the sphere. This can be seen easily as the resulting distribution is not radially symmetric.

Uniform in the Sphere

To generate random vectors in B^2 (i.e. in a disc) rejection from the enclosing rectangle is certainly the most natural and probably also the fastest method. This idea can be generalized to arbitrary dimension d. Unfortunately, as we have already mentioned above, the acceptance probability of the multivariate rejection method decreases quickly with dimension d, see Table 11.1. Thus Algorithm 11.7 (Uniform-Ball-Rejection) is slow for $d > 5$ and useless for $d > 10$.

For a more efficient approach for higher dimension d we can generate a random point $\mathbf{Y}$ uniformly distributed on the sphere S^{d-1}, get a random radius R, and return $\mathbf{X} = R\mathbf{Y}$. Such an algorithm performs linearly in dimension d when we use Algorithm 11.6 (Uniform-Sphere) for generating $\mathbf{Y}$. The density for the distribution of radius R is proportional to the area of the surface of a sphere of radius R, i.e. r^{d-1}. Thus the density of R is given by $d\, r^{d-1}$, for $0 \leq r \leq 1$, and 0 otherwise. Hence we can generate R by inversion as $U^{1/d}$. The details are collected in Algorithm 11.8 (Uniform-Ball) which is similar to the algorithms suggested by Deák (1978) and Rubinstein (1982); both references also describe algorithms using uniform spacings. Stefănescu

Algorithm 11.8 Uniform-Ball
Require: Dimension d.
Output: Random vector $\mathbf{X}$ uniformly distributed in ball B^d.
1: Generate random vector $\mathbf{Y}$ uniformly on S^{d-1}. /* use Algorithm 11.6 */
2: Generate $U \sim U(0,1)$.
3: $R \leftarrow U^{1/d}$.
4: **return** vector $\mathbf{X} = R\mathbf{Y}$.

(1990, 1994) suggests another method that is reported to be faster for high dimension.

By linear transformations the unit sphere can be transformed into arbitrary hyperellipsoids. As a linear transformation never destroys the uniform distribution but only changes its support we can use Algorithm 11.8 (Uniform-Ball) together with a linear transformation to generate random vectors uniformly distributed in arbitrary hyperellipsoids, see Rubinstein (1982) or Devroye (1986a, Sect. XI.2.4) for details.

11.2.2 Simplices and Polytopes

A *convex polytope* in $\mathbb{R}^d$ with vertices $\mathbf{v}_1, \ldots, \mathbf{v}_n$ is the set of all points that are representable as convex combinations of these vertices. Every point in this convex polytope can be written as

$$x = \sum_{i=1}^{n} a_i \mathbf{v}_i \quad \text{where } a_i \geq 0 \text{ and } \sum a_i = 1 \, .$$

The set of vertices is minimal for the generated polytope if no vertex can be written as a convex combination of the other vertices. If the set of vertices is minimal this implies that the vertices are in general position (i.e. no three points are on a line, no four points are on a plane etc.). A *simplex* in $\mathbb{R}^d$ is a convex polytope with $d+1$ vertices in general position. As d points in general position in $\mathbb{R}^d$ define a hyper-plane of dimension $d-1$, a simplex is the simplest polytope in $\mathbb{R}^d$ with non-zero d-dimensional volume. We define the basic simplex as

$$\Delta_d = \{(x_1, \ldots, x_d) : x_i \geq 0, \sum_{i=1}^{d} x_i \leq 1\} \, ,$$

i.e. as the simplex with vertices at the origin and points on the positive coordinate axes at distance one from the origin.

In the sequel we will show that we can use uniform spacings to easily generate a vector uniformly distributed in the basic simplex. By a linear transformation we can generalize this idea to arbitrary simplices. To give the details of the method we need the following definition. Let $0 \leq U_{(1)} \leq \ldots \leq U_{(n)} \leq 1$

be the order statistics of an iid. sample of n $U(0,1)$ random variates. Set $U_{(0)} = 0$ and $U_{(n+1)} = 1$ for convenience. Then the statistics S_i defined by

$$S_i = U_{(i)} - U_{(i-1)} \quad \text{for } 1 \leq i \leq n+1$$

are called the *uniform spacings* for this sample. Notice that the uniform spacings $(S_1, \ldots, S_{d+1})$ have the property that $S_i \geq 0$ for all $i = 1, \ldots, d+1$ and $\sum_{i=1}^{d+1} S_i = 1$. We have the following theorem.

Theorem 11.4. $(S_1, \ldots, S_d)$ *is uniformly distributed over the basic simplex* Δ_d.

Proof. It is known from the theory of order statistics that the vector $(U_{(1)}, \ldots, U_{(d)})$ is uniformly distributed over the simplex $\{(x_1, \ldots, x_d) : 0 \leq x_1 \leq \ldots \leq x_d \leq 1\}$. The transformation $S_1 = U_{(1)}$, and $S_i = U_{(i)} - U_{(i-1)}$ for $2 \leq i \leq d$ is obviously linear which implies that the uniform distribution is not changed. $\sum_{i=1}^{d} S_i = U_{(1)} + (U_{(2)} - U_{(1)}) + \ldots + (U_{(d-1)} - U_{(d)}) = U_{(d)} \leq 1$ show that the vector $(S_1, \ldots, S_d)$ is always in the basic simplex. Moreover, we obtain its vertices if one S_i is 1 and all others are 0 which completes the proof. □

For a detailed discussion on generating uniform spacings see Devroye (1986a, Chap. V). One approach is to generate a uniform sample of size d and then sort it, e.g. by means of bucket sort which has expected time $O(d)$. There is also a direct generation method without sorting based on the following relation between the exponential distribution and uniform spacings. As it is – in our opinion – the simplest approach to generate uniform spacings we are describing only this method here.

Theorem 11.5. *Let* $(E_1, \ldots, E_{d+1})$ *be a sequence of iid. exponential variates and define* $E = \sum_{i=1}^{d+1} E_i$. *Then* $(\frac{E_1}{E}, \ldots, \frac{E_{d+1}}{E})$ *are uniform spacings* $(S_1, \ldots, S_{d+1})$.

Proof. See Devroye (1986a, p.208). □

For the sake of easier reference we give the details as Algorithm 11.9 (Uniform-Spacings) as well.

We can use a linear transform of uniform spacings to generate random points uniformly over an arbitrary simplex. The necessary transformation is simple and based on the following theorem.

Theorem 11.6. *Let* $(S_1, \ldots, S_{d+1})$ *be the uniform spacings of a sample of size* $d+1$, *and let* $\mathbf{v}_1, \ldots, \mathbf{v}_{d+1}$ *be the vertices of a simplex in* $\mathbb{R}^d$. *Then the vector*

$$\mathbf{X} = \sum_{i=1}^{d+1} S_i \mathbf{v}_i$$

is uniformly distributed over the simplex.

11.2 Uniformly Distributed Random Vectors

Algorithm 11.9 Uniform-Spacings

Require: Dimension d.
Output: Uniform spacings $(S_1, \ldots, S_{d+1})$.
1: Generate a random vector $(E_1, \ldots, E_{d+1})$ of iid. exponential variates.
2: Compute $E \leftarrow \sum_{i=1}^{d+1} E_i$.
3: **for** $i = 1$ to $d+1$ **do**
4: $\quad S_i \leftarrow E_i/E$.
5: **return** vector $(S_1, \ldots, S_{d+1})$.

Algorithm 11.10 Uniform-Simplex

Require: Dimension d, vertices $\mathbf{v}_1, \ldots, \mathbf{v}_{d+1}$.
Output: Random vector $\mathbf{X}$ uniformly distributed in simplex.
1: Generate the uniform spacings S_i for $1 \leq i \leq d+1$. /* use Algorithm 11.9 */
2: $\mathbf{X} \leftarrow \sum_{i=1}^{d+1} S_i \mathbf{v}_i$.
3: **return** vector $\mathbf{X}$.

Proof. We know from Thm. 11.4 that the vector $\mathbf{S} = (S_1, \ldots, S_d)$ is uniformly distributed over the basic simplex. Using $S_{d+1} = 1 - \sum_{i=1}^{d} S_i$ it is not difficult to verify that the vector $\mathbf{X}$ can be rewritten as

$$\mathbf{X} = \sum_{i=1}^{d+1} S_i \mathbf{v}_i = \mathbf{v}_{d+1} + (S_1, \ldots, S_n)(\mathbf{v}_1 - \mathbf{v}_{d+1}, \mathbf{v}_2 - \mathbf{v}_{d+1}, \ldots, \mathbf{v}_d - \mathbf{v}_{d+1})$$

where the $\mathbf{v}_i$ are considered as column-vectors of a matrix. Thus vector $\mathbf{X}$ is obtained as a linear transformation of the vector $\mathbf{S}$. This shows that vector $\mathbf{X}$ is uniformly distributed. If we take $S_i = 1$ and all other $S_j = 0$ the vector $\mathbf{S}$ is transformed into the vertex $\mathbf{v}_i$. This shows that the vertices of the basic simplex are transformed into the vertices of the desired simplex which completes the proof. □

In Algorithm 11.10 (Uniform-Simplex) we combine Thms. 11.5 and 11.6 to obtain a surprisingly simple generator for random vectors uniformly distributed in arbitrary simplices.

The time complexity of Algorithm 11.9 is $O(d)$. This means for Algorithm 11.10 (Uniform-Simplex) that Step 1 executes with $O(d)$ operations whereas the linear combination of $d+1$ vectors in Step 2 requires $O(d^2)$ operations.

For $d = 2$ a simplex is a triangle. The generation of uniform random vectors in triangles, using a special case of Algorithm 11.10, is posed as Exercise 11.3.

A third approach for generating uniform spacings utilizes the conditional distribution method. In Example 11.1 we derived such an algorithm for triangles. It can be generalized to arbitrary dimension but is slower than Algorithm 11.9, see Ştefănescu (1985) for more details.

11.2.3 Polytopes

The situation changes if we consider the generation of uniform vectors in general convex polytopes instead of simplices. Despite their simple definitions (convex) polytopes are complex objects and there exist a lot of amazing nontrivial results. We refer the reader to the books of Grünbaum (1967) and Ziegler (1995) for a description of the rich structure of these interesting objects, a short survey is given by Henk, Richter-Gebert, and Ziegler (1997). Although we can write any point inside the polytope as a convex combination of its vertices we have no generalization of Thm. 11.6 to polytopes. The reason is that there is no linear transformation that transforms for example the unit square into an arbitrary convex quadrangle.

Triangulation

For convex polygons with n vertices in the plane (2-dimensional polytopes) we have a simple solution by means of *triangulation*: Choose a root vertex and join all remaining vertices with this vertex. Thus we get a partition into $n-2$ triangles. For generating a random point uniformly in the polygon, first sample one of the triangles with probabilities proportional to their areas. Then sample a point uniformly in this triangle and return the point. It is easy to see that the time complexity of the setup (i.e. triangulation) is $O(n)$ whereas sampling can be done in constant time.

The idea of triangulation, i.e. of decomposing the given polytope into simplices, also works for arbitrary dimensions d. Unfortunately, a triangulation of a convex polytope is not a simple task and the computing time explodes for higher dimensions. Rubin (1984) gives a survey of this problem and provides a decomposition algorithm. If we have all faces of the polytope given together with inclusion relations (called the *face lattice*, *poset* of all faces, or *incidence graph*) then this partitioning is quite trivial: Triangulate all 2-faces (faces of dimension 2) as described above, continue with all 3-faces by selecting a root vertex in each of these faces and connect this vertex to all triangles (2-simplices) to get a partition into 3-simplices. Continue with 4-faces, and so on. Another possibility is to use *complete barycentric subdivision*, i.e. use the collection of barycenters of each face (see e.g. Lee, 1997). For choosing one of these simplices at random we need the volume of each of these. Notice that for a simplex with vertices $\mathbf{v}_0, \mathbf{v}_1, \ldots, \mathbf{v}_d$ the volume is given by $|\det(\mathbf{V})|/d!$ where $\mathbf{V}$ denotes the matrix $(\mathbf{v_1} - \mathbf{v}_0, \ldots, \mathbf{v_d} - \mathbf{v}_0)$. The complexity of the setup (triangulation, computing volumes) depends on the number of faces of the given polytope which is known to be $O(n^{\lfloor d/2 \rfloor})$ (McMullen, 1970).

It is important to note that computing the face lattice of the polytope requires the computation of the convex hull of the given vertices. However, this is sensitive to round-off errors in higher dimensions. There exist some computer programs that can handle such computations, e.g. qhull (Barber and Huhdanpaa, 2002). We also remark that a convex polytope can equivalently be defined by half-spaces given as inequalities.

Grid Method

Other approaches for generating random points in a given polytope of dimension d utilize the rejection principle. As a simple bounding region a hyperrectangle $H = [l_1, u_1] \times \ldots [l_d, u_d]$ can be used However, as we have already seen in Algorithm 11.7 the rejection constant will decrease quickly for increasing dimension. Thus we divide each side into N_i intervals of length $(u_i - l_i)/N_i$ and get a decomposition of H. There are three types of grid rectangles, the good rectangle where we have immediate acceptance, the useless rectangles where we have immediate rejection, and the ambiguous rectangles which overlap both the acceptance and the rejection region, see Fig. 11.1.

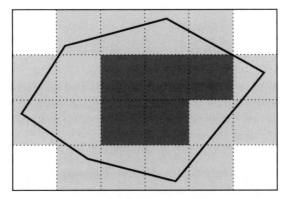

Fig. 11.1. Covering of a polygon by a grid. Grid rectangles with immediate acceptance (dark squares), rectangles outside (white squares), and ambiguous rectangles (gray squares)

Thus in the setup we compute the list G of all good rectangles and the list A of ambiguous rectangles. Algorithm 11.11 compiles the details based on the rejection method. The expected number of iterations is then the ratio between the volume of the polytope and the total volume of the grid rectangles in $G \cup A$. The speed of the algorithm, however, depends on the total volume of the ambiguous rectangles. Making this smaller requires a finer grid and thus the number of grid rectangles increases. The time complexity of the setup is about $O(N_1 \cdots N_d)$, i.e. it grows exponentially with dimension d if we keep the number of intervals for each side of H constant. For more details see Devroye (1986b).

Remark. It is obvious that the grid method works for any compact set C provided that we can compute whether a grid rectangle belongs to G, A, or neither of them. Thus we have stated Algorithm 11.11 in its general form.

It is also possible to improve the algorithm by adaptively decomposing ambiguous grid rectangles. This decreases the total volume of ambiguous rect-

Algorithm 11.11 Uniform-Grid

Require: Dimension d, compact set C.
Output: Random vector **X** uniformly distributed in C.
/* Setup */
1: Estimate bounding hyper-rectangle H.
2: Make grid.
3: Make list G of grid rectangles with immediate acceptance, and list A of ambiguous grid rectangles.
/* Generator */
4: **loop**
5:     Chose a rectangle $R \in G \cup A$ at random.
6:     Generate random point **X** uniformly in R.
7:     **if** $R \in G$ **then** /* immediate acceptance */
8:         **return** vector **X**.
9:     **if** $R \in A$ and $\mathbf{X} \in C$ **then**
10:        **return** vector **X**.

---

angles at the expense of even larger tables for the rectangle and the fact that the grid rectangles do not have the same volume any more.

In the next section we describe a third method for sampling uniformly in simple polytopes, called *sweep-plane algorithm*.

11.2.4 Sweep-Plane Method for Simple Polytopes

A *simple polytope* in $\mathbb{R}^d$ is a polytope where each vertex is adjacent to no more than d vertices. (Notice that every vertex of a polytope of dimension d is adjacent to at least d vertices.) The *sweep-plane technique* goes back to Hadwiger (1968) and has been applied to volume computation by Bieri and Nef (1983) and Lawrence (1991). The general idea of sweep-plane algorithms is to "sweep" a hyper-plane through a polytope, keeping track of the changes that occur when the hyper-plane sweeps through a vertex. Leydold and Hörmann (1998) have adopted this idea for generating random points. Then the plane is swept through the given simple polytope until a random fraction of the volume of the polytope is covered. The intersection of the sweep-plane with the polytope is again a simple polytope of smaller dimension. By recursion we arrive at a polytope of dimension 0, i.e. a single point. Although only derived for the convex case, the sweep plane algorithm also works for non-convex simple polytopes.

Sweep-Plane and Recursive Algorithm

In what follows let P denote a simple convex polytope in $\mathbb{R}^d$ and let **g** be some non-zero vector. We assume that $\|\mathbf{g}\| = 1$ and $\langle \mathbf{g}, \mathbf{x} \rangle$ is non-constant on every edge in P, where $\langle \cdot, \cdot \rangle$ denotes the scalar product. For a given **x** let $x = \langle \mathbf{g}, \mathbf{x} \rangle$ and denote the hyper-plane perpendicular to **g** through **x** by

$$\Phi(\mathbf{x}) = \Phi(x) = \{\mathbf{y} \in \mathbb{R}^d \colon \langle \mathbf{g}, \mathbf{y} \rangle = x\}$$

and its intersection with the polytope P with $Q(\mathbf{x}) = Q(x) = P \cap \Phi(x)$. $Q(x)$ again is a convex polytope (Grünbaum, 1967). Now move this sweep-plane $\Phi(x)$ through the domain P by varying x. Figure 11.2 illustrates the situation. The marginal density function $f_\mathbf{g}(x)$ along $\mathbf{g}$ of a uniform distribution on the polygon P is simply proportional to the volume $A(x)$ of $Q(x)$.

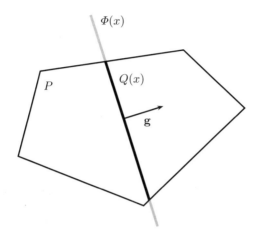

Fig. 11.2. Sweep-plane $\Phi(x)$ and cut polytope $Q(x)$

A generation algorithm based on the sweep-plane method can be seen as a generalization of the conditional distribution method for simple polytopes and works as following: Sample a random variate X from the marginal distribution and get the cut polytope $Q_{d-1} = Q(X)$. Now embed Q_{d-1} into a subspace $\mathbb{R}^{d-1}$ by eliminating the component y_M in $\mathbf{y} \in \mathbb{R}^d$, where $\mathbf{g}$ takes its maximum, i.e. $g_M = \max_{j=1,\ldots,d} g_j$ and

$$\mathbf{y} = (y_1, \ldots, y_M, \ldots, y_d) \mapsto \mathbf{y}^\circ = (y_1, \ldots, y_{M-1}, y_{M+1}, \ldots, y_d) \ .$$

Thus we get a polytope Q°_{d-1} of dimension $d-1$. Now apply the same procedure to the polytope Q°_{d-1} and get a polytope Q_{d-2} of dimension $d-2$ at random. Then we continue recursively until we arrive at a cut polytope Q_1 of dimension 1, i.e. a line segment, from which we easily can sample a random point uniformly.

At last notice that we can compute the random point $\mathbf{X} \in Q_{d-1}$ from $\mathbf{X}^\circ \in Q_{d-2} \subset Q^\circ_{d-1}$ by means of the fact that $\langle \mathbf{g}, \mathbf{X} \rangle = X$ for the sampled variate X.

$$\mathbf{X} = (X_1, \ldots, X_d) = (y_1, \ldots, y_{M-1}, X_M, y_{M+1}, \ldots, y_d)$$

with $\qquad X_M = \frac{X - \langle \mathbf{g}, \mathbf{y} \rangle}{g_M}$ \hfill (11.1)

where $\qquad \mathbf{y} = (y_1, \ldots, y_d) = (X_1^\circ, \ldots, X_{M-1}^\circ, 0, X_M^\circ, \ldots, X_{d-1}^\circ) \ .$

Sampling from the Marginal Distribution

The crucial part of this algorithm is to compute the marginal density and thus the volume $A(x)$ of $Q(x)$. Let $\Phi^-(x) = \{\mathbf{y} \in \mathbb{R}^d : \langle \mathbf{g}, \mathbf{y} \rangle \leq x\}$. Then $A(x) = V_{d-1}(P \cap \Phi(x)) = dV_d(P \cap \Phi^-(x))/dx$, where V_d denotes the d-dimensional volume. We denote the vertices of P by $\mathbf{v}_j \in \mathbb{R}^d$, $j = 1, \ldots, n$, and $v_j = \langle \mathbf{g}, \mathbf{v}_j \rangle$, such that
$$-\infty < v_1 \leq v_2 \leq \ldots \leq v_n < \infty\,.$$

The polytope P can be built up by *simple cones* at the vertices $\mathbf{v}_j$. Let $\mathbf{t}_1^{\mathbf{v}_j}, \ldots, \mathbf{t}_d^{\mathbf{v}_j}$ be nonzero vectors in the directions of the edges of P originated from $\mathbf{v}_j$, i.e. for each $i = 1, \ldots, d$ and every $\mathbf{x} \in P$, $\langle \mathbf{t}_i^{\mathbf{v}_j}, \mathbf{x} \rangle \geq 0$, see Fig. 11.3. We define $\gamma_i(\mathbf{v}_j) = \operatorname{sign}\langle \mathbf{t}_i^{\mathbf{v}_j}, \mathbf{g} \rangle$ and $\delta(\mathbf{v}_j) = \prod_{i=1}^d \gamma_i(\mathbf{v}_j)$. By our assumption $\langle \mathbf{t}_i^{\mathbf{v}_j}, \mathbf{g} \rangle \neq 0$. Then the vectors $\gamma_i(\mathbf{v}_j)\,\mathbf{t}_i^{\mathbf{v}_j}$ span the so called *forward cone* $C(\mathbf{v}_j)$ at $\mathbf{v}_j$, which is given by
$$C(\mathbf{v}_j) = \left\{ \mathbf{v}_j + \sum_{i=1}^n c_i \gamma_i(\mathbf{v}_j)\,\mathbf{t}_i^{\mathbf{v}_j} : c_i \geq 0 \right\}\,.$$

As can easily be seen, $\langle \mathbf{x}, \mathbf{g} \rangle \geq \langle \mathbf{v}_j, \mathbf{g} \rangle$ for all $\mathbf{x} \in C(\mathbf{v}_j)$. Moreover, $C(\mathbf{v}_j) \cap \Phi^-(x)$ is a simplex for $x > v_j = \langle \mathbf{v}_j, \mathbf{g} \rangle$ with volume
$$V_d(C(\mathbf{v}_j) \cap \Phi^-(x)) = (x - v_j)^d \frac{1}{d!} |\det(\mathbf{t}_1^{\mathbf{v}_j}, \ldots, \mathbf{t}_d^{\mathbf{v}_j})|\,\delta(\mathbf{v}_j) \prod_{i=1}^d \langle \mathbf{t}_i^{\mathbf{v}_j}, \mathbf{g} \rangle^{-1}\,.$$

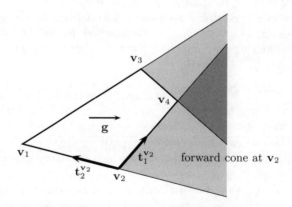

Fig. 11.3. Forward cone and Gram's relation

Let $\mathbf{1}_M$ denote the indicator function of the set $M \subseteq \mathbb{R}^d$. Then by a version of Gram's relation it follows that $\mathbf{1}_P = \sum_{j=1}^n \delta(\mathbf{v}_j)\,\mathbf{1}_{C(\mathbf{v}_j)}$ and consequently we find for the volume $V_d(P \cap \Phi^-(x)) = \sum_{j=1}^n \delta(\mathbf{v}_j)\,V_d(C(\mathbf{v}_j) \cap \Phi^-(x))$. Define

$$a_j = \frac{1}{(d-1)!} |\det(\mathbf{t}_1^{\mathbf{v}_j}, \ldots, \mathbf{t}_n^{\mathbf{v}_j})| \prod_{i=1}^{d} \langle \mathbf{t}_i^{\mathbf{v}_j}, \mathbf{g} \rangle^{-1}.$$

Using the fact that $\delta(\mathbf{v}_j)^2 = 1$ we arrive at $V_d(P \cap \Phi^-(x)) = \sum_{\substack{1 \le j \le n \\ v_j \le x}} \frac{a_j}{d}(x - v_j)^d$. Thus

$$A(x) = \frac{dV_d(P \cap \Phi^-(x))}{dx} = \sum_{\substack{1 \le j \le n \\ v_j \le x}} a_j(x - v_j)^{d-1}.$$

Using the binomial theorem we get

$$A(x) = \sum_{k=0}^{d-1} b_k^{(x)} x^k$$

where the coefficients

$$b_k^{(x)} = \binom{n-1}{k} \sum_{\substack{1 \le j \le m \\ v_j \le x}} a_j (-v_j)^{n-1-k} \tag{11.2}$$

are constant on the intervals $[v_{j-1}, v_j)$, i.e. $b_k^{(x)} = b_k^{(v_{j-1})}$ for all $x \in [v_{j-1}, v_j)$. On each of these intervals the marginal density function $f_{\mathbf{g}}(x) = A(x)$ is a polynomial of degree $d-1$ with both positive and negative coefficients.

To generate variates from these marginal distributions it is good luck that all of them are log-concave. This follows from the fact that the multivariate uniform distribution is log-concave itself together with the theorem that all marginal densities of a log-concave distribution are again log-concave (Prekopa, 1973, Thm. 8). Thus we can use any algorithm from Chap. 4. Since the coefficients $b_k^{(x)}$ change in every recursion step of the algorithm (and each time when we sample one random point $\mathbf{X}$) it seems most convenient to use Algorithm 4.1 (TDR) with adaptive rejection sampling for finding construction points. The performance of this algorithm can be improved when we first use the composition method and choose an interval $[v_{j-1}, v_j)$ at random with probabilities proportional to the volumes of $P \cap \{\mathbf{x} \in \mathbb{R}^d : v_{j-1} \le x \le v_j\}$ and use Algorithm 4.1 (TDR) to sample from the marginal distribution restricted to $[v_{j-1}, v_j)$.

Algorithm 11.12 (Uniform-Sweep-Plane) sketches the sweep-plane method. For more details especially on finding an appropriate vector $\mathbf{g}$, handling polytopes, and sampling from the marginal distribution we refer the reader to the original paper by Leydold and Hörmann (1998).

It is clear that computing the vertices of the cut polytope $Q(X)$ in each recursion step also requires the computation of the face lattice in a setup step. Compared to the triangulation method, the sweep plane method has lower marginal speed and only works for simple polytopes, but it also has faster

Algorithm 11.12 Uniform-Sweep-Plane

Require: Dimension d, simple polytope with vertices $\mathbf{v}_1, \ldots, \mathbf{v}_n$.
Output: Random vector $\mathbf{X}$ uniformly distributed in polytope.
1: **if** $d = 1$ **then**
2:     Generate $U \sim U(0,1)$.
3:     $\mathbf{X} \leftarrow U \mathbf{v}_1 + (1-U) \mathbf{v}_2$.
4:     **return** vector $\mathbf{X}$.
5: **else**
6:     Find a proper $\mathbf{g}$.
7:     Compute coefficients $b_k^{(v_j)}$ for marginal density. /* Equation (11.2) */
8:     Generate random variate X from marginal distribution.
9:     Compute cut polytope $Q(X)$ and its projection Q°.
10:    Call Algorithm Uniform-Sweep-Plane with dimension $d-1$ and polytope Q°. Receive $\mathbf{X}^\circ$.
11:    Compute $\mathbf{X}$ using (11.1).
12:    **return** vector $\mathbf{X}$.

setup for polytopes with many faces and works (with some minor modifications for the marginal distribution) also for simple non-convex polytopes where triangulation is difficult. Most important we also can use the sweep-plane technique for sampling from non-uniform distributions, see Sect. 11.3.

11.3 Multivariate Transformed Density Rejection

It is well known that we can generalize the rejection method easily to multivariate distributions (Sect. 11.1.2). The same is true for the main idea of transformed density rejection (Sect. 4.1).

We are given a multivariate distribution with differentiable (quasi-) density

$$f: D \to [0, \infty), \quad D \subseteq \mathbb{R}^d, \quad \text{with mode } \mathbf{m}.$$

For simplicity we assume $D = \mathbb{R}^d$. Completely analogous to the univariate case we say that f is T-concave if there exists a strictly monotonically increasing transformation T such that $\tilde{f}(\mathbf{x}) = T(f(\mathbf{x}))$ is concave for all $\mathbf{x}$ in the domain D of the distribution. In Sect. 4.2 we have discussed why we restrict our attention to the class of transformations T_c defined as

$$T_c(x) = \text{sign}(c)\, x^c \quad \text{and} \quad T_0(x) = \log(x).$$

For a T-concave quasi-density it is then easy to construct a hat $h(\mathbf{x})$. Choose N points $\mathbf{p}_i \in D \subseteq \mathbb{R}^d$ and take tangent hyper-planes $\ell_i(\mathbf{x})$ of the transformed density $\tilde{f}$ at these points, i.e.

$$\ell_i(\mathbf{x}) = \tilde{f}(\mathbf{p}_i) + \langle \nabla \tilde{f}(\mathbf{p}_i), (\mathbf{x} - \mathbf{p}_i) \rangle.$$

11.3 Multivariate Transformed Density Rejection

The hat is then the pointwise minimum of these tangents and we get h by transforming these hyper-planes back to the original scale,

$$\tilde{h}(\mathbf{x}) = \min_{i=1,\ldots,n} \ell_i(\mathbf{x}) \quad \text{and} \quad h(\mathbf{x}) = T^{-1}(\tilde{h}(\mathbf{x})) \,.$$

These are the similarities to the one-dimensional method but we have to overcome several problems as well. It is, for example, very hard to define a squeeze by just using the values of the transformed density in the design points. The regions around the design points where the local definition $\ell_i(\mathbf{x})$ of the hat function is used, is not an interval but a convex polytope. Thus we have to find generators for the hat distribution. We also have to cope with the "curse of dimensionality" which implies that we need a lot more design-points than in dimension one to obtain a close-fitting hat function.

There exist three algorithms based on multivariate TDR. These can be mainly distinguished by the way they decompose the multidimensional space into regions around a design point with a single hat function. In Leydold and Hörmann (1998) the hat is defined as described above as the minimum of the tangential hyper-planes. This implies that the regions around every design point form convex polytopes. In higher dimensions (starting with $d = 4$ or 5) the number of vertices of these polytopes explodes which makes the generation of random vectors from the dominating distribution very slow and numerically unstable (Sect. 11.3.2). Hörmann (2000) uses the special situation of $\mathbb{R}^2$ where it is simple to triangulate these convex polygons (Sect. 11.3.3). Leydold (1998) first decomposes the $\mathbb{R}^d$ into many cones rooted near the mode $\mathbf{m}$ and constructs the hat function inside each cone with a single design point. The resulting hat function is no longer continuous but the algorithm is much simpler than before as we have now cone-regions instead of general convex polytopes which also reduces the numerical problems. This algorithm works up to dimension 10 (Sect. 11.3.4).

In the sequel of this section we will discuss the main ingredients of these algorithms. For the sake of brevity we are omitting many of the implementation specific technical problems. The reader can find more technical details in the original papers and in the source code of our UNU.RAN library.

11.3.1 Multivariate T_c-Concave Distributions

Let us first consider the property of being multivariate T_c-concave. Functions on $\mathbb{R}^d$ have much more possible different shapes than functions over $\mathbb{R}$. This means that the condition of T_c-concavity is more restrictive for multivariate distributions. It follows directly from the definition that for any T-concave distribution all sets $S_k = \{\mathbf{x}\colon f(\mathbf{x}) \geq k\}$ must be either convex or empty. Devroye (1997a) calls such densities *convex unimodal*. Clearly convex unimodal distributions are a small group among all multivariate distributions but they are still more general than T_c-concave distributions.

Nevertheless, multivariate log-concave distributions are of great importance for statistics. T_c-concave distributions are interesting at least for the multivariate t-distribution which is considered to be of major importance in multivariate Monte-Carlo integration (Evans and Swartz, 2000). These integration problems are often restricted to a certain hyper-rectangle or polytope but in the literature there are no generators suggested to generate t-distributed (or even Gaussian) random vectors over polytopes. Thus the algorithms developed in this section are certainly useful to solve these generation problems. We also hope that, by facilitating the generation of random vectors with a given density, the methods proposed here will intensify the discussion about multivariate simulation (e.g. Agathokleous and Dellaportas, 1999).

To obtain hat functions with bounded integral over a cone the case $c \geq 0$ is never a problem, since then – analogously to the univariate case – the domain of the density f must be bounded. For negative c and unbounded domain we have to remember that the hat function has tail behavior like $x^{1/c}$. When we assume that the hat function remains the same over a cone the integral is only bounded if $x^{d-1} x^{1/c}$ converges faster to 0 than $1/x$. Thus the inequality $d - 1 + 1/c < -1$ must hold, i.e. $c > -1/d$.

11.3.2 TDR with Polytope Regions

Polyhedra and Marginal Densities of the Hat

If we make a direct generalization of TDR to multivariate distributions as described above we construct the pointwise minimum of several tangent hyperplanes as an upper bound for the transformed density. Transforming this bound back using T^{-1} we obtain a hat function for the density. Figure 11.4 shows a simple example for the bivariate normal distribution.

It is not difficult to see that the domain in which a particular tangent ℓ_i determines the hat function h is a convex polyhedron which may be bounded or not, see Fig. 11.5. The polyhedron P_i for design point $\mathbf{p}_i$ and its tangent $\ell_i(\mathbf{x})$ is given by

$$P_i = \{\mathbf{x} \in \mathbb{R}^d \colon h(\mathbf{x}) = T^{-1}(\ell_i(\mathbf{x}))\} = \bigcap_{j=1,\dots,N} \{\mathbf{x} \in \mathbb{R}^d \colon \ell_i(\mathbf{x}) \leq \ell_j(\mathbf{x})\} \,.$$

To avoid lots of indices we write $\mathbf{p}$, $\ell(\mathbf{x})$ and P without the index i if there is no risk of confusion. In the sequel we assume that P is a *simple* polyhedron, and that $\ell(\mathbf{x})$ is bounded and non-constant on every edge in P. We always can find touching points such that these restrictions hold if the transformed density is strictly concave, i.e. the determinant of its Hessian is negative almost everywhere.

To sample from the hat distribution restricted to P we apply the sweep-plane technique (Sect. 11.2.4). Notice that $\ell(\mathbf{x})$ is constant on every intersection of P with an affine hyper-plane perpendicular to the gradient $\nabla \ell = \nabla \tilde{f}(\mathbf{p})$. Define

11.3 Multivariate Transformed Density Rejection   267

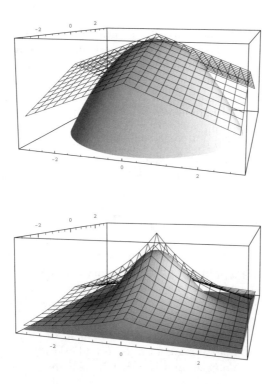

Fig. 11.4. Density (solid surface) and hat (grid) for bivariate normal distribution using four points of contact at $(\pm 0.5, \pm 0.5)$. Transformed (logarithmic) scale and original scale

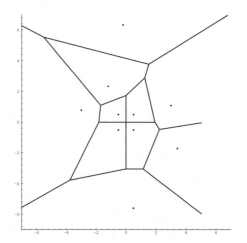

Fig. 11.5. TDR with ten design points and the resulting partition of the plane into ten polygons for the bivariate normal distribution

$$\mathbf{g} = -\frac{\nabla \tilde{f}(\mathbf{p})}{\|\nabla \tilde{f}(\mathbf{p})\|} \;.$$

As in Sect. 11.2.4 let again $x = \langle \mathbf{g}, \mathbf{x} \rangle$ and denote the sweep-plane by $\Phi(x)$ and its intersection with P by $Q(x)$. Our assumptions imply that the cut polytope $Q(x)$ is bounded. By setting

$$\alpha = \tilde{f}(\mathbf{p}) - \langle \nabla \tilde{f}(\mathbf{p}), \mathbf{p} \rangle \quad \text{and} \quad \beta = \|\nabla \tilde{f}(\mathbf{p})\| \tag{11.3}$$

we can write the hat function in the domain P as

$$h|_P(\mathbf{x}) = T^{-1}(\ell(\mathbf{x})) = T^{-1}(\alpha - \beta x) \;.$$

The marginal density $h_\mathbf{g}$ of the hat $h|_P$ along $\mathbf{g}$ is then

$$h_\mathbf{g}(x) = \int_{Q(x)} h(\mathbf{y})\,\mathrm{d}\mathbf{y} = A(x) \cdot T^{-1}(\alpha - \beta x)$$

where the volume $A(x)$ of $Q(x)$ is given by $A(x) = \sum_{k=0}^{d-1} b_k^{(x)} x^k$ with coefficients $b_k^{(x)}$ defined in (11.2).

For the special case of log-concave densities we have $T(x) = \log(x)$. Thus $T^{-1}(x) = \exp(x)$ and consequently

$$h_\mathbf{g}(x) = \exp(\alpha - \beta x) \sum_{k=0}^{d-1} b_k^{(x)} x^k \;.$$

$h_\mathbf{g}$ is again log-concave since it is the marginal density of the log-concave function $\exp(\ell(\mathbf{x}))|_P$ (Prekopa, 1973, Thm. 8). For the case $-1/d < c < 0$ the marginal density function is given by

$$h_\mathbf{g}(x) = (\beta x - \alpha)^{\frac{1}{c}} \sum_{k=0}^{d-1} b_k^{(x)} x^k \;.$$

Again this quasi-density is T_c-concave (Leydold and Hörmann, 1998). In this case, however, we must have $h|_P < 0$. To ensure this condition we always have to choose the mode $\mathbf{m}$ as construction point for a tangent plane ℓ. Notice, that the hat is then constant on P and we have to apply the method from Sect. 11.2.4 to this polytope. To sample from the marginal hat distribution h_P we can use any algorithm from Chap. 4. Since the coefficients $b_k^{(x)}$ change in every recursion step of the algorithm (and each time when we sample one random point $\mathbf{X}$) it seems most convenient to use Algorithm 4.1 (TDR) with adaptive rejection sampling for finding construction points. As suggested for the sweep-plane method in Sect. 11.2.4 the performance of the algorithm can be improved when we first use the composition method and choose an interval $[v_{j-1}, v_j)$ at random with probabilities proportional to the volumes

11.3 Multivariate Transformed Density Rejection

Algorithm 11.13 TDR-mv (Multivariate transformed density rejection)

Require: Dimension d, $c > -1/d$, T_c-concave quasi-density f, gradient ∇f; construction points $\mathbf{p}_1, \ldots, \mathbf{p}_N$ ($N \geq d$).
Output: Random vector $\mathbf{X}$ with density prop. to f.
/* Setup */
1: Compute ℓ_i and P_i for all $i = 1, \ldots, N$.
2: Compute coefficients $b_k^{(x)}$ for all polytopes.
3: Compute volumes A_i below hat in all P_i.
/* Generator */
4: **loop**
5:    Generate J with probability vector proportional to $(A_1, \ldots, A_N)$.
6:    Generate random variate X from marginal distribution $h|_{P_J}$.
7:    Compute cut polytope $Q(X)$ and its projection Q°.
8:    Call Algorithm 11.12 (Uniform-Sweep-Plane) with dimension $d-1$ and polytope Q°. Receive $\mathbf{X}^\circ$.
9:    Compute $\mathbf{X}$ using (11.1).
10:   Generate $V \sim U(0,1)$.
11:   **if** $V h(\mathbf{X}) \leq f(\mathbf{X})$ **then**
12:     **return** vector $\mathbf{X}$.

---

of $P \cap \{\mathbf{x} \in \mathbb{R}^d : v_{j-1} \leq x \leq v_j\}$ below the hat h and then use Algorithm 4.1 (TDR) to sample from the marginal hat distribution restricted to $[v_{j-1}, v_j]$.

For the sake of completeness we sketch the method in Algorithm 11.13 (TDR-mv), see Sect. 11.2.4 for the projection Q°. For the technical details, especially on handling polytopes and sampling from the marginal hat distributions we refer to the original paper by Leydold and Hörmann (1998).

Construction Points

If we want to use multivariate TDR in practice we have to choose construction points. In one dimension there are a number of possible techniques for finding appropriate points, see Sect. 4.4. For example, we can compute optimal points of contact for the tangents which makes it also possible to show that the execution time of the algorithm is uniformly bounded for a family of T-concave distributions.

In higher dimensions optimization is very difficult and out of scope for our algorithms. Instead the idea of adaptive rejection sampling (ARS, Sect. 4.4.5) seems to be the most useful one. Adapted to our situation it works in the following way: First take some points of contact (called starting points) which only must have the property that the volume below the hat $h(\mathbf{x})$ is bounded. Then start the generation of random variates with this hat until a point $\mathbf{x}$ is rejected. Use $\mathbf{x}$ to construct an additional tangent and thus a new hat and restart the generation of random points. Every rejected point is used as an additional design point until a certain stopping criterion is fulfilled, e.g. the maximal number N of design points or the aimed acceptance probability is reached. Using this rule the points of contact are chosen by a stochastic algorithm and it is clear that the multivariate density of the distribution of the next point of contact is proportional to $h(\mathbf{x}) - f(\mathbf{x})$. Thus with N tending towards infinity the hat function converges against the density with probability 1. We have the following simple observation which is a generalization of Thm. 4.16 to the multivariate case.

Theorem 11.7. *For a two times differentiable T-concave quasi-density function f with bounded domain in $\mathbb{R}^d$, and N design-points placed on a regular grid, the expected volume between hat and density is $O(N^{-2/d})$, i.e. we find for the rejection constant α,*

$$\alpha = 1 + O(N^{-2/d})\ .$$

Proof. As h and f are both two times differentiable functions with the same first-order Taylor-expansion in the design points, we have $|h(\mathbf{x}) - f(\mathbf{x})| = O(r^2)$ around each design-point $\mathbf{p}$, where $r = \|\mathbf{x} - \mathbf{p}\|$ is the the distance from the design point. Since we have N design points on a regular grid, the average radius is $r = O(N^{-1/d})$, which implies that the average distance $|h(\mathbf{x}) - f(\mathbf{x})| = O(N^{-2/d})$. As we have assumed a bounded domain D we get directly $\int_D |h(\mathbf{x}) - f(\mathbf{x})|\,d\mathbf{x} = O(N^{-2/d})$. □

The result of the theorem is another expression of the "curse of dimensionality". For $d = 1$ it is enough to multiply N by $2^{1/2} \approx 1.41$ to half the area between density and hat. In dimension 4 we have to multiply N by $2^2 = 4$ to obtain the same improvement, in dimension 10 we have to multiply even by $2^5 = 32$.

Using adaptive rejection sampling the design points do not form a regular grid. Nevertheless, it seems to be very likely that the results for adaptive rejection sampling are in average better than using a regular grid of design points. Thus we have $\alpha = 1 + O(N^{-2/d})$. This observation is clearly supported by our computational experiences.

The crucial step is to find the starting points such that the hat is bounded before we can run adaptive rejection sampling for finding more construction points. In the univariate case this was rather simple by choosing two points on either side of the mode. For multivariate densities with bounded domain this is also a simple task. Just take one or more points not too far away from the mode. For multivariate distributions with unbounded domain the problem can be more difficult since ℓ and thus the volume below the hat might not be bounded. Hörmann (2000) suggests to define a bounded auxiliary domain, that contains the mode and the main region of the distribution. Then use one starting point, if possible close to the mode, and start adaptive rejection sampling to find design points for a moderately good-fitting hat for the density restricted to the auxiliary domain. We have to find constructions points at least until the integral of the hat over the original domain is also bounded. These design points are then used as starting values for constructing a hat for the density over the unbounded domain.

Polyhedral Shaped Domains

Obviously we can restrict the domain D of density f to a simple polyhedron, which may be bounded or not. Then the intersection of D with the polyhedra P_i are again polytopes and the modifications to Algorithm 11.13 (TDR-mv) are straightforward.

Squeezes

It is possible to include squeezes in Algorithm 11.13 (TDR-mv) as well. To realize them the main problem is that we have to store the boundaries of these squeeze regions which again are polytopes. This makes the setup more expensive and requires more memory. Furthermore, in the rejection step of the algorithm we then have to find the right region. Even in the simplified versions of the sweep-plane method for dimension 2, described in Sect. 11.3.3 below, this is time-consuming and often slower than evaluating the density.

11.3.3 Bivariate Distributions

For bivariate distributions the method described in Sect. 11.3.2 above works as well. However, the situation is much simpler, since in $\mathbb{R}^2$ we have convex polygons instead of multidimensional polytopes (see Fig. 11.5) and it is easy to decompose such polygons into triangles. In Hörmann (2000) this observation has been used to design a universal generator for bivariate log-concave distributions. The setup of the generator is rather slow but the marginal execution time is fast. In the sequel we use the same notation as in Sect. 11.3.2. (However, we restrict our explanation here to the case $c = 0$, i.e. $T(x) = \log(x)$, for the sake of simplicity.)

The first steps, i.e. computing the polygons for the given construction points $\mathbf{p}_1, \ldots, \mathbf{p}_N$, remains the same, albeit less expensive than for higher dimensions. Next we make some changes to the method described in the previous section. After computing a polygon P it is now translated and rotated into a polgon P° such that the vertex $\mathbf{v}_0$ with the maximal value for the hat $h(\mathbf{x})$ is moved into the origin and the negative of the gradient of the transformed hat given by $-\nabla \ell = -\nabla f(\mathbf{p})$ points to the positive x_1-axis, i.e. we then have vector $\mathbf{g} = (1, 0)$ and the tangent plane has the form $\ell^\circ(\mathbf{x}) = \alpha - \beta x$, where $x = \langle \mathbf{x}, \mathbf{g} \rangle$ is just the first coefficient of $\mathbf{x} = (x_1, x_2)$. Notice that β is the length of the gradient of f at construction point $\mathbf{p}$, i.e. $\beta = \|\nabla f(\mathbf{p})\|$, and α is the value of the translated tangent plane in the origin. By these changes the generation step itself becomes simpler but we have to translate and rotate the generated random point back into the origininal scale afterwards.

We denote the components of $\nabla f(\mathbf{p})$ by t_1 and t_2. Then $\beta = \sqrt{t_1^2 + t_2^2}$ and using elementary algebra the matrix $\mathbf{R}$ for the required rotation is given by

$$\mathbf{R} = \begin{pmatrix} r_1 & -r_2 \\ r_2 & r_1 \end{pmatrix}, \quad \text{with } r_1 = \frac{-t_1}{\beta} \text{ and } r_2 = \frac{t_2}{\beta}, \text{ if } \beta \neq 0 \,.$$

Otherwise the hat is constant on polygon P and we choose $r_1 = 1$ and $r_2 = 0$. Next the translated polygon P° is triangulated such that each triangle has vertex $\mathbf{v}_0$, see Fig. 11.6. We can see that this triangulation is a simple and fast procedure and Hörmann (2000) shows that the total number of triangles grows only linearly with the number of design points, which is no longer true for dimensions larger than two.

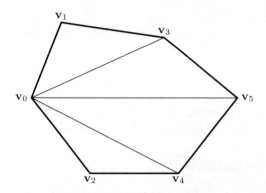

Fig. 11.6. Triangulation of a polygon. Vertex $\mathbf{v}_0$ is maximum of hat function

As shown in Fig. 11.7, each triangle is decomposed into two generator regions (G_1 and G_2) by the line l parallel to the x_2-axis through the vertex $\mathbf{v}_1$ that is closer to the x_2-axis. b denotes the distance between that line and the x_2-axis. For the generation step we also need the slopes k_0 and k_1 of the lines connecting the origin with the two vertices of the transformed triangle. The hat over the generator region is $h(x_1, x_2) = \exp(\alpha - \beta x_1)$. So we easily find for the marginal density

$$h_{x_1}(x) = (k_1 - k_0)\, x\, e^{\alpha - \beta x}$$

and for the volume below the hat

$$\text{vol} = \int_0^b h_{x_1}(x)\,dx = \frac{e^\alpha (k_1 - k_0)}{\beta} \left(e^{-\beta b} \left(-b - \frac{1}{\beta} \right) + \frac{1}{\beta} \right) .$$

The part of the triangle on the right hand side of line l (see Fig. 11.7) is the second generator region G_2. In the case of a bounded triangle the constants necessary for this generator region are computed analogously, except that in a first step $\mathbf{v}_2$ is translated into the origin and the triangle rotated into the opposite direction. This also implies that we have to replace β by $-\beta$.

To generate random pairs from the hat over a bounded generator region we must be able to sample from the marginal distribution $x \exp(-\beta x)$ in the interval $(0, b)$. This is trivial for the special case $\beta = 0$. For $\beta \neq 0$ it is best to generate X in the interval $(0, |\beta b|)$ with quasi-density $x \exp(x)$ or

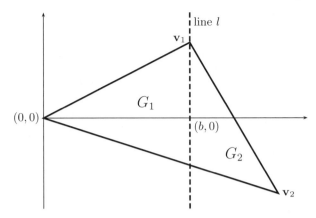

Fig. 11.7. Bounded case: The triangle is split by line l into two triangular generator regions G_1 (l.h.s.) and G_2 (r.h.s.). Line l is parallel x_2-axis through vertex $\mathbf{v}_1$

$x \exp(-x)$ depending on the sign of β and to return $X/|\beta|$. As the values of β and b vary at every call it is important to find a generator with a fast setup. After some experimentation we used a rejection algorithm with a fixed linear-constant-exponential hat for $x \exp(-x)$, and an exponential hat touching at $x_0 = 0.65\,|\beta b|$ for $x \exp(x)$. As the speed of this marginal generator has a considerable influence on the speed of the whole algorithm we use squeezes to accelerate the generation.

We can see in Fig. 11.5 that it is also possible that the region around a design point is an unbounded polygon. By triangulation this open polygon is decomposed into several bounded triangles but there remains an unbounded region, either a simple angle or an unbounded polygon consisting of two vertices and two lines towards infinity, see Fig. 11.8. From this region we can cut off a bounded generator region that is the same as the first generator region of a bounded triangle. After cutting off the bounded region the rest, which is the second generator region, can be described as a single angle and a parallel strip. (The parallel strip has width 0 for the case that the whole open polygon is only a single angle.) The second generator region G_2 of the unbounded case is rotated in the same way as the first region but the second vertex is translated into the origin. k_0 and k_1 are the respective slopes of the two lines l_1 and l_2 towards infinity and b is the x_2-coordinate of the intersection of the line l_1 with the x_2-axis (see Fig. 11.8).

For the generation of the marginal density in x_1-direction for an unbounded generator region we consider the region as divided into a parallel strip and an angle by the line p_2 through the point $(0, b)$ with slope k_2. Then it is easy to see that the marginal density is the mixture of an exponential density ($e^{-\beta x}$, parallel strip) and a gamma(2) density ($x\,e^{-\beta x}$, angle), both with scale parameter $\beta > 0$. The weights of the mixture depend on the volume below the hat for these two parts of the region. We get

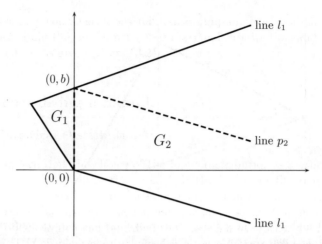

Fig. 11.8. Unbounded case: The polygon is split by the vertical line into the triangular generator regions G_1 (l.h.s.) and the "two-angle" G_2 bordered by l_1, the x_2-axis, and l_2 (r.h.s.). G_2 is furthermore decomposed by line p_2 into a simple angle and a parallel strip

$$\mathrm{vol}_{\mathrm{angle}} = \int_0^\infty h_{x_1}(x)\,\mathrm{d}x = e^\alpha\,|k_1 - k_0|/\beta^2 \;,$$

$$\mathrm{vol}_{\mathrm{strip}} = e^\alpha\,b/\beta\;.$$

Using these two volumes we can compute the probabilities of the mixture and get the marginal quasi-density for the open generator-region as

$$\frac{|k_1 - k_0|}{\beta}\,x\,e^{-\beta x} + b\,e^{-\beta x}\;.$$

It is simple to generate variates from this marginal density using the decomposition method: an exponential variate divided by β has to be returned with probability $p_0 = \beta b/(|k_1 - k_0| - \beta b)$, a gamma(2) variate divided by β with probability $1 - p_0$.

Up to now we have explained how to divide the domain of the distribution into generator regions, how to compute the volume below the hat for the different generator regions and how to sample from the marginal distribution along the x_1-axis. To sample from the hat over a generator region we have to add the remark that the conditional distribution of X_2 for given x_1 is uniform. Note that even for an unbounded polygon for any given x_1 the domain of X_2 is bounded, if the volume below the hat is bounded. Thus it is easy to generate random variates with density proportional to $h(x_1, x_2)$ over any of the regions by using the conditional distribution method. To decide between the different regions of the hat the volume between $h(x_1, x_2)$ and the (x_1, x_2)-plane for each generator region must be computed and the decomposition method can

11.3 Multivariate Transformed Density Rejection   275

Algorithm 11.14 TDR-Bivariate

Require: Log-concave quasi-density $f(x,y)$, gradient ∇f, domain of f; construction points $\mathbf{p}_1, \ldots, \mathbf{p}_N$.
Output: Random vector $\mathbf{X}$ with density prop. to f.
/* Setup */
1: Compute tangent planes for all $\mathbf{p}_i$.
2: Compute polytopes P_i for all $\mathbf{p}_i$.
3: **for all** polygons P_i **do**
4:     Triangulate P_i.
5:     Decompose all triangles into generator regions G_j.
6: **for all** generator regions G_j **do**
7:     Translate and rotate generator region G_j and compute all constants necessary for generation: α, β, g, k_0, and k_1.
8: Compute volumes A_j below the hat in all generator regions.
/* Generator */
9: **loop**
10:     Generate J with probability vector proportional to (A_j).
11:     **if** generator region G_J is bounded **then**
12:       Generate X from the marginal density $x \exp(-\beta x)$ in interval $(0, b)$.
13:       Generate $U \sim U(0,1)$.
14:       Set $Y \leftarrow X(k_0 + U(k_1 - k_0))$.
15:     **else** /* generator region unbounded */
16:       Generate X from the marginal density $\frac{|k_1 - k_0|}{\beta} x e^{-\beta x} + b e^{-\beta x}$.
17:       Generate $U \sim U(0,1)$.
18:       Set $Y \leftarrow X k_0 + U(b + X(k_1 - k_0))$.
19:     $Z \leftarrow \alpha - \beta X$. /* value of log of transformed hat for (X, Y) */
20:     Compute random point $\mathbf{X} = (X_1, X_2)$ by rotating and translating the pair (X, Y) back into original scale.
21:     Generate $V \sim U(0,1)$.
22:     **if** $\log(V) + Z \leq \log(f(X_1, X_2))$ **then**
23:       **return** vector $\mathbf{X}$.

---

be utilized to decide, which region should be used to generate the random variate.

These are the mathematical considerations necessary to implement two-dimensional TDR. Of course there remain a lot of computational difficulties: How are we going to store bounded and unbounded polytopes? How are we going to organize the necessary tables etc.? The algorithm description of Hörmann (2000) contains many details, a full documentation of a possible implementation is contained in UNU.RAN (Leydold et al., 2002). As this algorithm is rather long we are including here only a quite informal overview presented as Algorithm 11.14 (TDR-Bivariate). Hörmann (2000) shows that the total number of triangles that have to be stored is bounded by $4N$ and the total number of generator regions by $8N$. It is also shown that the time complexity of the setup is $O(N^2)$. In contrast the execution time of the generation part of Algorithm 11.14 (TDR-Bivariate) does not grow with N as long as we can guarantee that the generation of the index of the generator region is not influenced by N. Using Algorithm 3.2 (Indexed-Search) this is the case if the size of the guide table grows linearly with the total number of generator regions or – which is the same – linearly with N.

11.3.4 TDR with Cone Regions

Algorithm 11.13 (TDR-mv) in Sect. 11.3.2 works, but it is rather slow, since the domain of the density f is decomposed into polyhedra. This is due to the construction of the hat function, where we take the pointwise minimum of tangent hyper-planes. In this section we use a different approach: We first decompose the domain of the density into cones and then compute tangent hyper-planes in each of these. The resulting hat function is not continuous any more and the rejection constant is bounded from below. Moreover, the setup is slower. But on the other hand sampling from the hat distribution is much faster than in the original algorithm, and a version of derandomized adaptive rejection sampling (Sect. 4.4.6) can be applied. In the sequel we follow Leydold (1998). The notation is the same as in Sect. 11.3.2.

Simple Cones

A *simple cone* C in $\mathbb{R}^d$ (with its vertex in the origin) is an unbounded subset spanned by d linearly independent vectors:

$$\mathbf{t}_1, \ldots, \mathbf{t}_d \in S^{d-1}$$
$$C = \{\lambda_1 \mathbf{t}_1 + \cdots + \lambda_d \mathbf{t}_d : \lambda_i \geq 0\}.$$

$\mathbb{R}^d$ is now decomposed into such simple cones. In opposition to the procedure described above we now have to choose a proper point $\mathbf{p}$ in this cone C for constructing a tangent. In the whole cone the hat h is then given by this tangent. The method itself remains unchanged.

Using such a cone the intersection $Q(x)$ of the sweep plane $\Phi(x)$ with the cone C is bounded if and only if $\Phi(x)$ cuts each of the sets $\{\lambda \mathbf{t}_i : \lambda > 0\}$ for all $x > 0$, i.e. if and only if

$$\langle \mathbf{g}, \mathbf{t}_i \rangle > 0, \quad \text{for all } i. \tag{11.4}$$

For the volume $A(x)$ of $Q(x)$ we find

$$A(x) = \begin{cases} a\, x^{d-1} & \text{for } x \geq 0, \\ 0 & \text{for } x < 0, \end{cases}$$

where (again)

$$a = \frac{1}{(d-1)!} |\det(\mathbf{t}_1, \ldots, \mathbf{t}_d)| \prod_{i=1}^{d} \langle \mathbf{g}, \mathbf{t}_i \rangle^{-1}.$$

11.3 Multivariate Transformed Density Rejection

It is important to note that $A(x)$ does not exist if condition (11.4) is violated and that then the above formula is not valid any more. In close analogy to Sect. 11.3.2 we find for the marginal density and the area below the hat in cone C (for α and β see (11.3), p. 268)

$$h_{\mathbf{g}}(x) = a\, x^{d-1}\, T^{-1}(\alpha - \beta x)$$

and

$$H_C = \int_0^\infty h_{\mathbf{g}}(x)\, \mathrm{d}x = \int_0^\infty a\, x^{d-1}\, T^{-1}(\alpha - \beta x)\, \mathrm{d}x\ .$$

The intersection $Q(x)$ is always a $(d-1)$-simplex with vertices $\mathbf{v}_1,\ldots,\mathbf{v}_d$ in $\mathbb{R}^d$ given by

$$\mathbf{v}_i = \frac{x}{\langle \mathbf{g}, \mathbf{t}_i\rangle}\, \mathbf{t}_i\ .$$

Thus it is easy to sample a uniform random point from $Q(x)$ using Algorithm 11.10 (Uniform-Simplex).

The Choice of p

One of the main difficulties of this approach is the choice of the design point $\mathbf{p}$. Notice that $\mathbf{g}$ and thus a, α, β, and H_C depend on this point. Choosing an arbitrary $\mathbf{p}$ may result in a very large volume below the hat and thus in a very poor rejection constant.

Thus we have to find a construction point $\mathbf{p}$ such that condition (11.4) holds and the volume H_C below the hat is small. Searching for such a design point in the whole cone C with techniques for multidimensional minimization is cumbersome. Thus, for the sake of simplicity, we suggest to consider for the design point only the "center line" of C. Let $\bar{\mathbf{t}} = \frac{1}{d}\sum_{i=1}^d \mathbf{t}_i$ be the barycenter of the spanning vectors. Let $a(s)$, $\alpha(s)$, and $\beta(s)$ denote the corresponding parameters for $\mathbf{p} = s\,\bar{\mathbf{t}}$. Then we choose $\mathbf{p} = s\,\bar{\mathbf{t}}$ by minimizing the function

$$\mathcal{D}_A \to \mathbb{R},\quad s \mapsto \int_0^\infty a(s)\, x^{d-1}\, T^{-1}(\alpha(s) - \beta(s)\, x)\, \mathrm{d}x\ .$$

The domain $\mathcal{D}_A$ of this function is given by all points, where $\|\nabla \tilde{f}(s\bar{\mathbf{t}})\| \neq 0$ and where $A(x)$ exists, i.e. where $\mathbf{g}$ fulfills condition (11.4). For minimizing such a function standard methods can be used, e.g. Brent's algorithm (see e.g. Forsythe, Malcolm, and Moler, 1977). Notice that $\mathcal{D}_A$ is a subset of $[0,\infty)$ which might even be the empty set. In the latter case we have to decompose the cone into smaller cones (see below).

Log-Concave Densities

For log-concave densities with transformation $T(x) = \log(x)$ the marginal distribution follows a gamma law. The above formulae for marginal density, volume below hat, and the function that has to be minimized for construction point **p**, respectively, read

$$h_\mathbf{g}(x) = a\, x^{d-1} \exp(\alpha - \beta\, x) = a\, e^\alpha \cdot x^{d-1}\, e^{-\beta x}\,,$$

$$H_C = \int_0^\infty a\, x^{d-1} \exp(\alpha - \beta\, x)\, \mathrm{d}x = a\, e^\alpha\, \beta^{-d}\, (d-1)!\,,$$

$$\mathcal{D}_A \to \mathbb{R}, \quad s \mapsto \log(a(s)) + \alpha(s) - d\, \log(\beta(s))\,.$$

Triangulation

For the current variant of multivariate TDR we need a partition of the $\mathbb{R}^d$ into simple cones. It can be obtained by triangulating the unit sphere S^{d-1}. Each cone C is then spanned by a simplex $\Delta \subset S^{d-1}$, $C = \{\lambda \mathbf{t} : \lambda \geq 0, \mathbf{t} \in \Delta\}$. To avoid that $\mathcal{D}_A$ is the empty set and for having a good hat with a small rejection constant, some (or many) of these simplices must be small and "regular", i.e. the distances from the center to the vertices should be similar. Since we have to compute their volumes for computing the marginal densities, these simplices should have the same volumes. Thus the triangulation should have the following properties: (1) recursive construction; (2) $|\det(\mathbf{t}_1, \ldots, \mathbf{t}_d)|$ is easily computable for all simplices; (3) edges of a simplex have equal length. Although it is not possible to get such a triangulation for $d \geq 3$ we suggest an algorithm where we start with initial cones that are recursively split.

We get the initial simplices as the convex hull in S^{d-1} of unit vectors along the coordinate axes. As can easily be seen the resulting partition of the $\mathbb{R}^d$ is that of the arrangement of the hyper-planes orthogonal to the coordinate axes. Hence we have 2^d initial cones.

To get smaller cones we have to triangulate these simplices. Standard triangulations of simplices which are used for example in fixed-point computation (see e.g. Todd, 1976) are not appropriate for our purpose as the number of simplices increases too fast for each triangulation step. Instead we use a barycentric subdivision of the longest edges $(\mathbf{t}_i, \mathbf{t}_j)$ of the cone and split this edge at its barycenter, i.e.,

$$\mathbf{t}_\text{new} = \frac{\mathbf{t}_i + \mathbf{t}_j}{\|\mathbf{t}_i + \mathbf{t}_j\|}\,.$$

To get two smaller simplices replace vertex $\mathbf{t}_i$ by $\mathbf{t}_\text{new}$ for the first simplex and vertex $\mathbf{t}_j$ by $\mathbf{t}_\text{new}$ for the second one. We find for the volume of the new simplices

$$|\det(\mathbf{t}_1, \ldots, \mathbf{t}_\text{new}, \ldots, \mathbf{t}_d)| = \frac{1}{\|\mathbf{t}_i + \mathbf{t}_j\|}\, |\det(\mathbf{t}_1, \ldots, \mathbf{t}_d)|\,.$$

This triangulation is more flexible. Whenever we have a cone C, where the volume below the hat is too large or even unbounded we can split C and

11.3 Multivariate Transformed Density Rejection

Algorithm 11.15 TDR-Cone (Multivariate transformed density rejection with cones)

Require: Dimension d, log-concave quasi-density f, gradient ∇f, mode **m**.
Output: Random vector **X** with density prop. to f.
    /∗ Setup ∗/
1: Create initial cones.
2: Find touching points **p** and compute hat parameters if possible.
3: **repeat**
4:    **for all** cones C **do**
5:       **if** volume below hat H_C is unbounded or too large **then**
6:          Split cone.
7:          Find touching point **p** and compute hat parameters if possible.
8: **until** all cones computed and rejection constant sufficiently small.
    /∗ Generator ∗/
9: **loop**
10:    Select a random cone C by sampling from the probability vector proportional to the volumes below the hat (H_C) (use Algorithm 3.2, Indexed-Search).
11:    Generate random variate X from marginal distribution $h_{\mathbf{g}}(x)$.
12:    Compute cut simplex $Q(X)$.
13:    Generate **X** in simplex $Q(X)$ (use Algorithm 11.10, Uniform-Simplex).
14:    Generate $V \sim U(0,1)$.
15:    **if** $V\,h(\mathbf{X}) \leq f(\mathbf{X})$ **then**
16:       **return** vector **X**.

try again to find a proper touching point in both new cones. This can be continued using derandomized adaptive rejection sampling (Sect. 4.4.6) until the required rejection constant or the maximum number of cones is reached. In higher dimensions the latter is much more likely and an important stopping criterion since otherwise the computer runs out of memory.

Notice that it is not a good idea to use barycentric subdivision of the whole simplex (instead of dividing the longest edge only). This triangulation exhibits the inefficient behavior of creating long, skinny simplices (see remark in Todd, 1976).

Algorithm

The resulting algorithm is again rather long and too complicated to be describe in detail here. We thus refer the reader to the original paper by Leydold (1998) and give a sketch of the method as Algorithm 11.15 (TDR-Cone).

It is also possible to use this algorithm for a distribution where the domain is a strict subset of the $\mathbb{R}^d$. Then the cones are cut and we obtain pyramids instead. The cutting hyper-plane has to be perpendicular to the direction **g**. Thus the domain of the hat function is a strict superset of the domain of the density and we have to reject random points that fall outside of the domain of the density. This can be a problem in higher dimension since this fact may increase the rejection constant considerably.

11.4 Orthomonotone Densities

As the assumption of a log-concave or T_c-concave density is very strong for the multi-variate case it is important to develop universal algorithms for larger classes of multivariate distributions. The paper of Devroye (1997a) is the first and, up to now, only step into this very difficult direction. He presents several universal algorithms for various classes of distributions. We selected those that seem to be of highest practical importance.

These algorithms can be classified by three different principles for designing generators:

- Use global inequalities to design hat functions (Sect. 11.4.2).
- Divide the domain of the distribution into many hyper-rectangles and use a constant hat function for these rectangles; of course this requires a slow setup and large tables (Sect. 11.4.3).
- Use conditional densities to construct hat functions that touch the density along the axes (Sect. 11.4.4). In this section we also include new algorithms that are not presented in the original paper by Devroye (1997a).

Before we start with the presentation of the algorithms we have to find a definition of unimodality for higher dimensional distributions. However, following Devroye (1997a), we only present generators for the class of multivariate distributions with *orthounimodal* densities on $[0, \infty)^d$, i.e. *orthomonotone* densities.

11.4.1 Notions of Unimodality

There are different possibilities to generalize unimodality to higher dimensions. For the definitions given below and other notions like monotone-, axial-, linear-, and α-unimodality see Dharmadhikari and Joagdev (1988).

A density f is called *convex unimodal* (at the origin) if all sets $S_k = \{\mathbf{x}: f(\mathbf{x}) \geq k\}$ are either empty or are convex sets (that contain the origin). Clearly all T_c-concave densities are also convex unimodal. Convex unimodal densities do not form a robust class. Dharmadhikari and Joagdev (1988) have shown that lower-dimensional marginals of convex unimodal densities need not be convex unimodal. On the other hand convex unimodality is not influenced by rotations of the axes.

A density is called *orthounimodal* at the origin if all one-dimensional full conditional densities (parallel to the the axes) are unimodal with mode at 0. A mixture of orthounimodal densities with the same mode is orthounimodal again and all lower-dimensional marginals of orthounimodal densities are also orthounimodal. So this is quite a robust distribution family, but it is not closed under rotation of the axis. Therefore, if we compare it with convex unimodality, it is obvious that we have orthounimodal densities that are not convex unimodal and also convex unimodal densities that are not orthounimodal. On each of the 2^d quadrants of $\mathbb{R}^d$ an orthounimodal density is called *orthomonotone*.

If $\mathbf{Z} = (Z_1, \ldots, Z_d)$ is a random vector with $P(Z_i \leq 0) = 0$ and $(U_1, \ldots, U_d)$ are iid. uniform random variates independent of $\mathbf{Z}$, then $\mathbf{X} = (Z_1 U_1, Z_2 U_2, \ldots, Z_d U_d)$ is called a *block monotone* random variable (Shepp, 1962). This is a generalization of monotonicity based on Khinchine's Theorem (see e.g. Feller, 1971) stating, that every one-dimensional unimodal distribution can be written as an independent mixture UY, of a uniform random variate U and an arbitrary random variate Y. If $\mathbf{Z}$ is fixed, $\mathbf{X}$ is uniform between the origin and $\mathbf{Z}$ and therefore orthounimodal. If $\mathbf{Z}$ is random, $\mathbf{X}$ is a mixture of orthomonotone densities and therefore orthomonotone itself. Thus block monotonicity implies orthomonotonicity, the converse, however, is false.

A density f is *star-unimodal* at the origin if for any vector $\mathbf{x}$, $f(t\mathbf{x})$ is non-increasing as t varies from 0 to ∞. Clearly all orthounimodal densities are also star-unimodal.

Star-unimodal seems to be too general to design universal generators; for this task orthounimodality is best suited. Also many of the multivariate special distributions suggested in the literature are either orthounimodal, or orthounimodal after some linear transformation. If the mode is known we can decompose an orthounimodal density into orthomonotone parts on the 2^d quadrants. So we discuss only the basic problem here to generate from orthomonotone densities on the first quadrant.

Examples of orthomonotone densities include

$$f(x_1, \ldots, x_d) = \prod_{i=1}^d f_i(x_i), \quad x_i \geq 0, \ 1 \leq i \leq d$$

where all f_i are non-increasing, and

$$f(x_1, \ldots, x_d) = C e^{-\prod_{i=1}^d g_i(x_i)}, \quad x_i \geq 0, \ 1 \leq i \leq d$$

where all g_i are increasing.

11.4.2 Generators Based on Global Inequalities

The methods in this section do not require that f is analytically known. We only need a black-box that can evaluate the density together with a piece of additional information. These algorithms are multidimensional generalizations of several of the one-dimensional algorithms explained in Sect. 6.1 of this book. There we have developed universal algorithms for monotone distributions on $\mathbb{R}^+$ and on $(0, 1)$. It turned out that even for the class of bounded monotone densities on $(0, 1)$ we have not been able to design a uniformly fast algorithm. So we cannot expect to obtain such an algorithm in higher dimensions.

One algorithm given by Devroye (1997a) requires the knowledge of the moments of all one-dimensional marginals, or the knowledge of a moment of the sum of the random vector, or the knowledge of the moment generating function. As we do not think that such information is readily available in practice we omit these algorithms here and refer the interested reader to the original paper. Instead we focus our interest on orthomonotone densities on the unit cube. Note that – using a linear transform – this algorithm can be used for any orthomonotone density with known compact support.

Densities on the Unit Cube

The naive rejection algorithm using a constant hat of height $f(\mathbf{0})$ is valid for all monotone densities over the unit cube. Its validity is not influenced by the notion of monotonicity we use and no additional information is necessary. The expected number of iterations of the naive algorithm is equal to $f(\mathbf{0})/A_f$, where $A_f = \int f(\mathbf{x})\,\mathrm{d}\mathbf{x}$ denotes the volume below the quasi-density. Therefore it is very high for most distributions especially in higher dimensions. As in dimension one we can improve this algorithm when we know A_f, the volume below the density. The definition of orthomonotonicity implies that for any fixed point $\mathbf{x} = (x_1,\ldots,x_d)$, $f(\mathbf{y})$ is not smaller than $f(\mathbf{x})$ for all $\mathbf{y}$ in the d-dimensional interval $[\mathbf{0}, \mathbf{x}] = \{\mathbf{y}\colon 0 \leq y_i \leq x_i\}$, which has volume $\prod_{i=1}^{d} x_i$. Thus we have

$$f(\mathbf{x}) \prod_{i=1}^{d} x_i \leq \int f(\mathbf{y})\,\mathrm{d}\mathbf{y} = A_f$$

which can be rewritten as

$$f(\mathbf{x}) \leq \min\left(f(\mathbf{0}), \frac{A_f}{\prod_{i=1}^{d} x_i}\right) =: h(\mathbf{x}) .$$

We can see that this inequality is a multivariate extension of Thm. 6.3. Unfortunately the generalization of the generation algorithm for the hat function is not so simple as we have no multidimensional inversion method and the conditional distribution method is not simple even for $d = 2$ (see Exercise 11.5). For higher d it seems to become intractable. Devroye (1997a) uses the ingenious idea to transform the random vector $\mathbf{Z} = (Z_1,\ldots,Z_d)$ with density $h(\mathbf{z})$ over the unit cube into the vector $\mathbf{Y} = (Y_1,\ldots,Y_d) = (-\log Z_1,\ldots,-\log Z_d)$. Using the transformation theorem for multivariate densities we can see that the vector $\mathbf{Y}$ has density

$$f_\mathbf{Y}(\mathbf{y}) = \min\left(A_f, f(\mathbf{0}) e^{-\sum_{i=1}^{d} y_i}\right), \qquad y_i \geq 0 .$$

Devroye defines the *platymorphous* density with parameters $b \geq 1$ and d as

$$h_{b,d}(\mathbf{y}) = \min\left(1, b\,e^{-\sum_{i=1}^{d} y_i}\right), \qquad b \geq 1, \quad y_i \geq 0.$$

Thus $f_\mathbf{Y}(\mathbf{y})$ is a multiple of $h_{b,d}(\mathbf{y})$ with $b = f(\mathbf{0})/A_f$. An algorithm for sampling from such a distribution is given in Algorithm 11.16 (Platymorphous). It is the basis for Algorithm 11.17 (Monotone-Multivariate), a simple rejection algorithm for bounded orthomonotone densities over the unit cube.

11.4 Orthomonotone Densities

Algorithm 11.16 Platymorphous

Require: Density parameter $b \geq 1$ and dimension d.
Output: Random vector $\mathbf{X}$ with platymorphous density $h_{b,d}$.
1: Generate random variate Y from log-concave density $\xi(y) = \frac{y^{d-1}}{(d-1)!} \min(1, b\,e^{-y})$
   with mode at $\max(\log b, d-2)$. /∗ see Chap. 4 ∗/
2: Generate uniform spacings $(S_1, \ldots, S_d)$
   (use Algorithm 11.9 (Uniform-Spacings) with dimension $d-1$).
3: **return** vector $(YS_1, \ldots, YS_d)$.

Algorithm 11.17 Monotone-Multivariate

Require: Bounded and orthomonotone quasi-density $f(\mathbf{x})$ on $[0,1]^d$,
   upper bound $\breve{A}_f \geq A_f = \int f(\mathbf{x})\,d\mathbf{x}$.
Output: Random vector $\mathbf{X}$ with density prop. to f.
1: **loop**
2:   Generate $\mathbf{Y}$ with platymorphous density $h_{f(\mathbf{0})/\breve{A}_f,d}$.
3:   Set $\mathbf{X} \leftarrow (e^{-Y_1}, \ldots, e^{-Y_d})$.
4:   Generate $V \sim U(0,1)$.
5:   **if** $V \min\left(f(\mathbf{0}), \frac{\breve{A}_f}{\prod_{i=1}^d X_i}\right) \leq f(\mathbf{X})$ **then**
6:     **return** vector $\mathbf{X}$.

The area below the platymorphous density $h_{b,d}(\mathbf{x})$ is $\sum_{i=0}^d \frac{\log^i b}{i!}$. Thus the expected number of iterations in Algorithm 11.17 is given by

$$\mathrm{E}(I) = \frac{A_h}{A_f} = \sum_{i=0}^d \frac{\log^i(f(\mathbf{0})/A_f)}{i!}.$$

Clearly $\mathrm{E}(I)$ is the sum of the first $d+1$ terms of the Taylor series of $f(\mathbf{0})/A_f = \exp(\log(f(\mathbf{0})/A_f))$. Hence $\mathrm{E}(I)$ will never exceed $f(\mathbf{0})/A_f$, which is the rejection constant for naive rejection from a constant hat. If d is much smaller than $\log(f(\mathbf{0})/A_f)$, then $\mathrm{E}(I)$ is much smaller than $f(\mathbf{0})/A_f$. For fixed dimension d, $\mathrm{E}(I)$ grows for $f(\mathbf{0}) \to \infty$ like $(\log^d f(\mathbf{0}))/d!$.

It remains to explain the generator for the platymorphous distributions (Algorithm 11.16). Its density $h_{b,d}$ only depends on the sum of all x_i. Therefore it has a simple shape. For example, for the special case $b=1$ we have $\log h_{1,d}(\mathbf{x}) = \min(0, -\sum x_i)$, i.e. the graph of its logarithm is simply the pointwise minimum of two hyper-planes. Therefore $h_{b,d}(\mathbf{x})$ is obviously constant over all sets $C_y = \{\mathbf{x} | \sum x_i = y \text{ and } x_i \geq 0\}$. Thus density $\xi(y)$ for $y = \sum x_i$ is given by

$$\xi(y) = \frac{y^{d-1}}{(d-1)!} \min(1, b\,e^{-y}).$$

$\xi(y)$ is called the generator of this distribution. If we generate a random variate Y from this density and $(S_1, \ldots, S_d)$ as uniform spacings (i.e. the spacings of $d-1$ iid. $U(0,1)$ random numbers, see Sect. 11.2.2), then it is clear that the vector $\mathbf{X} = (YS_1, \ldots, YS_d)$ has sum Y and falls into the first quadrant. As Y was generated such that it follows the distribution of $\sum x_i$ it is also quite obvious that the vector $\mathbf{X}$ follows the platymorphous distribution. For a formal proof of the formula for $\xi(y)$ above and for the correctness of the generation method described here a result about Liouville distributions proven by Sivazlian (1981) is used (see Devroye, 1997a). To sample from the platymorphous distribution notice that ξ is log-concave and we can use generators based on transformed density rejection (Chap. 4). Depending on parameter b the mode is either in the "gamma-part" and thus $d-2$ which is the case when $\log b < d - 2$. Or, when $\log b \geq d - 2$, the mode of ξ is simply $\log b$.

11.4.3 Table Methods

It is easy to generalize the "Ahrens method" (Sect. 5.1) that uses constant hat functions and squeezes over many small intervals to higher dimensions d. If we decompose the domain of the distribution into many d-dimensional intervals orthomonotonicity is exactly the notion of multidimensional monotonicity that we need for this method. Using the definition of orthomonotonicity we have the trivial bounds $f(\mathbf{u}) \leq f(\mathbf{x}) \leq f(\mathbf{l})$ for any vector $\mathbf{x}$ in the interval $(\mathbf{l}, \mathbf{u}) = (l_1, u_1) \times \ldots \times (l_d, u_d)$, with $l_i \leq u_i$ for $i = 1, \ldots, d$. So we can easily calculate the necessary constant hat and constant squeeze for arbitrary intervals. Of course everything is simplest if we consider the decomposition of the unit-cube $[0,1]^d$ into $N = N_1^d$ cubes of equal size (see Algorithm 11.18 (Ahrens-Multivariate) for the details). It is also possible to generalize the global bound for $\mathrm{E}(I)$ given in Thm. 5.1 to higher dimensions. We have the following theorem (a generalization of a result by Devroye, 1997a).

Theorem 11.8. *The expected number of iterations for sampling one random vector using Algorithm 11.18 (Ahrens-Multivariate) is given by*

$$\mathrm{E}(I) \leq \frac{f(\mathbf{0})}{A_f} \frac{N_1^d - (N_1 - 1)^d}{N_1^d} + 1 = \frac{f(\mathbf{0})}{A_f}\left(1 - \left(1 - \frac{1}{N_1}\right)^d\right) + 1 \; .$$

Proof. We number our N_1^d cells and denote the lower left vertex by $\mathbf{l}_i$, the upper right vertex by $\mathbf{u}_i$. Then we have

$$A_h = \sum_{i=0}^{N_1^d-1} f(\mathbf{l}_i)/N_1^d = \sum_{i=0}^{N_1^d-1} (f(\mathbf{l}_i) - f(\mathbf{u}_i))/N_1^d + \sum_{i=0}^{N_1^d-1} f(\mathbf{u}_i)/N_1^d \; .$$

Now it is clear that for some j we have $\mathbf{l}_i = \mathbf{u}_j$. Only for the case that $\mathbf{l}_i$ has at least one zero component we cannot find such a $\mathbf{u}_j$. So we have a total number of $N_1^d - (N_1 - 1)^d$ cells where $f(\mathbf{l}_i)$ is not "telescoped" away. But for all these cells $f(\mathbf{l}_i)$ can be $f(\mathbf{0})$. We also have the trivial bound $\sum_i f(\mathbf{y_i})/N_1^d \leq A_f$. So putting these arguments together we get

Algorithm 11.18 Ahrens-Multivariate

Require: Bounded and orthomonotone quasi-density $f(\mathbf{x})$ on $[0,1]^d$;
number N_1 of decompositions in each dimension (thus we have $N = N_1^d$ cells).
Output: Random vector $\mathbf{X}$ with density prop. to f.
/* Setup */
1: Store $f(\mathbf{x})$ for $\mathbf{x} = (i_1, i_2, \ldots, i_{N_1})/N_1$ for all integers $i_k \in \{0, \ldots, N_1\}$.
/* Generator */
2: **loop**
3:     Generate discrete random vector $\mathbf{D} \in \{0, 1, \ldots, N_1 - 1\}^d$ with probability vector prop. to $f(\mathbf{D}/N_1)$. /* use indexed search (Sect. 3.1.2) */
4:     Generate d independent $U(0,1)$ variates $U_1, \ldots, U_d$.
5:     Set $\mathbf{X} \leftarrow ((D_1 + U_1)/N_1, \ldots, (D_d + U_d)/N_1)$.
6:     Generate $V \sim U(0,1)$.
7:     **if** $V f(\mathbf{D}/N_1) \leq f((\mathbf{D} + (1, \ldots, 1))/N_1)$ **then** /* squeeze test */
8:         **return** $\mathbf{X}$.
9:     **if** $V f(\mathbf{D}/N_1) \leq f(\mathbf{X})$ **then**
10:         **return** vector $\mathbf{X}$.

$$\mathrm{E}(I) = A_h/A_f \leq \frac{f(\mathbf{0})}{A_f} \frac{N_1^d - (N_1 - 1)^d}{N_1^d} + 1$$

as proposed. □

The above bound mainly shows that this simple algorithm needs huge tables to obtain a sensible guaranteed performance even for moderate dimension d. $\mathrm{E}(I)$ is improved over naive rejection by the factor $\frac{N_1^d - (N_1-1)^d}{N_1^d} + 1/f(\mathbf{0})$ which is approximately d/N_1 for $N_1 \gg d$. But this means that e.g. for $d = 5$ and $N_1 = 10$ we make a decomposition into $N = 10^5$ subcells to get (in the worst case) an improvement by a factor of only $1/2$. In other words, in terms of the total N, we have $A_h/A_f = O(N^{-1/d})$. This is poor compared with the result for multivariate TDR, $A_h/A_f = O(N^{-2/d})$ (see Thm. 11.7). So for $d > 3$ we should not expect too much from the multivariate table method.

As this general bound is so poor it is of course worthwhile to think how we could improve the performance of Algorithm 11.18 (Ahrens-Multivariate). First we could allow cells of different sizes. However, we need an extra amount of book-keeping as we have to store the left lower vertex and the size of all cells. We also need a strategy to find an appropriate decomposition of the hypercube $[0,1]^d$ into cells. Optimizing the choice of the cells seems to be impossible in higher dimensions. A practical and simple solution is using derandomized adaptive rejection sampling (Sect. 4.4.6): Start with cells of equal sizes. Whenever the difference between hat and squeeze over the cell (i.e. $f(\mathbf{l}_i) - f(\mathbf{u}_i)$) is bigger than a certain threshold then split the cell into 2^d smaller cells in the obvious way. This trick can improve the performance of the algorithm considerably especially if there are large regions where the density is flat. (In Sect. 5.1 we have seen that for the one-dimensional case $d = 1$ using this rule leads to asymptotically optimal points if the threshold goes to 0.) We also could use adaptive rejection sampling (Sect. 4.4.5). However, the repeated recomputation of necessary guide tables makes the adaptive steps slow. Using such a technique the improvements in terms of E(I) for a given number of cells can be impressive; but it seems impossible that they can improve the $O(N^{-1/d})$-law for general densities.

11.4.4 Generators Based on Inequalities Involving Conditional Densities

It can be useful to construct hat functions by inequalities that include conditional densities if we can compute the marginals of these hat functions. Of course we have to find generators for these marginals as well. But simple and easy to check conditions on the required conditional distributions (in most cases log-concavity) guarantee that we can easily generate from the hat distribution without using the analytic form of the density. Adding this new idea makes the presented algorithms similar universal as those based on global inequalities.

Bounding by Product Densities

Writing $f_i(u) = f(0,\ldots,0,u,0,\ldots,0)$ (with u in the i-th position) for the full conditional density given that all $X_j = 0$ for $j \neq i$, we get the simple inequality

$$f(\mathbf{x}) \leq \min_{1 \leq i \leq d} f_i(x_i) \leq \prod_{i=1}^{d} f_i^{1/d}(x_i) =: h(\mathbf{x})$$

which can be utilized for a rejection algorithm, provided that the volume below the hat $h(\mathbf{x})$ is bounded. Then it remains to find a method to generate random variates from the (quasi-)densities $f_i^{1/d}$. Corollary 4.7 tells us that for a T_c-concave density f^a is $T_{(c/a)}$-concave. Thus if $f_i^{1/d}$ is T_{c_1}-concave for

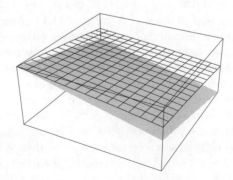

Fig. 11.9. Density $f(\mathbf{x}) = 2 - x_1 - x_2$ on $[0,1]^2$ (solid surface) and hat $h(\mathbf{x}) = \sqrt{(2-x_1)(2-x_2)}$ (grid) used in Algorithm 11.19 (Orthomonotone-I)

Algorithm 11.19 Orthomonotone-I

Require: Bounded and orthomonotone quasi-density $f(\mathbf{x})$ on $\mathbb{R}^d$; all d full conditional densities $f_i(u) = f(0,\ldots,0,u,0,\ldots,0)$ (with u in the i-th position) are log-concave.
Output: Random vector $\mathbf{X}$ with density prop. to f.
/* Setup */
1: Setup generators for the log-concave densities $f_i^{1/d}$.
/* Generator */
2: **loop**
3:    **for** $i = 1,\ldots,d$ **do**
4:       Generate X_i with log-concave density $f_i^{1/d}$.
5:    Generate $V \sim U(0,1)$.
6:    **if** $V \prod_{i=1}^d f_i^{1/d}(X_i) \leq f(\mathbf{X})$ **then**
7:       **return** vector $\mathbf{X}$.

some $c_1 > -1$ it follows that f_i must be T_c-concave for some $c > -1/d$. Thus we are not sacrificing much and obtain the easiest conditions not depending on d if we assume that all f_i are log-concave as this implies (using Corollary 4.7) that $f_i^{1/d}$ is log-concave as well. Then we can use one of the universal Algorithms for log-concave distributions described in Sects. 4 and 6.3. In combination with the product form of the hat function we have no problem to generate random vectors from the multidimensional hat function. Algorithm 11.19 (Orthomonotone-I) contains the details.

It is obvious that we have $\mathrm{E}(I) = \prod_i \int f_i^{1/d}(u)/A_f \, du$ for the expected number of iterations. This number strongly depends on the shape of the distribution. To give a simple example $\mathrm{E}(I)$ is one for the product of uniform densities, $d^{d/2}$ for the product of half-normal densities and d^d for the product of exponential densities (see Exercises 11.6, 11.7, and 11.8). We also observe that the hat function touches the density in general only in the mode. Figure 11.9 shows the linear density $f(\mathbf{x}) = 2 - x_1 - x_2$ on the unit square together with the hat used by Algorithm 11.19 (Orthomonotone-I).

Symmetric Densities

We say a density f is symmetric in all x_i if

$$f(\ldots, x_j, \ldots, x_k, \ldots) = f(\ldots, x_k, \ldots, x_j, \ldots) \text{ for all } j, k.$$

Such densities form an important subclass of orthomonotone densities. Writing $g(u) = f(u, 0, \ldots, 0)$ and using the obvious inequality

$$f(\mathbf{x}) \leq g\left(\max_i x_i\right) =: h(\mathbf{x})$$

we have a hat function for a rejection algorithm. The hat is obviously touching the density above all axes; see Fig. 11.10 for an example with quasi-density $f(\mathbf{x}) = \max(2 - x_1 - x_2, 0)$. The hat function is constant on the set

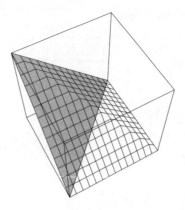

Fig. 11.10. Density $f(\mathbf{x}) = \max(2 - x_1 - x_2, 0)$ on $[0,2]^2$ (solid surface) and hat $h(\mathbf{x}) = \max(2 - \max(x,y), 0)$ (grid) used in Algorithm 11.20 (Orthomonotone-II)

$$H = \{\mathbf{x} : x_i \geq 0 \text{ and } \max_i x_i = k\}$$

which consists of half of the surface of the hypercube defined by the origin and the point $(k, \ldots, k)$ and therefore has $d-1$ dimensional volume $d\,k^{d-1}$. This implies that the one-dimensional distribution of $\max_i(X_i)$ has density

$$\xi(u) = d\,u^{d-1} g(u)\,.$$

Using this result we obtain for the expected number of iterations

$$\mathrm{E}(I) = \frac{A_h}{A_f} = \frac{d}{A_f} \int_0^\infty t^{d-1} f(t, 0 \ldots, 0)\,\mathrm{d}t\,.$$

Of course the rejection algorithm for symmetric densities only works if the volume below the hat and equivalently the area below $\xi(u)$ is bounded. Considering the tail behavior we can see that the area below the hat is only bounded if the tails of $g(u) = O(u^{-1/c})$ for $c > -1/d$ which is the same condition as for Algorithm 11.19 (Orthomonotone-I). This is amazing as we here use symmetry which was not assumed above. To facilitate the conditions we assume that $g(u)$ is log-concave in the sequel. This obviously implies that $\xi(u)$ is log-concave as well.

To generate from the hat distribution we start by sampling $Y = \max_i(X_i)$ which has density $\xi(u)$. Then we have to generate a vector uniformly on the set H, which is not difficult. First we randomly generate the index i of the variable for which the maximum Y will be attained. All other elements of the vector $\mathbf{X}$ are uniformly distributed between 0 and Y. Algorithm 11.20 (Orthomonotone-II) contains the details.

The hat function is touching the density along the axes and has constant values over half of the surfaces of the hypercubes with one vertex the origin and the second vertex on the line through the origin in the direction

Algorithm 11.20 Orthomonotone-II

Require: Bounded and orthomonotone quasi-density $f(\mathbf{x})$ on $\mathbb{R}^d$, symmetric in all x_i, $f(u, 0, \ldots, 0)$ is log-concave.
Output: Random vector $\mathbf{X}$ with density prop. to f.
  /* Setup */
1: Setup generator for the log-concave density $u^{d-1} f(u, 0, \ldots, 0)$.
  /* Generator */
2: **loop**
3:   Generate Y with log-concave density $u^{d-1} f(u, 0, \ldots, 0)$.
4:   Generate J uniformly on $\{1, \ldots, d\}$. /* J is index of maximal component */
5:   Set $X_J \leftarrow Y$.
6:   **for** $i = 1, \ldots, d, i \neq J$ **do**
7:     Generate $U \sim U(0, 1)$.
8:     $X_i \leftarrow UY$.
9:   Generate $V \sim U(0, 1)$.
10: **if** $V f(Y, 0, \ldots, 0) \leq f(\mathbf{X})$ **then**
11:    **return** vector $\mathbf{X}$.

$(1, \ldots, 1)$. Due to this shape we expect a good fit of this hat, especially for symmetric densities where the contour lines form convex sets containing the origin. As the hat is touching along the axes we expect that Algorithm 11.20 (Orthomonotone-II) has a better fit for symmetric densities than the hat used for Algorithm 11.19 (Orthomonotone-I). This can be easily proved as for symmetric densities Algorithm 11.19 has hat function

$$h(\mathbf{x}) = \left(\prod_i g(x_i)\right)^{1/d} \geq g\left(\max_i x_i\right)$$

which is obviously bigger than the hat function of Algorithm 11.20 as $g(x)$ is monotonically decreasing. Equality only occurs for the case that all x_i are equal. Nevertheless, for both algorithms the expected number of iterations $\mathrm{E}(I)$ is bounded for the same densities. To compare the values of $\mathrm{E}(I)$ we have to compare $\left(\int g^{1/d}(t)\, dt\right)^d$ and $d \int t^{d-1} g(t)\, dt$. If $g(t)$ is the exponential density we have $d!$ for Algorithm 11.20 and d^d for Algorithm 11.19 (see Exercise 11.6). For the normal distribution (see Exercise 11.7), the gain is no longer that spectacular but still about 1.25 for $d = 2$, 2.51 for $d = 4$, and 32.64 for $d = 10$. It is clear that for the uniform distributions both hat functions and thus both algorithms are identical.

An additional advantage of Algorithm 11.20 (Orthomonotone-II) over Algorithm 11.19 (Orthomonotone-I) is that it needs only one instead of d calls to the one-dimensional universal log-concave generator.

Densities with Compact Support

We can easily construct a hat touching the density in the direction of the vector $(1, \ldots, 1)$ using the obvious inequality

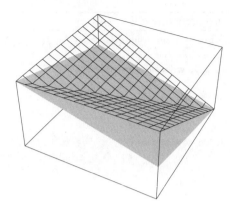

Fig. 11.11. Density $f(\mathbf{x}) = 2 - x_1 - x_2$ on $[0,1]^2$ (solid surface) and hat $h(\mathbf{x}) = 2 - \min(x,y) - \min(x,y)$ (grid) used in Algorithm 11.21 (Orthomonotone-III)

$$f(\mathbf{x}) \leq g(\min_i x_i)$$

where $g(u) = f(u, \ldots, u)$. The contour lines of this hat function are the sets $\{\mathbf{x}\colon \min_i x_i = k, k > 0\}$ which are not bounded. Thus we can use this algorithm only for densities with bounded domain. Without loss of generality we will assume that the domain of the density f is the unit cube $[0,1]^d$. Then the level sets have the form $\{\mathbf{x}\colon 0 \leq x_i \leq 1$ and $\min_i x_i = k, k > 0\}$; see Fig. 11.11 for an example with density $f(\mathbf{x}) = \max(2 - x_1 - x_2, 0)$. Due to the shape of the hat function this algorithm is especially suited for distributions in $\mathbb{R}^2$ that have contour lines shaped approximately like the graph of the function $1/x$.

To design the rejection algorithm we need the density of the random variate $Y = \min x_i$ which is

$$\xi(u) = d\,(1-u)^{d-1} g(u) \quad \text{for} \quad 0 \leq u \leq 1\,.$$

It should be obvious now how close this algorithm is to the algorithm for symmetric densities described above. But as we only consider densities on the unit cube we need no restriction for the conditional density along the diagonal. So we only assume that it is monotone which easily follows from orthomonotonicity anyway. This implies that $\xi(u)$ is monotone as well. We leave the details of the deviation to the reader and only present the full description as Algorithm 11.21 (Orthomonotone-III). The expected number of iterations can be computed as

$$\mathrm{E}(I) = \frac{A_h}{A_f} = \frac{d}{A_f} \int_0^1 t^{d-1} f(t, \ldots, t)\,\mathrm{d}t\,.$$

To compare $\mathrm{E}(I)$ for a simple example we have used the density $f(\mathbf{x}) = (2 - 2x_1)(2 - 2x_2)$ on the unit square ($A_f = 1$). For the naive rejection algorithm we find $\mathrm{E}(I) = 4$, Algorithm 11.17 (Monotone-Multivariate)

11.4 Orthomonotone Densities

Algorithm 11.21 Orthomonotone-III

Require: Bounded and orthomonotone quasi-density $f(\mathbf{x})$ on $[0,1]^d$.
Output: Random vector $\mathbf{X}$ with density prop. to f.
/* Setup */
1: Setup generators for monotone density $\xi(u) = (1-u)^{d-1} f(u,\ldots,u)$ with $0 < u < 1$ (use e.g. Algorithm 6.1 (Monotone-01) or 5.1 (Ahrens)).
/* Generator */
2: **loop**
3:    Generate Y with density $\xi(u) = (1-u)^{d-1} f(u,\ldots,u)$ on $[0,1]$.
4:    Generate J uniformly on $\{1,\ldots,d\}$. /* J is index of maximal component */
5:    Set $X_J \leftarrow Y$.
6:    **for** $i = 1,\ldots, d, i \neq J$ **do**
7:      Generate $U \sim U(0,1)$.
8:      $X_i \leftarrow Y + (1-Y)U$.
9:    Generate $V \sim U(0,1)$.
10:    **if** $V f(Y,\ldots,Y) \leq f(\mathbf{X})$ **then**
11:      **return X**.

gives 3.34, Algorithm 11.21 (Orthomonotone-III) 2, and for Algorithm 11.20 (Orthomonotone-II), which is using the symmetry, we get 4/3. If we look at the contour lines of the density, which are not too different from circles with center origin, we can understand, why Algorithm 11.20 (Orthomonotone-II) performs best.

Somebody could argue that our values of E(I) do not tell the full truth as there is a universal rejection algorithm for one-dimensional distributions hidden within the algorithms of this chapter and we have not included this in our analysis. This objection is correct but of little practical relevance as we have (very) fast algorithms for univariate log-concave distributions available.

Using Several Conditional Densities

When we compare hats and characteristics of different algorithms of this section it seems to be a good idea to combine different hat functions. For example, we would get a hat touching an arbitrary orthomonotone density at all axes using the trivial inequality

$$f(\mathbf{x}) \leq \min(f_1(x_1), f_2(x_2), \ldots, f_d(x_d))$$

where $f_i(x_i) = f(0, \ldots, 0, x_i, 0, \ldots, 0)$ denotes the full conditional density.

However, this idea cannot be used directly as we cannot solve the problem of computing the domain of the different parts of the hat function unless f has a simple analytical form. Nevertheless, we can obtain usable algorithms if we partition the first quadrant into simple parts first and then use a single inequality and thus a single hat for every region, using inequalities that we can easily get from the orthomonotonicity property.

Let $\mathbf{b} = (b_1, \ldots, b_d)$ be a vector with nonnegative components $b_i \geq 0$, that defines a line through the origin. For an arbitrary point $\mathbf{x} = (x_1, \ldots, x_d)$ we have
$$f(x) \leq f\left(\min_i \left(\frac{x_i}{b_i}\right) \mathbf{b}\right)$$
where we set $x_i/b_i = \infty$ if $b_i = 0$. For example, for $\mathbf{b} = (1, 0, \ldots, 0)$ we get $f(\mathbf{x}) \leq f(x_1, 0, \ldots, 0) = f_1(x_1)$ and for $\mathbf{b} = (1, \ldots, 1)$ we obtain $f(\mathbf{x}) \leq f(\min x_i, \ldots, \min x_i)$, i.e. we get the inequalities that we have used for Algorithms 11.20 and 11.21. If we want to design an algorithm for densities on unbounded domains we have to use at least the d unit vectors $\mathbf{e}_i$ as vectors $\mathbf{b}$. If we only use these we get the simplest algorithm in this new class. The first quadrant is decomposed into d regions $R_i = \{\mathbf{x}|\max_j x_j = x_i\}$. For each of the R_i the hat is touching f along the x_i axis. Thus Algorithm 11.22 (Orthomonotone-IV) below can be seen as a generalization of the algorithm for symmetric densities above.

The marginal quasi-density of the hat for every R_i is $\xi_i(u) = u^{d-1} f_i(u)$. It could also be used to compute the area below the hat for the different R_i. Unfortunately this includes a numerical integration problem and would therefore lead to a very slow setup. It is therefore easier to compute the area below the hat for the d rejection algorithms that we will use to generate from the marginal densities of the hat function. We can use these probabilities to decide randomly which R_i should be chosen to make the next try to generate the random vector. Clearly we need a condition on the conditional densities to guarantee that the volume below the hat is bounded and we need universal algorithms to sample from the densities ξ_i. Again T_c-concave with $c > -1/d$ would be sufficient to obtain a T_c-concave density $\xi(u)$ but we restrict our attention to log-concave densities. For the expected number of iterations, when we – as above – do not include the number of trials in the one dimensional algorithm to generate the marginal density, we have:

$$\mathrm{E}(I) = \sum_{i=1}^{d} \int_0^\infty u^{d-1} f_i(u)/A_f \, du \ .$$

A first possible refinement is to define the regions R_i of the different definition of the hat depending on the density f such that $\mathrm{E}(I)$ is minimized. If the domain of f is the whole first quadrant this is not difficult. We use a vector $\mathbf{a} = (a_1, \ldots, a_d)$ as the "borderline" of all regions $R_i = \{x| \max_j \frac{x_j}{a_j} = \frac{x_i}{a_i}\}$. Then the marginal densities of the hat for R_i is denoted by ξ_i and has to be multiplied by the factors $k_i = \prod_{j=1}^{d} \frac{a_j}{a_i}$. Like in Algorithm 11.22 (Orthomonotone-IV) we denote the area below the hat of the generator for the different regions by p_i. We want to minimize the total volume below the hat which is given by

$$\mathrm{Vol}_{\text{total}}(\mathbf{a}) = \left(\prod_{i=1}^{d} a_i\right) \left(\sum_{i=1}^{d} \frac{p_i}{a_i^d}\right) \ .$$

11.4 Orthomonotone Densities

Algorithm 11.22 Orthomonotone-IV

Require: Bounded and orthomonotone quasi-density $f(\mathbf{x})$ on $\mathbb{R}^d$; all d full conditional densities $f_i(u) = f(0, \ldots, 0, u, 0, \ldots, 0)$ (with u in the i-th position) are log-concave.
Output: Random vector $\mathbf{X}$ with density prop. to f.
 /* Setup */
1: Setup generators for log-concave densities $\xi_i(u) = (1-u)^{d-1} f_i(u)$.
2: Compute volume below the hat for the d one-dimensional rejection algorithms above and store these as vector $\mathbf{P} = (p_1, \ldots, p_d)$.
3: Compute the optimal vector $\mathbf{a}$ setting $a_1 \leftarrow 1$, $a_i \leftarrow (p_i/p_1)^{1/d}$.
 (Or use vector $\mathbf{a} = (1, \ldots, 1)$ instead.)
4: Compute $\tilde{\mathbf{P}}$ setting $\tilde{p}_i \leftarrow p_i \left(\prod_{j=1}^{d} a_j \right) / a_i^d$
 /* Generator */
5: **loop**
6:   Generate J on $\{1, \ldots, d\}$ with probability vector prop. to $\tilde{\mathbf{P}}$.
7:   Make one trial to generate Y with density $\xi_J(u) = (1-u)^{d-1} f_J(u)$ using the rejection algorithm.
8:   **if** Y was accepted in the first trial **then**
9:     Set $X_J \leftarrow Y$.
10:     **for** $i = 1, \ldots, d, i \neq J$ **do**
11:       Generate $U \sim U(0,1)$.
12:       $X_i \leftarrow UY a_i / a_J$.
13:     Generate $V \sim U(0,1)$.
14:     **if** $V f_J(Y) \leq f(\mathbf{X})$ **then**
15:       **return** $\mathbf{X}$.

Setting all partial derivatives of $\mathrm{Vol}_{\mathrm{total}}$ to zero we obtain d equations of the form

$$\sum_{i=1}^{d} \frac{p_i}{a_i^d} = \frac{dp_j}{a_j^d}, \quad \text{for } j = 1, \ldots, d.$$

As the left hand sides of these equations are all equal, the right hand sides must be equal as well. Without loss of generality we may assume that $a_1 = 1$ and thus we get the optimal solution for $\mathbf{a}$ as

$$a_i = \left(\frac{p_i}{p_1} \right)^{1/d}.$$

For this vector $\mathbf{a}$, $\mathrm{Vol}_{\mathrm{total}}$ is d times the geometric mean of the p_i,

$$\mathrm{Vol}_{\mathrm{total}} = d \left(\prod_{i=1}^{d} p_i \right)^{1/d}.$$

If some of the p_i are clearly different we can therefore gain a lot by using this refinement, but if f is for example restricted to the unit cube we get practical problems with using a vector $\mathbf{a}$ different from $(1, \ldots, 1)$. Algorithm 11.22

(Orthomonotone-IV) gives the necessary details for both variants of the algorithm. For the expected number of iterations we get for the optimal choice of **a**

$$\mathrm{E}(I) = d \left(\prod_{i=1}^{d} \int_0^\infty u^{d-1} f_i(u) \, \mathrm{d}u \right)^{1/d} / A_f \, .$$

11.5 Computational Experience

In Sects. 11.3 and 11.4 we have presented quite a few different universal rejection algorithms for multivariate distributions. The methods of Sect. 11.3 are applicable to log-concave distributions and they require a time consuming and complicated setup. In contrast the much simpler algorithms of Sect. 11.4 were designed for orthomonotone distributions. So the decision for an algorithm mainly depends on the known properties of the distribution we want to sample of. Nevertheless, the reader may be interested in approximate execution times of these algorithms. So we briefly communicate the computational experience we made with our C-implementations. We emphasize that these results depend not only on the computer, uniform generator, compiler and implementation but above all on the density we use for the experiments. Table 11.2 contains our timings for the density of independent half-normal variates with standard deviation 0.2. (Recall that the relative generation time is the generation time divided by the generation time of one exponential variate.)

Table 11.2. Relative average generation times for a sample of size $n = 10^5$ of d-tuples of independent half-normal variates

	Dimension		
	2	4	8
Algorithms for log-concave distributions			
11.14 (TDR-Bivariate)	2	–	–
11.15 (TDR-Cone)	2.5	4	26
Algorithms for orthomonotone distributions on $[0,1]^d$			
naive rejection	49	1180	$5 \cdot 10^5$
11.17 (Monotone-Multivariate)	45	780	$2 \cdot 10^5$
11.18 (Ahrens-Multivariate), $N_1 = 10$	10	21	–
11.21 (Orthomonotone-III)	23	605	$4 \cdot 10^5$
Algorithms for orthomonotone distributions on $\mathbb{R}_+^d$			
11.19 (Orthomonotone-I)	12	145	$1 \cdot 10^5$
11.20 (Orthomonotone-II)	7	22	605
11.22 (Orthomonotone-IV)	15	36	1140

As expected we can see from Table 11.2 that the generation time rapidly increases with dimension. Algorithms 11.15 and 11.18 that are fastest for dimension 4 need large tables that grow rapidly with the dimension d. For Algorithm 11.18 with a partition into $N = N_1^d = 10^d$ hypercubes this increase is so fast that we were not able to use it for dimension 8. Algorithm 11.15 works fine for our simple distribution and dimension 8 but reaches, depending on the distribution, its memory limits for dimensions between 10 and 14. All other algorithms are much slower and the considerable differences in the execution times shown by Table 11.2 should not be considered as too important. As we have discussed when presenting the different Algorithms in Sect. 11.4 these times strongly depend on the distribution. In addition the different algorithms are well suited for very different distributions. To understand the poor performance for Algorithms 11.17 and 11.21 we should not overlook that the density of the normal distribution with standard deviation 0.2 is very close to 0 on the interval (0.5,1). Thus this distribution is a comparatively difficult distribution for algorithms designed for bounded domain. Using standard deviation 1 instead of 0.2 would result in much shorter execution times.

What is a possible general conclusion from these results? For dimensions higher than 4 it is by no means clear if we can find a good working generation algorithm for a desired distribution. For log-concave distributions the situation is better but even then we should not hope to reach more than dimension 10.

11.6 Multivariate Discrete Distributions

For the generation of discrete random vectors we can utilize the general principles for discrete random variates (Sect. 3) and for continuous random vectors (Sect. 11.1). Necessary modifications and possible problems for the discrete case are given below. However, universal algorithms for multivariate discrete distributions are not discussed in the literature. Thus we are just considering how these principles can be applied to generate from multivariate discrete distributions. The resulting algorithms must be used with care since they can be extremely slow and/or require a lot of memory when they are applied imprudently to a particular distribution.

For the sake of simplicity we only consider random vectors of nonnegative integers. Thus the domain of all distributions considered is a subset of $\mathbb{N}_0^d$.

11.6.1 The Conditional Distribution Method

The conditional distribution method remains exactly the same as explained in Sect. 11.1.1 for continuous distributions. Even the notation can be left unchanged but we interpret f now as a probability mass function (pmf). For dimension d we again have to know all $d - 1$ *marginal distributions* $f_{1,2,\ldots,i}(x_1, x_2, \ldots, x_i)$ of the first i coordinates $(X_1, X_2, \ldots, X_i)$, for $i < d$. Then we can write the joint pmf for the random vector $\mathbf{X} = (X_1, \ldots, X_d)$ as

$$f(x_1, \ldots, x_d) = f_1(x_1) \, f_2(x_2|x_1) \, f_3(x_3|x_1, x_2) \cdots f_d(x_d|x_1, \ldots, x_{d-1})$$

where the *conditional distributions* are computed by

$$f_i(x_i|x_1, \ldots, x_{i-1}) = \frac{f_{1,\ldots,i}(x_1, x_2, \ldots, x_i)}{f_{1,2,\ldots,i-1}(x_1, x_2, \ldots, x_{i-1})} \, .$$

So, if we know a closed form representation for the marginal distributions, Algorithm 11.1 (Conditional-Distribution-Method) reduces the d-dimensional generation problem into d 1-dimensional problems. For most distributions we do not have this representation of the marginals available, but if all components have bounded domain the required marginal distributions could be calculated by summation. If d or the size of the domain are not small this idea results in huge tables for storing the required marginal probabilities. The setup has time complexity linear in the total size of the domain of the multivariate distribution and can thus become very slow.

For the one-dimensional generation step required in the conditional distribution method it is natural to use discrete inversion as explained in Algorithm 3.1 (Sequential-Search). The speed-up presented as Algorithm 3.2 (Indexed-Search) is easily applicable for the first component. For the other components it is not possible to use guide tables in practice as this would require an extra guide table for each possible value of X_1, for each pair (X_1, X_2), and so on. The combination of the conditional distribution method with sequential search requires d uniform random variates; the expected number of iterations I, i.e. the expected number of comparisons in the sequential search of the sampling part is equal to

$$\mathrm{E}(I) = d + \sum_{i=1}^{d} \mathrm{E}(X_i) \, .$$

A proof for $d = 2$ is given by Kemp and Loukas (1978), for general d see Exercise 11.9. By using a guide table of moderate size for generating the first component, the expected number of comparisons is reduced by approximately $\mathrm{E}(X_1) - 1$. Whether this is worthwhile depends on the distribution.

For most distributions the main problem of the algorithm discussed above is – as already mentioned – the fact that we have to calculate and store the marginal distributions which leads to a (very) slow setup and huge tables. The situation is different if we have a closed form representation for the marginal distribution. We demonstrate such a case by means of the multinomial distribution.

Example 11.9 (Multinomial Distribution).
Consider the multinomial distribution with parameters p_i, $i = 1, 2, \ldots, d$, forming a probability vector, and parameter n, which is a positive integer. For the joint pmf we have the well-known formula

$$f(x_1, x_2, \ldots, x_d) = \frac{n! \prod_{i=1}^{d} p_i^{x_i}}{\prod_{i=1}^{d} x_i!} \,.$$

From the genesis of the multinomial distribution it is clear that the component X_i follows a binomial law with parameters n and p_i. Given $X_1 = x_1$ the conditional distribution of the vector $(X_2, X_3, \ldots, X_d)$ is again multinomial with new parameters $p_i/(1 - p_1)$, $i = 2, 3, \ldots, d$, and $n - x_1$. Thus we can see that the required conditional distributions $f_i(x_i|x_1, x_2, \ldots, x_{i-1})$ follow a binomial distribution with $\tilde{n} = n - \sum_{j=1}^{i-1} x_j$ and $p = p_i / \left(1 - \sum_{j=1}^{i-1} p_j\right)$. Using the conditional distribution method we can compile the following generator for the multinomial distribution:

Require: Parameters n and p_i, $i = 1, 2, \ldots, d$, forming a probability vector.
Output: Discrete random vector from the multinomial distribution.
1: Set $p_s \leftarrow 1$ and $\tilde{n} \leftarrow n$.
2: **for** $i = 1$ to $d - 1$ **do**
3:    Generate binomial random variate X with parameters $\tilde{n}$ and p_i/p_s.
4:    Set $X_i \leftarrow X$, $\tilde{n} \leftarrow \tilde{n} - X$, and $p_s \leftarrow p_s - p_i$.
5: Set $X_d \leftarrow \tilde{n}$.
6: **return** vector $\mathbf{X} = (X_1, X_2, \ldots, X_d)$.

The parameters of the binomial variate generated in Step 3 change at every call. Thus a table method is not suitable. But for small n Algorithm 3.1 (Sequential-Search) and for moderate or large n Algorithm 10.4 (Automatic-Rejection-Inversion) can be used. A special algorithm for the binomial distribution (see Hörmann, 1993, and the references given there) will gain the best possible speed. When a uniformly fast binomial generator (like e.g. Algorithm 10.4, Automatic-Rejection-Inversion) is used the generation time for multinomial random vectors is linear in dimension d and uniformly bounded for n.

11.6.2 Transformation into a Univariate Distribution

We have seen in the last subsection that the very fast table methods (Algorithm 3.2, Indexed-Search, and 3.3, Alias-Sample) are only of limited use for the conditional distribution method. But they can be utilized for discrete vector generation if we remember the fact that there is a one-to-one correspondence between the space of all integer vectors ($\mathbb{N}_0^d$) and the integers ($\mathbb{N}_0$). This observation can be used to design inversion algorithms for discrete multivariate standard distributions (Kemp and Loukas, 1978; Loukas and Kemp, 1983, 1986; Loukas, 1984). Devroye (1986a) calls the mapping from the d-tuples into the nonnegative integers a "coding function". The inverse function is called the decoding function.

As a simple example let us consider a distribution with 2-dimensional domain $\{(x_1, x_2) \in \mathbb{N}_0^2 : 0 \leq x_1 \text{ and } 0 \leq x_2 < M\}$. We can easily see that the coding function

$$\mathcal{C}(x_1, x_2) = Mx_1 + x_2$$

corresponds to a "column wise" visit through all pairs of the domain. This is exactly the way two-dimensional arrays are stored in many computer languages. The decoding function is very simple:

$$x_1(\mathcal{C}) = \lfloor \mathcal{C}/M \rfloor \qquad x_2(\mathcal{C}) = \mathcal{C} \bmod M \,.$$

So we can easily use Algorithms 3.2 or 3.3 to generate pairs from discrete joint distributions when both components have bounded domain. If only the second component has bounded domain Algorithm 10.1 (Indexed-Search-Tail) must be used.

If both components have unbounded domains then we can construct a coding function that visits all integer pairs in the first quadrant in cross diagonal fashion:

$$h(x_1, x_2) = \frac{(x_1 + x_2)(x_1 + x_2 - 1)}{2} + x_1 \,.$$

It is also not too difficult to generalize both ideas mentioned above to higher dimensions. However, we face the practical problem that the table sizes explode exponentially with increasing dimension d. Of course we can also try sequential search if the tables become too large. Kemp and Loukas (1978) discuss the performance of sequential search for different coding functions. We have for the number of comparisons (i.e. the number of iterations in the sequential search)

$$\mathrm{E}(I) = \mathrm{E}(h(X_1, X_2, \ldots, X_d)) + 1 \,.$$

For our first example of the coding function this results in

$$\mathrm{E}(I) = M\,\mathrm{E}(X_1) + \mathrm{E}(X_2) + 1 \,,$$

and for the second we get

$$\mathrm{E}(I) = \mathrm{E}((X_1 + X_2)(X_1 + X_2 - 1)/2 + X_1 + 1) \,.$$

Both are much larger than the result for the conditional distribution method in the last section, but on the other hand sequential search combined with a coding function does not require any setup.

When implementing such an algorithm one must never forget that, when $\mathrm{E}(I)$ is large, sequential search requires adding up many small numbers which might result in an infinite loop due to round-off errors. Of course this serious problem can also occur when tables have to be calculated in a setup.

As any coding function is destroying the natural structure of the joint distribution in a largely arbitrary way, unimodality or other nice properties of the joint distribution are not preserved. Therefore we do not recommend to use the rejection method in combination with coding functions. Rejection should always be applied on the "natural" domain of the distribution.

11.6.3 Rejection Method

The advantage of rejection for discrete distributions (Sect. 3.3) is the short setup but on the other hand sampling by rejection is comparatively slow. Moreover, as an automatic method it is only applicable to probability mass functions with special properties. They must, e.g., be monotone or log-concave. For many multivariate discrete distributions the number of different tuples in the domain is so high that a table method of Sect. 11.6.2 is out of question and sequential search becomes very slow (and numerically unstable). Consider, for example, the multinomial distribution with $n = 100$ and dimension $d = 5$. It has 10^8 5-tuples when using naive coding. In Example 11.9 we have explained a fast generator for this distribution. But this generator is based on the conditional distribution method and thus on the fact, that for this special distribution we can easily derive the marginal distributions. A generator for the multinomial distribution based on a coding function and sequential search would be very slow even for moderate values of n and d. For other multivariate discrete distributions the situation is similar and the rejection method should be better suited for them. Here we "only" suffer from the "curse of dimensionality" which is not influenced by the size of the domain.

In Sect. 3.3 three variants for discrete rejection are presented from which continuous rejection with the histogram function and rejection from a discrete hat distribution seem to be applicable to multivariate discrete distributions. But it is not clear which definition of a histogram should be used such that the different multivariate rejection algorithms of Sects. 11.3 and 11.4 can be applied. These problems are – up to our knowledge – not covered in the literature and require separate research. Therefore we only consider the easiest case of orthomonotone distributions here.

The definition of orthomonotonicity can directly be applied to discrete distributions. Thus a joint discrete pmf $f(x_1, x_2, \ldots, x_d)$ is orthomonotone on $\mathbb{N}_0^d$ if all full conditional distributions are monotone on $\mathbb{N}_0$. It is not difficult to see that the histogram function

$$\tilde{f}(x_1, x_2, \ldots, x_d) = f(\lfloor x_1 \rfloor, \lfloor x_2 \rfloor, \ldots, \lfloor x_d \rfloor)$$

is an orthomonotone density. So we can use a slightly changed version of Algorithm 11.18 (Ahrens-Multivariate) to generate a random vector $\tilde{\mathbf{X}}$ with density $\tilde{f}$. The integer vector $\mathbf{X} = \lfloor \tilde{\mathbf{X}} \rfloor$ clearly has joint pmf f as desired. Notice that for a discrete distribution with domain $\{0, 1, 2, \ldots, M-1\}^d$ the histogram function has domain $[0, M)^d$. Consequently the design parameter N_1 should be a divisor of M, as this implies that the $N = N_1^d$ multidimensional subrectangles have integer borders, which implies the best possible fit of the hat function (for a fixed number of rectangles).

Algorithms 11.19, 11.20, 11.21, and 11.22 that do not require (large) tables could also be used in a similar way for rejection with the histogram function. But as all these algorithms need generation from the full conditional distributions it is certainly better to generate from the discrete full conditional distribution by rejection or sequential search. Then it is possible to construct a discrete hat distribution and use rejection for that discrete hat. The details of such a method also require additional investigations.

11.7 Exercises

Exercise 11.1. Use the conditional distribution method to design an algorithm for the uniform distribution over the set $\{(x_1, x_2): x_1, x_2 \geq 0 \text{ and } x_1^2 + x_2^2 \leq 1\}$.

Exercise 11.2. Formulate the two algorithms for the multinormal distribution explained in Sect. 11.1.6 for the two-dimensional normal distribution with $\sigma_1 = \sigma_2 = 1$ and arbitrary ρ between -1 and 1. Check if the two algorithms are producing the same random vectors.

Exercise 11.3. Specialize Algorithm 11.10 (Uniform-Simplex) for generating uniform vectors only in arbitrary triangles in $\mathbb{R}^2$. Which is the easiest way to generate the uniform spacings S_i for this special case?

Exercise 11.4. Prekopa (1973) has proven that all marginals of log-concave distributions are log-concave as well. Explain, how one can use this fact and the conditional distribution method to design a universal algorithm for log-concave distributions. What informations are necessary from the multivariate distributions? Which universal algorithms for one-dimensional distributions are useful for this problem?

Exercise 11.5. Find a generator for the density $f(x, y) = \min\left(c, \frac{1}{xy}\right)$ which can be used as hat function for the two-dimensional version of Algorithm 6.1 (Monotone-01).
Hint: Use the equation $c = 1/(xy)$ to find the border of the two different regions of the hat-function. Then compute the marginal density $f_X(x)$ and design a generator for that marginal density. Proceed with the conditional distribution method to finish the algorithm design.

Exercise 11.6. Compare the expected number of iterations $E(I)$ of Algorithm 11.19 (Orthomonotone-I) and 11.20 (Orthomonotone-II) for the density $f(x) = \prod_{i=1}^{d} e^{-x_i}$.

Exercise 11.7. Compare the expected number of iterations $E(I)$ of Algorithm 11.19 (Orthomonotone-I) and 11.20 (Orthomonotone-II) for the product of half-normal densities.

Exercise 11.8. Compute the expected number of iterations $E(I)$ of Algorithm 11.19 (Orthomonotone-I) for the density $f(x) = \prod_{i=1}^{d} e^{-x_i^r}$ with $x \geq 0$ and $r > 0$. (The special cases $r = 1, 2$, and $r \to \infty$ are the exponential, normal, and uniform distribution, resp.)
Hint: $A_f = \Gamma(1 + 1/r)$.

Exercise 11.9. Show that generating discrete random vectors using a combination of the conditional distribution method with sequential search requires d uniform random variates and the expected number of comparisons is given by

$$E(I) = d + \sum_{i=1}^{d} E(X_i) .$$

Hint: Use the result that for sequential search $E(I) = E(X) + 1$. Start with $d = 2$ and the fact that the weighted sum of conditional expectations is equal to the unconditional expectation. Then continue by induction.

Exercise 11.10. Compile an algorithm for generating discrete random vectors that uses the conditional distribution method together with sequential search.

Part V

Implicit Modeling

12

Combination of Generation and Modeling

Up to now we have assumed that we have a full characterization of the distribution we want to sample from. In most cases the distribution was characterized by its density and we tried to find universal algorithms that are able to sample exactly from that distribution. This is a well posed and often interesting mathematical problem but for the simulation practitioner the problem starts earlier: He has to decide about the input distribution before he can think about generating the variates. There are quite a few different approaches to tackle this task. They depend on the information we have available from the stochastic input we should model. For example, data from the real system might be available. Then we should sample random variates that have the "same property" as the observed sample. Strictly speaking questions of this kind are not the subject of a book dealing with random variate generation. But due to its great practical importance we decided to include this chapter.

Typically we can distinguish between the case that a sample from the real system is available and the case that the user has some assumptions about the distribution (like moments, percentiles, etc.) that do not give a full characterization of the distribution. In both cases the problem we have to solve consists of a modeling step and of a generation step. We will mainly concentrate on the former problem assuming that a sample is available. In Sect. 12.1 it is demonstrated that using the simple idea of resampling with noise results in a very useful algorithm. It is generalized to higher dimensions in Sect. 12.2.

For the latter problem we only provide some hints to the literature in Sect. 12.3 and discuss how to sample from distributions with given moments in more detail in Sect. 12.4. A similar topic is the problem of generating random vectors with given marginals and correlation matrix. The modeling step here consists of deciding about the joint distribution. This problem seems to be of special practical importance and attracted quite an interest in recent papers. So we are including here general considerations concerning this problem together with different algorithms that can generate from vectors with specified one-dimensional marginals and specified correlation as Sect. 12.5.

This problem, of course is closely related to sampling from non-Gaussian time series where only the one-dimensional marginal distribution and a vector of autocorrelations is given (Sect. 13.2).

12.1 Generalizing a Sample

12.1.1 Empirical Distributions

It is well known that the choice of the input distribution is a crucial task for building a stochastic simulation model. If the inputs of the real system we are interested in are observable, it is possible to collect data. In this case the choice of the input distribution for the stochastic simulation model is a statistical problem, which can be called the *modeling of probability distributions from data*. The problem can be solved in a parametric approach by estimating the parameters of a suitable standard distribution or in a non-parametric approach by estimating the unknown distribution. If the size of the available sample is not too small, we are convinced that due to its greater flexibility the non-parametric approach has more advantages unless there are profound a priori reasons (e.g. of physical nature) favoring a certain standard distribution. The task of fitting a distribution to a given sample and then generating variates from that distribution is called *"generating variates from empirical distributions"* or *"generalizing a sample"* in the simulation literature (see e.g. Bratley, Fox, and Schrage, 1987; Law and Kelton, 2000). As these names indicate, the problem of estimating (or modeling) the input distribution is often hidden behind a procedure to generate random variates from data. Therefore no discussion of the quality of the estimation of the different methods is given.

For the parametric approach many considerations about the properties of the chosen distribution have been discussed in papers describing the parametric fit of large flexible distribution families. See Swain, Venkatraman, and Wilson (1988) for the fitting of "Johnson's translation system", Wagner and Wilson (1996) for fitting univariate Bézier distributions, and Avramidis and Wilson (1994) for estimating a polynomial filter to enhance the fit of an arbitrary distribution to a given sample.

For the non-parametric point of view there is a developed statistical theory called density estimation that discusses the optimal estimation of distributions. Especially *kernel density estimation* is well suited for modeling input distributions, as variate generation from these estimates is very simple. This was already observed in the monographs by Devroye and Györfi (1985), Devroye (1986a), and Silverman (1986) but seems to be widely unknown in the simulation literature. Therefore we compare the theoretical properties of different methods of generating random variates from data and will demonstrate with simple examples that the choice of the method can have an influence on simulation results. This section follows mainly the presentation of Hörmann and Bayar (2000), the main idea was also presented in Hörmann and Leydold (2000).

Barton and Schruben (2001) point out an important fact about running simulations with empirical distributions in practice. It must not be overlooked that the confidence intervals computed for performance parameters are computed conditional on the chosen input distribution which also depends on the resampling method we are using for the empirical distribution. We must not assume that the usual confidence intervals also account for the uncertainty that comes from the limited number of observations from the real system. That is the reason why Barton and Schruben (2001) discuss resampling methods together with output analysis. They discuss how to compute confidence intervals for unknown system parameters also taking into account the uncertainty about the true input distribution. A similar problem arises if we estimate the parameters of an input distribution from data. But for this case it is much easier to understand the statement: "Assuming the service time follows a gamma distribution with $\alpha = 3.2$ and $\beta = 2$ we obtain the following system characteristics." And it is also easier to try different scenarios by just changing the parameters of the chosen input distribution.

12.1.2 Sampling from Empirical Distributions Using Kernel Density Estimation

We are given a random sample of size n, denoted by $X_1, X_2, \ldots, X_n$, with sample standard deviation s. Of course the simplest method of sampling from the empirical distribution defined by this sample is *naive resampling*. We just draw numbers from the sample randomly. If the sample is based on a continuous random variable this method has the obvious drawback that only a small number of different values can be generated.

There is a simple modification of naive resampling called *smoothed bootstrap* in the statistic literature (see Shao and Tu (1995, Sect. 3.5) and the references given there). Instead of only resampling, some white noise (i.e. a continuous random variable with expectation 0 and small variance) is added to the resampled numbers. It is not difficult to see, that smoothed bootstrap is the same as generating random variates from a density estimate using the kernel method. The advantage is that it is not even necessary to compute the density estimate. Algorithm 12.1 (Resampling-KDE) contains the formal statement. The ingredients are explained below.

The density of the random noise distribution W is called *kernel* and is denoted by $k(x)$. Clearly $k(x)$ must be a density function and should be symmetric around the origin. As we want to adjust the variance of the random noise to the given data we introduce the scale (smoothing) parameter b. The random variable bW has density $k(x/b)/b$. The random variate Y generated by Algorithm 12.1 (Resampling-KDE) is the equiprobable mixture of n noise distributions, each centered around one of the sample points. This implies that the density of Y (denoted by f_Y) is the sum of n translated versions of

Algorithm 12.1 Resampling-KDE (Kernel density estimation)

Require: Sample $X_1, \ldots, X_n$;
  kernel distribution (Gaussian or boxcar or any from Table 12.1).
Output: Random variate from density estimate $\hat{f}(x)$.
  /* Setup */
1: Compute standard deviation s and interquartile range R of the sample.
2: Set $b \leftarrow 1.364\, \alpha(k) \min(s, R/1.34)\, n^{-1/5}$.
   (See Table 12.1 for kernel parameter $\alpha(k)$.)
  /* Generator */
3: Generate $U \sim U(0,1)$.
4: Set $J \leftarrow \lfloor nU \rfloor + 1$. /* J is uniformly distributed on $\{1, 2, \ldots, n\}$. */
5: Generate W from the kernel distribution.
6: **Return** $Y = X_J + bW$.

$k(x)$ multiplied by $1/n$. Hence we have

$$\hat{f}(x) = f_Y(x) = \frac{1}{nb} \sum_{i=1}^{n} k\left(\frac{x - X_i}{b}\right).$$

Notice that in density estimation this density f_Y is the kernel estimate of the unknown density of the observed sample and therefore denoted with $\hat{f}$.

Of course there remains the question of the choice of the standard deviation b (called *bandwidth* or *window width* in density estimation) and the choice of the kernel function $k(x)$. Here we can use the results of the theory of density estimation as presented, e.g., in Silverman (1986) or Wand and Jones (1995). Silverman (1986) suggests a simple and robust formula for the bandwidth b. It is based on an asymptotic result and minimizes the mean integrated squared error using the sample standard deviation s if the unknown distribution of the data is Gaussian. However, the calculated b is often too large when the unknown distribution is non-Gaussian. Therefore s is replaced by $\min(s, R/1.34)$ (where R denotes the interquartile range) to avoid over-smoothing. We arrive at

$$b = 1.364\, \alpha(k) \min(s, R/1.34)\, n^{-1/5} \tag{12.1}$$

where $\alpha(k)$ is a constant that only depends on the used kernel which can be computed as $\alpha(k) = \sigma_k^{-4/5} \left(\int k(t)^2 dt\right)^{1/5}$; σ_k denotes the standard deviation of the kernel. There are numerous much more complicated ways to determine b published in the literature. For an overview we refer to Devroye (1997b), where the L_1-errors (i.e. the mean integrated absolute error) of many different bandwidth selection procedures are compared. The method we use is a mixture of the methods called "reference: L_2, quartile" and "reference: L_2, std. dev" therein. The results of the simulation study show that with the exception of some very strangely shaped multi-modal distributions the performance of this very simple choice of b is really not bad.

The last question that has to be solved before we can use Algorithm 12.1 is the choice of the kernel. Asymptotic theory shows that the mean integrated squared error (MISE) is minimal for the Epanechnikov kernel $k(x) = (1 - x^2)3/4$ but some other kernels have almost the same efficiency. Therefore we can choose the kernel by considering other properties, e.g. the speed and simplicity of our generation algorithm. In that respect the rectangular kernel (i.e. uniformly distributed noise) is of course the best choice, but it has the theoretical drawback that the estimated density is not continuous. Due to the nice statistical interpretation we prefer Gaussian noise. Table 12.1 gives a list of various kernels with different efficiency and tail behavior. The efficiency is measured by the ratio of the mean integrated squared error of the kernel to that of the Epanechnikov kernel.

Table 12.1. Some kernels with their parameters

	$k(x)$		$\alpha(k)$	σ_k^2	Efficiency
Epanechnikov	$\frac{3}{4}(1-x^2)$	$\|x\|<1$	1.7188	0.2	1.000
Quartic (biweight)	$\frac{15}{16}(1-x^2)^2$	$\|x\|<1$	2.0362	0.14285	0.994
triweight	$\frac{35}{32}(1-x^2)^3$	$\|x\|<1$	2.3122	1/9	0.987
triangular	$1-\|x\|$	$\|x\|<1$	1.8882	1/6	0.986
Gaussian	$\frac{1}{\sqrt{2\pi}}e^{-x^2/2}$		0.7764	1	0.951
boxcar (rectangular)	$1/2$	$\|x\|<1$	1.3510	1/3	0.930
logistic	$\frac{1}{2+e^{-x}+e^x}$		0.4340	3.2899	0.888
t-7 distribution	$\frac{k}{(7+x^2)^4}$		0.6666	7/5	0.878
t-5 distribution	$\frac{k}{(5+x^2)^3}$		0.6174	5/3	0.834
t-3 distribution	$\frac{k}{(3+x^2)^2}$		0.4802	3	0.674

Algorithm 12.1 (Resampling-KDE) guarantees that the density function of the empirical distribution approximates the density of the unknown true distribution as good as possible with respect to the mean integrated squared error. But on the other hand the variance of the empirical distribution is always larger than the variance s^2 of the observed sample. This can easily be seen, since the variance of the empirical distribution is given by $s^2 + b^2\sigma_k^2$, where σ_k^2 denotes the variance of the kernel. This can be a disadvantage in simulations that are sensitive against changes of the variance of the input distributions. To overcome this problem Silverman (1986) suggests rescaling of the empirical distribution such that it has the same variance as the sample. Thus we introduce the correcting factor $\sigma_e^2 = 1 + b^2\sigma_k^2/s^2$, which is simply the ratio of the two variances. Algorithm 12.2 (Resampling-KDEvc) contains the necessary details.

Remark. Positive random variables are interesting for many applications. However, Algorithm Resampling-KDE can cause problems for such applications as it could also generate negative variates. The easiest way out is the so-called mirroring principle. Instead of a negative number Y simply return $-Y$. Unfortunately the mirroring principle disturbs the variance correction. They

Algorithm 12.2 Resampling-KDEvc (Kernel density estimation, variance corrected)

Require: Sample $X_1, \ldots, X_n$;
  kernel distribution (Gaussian or rectangular or any from Table 12.1).
Output: Random variate from variance corrected density estimate $\hat{f}(x)$.
  /* Setup */
1: Compute mean $\bar{x}$, standard deviation s and interquartile range R of the sample.
2: Set $b \leftarrow 1.364\, \alpha(k) \min(s, R/1.34)\, n^{-1/5}$.
   (See Table 12.1 for kernel parameter $\alpha(k)$.)
3: Set $\sigma_e \leftarrow \sqrt{1 + b^2 \sigma_k^2/s^2}$. (See Table 12.1 for the variance σ_k^2 of the kernel.)
  /* Generator */
4: Generate $U \sim U(0, 1)$.
5: Set $J \leftarrow \lfloor nU \rfloor + 1$. /* J is uniformly distributed on $\{1, 2, \ldots, n\}$. */
6: Generate W from the kernel distribution.
7: **Return** $Y = \bar{x} + (X_J - \bar{x} + bW)/\sigma_e$.

---

can be used together but the resulting empirical distribution has a smaller variance than the sample. This can be a practical problem if the sample of a positive distribution has many values close to zero.

It is also possible to simply reject all negative numbers generated to ensure nonnegativity of the output. In density estimation the mirroring principle is generally assumed to result in better estimates as the use of rejection implies that the weight of observations close to 0 is only about half the weight of other observations.

Remark. Algorithms 12.1 (Resampling-KDE) and 12.2 (Resampling-KDEvc) also work with kernels that have heavier tails than the normal distribution, for example a density of the t-distribution or of the logistic distribution. For density estimation applications these kernels are not interesting as they have numerical disadvantages compared to Gaussian or rectangular kernels. For generating empirical distributions they could be interesting as the empirical distribution has the same tail behavior as the kernel. Thus using heavier tailed kernels allows to generate distributions that have a behavior like the given sample for the main part of this distribution together with the desired tail behavior that can be reached selecting the respective kernel from Table 12.1.

Remark. We tried the different kernels suggested in the literature and kernels with heavier tails in some of the below simulation studies (see Sect. 12.1.5) and found that the results were practically the same for all different kernels. Therefore we are only reporting the results of the Gaussian kernel although heavier tailed kernels could be of interest for certain applications.

12.1.3 Linear Interpolation of the Empirical Distribution Function

A different approach for generating random variates from an unknown distribution is based on the empirical cumulative distribution function. It is pro-

Algorithm 12.3 Resampling-ELK (Linear interpolation of empirical cdf)

Require: Sample $X_1, \ldots, X_n$.
Output: Random variate from empirical cdf obtained by linear interpolation.
  /* Setup */
1: Sort sample, $X_{(1)} \leq \ldots \leq X_{(n)}$.
  /* Generator */
2: Generate $U \sim U(0,1)$.
3: Set $J \leftarrow \lfloor (n-1)U \rfloor + 1$ and $V \leftarrow (n-1)U - J + 1$.
4: Set $X \leftarrow X_{(J)} + V(X_{(J+1)} - X_{(J)})$.
5: **Return** X.

---

posed in many textbooks on simulation. Law and Kelton (2000) and Banks, Carson, and Nelson (1999) suggest using a linear interpolation of the empirical cdf as the cdf for generating random variates. Let $X_{(1)} \leq \ldots \leq X_{(n)}$ be the ordered sample of the unknown distribution. Then we use the cdf

$$\hat{F}_e(x) = \begin{cases} 0 & \text{if } x < X_{(1)}, \\ \frac{1}{n-1}\left(i + \frac{x - X_{(i)}}{X_{(i+1)} - X_{(i)}}\right) & \text{if } X_{(i)} \leq x < X_{(i+1)} \text{ for } i = 1, \ldots, n-1, \\ 1 & \text{if } x \geq X_{(n)}. \end{cases}$$

Algorithms for sampling from that linear interpolation of the empirical cdf using the inversion method are simple to implement. We call the algorithm described in Law and Kelton (2000) "empirical Law Kelton" (ELK), see Algorithm 12.3 (Resampling-ELK). This algorithm, however, only generates random points between the minimum and maximum of the sample. This can have a significant impact on the result of a simulation run. For example, large service times can have considerable consequences for a queuing model. Therefore Bratley et al. (1987) added an exponential tail on the right side of the empirical distribution. This is done by fitting an exponential distribution to the right most data points of the sample. We call this method "empirical Bratley Fox Schrage" (EBFS).

12.1.4 Comparison of Methods

Expectation and Variance

One important concern for many simulations are the expectation and the variance of the input distribution. The best we can hope to reach is that the expectation and the variance of the empirical distribution are equal to the sample mean and sample variance of the observed sample as these are the best available estimates for the unknown values. All methods produce random variates that have as expectation the sample mean. Only for ELK the result is slightly different.

Concerning the variance the situation is more complicated: For kernel density estimation we know that the variance of the empirical distribution is larger than the sample variance. A simple calculation shows that the variance of the empirical distribution of Algorithm 12.1 (Resampling-KDE) is $s^2\left((n-1)/n + b^2\sigma_k^2\right)$. For the Gaussian kernel and the choice of b given in formula (12.1) we have $s^2\left(1 - 1/n + 1.058 n^{-2/5}\right)$. For example, for sample size $n = 100$ this factor is 1.164 which shows that it may be sensible to consider the variance corrected version Algorithm 12.1 (Resampling-KDE) which has by design the same variance as the sample for any sample size. If we fit a standard distribution to the data the variance of the fitted distribution is exactly the same as the sample variance if we use parameter estimation based on moment matching. If we use a maximum likelihood estimation for fitting the parameters the variance of the fitted distribution and the sample variance need not be the same but they are close together.

For the two methods based on linear approximation of the cdf (ELK and EBFS) there is no simple formula for the variance of the empirical distribution as the variance depends on the sample. So we computed the variance of the empirical distribution in a small simulation study. It is not surprising that our results showed that for ELK the variance of the empirical distribution is always smaller than the sample variance, for EBFS the variance was due to the added tails always bigger. It was astonishing for us to observe that the factor Var(emp. distr.)/s^2 is strongly influenced by the shape of the distribution. For samples of size 100 we observed factors up to 1.12 for EBFS and down to 0.91 for ELK.

The Fine Structure of the Empirical Distribution

Thanks to L. Devroye we have a theoretical result about the quality of the local approximation of the unknown density by the methods ELK and EBFS (see Bratley et al., 1987, p.132). He has shown that for sample size $n \to \infty$ the density of the empirical distribution does not converge pointwise against the true distribution. In contrast to this poor behavior we know that for method KDE the estimated density does not only converge but we even have (approximately) minimized the mean integrated squared error. Method KDEvc is not as good in that sense but asymptotically it is optimal as well as KDEvc coincides in that case with KDE.

Figure 12.1 illustrates the consequences of this theoretical result. It compares the empirical density of method ELK and of method KDE for a "well behaved" sample of size 20 (from a triangular distribution). When looking at Fig. 12.1 we can also understand that the high peaks in the ELK-density occur when two sample points are comparatively close together and this happens in practically all samples.

It is probably instructive to add here that the linear interpolation of the empirical cdf results in a much better method if we use data grouped in fixed intervals. In this case the density of the empirical distribution is an estimate for the unknown density, the so-called *histogram estimate*. Law and Kelton

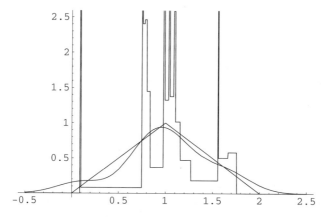

Fig. 12.1. A triangular density with the empirical densities of ELK (step function) and KDE

(2000) and Banks et al. (1999) mention this method for the case that only grouped data are available.

The Distance Between the Theoretical and the Empirical Distributions

There are different distance measures suggested in the literature to compute the distance between two distributions. In density estimation the L_2-difference (integrated squared difference) and the L_1-difference (integrated absolute difference) are of major importance. Another possibility is to use the L_1-difference or the L_2-difference between the two cdfs. As method KDE is based on estimating the density whereas ELK and EBFS are based on an approximation of the cdf we thought that all of these measures would favor automatically one group of the algorithms discussed here. Therefore we decided to use a third class of distance measures for our comparison: The test-statistics of three well known goodness-of-fit tests, the Chi-square test, the Kolmogorov-Smirnov test, and the Anderson-Darling test.

We are given a sample of size n from a fixed standard distribution. Then we generate a sample of size m of the empirical distribution. We use a goodness-of-fit test to see if this sample of size m is close to the fixed standard distribution. For the Chi-square test we took the number of equiprobable classes as $\sqrt{m}$. Then the test-statistic χ^2 divided by m converges for $m \to \infty$ towards

$$\int (f_n(x) - f(x))^2 / f(x) \, dx$$

where f_n denotes the density of the empirical distribution of a sample of size n. This is a weighted squared difference between the empirical and the theoretical density. As the weight is $1/f(x)$ the Chi-square test is more sensitive to deviations in the tails than to deviations in the center of the distribution.

The Kolmogorov-Smirnov test (KS-test) measures the maximal absolute difference between the empirical cdf and the theoretical cdf. The power of the KS-test for deviations in the tails is known to be low.

The Anderson-Darling test on the other hand was designed to detect discrepancies between the tails of the distributions. Like the KS-test it compares the theoretical and the empirical cdf but it uses a weighted L_2-difference to compare the cdfs. The test statistic is

$$A_m^2 = m \int (F_m(x) - F(x))^2 f(x)/(F(x)(1 - F(x))) \, dx$$

where F_m denotes the empirical cdf of the sample.

Our simulation experiment for comparing the distance between the theoretical and the different empirical distributions was organized as follows: For each random sample of size $n = 100$ generated from theoretical distributions, the empirical distribution is computed by means of the different methods. Then from each of these empirical distributions 40 random samples of size $m = 3000$ are generated and the test statistics are computed. The computed test statistics are then averaged over all 40 samples. The result can be interpreted as a (stochastic) distance measure between the theoretical distribution and the empirical distribution. It is like the integrated squared error (ISE) or the integrated absolute error a distance measure between the empirical distribution constructed from one sample and the theoretical distribution. We repeat this procedure 100 times and means and standard errors are computed. Our results with four distributions are shown in Table 12.2. We also compare with "fitting of standard distributions" (FSD) and "naive resampling" without adding noise (Naive).

If we compare our averages with the critical values we can see that all averages are in the critical region. This is not surprising as the empirical distribution was constructed based only on a sample of size 100 of the correct distribution. A larger sample from the empirical distribution cannot have exactly the same properties as a sample from the correct distribution. Nevertheless, we can interpret the observed values as distance measure between the empirical and the theoretical distribution. We see that the method chosen for constructing the empirical distribution has an important influence on the results. In the chi-square and in the Kolmogorov-Smirnov tests methods KDE and KDEvc perform considerably better than EBFS, ELK and naive resampling. For the Anderson-Darling tests the differences are small but even there KDEvc performs best. The results do not only show that kernel density estimation performs better than the other methods, it also shows that the variance corrected method performs in almost all cases better than the original version. We are also surprised that for these three very different distance measures and for four quite different distributions the same method for constructing the empirical distribution is best or close to best in all cases.

12.1 Generalizing a Sample   315

Table 12.2. Average test statistics and standard errors (in brackets) for different resampling methods (Naive denotes naive resampling without adding noise; FSD denotes fitting the gamma and normal distribution, resp.). The critical values (5%) are 71.0 for χ^2-statistics, 24.8×10^3 for KS-statistics, and 2.5 for AD-statistics

	EBFS	ELK	Naive	KDE	KDEvc	FSD
Theoretical distribution: Gamma(2)						
χ^2-mean	1218 (24)	1270 (25)	1642 (31)	319 (10)	241 (7)	110 (6)
KS-mean$\times 10^3$	79 (2)	79 (2)	84 (2)	56 (2)	54 (2)	44 (2)
AD-mean	27 (2)	27 (2)	27 (2)	31 (2)	24 (2)	17 (2)
Theoretical distribution: Gamma mixture G(2)&G(6)						
χ^2-mean	1196 (23)	1256 (24)	1651 (28)	268 (6)	230 (6)	220 (5)
KS-mean$\times 10^3$	84 (3)	84 (3)	88 (3)	57 (2)	63 (2)	68 (2)
AD-mean	32 (2)	32 (2)	32 (2)	30 (2)	28 (2)	28 (2)
Theoretical distribution: Normal(0,1)						
χ^2-mean	1290 (26)	1240 (23)	1633 (30)	220 (12)	155 (7)	113 (7)
KS-mean$\times 10^3$	82 (2)	83 (2)	87 (2)	59 (2)	54 (2)	46 (2)
AD-mean	30 (2)	30 (2)	31 (2)	30 (2)	22 (2)	19 (2)
Theoretical distribution: Normal Mixture N(0,1)&N(3,1)						
χ^2-mean	1261 (29)	1229 (26)	1601 (33)	350 (18)	209 (9)	513 (7)
KS-mean$\times 10^3$	80 (3)	80 (3)	80 (3)	62 (2)	65 (3)	92 (2)
AD-mean	32 (3)	31 (3)	31 (2)	35 (3)	26 (2)	44 (2)

It is clear that the method of fitting a standard distribution (FSD) performs best if we fit the correct distribution. But even then Table 12.2 shows that the results for KDEvc is in most cases close to those of FSD.

Summarizing the comparison of the different methods strongly favors the two methods based on kernel density estimation. We think it is justified at this point to recommend method KDEvc in all cases, where there are no strong arguments for fitting a standard distribution. Nevertheless, one might still think that the differences between the different methods are so small that it is not necessary to bother about. In the next section we therefore compare the influence of the different methods on the results of three small simulation studies.

12.1.5 Computational Experience for Three Simple Simulation Examples

First we want to stress that the methods of this section are simple and fast sampling methods. Experiments with the UNU.RAN implementations of Algorithms 12.1 (Resampling-KDE) and 12.2 (Resampling-KDEvc) have shown that they have practically the same speed. Of course the speed depends on the method used for sampling from the kernel density; using the boxcar kernel (i.e. the uniform distribution) both algorithms were in our experiments as fast as sampling from the exponential distribution by inversion. For the Gaussian kernel the algorithms were about 20% slower.

Changing the method of modeling the empirical distribution is not more than changing the fine structure and perhaps slightly the variance of the input distribution of a simulation model. It is to be expected that many simulations, which have as output averages of a large number of input random variables, are not very sensitive to small changes in the fine structure of the input distribution. For example, it is known that the average waiting time for the M/G/1 queue is only influenced by the expectation and the variance of the service time distribution and not by its shape. It is even better known that the distribution of the sample mean of a large sample is always very close to normal. The parameters of that normal distribution are again only influenced by the expectation and the variance of the underlying distribution and not by its shape. These are arguments why the choice of the method will not have a big influence on many simulation results. Nevertheless, we try to get some insight into this question by looking at three examples.

The first simulation model we have considered is the M/G/1 queue. We include this example here as the M/G/1 queue is well known and understood (see any introductory book on operations research or, e.g., Hall, 1991). The inter-arrival times are taken exponential with expectation 1, the service times are modeled from samples of different gamma distributions, using the different empirical methods described above. Then we have simulated the model starting with an empty system and observed the average waiting time (avw) and the maximal number in queue (maxniq). We know that by the Pollaczek-Khintchine formula (see e.g. Hall, 1991) avw can be computed directly from the mean and variance of the service time distribution but for maxniq the shape of the service time distribution could have some influence. So we repeated the experiment for several different samples of the theoretical service-time distribution to get an average over different samples. The results given in Table 12.3 only show that there is little to choose between the different methods to fit an empirical distribution. All methods have about the same performance and they rarely differ more than one standard error. This is true for avw as well as for maxniq. The second clear result is that the size of the error when using an empirical instead of the correct distribution strongly depends on ρ, the utilization factor of the server. Again the result for avw can be explained by the Pollaczek-Khintchine formula. Notice that for $\rho = 0.4$ and a sample of size $n = 100$ available, the errors are much smaller than those for $\rho = 0.9$ and $n = 500$. We also note that Shanker and Kelton (1991) have made a similar study for the M/G/1 queue to demonstrate the consequences of generating data from incorrect distributions.

Due to the very small differences between the methods for the M/G/1-queue we looked for simulation examples that are influenced by the fine struc-

Table 12.3. M/G/1-queue: Average error and its standard deviation (in brackets) of average waiting time (avw) and maximal number in queue (maxniq). All errors and standard deviations were multiplied by 100. ρ denotes the server utilization, α the shape parameter of the gamma distributed service times, n the size of the sample from the theoretic distribution used to construct the empirical distribution

	ρ	α	n	EBFS		ELK		Naive		KDE		KDEvc	
avw	0.9	10	100	138	(21)	132	(19)	138	(21)	137	(21)	137	(21)
maxniq	0.9	10	100	341	(42)	332	(37)	343	(40)	340	(41)	341	(40)
avw	0.9	10	500	52	(4)	53	(4)	52	(4)	53	(4)	52	(4)
maxniq	0.9	10	500	138	(11)	143	(10)	140	(11)	142	(11)	137	(10)
avw	0.4	2	100	3	(0.3)	3	(0.2)	3	(0.2)	3	(0.3)	3	(0.3)
maxniq	0.4	2	100	47	(4)	38	(3)	36	(3)	37	(3)	37	(3)
avw	0.4	2	500	1.5	(0.1)	1.5	(0.1)	1.5	(0.1)	1.5	(0.1)	1.5	(0.1)
maxniq	0.4	2	500	21	(2)	18	(1)	17	(1)	18	(1)	19	(1)
avw	0.4	10	100	1.0	(0.1)	1.0	(0.1)	1.0	(0.1)	1.0	(0.1)	1.0	(0.1)
maxniq	0.4	10	100	13	(1)	12	(1)	11	(1)	12	(1)	11	(1)
avw	0.4	10	500	0.5	(0.03)	0.5	(0.03)	0.5	(0.03)	0.5	(0.03)	0.5	(0.03)
maxniq	0.4	10	100	6	(0.4)	5	(0.4)	5	(0.4)	5	(0.4)	6	(0.4)

ture of the distribution. Thus we have made the following experiment: We have taken a sample of size 50 of a gamma distribution and computed the maximal and the minimal distance between two neighboring points. We then replace the true distribution by empirical distributions constructed from samples of sizes $n = 100$ and $n = 500$ drawn from the true distribution. We repeated each experiment 10^5 times. The results are given in Table 12.4.

Table 12.4. Average minimal and maximal distances, standard deviations in brackets

	n	EBFS		ELK		Naive		KDE		KDEvc		Correct	
Theoretical distribution: Gamma(2)													
$\min \times 10^5$	100	55	(0.7)	54	(0.7)	0	(0)	172	(2)	168	(2)	163	(2)
$\min \times 10^5$	500	76	(1)	77	(1)	12	(0.6)	173	(2)	167	(2)	163	(2)
$\max \times 100$	100	196	(2)	125	(1)	147	(1)	138	(1)	129	(1)	150	(1)
$\max \times 100$	500	172	(2)	139	(1)	148	(1)	146	(1)	141	(1)	150	(1)
Theoretical distribution: Gamma(20)													
$\min \times 10^5$	100	209	(3)	199	(3)	0	(0)	679	(7)	624	(6)	628	(6)
$\min \times 10^5$	500	306	(4)	304	(4)	53	(3)	662	(7)	628	(6)	628	(6)
$\max \times 100$	100	706	(5)	291	(1)	342	(2)	339	(2)	310	(2)	323	(2)
$\max \times 100$	500	482	(4)	314	(2)	331	(2)	338	(2)	324	(2)	323	(2)

The interpretation of Table 12.4 with respect to the minimal distance is simple. Naive resampling is useless if the fine structure of the distribution is of any importance, even though the sample generated from the empirical distribution had only size $m = 50$ whereas $n = 100$ or even $n = 500$ data points were available. The second observation is that the fine structure of EBFS and ELK are only slightly better whereas those of KDE and KDEvc are much better with results close to the results using the correct distribution. It is interesting that the results of the variance corrected method are better than those of the standard method. If we look at the results for the maximal distance we see that naive resampling works surprisingly well whereas EBFS works bad, although it assumes exponential tails, which should be an advantage. KDE and KDEvc again show good results.

Our last example can be interpreted as part of a computer system simulation. Two processes work with the same file. They start at the same time and end after ten hours. The time between two file accesses follows the same distribution (gamma(10, 0.1)). We want to compute the probability that the two processes try to access the file at "almost the same time". "Almost the same time" means that the time difference is smaller than a given tolerance. What happens in this example if again the true gamma distribution is replaced by an empirical distribution which is constructed from a sample from the correct distribution? Our results are given in Table 12.5.

Table 12.5. Estimated probability of file access at the same time multiplied with 10^6, standard deviation of estimates in brackets

n	tol	EBFS	ELK	Naive	KDE	KDEvc	Correct
100	10^{-5}	306 (25)	282 (24)	10328 (143)	202 (20)	240 (22)	167 (10)
500	10^{-5}	242 (22)	248 (22)	2156 (66)	186 (19)	214 (21)	167 (10)
100	10^{-4}	2592 (72)	2628 (72)	12142 (155)	1960 (63)	1978 (63)	1898 (30)
500	10^{-4}	2278 (67)	2234 (67)	3956 (89)	2044 (64)	2026 (64)	1898 (30)

The results of Table 12.5 are not surprising. As we have seen in Table 12.4 the empirical distributions constructed by KDE and KDEvc have about the same behavior as the true distribution. EBFS and ELK are considerably worse whereas naive resampling is totally useless for these examples. We think that this experiment is quite interesting as it shows how simulation results can be influenced by the choice of the resampling method in the case of rare events.

The first conclusion from the above investigations is in our opinion that methods that generate random variates directly from data are important and useful tools in simulation studies. They are easy to use and more flexible than fitting standard distributions to data. If there are enough, say at least around 100, data available resampling methods should be used whenever there are no a priori reasons for using a certain standard distribution. The second conclusion is even more obvious. Use kernel density estimates to construct the empirical distribution function. We hope that we have demonstrated that this method should not be unknown to simulation practitioners. Sampling from kernel density estimates is a simple task. There is mathematical theory that shows the good theoretical behavior of these estimates, and the empirical results of this section confirm that these good theoretical properties can lead to more accurate results in simulation studies. A question not fully solved here is, whether the variance correction should be used. We think that the results presented here clearly favor the variance corrected version (KDEvc) but there are applications where the original version could perform better. Bal (2002) concludes from his simulation results that for quantile estimations the variance corrected version should be applied.

12.2 Generalizing a Vector-Sample

When modeling the stochastic input of a simulation model we often have the problem of dependencies between different input values. One of the most common modeling errors is to overlook these dependencies. If enough data are available we can use a method to generalize a sample of vectors. This can be of great practical importance as there are very few methods that can be easily used to model and generate random vectors from data. Simple algorithms used in practice include naive resampling and using the multinormal distribution that has the same expectation and covariance matrix as the given sample. In the literature Taylor and Thompson (1986) describe a method that samples from a mixture of nearest neighbor and kernel density estimate, which is also included in the IMSL library. In this section we describe a much simpler method which is solely based on kernel density estimation to resample from the given data. As it is always the case the difficulty of estimation increases with dimension d and we need really large samples to obtain acceptable estimates of the unknown density if d is greater than 3 or 4.

12.2.1 Sampling from Multidimensional Kernel Density Estimates

The idea of resampling with noise remains the same as in dimension one and we can leave Algorithm 12.1 (Resampling-KDE) unchanged and interpret it as a vector algorithm by just taking X_J and W as vectors. However, we have to specify the full variance-covariance matrix of the noise W. This problem is more difficult than in dimension one and there is no generally accepted method recommended in the literature. Silverman (1986) explains a "quick and dirty" version that is based on the following idea: We can transform the data of the sample such that they have unit covariance matrix. Then we can use (as for dimension one) a simple reference method to find the smoothing parameter b of a radial symmetric kernel. The resulting b is optimal if the

Algorithm 12.4 Resampling-MVKDEvc

Require: Sample $\mathbf{X}_1, \ldots, \mathbf{X}_n$ of vector in $\mathbb{R}^d$.
Output: Random vector from density estimate using a Gaussian kernel.
/* Setup */
1: Compute mean vector $\bar{\mathbf{x}}$ and covariance matrix $\mathbf{S}$ of the sample.
2: Compute Cholesky-factor $\mathbf{L}$ of $\mathbf{S}$.
   /* use Algorithm 11.4 (Cholesky-Decomposition) */
3: Compute $b \leftarrow \left(\frac{4}{(d+2)n}\right)^{\frac{1}{d+4}}$ and $c_b \leftarrow 1/\sqrt{1+b^2}$.
   /* Generator */
4: Generate $U \sim U(0,1)$.
5: Set $J \leftarrow \lfloor nU \rfloor + 1$. /* J is uniformly distributed on $\{1, 2, \ldots, n\}$. */
6: Generate random vector $\mathbf{W}$ of d independent normal variates.
7: Set $\mathbf{Y} \leftarrow \bar{\mathbf{x}} + (\mathbf{X}_J - \bar{\mathbf{x}} + \mathbf{L}(b\mathbf{W}))c_b$.
8: **Return** random vector $\mathbf{Y}$.

---

vector-data come from a multinormal distribution. Using the Gaussian kernel we have

$$b = \left(\frac{4}{(d+2)n}\right)^{\frac{1}{d+4}}$$

where n denotes the sample size and d the dimension of the vectors. After adding the noise $b\mathbf{W}$ to the chosen transformed data point we transform them back to the original covariance structure.

It is not difficult to see that instead of transforming the data we can equivalently use a kernel that has the same covariance matrix as the data and is multiplied by the smoothing parameter b. As in dimension one the above b is optimal when we assume multinormal distributed data and it is over-smoothing the data in many cases where the unknown distribution is multi-modal or skewed. However, there is no simple way to correct this over-smoothing in higher dimensions as it was possible in dimension one. Therefore it can be sensible to use a smaller b (for example half of the calculated value) if we assume that the unknown distribution is strongly non-normal. As for the one-dimensional case the variance of this empirical distribution is too large. Thus we need a correcting factor c_b. The details are collected in Algorithm 12.4 (Resampling-MVKDEvc).

We know that the rough estimate of b can be far from optimal for many distributions. This is the very reason that we suggest to use the normal kernel and the variance corrected version here, as we can make the following considerations. Algorithm 12.4 (Resampling-MVKDEvc) is identical to naive resampling when we use the bandwidth $b = 0$ for the noise. On the other hand the estimated distribution converges to the kernel distribution (i.e. multinormal) with the same variance-covariance matrix as the sample if $b \to \infty$. Thus for very large values of b Algorithm 12.4 is almost identical to fitting a multi-normal distribution to the data. Fitting the multinormal distribution (the case $b \to \infty$) is optimal if the (unknown) distribution is normal, naive resampling (the case $b = 0$) is optimal if the unknown distribution is not continuous. In other cases it seems obvious that values of b between 0 and ∞ lead to a better approximation of the unknown distribution. So even if our guess of b is far from optimal it is still likely that it is better than using $b = 0$ or $b = \infty$. Therefore Algorithm 12.4 with an intermediate value of b should have better properties than naive resampling or fitting the normal distribution for most continuous distributions.

Addressing the performance characteristics of Algorithm 12.4 (Resampling-MVKDEvc), we can see that we have to store all sample vectors, the sample mean and the Cholesky factor; thus the memory requirements are in total at least $(n+1)d+d(d+1)/2$ floating point numbers. The computing time for the setup is linear in n and cubic in dimension d due to the Cholesky factorization. The generation part requires d independent normal variates and $O(d^2)$ floating point operations. So this is a fast generation algorithm for random vectors and its speed depends only quadratic on the dimension. We could even consider to generate a large random sample from a given distribution by some other (slower) method first, perhaps even using a MCMC algorithm as explained in Chap. 14 and to store that sample. Then we can continue with the resampling algorithm as it is probably much faster. The problem here, however, is that we need rapidly increasing sample sizes for increasing d to obtain a good estimate for the distribution which makes this idea again infeasible for d larger than 5.

12.2.2 Computational Experience

Algorithm 12.4 (Resampling-MVKDEvc) has the advantage that it is based on a well understood method of density estimation whereas no theory is available to show the approximation properties of the method of Taylor and Thompson (1986). Another practical disadvantage of the latter method is that sampling is much slower than for the kernel Method and that setup time and storage requirements explode for large samples and higher dimensions.

For Algorithm 12.4 the greatest computational effort is the generation of the multivariate noise distribution using Algorithm 11.3 (Multinormal-Cholesky); the random selection of data points is using the discrete uniform distribution and is therefore very fast. Our timing experiments with the UNU.RAN implementation of Algorithm 12.4 have shown that for dimensions $d = 2$ to 10 the execution times are at most 20 percent slower than the generation of d iid. exponential random variates by inversion. If the sample used is bigger than $n = 10^4$ we can observe a slow-down, probably due to cache effects as a large array is randomly accessed. Nevertheless, these empirical results underline that resampling with noise is a very fast method to generate random vectors.

12.3 Modeling of Distributions with Limited Information

Sometimes it is necessary to model an input distribution without any data from the real system. In this case we can have to use some subjective information about the distribution, like some moments or percentiles guessed by an "expert". If we have information (or make assumptions) about the first four moments of the input distribution we can use Algorithm 12.6 (4-Moments-Continuous) below to generate random variates that have the assumed moments.

DeBrota, Dittus, Roberts, and Wilson (1989) describe a software where a Johnson bounded distribution is chosen to match such informations. The Johnson bounded distribution has cdf

$$F(x) = \Phi_N \left(\gamma + \delta \log \left(\frac{x - \xi}{\lambda - (x - \xi)} \right) \right)$$

where Φ_N is the standard normal cdf, γ and δ are shape parameters, ξ is the location, and λ the scale parameter. The user has to provide an upper and lower bound for the distribution. In addition the mode and the standard deviation, the mode and a percentile, two percentiles or comparable information are required. Once the parameters have been identified random variate generation is no problem as we can easily invert the cdf.

Wagner and Wilson (1996) describe univariate Bézier distributions, which are a special case of spline curves that model the cdf. Thus a Bézier distribution can be fit if we have information about several percentiles. It is also possible to specify the first four moments and to construct the Bézier distribution such that it matches these moments. Bézier random variates are then generated by numerical inversion.

It can be of practical importance to consider the case where we have a small sample and some additional information about the distribution available. The Bézier distribution is a useful tool for this situation as well.

Note that Algorithms 12.1 (Resampling-KDE) and 12.2 (Resampling-KDEvc) can also be used for similar problems. For example, if we have a small sample together with the information that the tails of the distribution are exponential we can use these algorithms with a kernel density that has exponential tails. If we have a sample together with information about mean μ and standard deviation σ of the distribution we can use a slightly changed version of Algorithm 12.2 (Resampling-KDEvc) that generates random variates by resampling with noise and standardizes the resulting distribution such that it has the desired values for μ and σ.

For more references and information about software for this kind of generation problems see Nelson and Yamnitsky (1998).

12.4 Distribution with Known Moments

Sometimes it can be of practical importance to generate random variates from a distribution where just a few moments are known. It cannot be expected that these moments uniquely determine the distribution. See Fig. 12.3 on p. 328 for an example of two distinct distributions that have all moments $\mu_r = \int x^r f(x)\,dx$ for $r = 1, 2, \ldots$ in common. Therefore just a few moments can certainly not fully determine a distribution. Nevertheless, there may be applications for an algorithm that generates random variates with given moments. Therefore following the ideas of Devroye (1986a, Sect. XIV.2.4) we include a simple algorithm that constructs and generates from densities that match the first four moments.

12.4.1 Matching the First Four Moments

For a given density f with mean μ and standard deviation σ, the density of the standardized distribution is given by $\sigma f(\mu + \sigma x)$. Thus we can easily compute the *standardized moments* $\tilde{\mu}_i = \int x^i \sigma f(\mu + \sigma x)\,dx$ and obtain

$$\tilde{\mu}_3 = \frac{\mu_3 - 3\mu_2\mu_1 + 2\mu_1^3}{\sigma^3} \quad \text{and} \quad \tilde{\mu}_4 = \frac{\mu_4 - 4\mu_3\mu_1 + 6\mu_2\mu_1^2 - 3\mu_1^4}{\sigma^4}.$$

To generate random variates from a distribution that matches the first four given moments we generate a distribution Z with mean zero and unit variance which matches the standardized moments $\tilde{\mu}_3$ and $\tilde{\mu}_4$ first. Then we return $\mu + \sigma Z$ which has the desired first four moments. Hence we need a method to generate distributions with mean zero and unit variance and given moments μ_3 and μ_4. A special case of Thm. 2.1 of Devroye (1986a, Sect. XIV.2) gives the necessary and sufficient condition that a distribution with the given first four moments exists as

$$\begin{vmatrix} 1 & \mu_1 & \mu_2 \\ \mu_1 & \mu_2 & \mu_3 \\ \mu_2 & \mu_3 & \mu_4 \end{vmatrix} \geq 0.$$

Setting $\mu_1 = 0$ and $\mu_2 = 1$ and computing the determinant results in the condition

$$\mu_4 \geq \mu_3^2 + 1. \tag{12.2}$$

Thus in the (μ_3, μ_4)-plane all feasible pairs lie above or on the limiting parabola.

To construct a simple algorithm it is useful to consider the standardized Bernoulli distribution with parameter q. It is a two-atomic distribution which takes the value $-q/\sqrt{q(1-q)}$ with probability $1-q$ and the value $(1-q)/\sqrt{q(1-q)}$ with probability q. Straightforward calculations show that this distribution has the moments

$$\mu_3 = \frac{1 - 2q}{\sqrt{q(1-q)}} \quad \text{and} \quad \mu_4 = \frac{1 - 3q + 3q^2}{q(1-q)}.$$

It turns out that $\mu_3^2 + 1 = \mu_4$ for the standardized Bernoulli distribution. Consequently all pairs (μ_3, μ_4) fall on the limiting parabola and by letting q vary from 0 to 1, all points on the parabola are obtained. Thus for a given arbitrary value of $\mu_4 \geq 1$ we can find the parameters q of the two corresponding standardized Bernoulli distributions by solving the equation of μ_4 above with respect to q. We obtain

$$q_{1,2} = 0.5\left(1 \pm \sqrt{\frac{\mu_4 - 1}{\mu_4 + 3}}\right).$$

Note that $q_2 = 1 - q_1$. To have a free choice for μ_3 as well we can utilize the general observation that the mixture of two distributions X and Y where X is taken with probability p and Y with probability $1 - p$ has moments $\mu_i = p\mu_i(X) + (1-p)\mu_i(Y)$. We have seen above that for the standardized Bernoulli distribution all pairs (μ_3, μ_4) fall on the limiting parabola. Thus for fixed μ_4 the standardized Bernoulli distributions with parameters q_1 and q_2 have maximal and minimal feasible moment $\mu_3 = \pm\sqrt{\mu_4 - 1}$, respectively. Hence taking a mixture of these two distributions we can obtain all possible values of μ_3 without changing μ_4. To obtain the mixture probability p for choosing the distribution with parameter q_1 we have to solve the equation $(p - (1-p))\sqrt{\mu_4 - 1} = \mu_3$ and find

$$p = \frac{1}{2}\left(1 + \frac{\mu_3}{\sqrt{\mu_4 - 1}}\right).$$

Summarizing we have derived the necessary formulas to compute q_1 and p. The resulting mixture of two standardized Bernoulli distributions with parameters q_1 and $1 - q_1$ is a discrete distribution with only 4 atoms but it has mean zero and unit variance and μ_3 and μ_4 as desired. Of course it is very simple to generate variates from such a distribution. We collect details of one possible variant as Algorithm 12.5 (4-Moments-Discrete). For most applications this distribution is probably not a good solution to the moment problem. Nevertheless, we will demonstrate below that this very simple distribution is the main part of an algorithm to construct and generate continuous distributions with given first four moments.

12.4.2 Unimodal Densities and Scale Mixtures

Khinchine's theorem tells us that the product UY of two independent random variates $U \sim U(0,1)$ and Y always results in a unimodal distribution and vice versa that for any unimodal distribution X there exists a random variate Y such that $X = UY$.

If U is non-uniform, then $X = UY$ is called a *scale mixture*. Such products are useful here as $\mathrm{E}(X^i) = \mathrm{E}(Y^i)\mathrm{E}(U^i)$. If U is a fixed distribution with moments ν_i and we want X to have moments μ_i then it is obvious that Y must have moments μ_i/ν_i. Thus if we can generate an arbitrary Y with moments

12.4 Distribution with Known Moments   325

Algorithm 12.5 4-Moments-Discrete

Require: First four moments $(\mu_1, \mu_2, \mu_3, \mu_4)$; moments must be feasible.
Output: Discrete random variate with four atoms and desired moments.
    /* Setup */
1: Set $\sigma^2 \leftarrow \mu_2 - \mu_1^2$.
2: **if** $\sigma^2 \leq 0$ **then**
3:     **abort** (μ_2 not feasible).
4: Set $\sigma \leftarrow \sqrt{\sigma^2}$.
5: Set $\tilde{\mu}_3 \leftarrow (\mu_3 - 3\mu_2\mu_1 + 2\mu_1^3)/\sigma^3$ and $\tilde{\mu}_4 \leftarrow (\mu_4 - 4\mu_3\mu_1 + 6\mu_2\mu_1^2 - 3\mu_1^4)/\sigma^4$.
6: **if** $\tilde{\mu}_4 - 1 \leq 0$ **then**
7:     **abort** (μ_4 not feasible).
8: Set $q \leftarrow 0.5\left(1 + \sqrt{(\tilde{\mu}_4 - 1)/(\tilde{\mu}_4 + 3)}\right)$ and $p \leftarrow 0.5\left(1 + \tilde{\mu}_3/\sqrt{\tilde{\mu}_4 - 1}\right)$.
9: **if** $p < 0$ or $p > 1$ **then**
10:     **abort** (μ_3 not feasible).
    /* Locations x_i and probabilities p_i of the discrete distribution */
11: Set $x_0 \leftarrow \mu_1 + \sigma q/\sqrt{q(1-q)}$ and $p_0 \leftarrow p(1-q)$.
12: Set $x_1 \leftarrow \mu_1 - \sigma(1-q)/\sqrt{q(1-q)}$, and $p_1 \leftarrow pq$.
13: Set $x_2 \leftarrow \mu_1 + \sigma(1-q)/\sqrt{q(1-q)}$, and $p_2 \leftarrow (1-p)q$.
14: Set $x_3 \leftarrow \mu_1 - \sigma q/\sqrt{q(1-q)}$, and $p_3 \leftarrow (1-p)(1-q)$.
    /* Generator */
15: Generate $U \sim U(0,1)$.
16: **if** $U \leq p_0$ **return** x_0.
17: **else if** $U \leq p_0 + p_1$ **return** x_1.
18: **else if** $U \leq p_0 + p_1 + p_2$ **return** x_2.
19: **else return** x_3.

μ_i/ν_i the scale mixture YU will have the desired moments and a continuous distribution if U is continuous. Note that it is possible for a feasible moment sequence μ_i that μ_i/ν_i is not feasible. This means that for the given U the scale mixture might not include a density with the desired moments.

It is not useful to take U as a symmetric standardized distribution (like the standard normal distribution) as this implies that $\nu_3 = 0$. For such a U all scale mixtures are symmetric themselves and no skewed distributions can be constructed. It is also not convenient to use a distribution U where ν_4 is much bigger than ν_3 as this implies that the scale mixture family only contains distributions with very large μ_4. This is the reason why taking for example U the exponential distribution (i.e. $\nu_i = i!$) is not useful in most practical situations. In Table 12.6 we include simple examples for U that seem to be useful to us. Using the condition

$$\begin{vmatrix} 1 & \mu_1/\nu_1 & \mu_2/\nu_2 \\ \mu_1/\nu_1 & \mu_2/\nu_2 & \mu_3/\nu_3 \\ \mu_2/\nu_2 & \mu_3/\nu_3 & \mu_4/\nu_4 \end{vmatrix} \geq 0$$

it is not difficult to construct the feasible region in the (μ_3, μ_4) plane for standardized scale mixtures (as posed in Exercise 12.3). Algorithm 12.6 (4-

Table 12.6. Possible examples for U in the scale mixture (Z denotes a standard normal variate)

Distribution	ν_1	ν_2	ν_3	ν_4	Generator
Uniform $U(0,1)$	1/2	1/3	1/4	1/5	trivial
Trapezoidal on $(-1,2)$	1/2	2/3	3/4	16/15	$U(-1,1)+U(0,1)$
Normal $N(1,1)$	1	2	4	10	$Z+1$
Normal $N(1,0.5)$	1	1.25	1.75	2.6875	$0.5\,Z+1$
Normal $N(1,\sigma)$	1	$1+\sigma^2$	$1+3\sigma^2$	$1+6\sigma^2+3\sigma^4$	$\sigma Z+1$

Algorithm 12.6 4-Moments-Continuous

Require: First four moments $(\mu_1,\mu_2,\mu_3,\mu_4)$, moments must be feasible; mixture distribution U, see Table 12.6.
Output: Random variate with desired moments.
/∗ Setup ∗/
1: Set $\tilde{\mu}_i \leftarrow \mu_i/\nu_i$, for $i=1,2,3,4$.
2: Call the setup of Algorithm 12.5 (4-Moments-Discrete) with moments $\tilde{\mu}_i$.
  (If moments $\tilde{\mu}_i$ are not feasible for chosen mixture distribution, try mixture distributions $U_n \sim N(1, 2^{-n})$ recursively for $n=1,2,\ldots$ till we obtain feasibility.)
/∗ Generator ∗/
3: Generate random variate Y with moments $\tilde{\mu}_i$, use Algorithm 12.5.
4: Generate a random variate U from the mixture distribution.
5: **return** YU.

Moments-Continuous) gives the details of how to generate random variates with the desired first four moments for these scale mixtures. Figure 12.2 shows the resulting densities for different desired moments. We can see that uniform U leads to staircase shaped densities, a trapezoidal U to piecewise linear densities and normal $U \sim N(1,1)$ to very smooth bell-shaped densities. For $N(1,\frac{1}{2})$ the resulting scale mixture has a multi-modal density.

It is natural to ask here which mixture distribution U leads to the largest possible feasible region in the (μ_3,μ_4)-plane. If U is the constant 1 then the mixture $X=UY$ is equal to Y. Clearly this is not what we aimed at when introduced mixture distributions. But this consideration makes clear that a sequence of distributions U_n converging to the constant 1 will lead to increasing regions of feasibility. Thus, for example, $U_n \sim N(1,2^{-n})$ can be used for U to obtain mixture distributions for pairs (μ_3,μ_4) that are close to the general feasibility border given in condition (12.2). We can see in the bottom rows of Table 12.6 how the moments ν_i for U normal with $\mu=1$ approach 1 when σ becomes smaller. This implies that the feasible region of the scale mixture approaches to the feasible region for arbitrary distributions. On the other hand, however, we can see in Fig. 12.2 that small values of σ result in multi-modal densities for the scale mixture.

12.4 Distribution with Known Moments 327

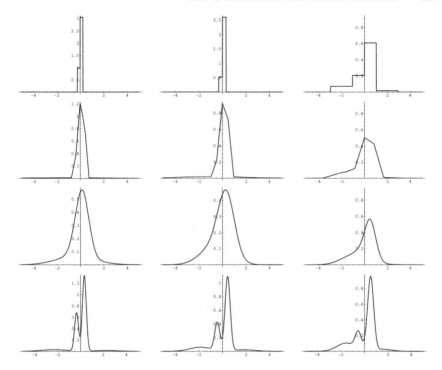

Fig. 12.2. Densities of different scale mixtures: $U \sim U(0,1)$ (top row), U trapezoidal (second row), U normal with $\mu = 1$ and $\sigma = 1$ (third row), and U normal with $\mu = 1$ and $\sigma = 0.5$ (bottom row). The moments $\mu_1 = 0$, $\mu_2 = 1$, and $\mu_3 = 0.9$ are the same for all scale mixtures; μ_4 is 10 (left column), 6 (middle column), and 4 (right column), resp.

Remark. Poirion (2001) has extended these simulation methods for random variates with given moments to d-dimensional random vectors using the conditional distribution method.

12.4.3 The Moment Problem

The classical moment problem poses the question whether for a given sequence of moments there exists a corresponding distribution and whether this distribution is unique. A very nice illustration that a moment sequence is not always uniquely determining the density is due to Godwin (1964). He shows that the following two densities

$$f_1(x) = \frac{1}{4}\exp\left(-\sqrt{|x|}\right) \quad \text{and} \quad f_2(x) = \frac{1}{4}\exp\left(-\sqrt{|x|}\right)\left(1 + \cos\left(\sqrt{|x|}\right)\right)$$

have the same moment sequence but are obviously very different, see Fig. 12.3.

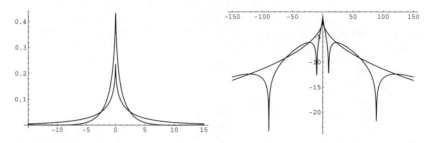

Fig. 12.3. Two densities with the same moment sequence. Original scale (l.h.s.) and logarithmic scale (r.h.s.)

For a discussion of the moment problem and necessary and sufficient conditions for uniqueness see Devroye (1986a, Sect. XIV). These conditions also imply that for the case that f is a density with compact support the moment sequence uniquely determines the density. Devroye (1989) demonstrates how the given moment sequence together with Legendre polynomials and Legendre series can be used to obtain a series representation of the density. Applying the series method results in an algorithm that is similar to Algorithm 9.9 (Fourier-Coefficients). Unfortunately the error bound for the Legendre series includes quantities that are usually not available if only the moment series is known but the density is not known explicitly. Thus this idea seems to be of limited practical use and we do not include the details here.

12.4.4 Computational Experience

The algorithms of this section are quite simple and fast. It is also easy to test them by computing the four empirical moments of a large generated sample.

We coded Algorithm 12.5 (4-Moments-Discrete) that turned out to be a bit faster than generating the exponential distribution by inversion. For Algorithm 12.6 (4-Moments-Continuous) the speed of course depends on the mixing distribution chosen. Our code has been between 1.5 and 2.5 times slower than generating the exponential distribution by inversion.

12.5 Generation of Random Vectors where only Correlation and Marginal Distributions are Known

There are situations in simulation practice where users want to generate vectors from a distribution with given continuous marginal distributions and given correlation matrix. For example Clemen and Reilly (1999) discuss applications in decision and risk analysis, Lurie and Goldberg (1998) describe applications in cost analysis, Hill and Reilly (1994, 2000) describe the generation of test problems for an algorithm. All these authors of course know and state that a multivariate distribution is not fully characterized by the

12.5 Random Vectors where only Correlation and Marginals are Known

marginal distribution and the correlation matrix but they do not check if or how their simulation results are influenced by the different possible shapes of the joint distribution. Therefore we want to stress here that we have the freedom of choice for the multivariate distribution. For example, we can take a distribution which allows easy generation and generate vectors in dimensions up to several hundred. However, the freedom of choice implies that we also have to take a modeling decision and it is possible (or even likely) that this decision can have an influence on our simulation result. To stress this aspect of the problem, that can be easily overlooked, we have not included this section in Chap. 11 on the generation of random vectors but here in the chapter on the combination of generation and modeling. Note that we will present some of the methods of this chapter also in the context of non-Gaussian stationary time series in Sect. 13.2.2. (Being pedantic that section should be part of this chapter.)

There are several different notions of correlation suggested in the statistical literature. The best known among them is certainly the *Pearson correlation coefficient* for two random variates X and Y, given by

$$\rho(X,Y) = \frac{\text{Cov}(X,Y)}{\sigma_X \sigma_Y}$$

that is natural for linear regression and for the multinormal distribution. ($\sigma_X = \sqrt{\text{Var}(X)}$ denotes the standard deviation of X.) It has the important disadvantage that it is not invariant under monotone transformations. Therefore we will mainly consider the *rank correlation* (also called *Spearman's rho*) that can be defined as

$$\text{r}(X,Y) = \rho(F_x(X), F_y(Y))$$

where F_x and F_y are the respective marginal distribution functions of X and Y. Another useful correlation is *Kendall's* τ: For two independent pairs (X_1, Y_1) and (X_2, Y_2) of the distribution (X, Y) we have the definition

$$\tau(X,Y) = 2\text{Prob}((X_1 - Y_1)(X_2 - Y_2) > 0) - 1 \ .$$

It is well known that both r and τ are invariant under monotone transformations.

It is useful to observe here that we can obtain arbitrary marginals if we transform a multivariate distribution component-wise with monotone transforms. We can for example transform a multivariate distribution with normal marginals component-wise with Φ_N, the distribution function of the normal distribution, to obtain uniform marginals. The distribution function of a random pair with uniform marginals is called a *copula*. Transforming a random vector with uniform marginals component-wise with arbitrary inverse cdfs we can obtain arbitrary marginal distributions. The advantage of the rank correlation r and Kendall's τ is that they remain invariant under such monotone transformations.

We restrict ourselves here to the case of continuous marginals. First we discuss the simple bivariate case where solutions to the generation problem are known at least since Mardia (1970). We discuss different distribution families of pairs with normal and uniform marginals. Based on these families we present three different algorithms to generate pairs with given marginals and rank correlation. By means of illustrations we also try to give the reader some insight into the differences between the methods that are based on different joint distributions. For the multidimensional problem we present the standard method in the literature (suggested already by Li and Hammond, 1975). It is based on the component-wise transformation of a multinormal vector. Following Cario and Nelson (1997) we call the resulting distribution NORTA (normal to anything). It is well known that NORTA distributions do not allow for all possible covariance matrices in higher dimensions. Thus we also give references to recent results of Kurowicka and Cooke (2001) and Ghosh and Henderson (2002a) that suggest multivariate distributions that allow for arbitrary positive definite variance-covariance matrices.

12.5.1 The Bivariate Case

As the bivariate problem is much easier than the multivariate we take it as the starting point. We present different methods to construct bivariate distributions with given marginals. Contour plots of the densities of these distributions also help us to see how much freedom for modeling is left if we only specify the marginals and the correlation.

A very simple method to obtain a joint distribution with given marginals and correlation is based on the mixing of independent and highly correlated pairs. The generation of independent pairs is no problem. The generation of pairs with rank correlation $r = 1$ (i.e. maximal Pearson correlation $\rho_{\max}$) can be done easily if we use the inversion method and the same uniform random numbers for both random variates. For $r = -1$ we use antithetic variates, i.e. take again inversion but use $1-U$ instead of U for the second random variate. Then, taking the independent pair with probability $1-p$ and the correlated one with probability p, results in a bivariate distribution with the given marginals and rank correlation $r = p$ or $r = -p$. For the Pearson correlation we get $\rho = p\,\rho_{\max}$ and $\rho = p\,\rho_{\min}$, resp., where $\rho_{\max}$ and $\rho_{\min}$ denote the maximal and minimal possible correlation that depends on the marginal distributions. This method has been known since the seventies of the last century and has been generalized to higher dimensions by Hill and Reilly (1994). For two dimensions it is very simple but the resulting distribution does not have a density and can be considered as strange. Therefore this distribution is not seen as a good solution to this problem in the simulation literature. We call this distribution *zero-maximal correlation mixture*. For uniform marginals it is presented as Algorithm 12.7 (Zero-Maximal-Bivariate).

Note that we can obtain a similar algorithm if we mix pairs with maximal negative and maximal positive correlation. The resulting distribution is called

12.5 Random Vectors where only Correlation and Marginals are Known

Algorithm 12.7 Zero-Maximal-Bivariate (Correlation mixture for uniform marginals)

Require: Correlation ρ (which is in this case equal to rank correlation r).
Output: Random pair from the zero-maximal correlation mixture, which has uniform marginals and correlation ρ.
1: Generate $U \sim U(0,1)$ and $W \sim U(0,1)$.
2: **if** $W < |\rho|$ **then** /* maximal (minimal) correlation */
3:    **if** $\rho > 0$ **then**
4:       **return** pair (U, U).
5:    **else**
6:       **return** pair $(U, 1 - U)$.
7: **else** /* no correlation (independent) */
8:    Generate $V \sim U(0,1)$.
9:    **return** pair (U, V).

Frechet distribution. The Frechet copula (i.e. bivariate distribution with uniform marginals) has all mass concentrated on the two diagonals and the mass is constant for each diagonal respectively.

We continue with an example of different bivariate distributions with normal marginals.

Example 12.1. We have explained in Sect. 11.1.6 how we can generate random vectors from the multinormal distribution with arbitrary variance-covariance matrix. As the multinormal distribution has normal marginals this is one possible solution to the problem but it is not the only one. To demonstrate this we define the following bivariate normal-mixture distribution with parameters p, ρ_1 and ρ_2. With probability p we have a normal distribution with mean zero and variance-covariance matrix Σ_1 and with probability $1 - p$ one with mean zero and variance-covariance matrix Σ_2, where

$$\Sigma_1 = \begin{pmatrix} 1 & \rho_1 \\ \rho_1 & 1 \end{pmatrix} \quad \text{and} \quad \Sigma_2 = \begin{pmatrix} 1 & \rho_2 \\ \rho_2 & 1 \end{pmatrix}.$$

It is obvious that all members of this mixture have standard normal marginals and mean 0, the correlation is $\rho = p\rho_1 + (1-p)\rho_2$. So even in this very simple and comparatively small normal mixture family we have infinitely many members that have standard normal marginals and the same ρ. Figure 12.4 shows four different examples for $\rho = 0.3$.

Copulae (Bivariate Distributions with Uniform Marginals)

We continue with the case of uniform marginals, as copulae are the most important special case. Random pairs with uniform marginals can be easily transformed into arbitrary marginals using the inverse cdf of the desired marginal distribution. For copulae the rank correlation r and the Pearson correlation ρ coincide. It is not trivial to find a continuous bivariate distribution

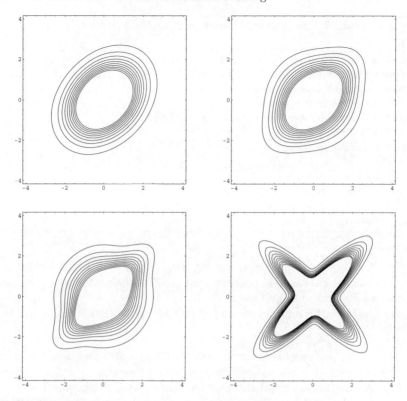

Fig. 12.4. Four different distributions with standard normal marginals and correlation $\rho = 0.3$. Multinormal distribution (upper left), mixture with $p = 0.5$, $\rho_1 = 0.6$, $\rho_2 = 0$ (upper right), mixture with $p = 3/8$, $\rho_1 = 0.8$, $\rho_2 = 0$ (lower left), and mixture with $p = 2/3$, $\rho_1 = 0.9$, $\rho_2 = -0.9$ (lower right)

with uniform marginals that can take all possible correlations between -1 and 1. Devroye (1986a) gives several examples of bivariate distribution families with uniform marginals, among them the uniform version of Morgenstern's family with parameter $a \in [-1, 1]$ and density

$$f(x, y) = 1 + a(2x - 1)(2y - 1), \quad \text{on } [0, 1]^2.$$

However, for this distribution the correlation is $\rho = a/3$ and therefore restricted to the interval $[-1/3, 1/3]$.

A family with uniform marginals that includes all possible correlations is the multinormal distribution transformed such that the marginals become uniform. To be more precise we write (Y_1, Y_2) for a multinormal random vector with covariance matrix

$$\Sigma_Y = \begin{pmatrix} 1 & \rho_Y \\ \rho_Y & 1 \end{pmatrix}.$$

Then the random vector with uniform marginals is defined as

12.5 Random Vectors where only Correlation and Marginals are Known

$$(U_1, U_2) = (\Phi_N(Y_1), \Phi_N(Y_2)),$$

where $\Phi_N(\cdot)$ denotes the cdf of the standard normal distribution.

In close analogy to the name NORTA we call this distribution NORTU (normal to uniform). In the literature it is also called the *normal copula*. Using a result of Kruskal (1958) we get the following formula for the correlation of the NORTU distribution:

$$\rho(U_1, U_2) = r(U_1, U_2) = r(Y_1, Y_2) = \frac{6}{\pi} \arcsin\left(\frac{\rho_Y}{2}\right).$$

For uniform marginals the rank correlation is the same as the Pearson correlation and it is not difficult to see from the formula for the correlation above that it can take all values between -1 and 1. Thus for NORTU distributions ρ can take all values between -1 and 1. For a desired rank correlation r we can compute

$$\rho_Y = 2\sin\left(\frac{\pi r}{6}\right). \tag{12.3}$$

Then we generate a normal pair with correlation ρ_Y and transform it component-wise with Φ_N. Algorithm 12.8 (NORTU-Bivariate) contains the details for easy reference. Notice that – as also pointed out in Sect. 13.2.2 – this transformation has some nice properties: If r $= -1$, 0, or 1 then ρ results in $\rho = -1$, 0, and 1, respectively. It is an odd, strictly monotone function, so that $|\rho| > |r|$ except at -1, 0, or 1. The maximum deviation measured as $|\rho - r|$ is observed at $|r| = 0.5756$ and amounts only to 0.0181.

Remark. If we want to use Kendall's τ there is a similar formula telling us which ρ_Y is necessary for the bivariate normal distribution to obtain the desired value for τ:

$$\rho_Y = \sin\left(\frac{\pi \tau}{2}\right). \tag{12.4}$$

In the following algorithms we are only stating the formulas for r, which is the same as ρ for uniform marginals. Using the above formula we could formulate all of the algorithms for given τ instead of r as well. Note that this transform has the same nice properties but the distance of $|\rho - \tau|$ is much bigger than that of $|\rho - r|$; the maximal deviation is observed at $|\tau| = 0.5607$ and is equal to 0.2105.

Figure 12.5 shows the contour plot of different joint distributions that may occur in the above methods (models).

Ghosh and Henderson (2002a) have suggested "chessboard" distributions to model multivariate distributions with uniform marginals and arbitrary correlation structure. For the simple case $d = 2$ the unit square is decomposed into n^2 equally sized squares and chessboard distributions have bivariate densities constant over any of these squares. If we only consider chessboard distributions that have density $f(x, y) = n$ for the sub-squares along the first (or second) diagonal and zero elsewhere then it is not too difficult to compute

Algorithm 12.8 NORTU-Bivariate

Require: Correlation ρ (which is in this case equal to rank correlation r).
Output: Random pair from the NORTU distribution, which has uniform marginals and correlation ρ.
/* Setup */
1: Set $\rho_Y \leftarrow 2\sin\left(\frac{\pi\rho}{6}\right)$.
/* Generator */
2: Generate standard normal variates Y_1 and Y_2.
3: Set $Y_2 \leftarrow \rho_Y Y_1 + \sqrt{1-\rho_Y^2}\, Y_2$.
/* (Y_1, Y_2) is multinormal distributed with variances 1 and correlation ρ_Y */
4: **return** random pair $(\Phi_N(Y_1), \Phi_N(Y_2))$.

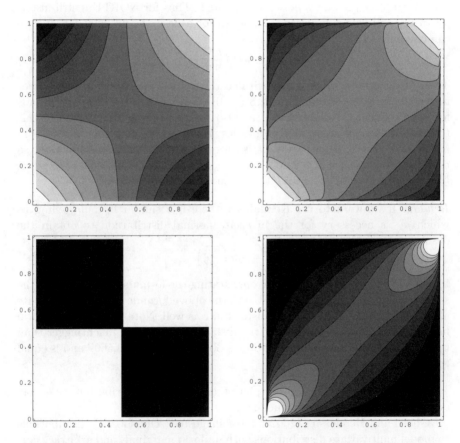

Fig. 12.5. Contour plots of four different distributions with uniform marginals. Morgenstern distribution with density $f(x,y) = 1 + 0.9\,(2x-1)(2y-1)$ and $\rho = 0.3$ (upper left), NORTU distribution with $\rho = 0.3$ (upper right), simple "chessboard" distribution with $\rho = 0.75$ (lower left), and NORTU distribution with $\rho = 0.75$ (lower right)

12.5 Random Vectors where only Correlation and Marginals are Known

Algorithm 12.9 Chessboard-Bivariate

Require: Correlation ρ (which is in this case equal to rank correlation r).
Output: Random pair from a chessboard distribution which has uniform marginals and correlation ρ.
/* Setup */
1: Set $\tilde{n} \leftarrow \left\lfloor \sqrt{1/(1-|\rho|)} \right\rfloor$.
2: Set $\rho_1 \leftarrow 1 - 1/\tilde{n}^2$ and $\rho_2 \leftarrow 1 - 1/(\tilde{n}+1)^2$.
3: Set $p \leftarrow (|\rho| - \rho_2)/(\rho_1 - \rho_2)$. /* p is the mixing probability between the chess board distribution with respective $\tilde{n}$ and $\tilde{n} + 1$ squares along the diagonal. */
/* Generator */
4: Generate $U, V, W \sim U(0,1)$.
5: **if** $W < p$ **then**
6:   Set $V \leftarrow (\lfloor \tilde{n} U \rfloor + V)/\tilde{n}$.
7: **else**
8:   Set $V \leftarrow (\lfloor (\tilde{n}+1)U \rfloor + V)/(\tilde{n}+1)$.
9: **if** $\rho < 0$ **then**
10:   Set $V \leftarrow 1 - V$.
11: **return** random pair (U, V).

---

that these distributions have $\rho = 1 - 1/n^2$ (or $\rho = -1 + 1/n^2$) and uniform marginals. For $n = 2$ we find $\rho = 0.75$ (or -0.75.) Thus we can easily obtain a correlation between 0.75 and -0.75 if we use the chessboard distribution with $n = 2$ and density-value $2p$ for the two squares touching at $(0,0)$ and $(1,1)$ and density-value $2 - 2p$ for the other two squares. Then it is a straightforward calculation that $\rho = -0.75 + 1.5\,p$.

If the required correlation ρ is in absolute value greater than $3/4$ but smaller than 1 we can determine $\tilde{n}$, the largest n with $1 - 1/n^2 \leq |\rho|$, by

$$\tilde{n} = \left\lfloor \sqrt{\frac{1}{1-|\rho|}} \right\rfloor .$$

The required distribution can then be constructed as a mixture of the two chessboard distributions with $n = \tilde{n}$ and $n = \tilde{n} + 1$. The required n can become quite large but this is no problem for the generation method described in Algorithm 12.9 (Chessboard-Bivariate). For this algorithm it is useful to note that the mixture procedure used for $|\rho| > 0.75$ can also be used for $|\rho| \leq 0.75$. We then have $\tilde{n} = 1$ and the mixture is between the chessboard distribution with $n = 2$ and the one with $n = 1$ which is the independent case. The resulting distribution is a chessboard distribution with $n = 2$. It has the desired ρ and is the same distribution as the one described in the last paragraph.

There are other copulae suggested in the literature to generate random pairs with given correlation and uniform marginals: Meuwissen and Bedford (1997) suggest to use the minimal informative distribution that is interesting from a theoretic point of view. Unfortunately its generation is not easy.

Kurowicka, Misiewicz, and Cooke (2001) (see also Kurowicka and Cooke, 2001) suggest an elliptically contoured copula that is easy to generate and is especially well suited to model higher-dimensional distributions. Its behavior is clearly different from the copulae explained so far as for the case $\rho = 0$, X and Y are not independent and most of its conditional distributions are U-shaped with two poles.

Of course a simple copula is also the Frechet-distribution which is the mixture of maximal and minimal correlated pairs. But it is not continuous and its joint distribution concentrates the mass only on the two diagonals of the unit square. So its conditional distributions consist of two point masses only. Moreover, for $\rho = 0$, X and Y are not independent.

Arbitrary Marginals

Johnson and Tenenbein (1981) introduce a general principle called weighted linear combination method to construct bivariate families where the rank correlation can take all values between -1 and 1. They apply the method to normal, uniform, exponential, and double exponential marginals.

Another possibility is to generate a pair with uniform marginals and desired rank correlation first using one of the three Algorithms 12.7, 12.8, or 12.9. We want to stress again here that the decision between these three algorithms is a modeling decision as it is a decision between three different joint distributions. It is also possible that none of the three distributions is suitable for a simulation problem. Using Algorithm 12.7 (Zero-Maximal-Bivariate) implies that pairs are generated from a distribution that mixes independent and maximal correlated pairs. This distribution is considered to be strange and is not suited for many simulation models in the literature. Using Algorithm 12.8 (NORTU-Bivariate) implies that the final vector follows the so called NORTA distribution, which is a transformed multinormal distribution. Finally, using Algorithm 12.9 (Chessboard-Bivariate) means that the final pair follows a transformed chessboard distribution.

In the final step the pair with uniform marginals is transformed with the inverse cdf of the desired marginal distribution which guarantees that the generated pair has the correct marginal distribution. As the rank correlation is invariant under monotone transforms we can generate pairs with arbitrary marginals and desired rank correlation. Algorithm 12.10 (Arbitrary-Marginals) collects the details.

If we want to generate pairs with arbitrary fixed marginals and given Pearson correlation the problem is more delicate as the Pearson correlation is not invariant under monotone transforms. Cario and Nelson (1997) explain in detail what can be done to solve this problem numerically. We are not repeating their considerations here as we are convinced that the Pearson correlation is simply not suited for marginals that are neither normal nor uniform.

12.5 Random Vectors where only Correlation and Marginals are Known 337

Algorithm 12.10 Arbitrary-Marginals

Require: Rank correlation r, cdfs F_1 and F_2 of the two marginals distributions.
Output: Random pair with desired rank correlation and marginals.
  /* Setup */
1: Choose one of the Algorithms 12.7, 12.8, or 12.9 for generating uniform copula and run setup. Notice that this is a modeling decision.
  /* Generator */
2: Generate random pair (U, V) with uniform marginals and correlation r.
3: Compute $X \leftarrow F_1^{-1}(U)$ and $Y \leftarrow F_2^{-1}(V)$ (numerically).
4: **return** random pair (X, Y).

The Distribution of $X + Y$ and $X - Y$

We have seen how many different shapes of the distribution are possible for a random pair (X, Y) with fixed marginal distributions and fixed correlation. The possible influence of these different shapes on simulation results are quite unclear. To investigate this influence let us consider the respective distributions of $X + Y$ and $X - Y$. The distribution of $X + \alpha Y$ for all real α would result in a full characterization of the joint density. For $\alpha = 0$ and $\alpha \to \infty$ the distribution is equivalent to the marginal distributions of X and Y respectively. Thus it makes sense to consider the cases $\alpha = \pm 1$ that include additional information about the joint distribution.

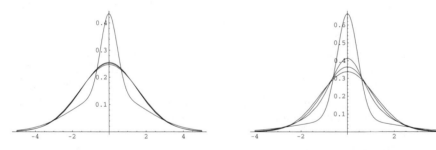

Fig. 12.6. The densities of $X + Y$ (l.h.s.) and $X - Y$ (r.h.s.) for the four joint distributions of Fig. 12.4

As the mean value and the variance of $X + Y$ and $X - Y$ are fully determined by the mean and variance of the marginal distributions together with correlation $\rho(X, Y)$ it is clear that they are not changed by the shape of the two-dimensional distribution. But the shapes of the distributions of $X + Y$ and $X - Y$ are strongly influenced. For example, for normal distributed marginals, $X + Y$ and $X - Y$ follow a normal distribution if we decided to take the multinormal distribution. If we use the normal mixture family introduced in Example 12.1 (see Fig. 12.4) we get a mixture of two normals with the same mean but possibly very different standard deviations, see Fig. 12.6. For the chessboard distribution with $n = 2$ and ρ close to 0.75 we even get a clearly bimodal distribution for $X + Y$. These examples should make clear that it is necessary for simulation modeling to assess the shape of the two-dimensional distribution as well. The possible differences between the distributions of $X + Y$ and $X - Y$ also imply that it is no problem to find examples of simulations where the use of different joint distributions with the same marginals and correlations leads to different results (see Exercises 12.4, 12.6 and 12.6). If we do not have enough information available to model that shape it is at least necessary to investigate the sensitivity of our simulation to different shapes of the joint distribution. Only specifying the marginals and the correlation can be compared with specifying a one dimensional distribution by only fixing its mean and variance.

To give the potential user a possibility to assess the difference between the pairs generated by Algorithms 12.7 (Zero-Maximal-Bivariate), 12.8 (NORTU-Bivariate), or 12.9 (Chessboard-Bivariate) we consider the distribution of $X + Y$ and $X - Y$ of random pairs (X, Y) with the respective joint distributions. To start with, let us consider the case $\rho = 0$. It is easy to see that then all three methods generate pairs from the same joint distribution, the uniform distribution over the unit square. In this case the distributions of $X + Y$ and $X - Y$ are symmetric triangular distributions on $(0, 2)$ and $(-1, 1)$, respectively. For $\rho = 1$ the distribution of $X + Y$ is uniform on $(0, 2)$, that of $X - Y$ is a point distribution with mass one at 0. For the values of ρ between 0 and 1 the joint distributions are quite different which results also in different shapes for the distributions of $X + Y$ and $X - Y$. Figs. 12.7, 12.8, and 12.9 show these shapes for three different values of ρ.

If data are available we could compare these densities with the empirical densities to take a decision which Algorithm fits best. But if data-pairs are available then we think that in most cases using resampling methods like Algorithm 12.4 (Resampling-MVKDEvc) is better than estimating the marginals and the correlation from the data. Merely estimating these characteristics means that we do not take full advantage of the information contained in the observed pairs.

In the absence of data the very strange distribution of $X - Y$ in Fig. 12.7 can be seen as a support of the general opinion in the simulation community that Algorithm 12.7 (Zero-Maximal-Bivariate) is leading to strange distributions. On the other hand the distribution of $X + Y$ in Fig. 12.9 shows that at least that sort of chessboard distributions we have suggested here are also behaving quite strange. Figure 12.8 shows that the NORTU distribution seems to be closest to what people may assume to be a well behaved distribution. Of course the final decision has to depend on the model we are trying to build. If we have no possibility to get an insight which joint distribution fits best to our model we suggest at least to try out all three variants when using Algorithm 12.10 (Arbitrary-Marginals). Then by comparing the results of the simulation we can get a first impression if and how our simulation results are influenced by the shape of the joint input distribution.

12.5 Random Vectors where only Correlation and Marginals are Known   339

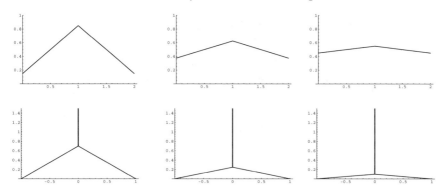

Fig. 12.7. Zero-maximal correlation mixture distribution (Algorithm 12.7): densities of $X + Y$ (upper row) and $X - Y$ (lower row) for $\rho = 0.3$ (left), $\rho = 0.75$ (middle), and $\rho = 0.9$ (right). (The point-mass is shown as a spike)

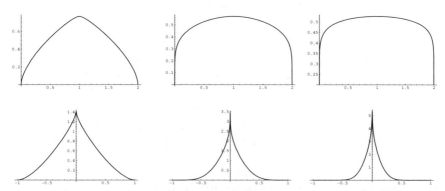

Fig. 12.8. NORTU distribution (Algorithm 12.8): densities of $X + Y$ (upper row) and $X - Y$ (lower row) for $\rho = 0.3$ (left), $\rho = 0.75$ (middle), and $\rho = 0.9$ (right)

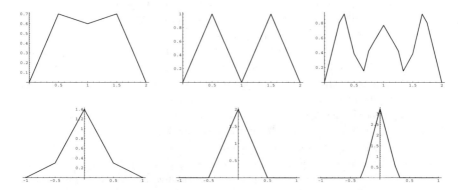

Fig. 12.9. Chessboard distribution (Algorithm 12.9): densities of $X + Y$ (upper row) and $X - Y$ (lower row) for $\rho = 0.3$ (left), $\rho = 0.75$ (middle), and $\rho = 0.9$ (right)

12.5.2 The Multidimensional Problem

The problem of modeling distributions with given marginals and correlation matrix becomes more difficult in higher dimensions. It is easy to reformulate Algorithm 12.10 (Arbitrary-Marginals) combined with Algorithm 12.8 (NORTU-Bivariate) for higher dimensions using NORTA distributions; see Algorithm 12.11 (NORTA-Multivariate). But it might happen that, for a given (positive definite) rank correlation matrix Σ_r, the corresponding matrix Σ_Y that we need for generating the multinormal $\mathbf{Y}$-vector is not positive-definite. Thus no multinormal distribution with the desired rank correlation Σ_r exists. Li and Hammond (1975) give the example of the correlation matrix Σ_r below where the corresponding matrix Σ_Y, obtained by applying Transformation (12.3) to all correlations, is not positive definite:

$$\Sigma_r = \begin{pmatrix} 1 & -0.4 & 0.2 \\ -0.4 & 1 & 0.8 \\ 0.2 & 0.8 & 1 \end{pmatrix}, \quad \Sigma_Y = \begin{pmatrix} 1 & -0.416... & 0.209... \\ -0.416... & 1 & 0.813... \\ 0.209... & 0.813... & 1 \end{pmatrix}.$$

Thus for this positive definite correlation matrix there exists no NORTA distribution. Ghosh and Henderson (2002a) have shown that the combinations of zero and maximal correlation as suggested in Hill and Reilly (1994) is not able to obtain the rank correlation matrix Σ_r as well. So the three-dimensional version of Algorithm 12.7 (Zero-Maximal-Bivariate) is not able to sample from a distribution with Σ_r as above. But Ghosh and Henderson (2002a) have constructed a chessboard distribution with rank correlation matrix Σ_r. They even prove that by means of solving a (possibly very large) linear programming problem it is possible to construct a chessboard distribution that approximates any feasible rank correlation matrix arbitrarily close.

Thus among the three joint distributions that we have discussed for random pairs, only chessboard distributions are suitable for higher dimensional distributions with arbitrary correlation structure. However, to construct such a distribution a linear programming problem has to be solved (Ghosh and Henderson, 2002a). On the other hand, the NORTA idea remains simple in higher dimensions. It works well for most correlation matrices in dimension $d = 3$ unless the smallest eigenvalue of the correlation matrix is very close to zero. Unfortunately the situation changes if dimension d increases. Kurowicka and Cooke (2001) present a simulation study using vines (a new graphical concept to model correlations of random vectors introduced by Bedford and Cooke, 2002, 2001) to sample from random positive definite matrices. This study shows that the proportion of feasible rank correlation matrices for the multinormal distribution goes to 0 rapidly with d increasing; see Table 12.7 column (C). We extended their study by generating the positive definite matrices in two different ways. Thus altogether we compare the following three methods:

(A) Following the suggestion of Devroye (1986a, p.605) we have generated a $(d \times d)$-matrix $\mathbf{H}$ of iid. normal variates and normalized the rows of this

12.5 Random Vectors where only Correlation and Marginals are Known

matrix to length one. Then the matrix $\mathbf{HH'}$ is a random positive definite correlation matrix.

(B) We have generated symmetric matrices with coefficients uniformly distributed on $(-1,1)$ and accepted only the positive definite ones. Unfortunately this method becomes too slow for dimension $d > 8$.

(C) The method of Kurowicka and Cooke (2001) using vines. For information about vines see Bedford and Cooke (2002, 2001).

Table 12.7 gives the proportion of feasible rank correlation matrices for the multinormal distribution. The results for methods (A), (B), and (C) are quite different as we sampled randomly from different distributions. Nevertheless, from a practical point of view we come to the same conclusion. For NORTA distributions the proportion of feasible rank correlation matrices is decreasing with dimension d. For a fixed desired rank correlation matrix with $d > 4$ it can easily happen that this matrix is not feasible for the NORTA distribution.

Table 12.7. Percentage of feasible rank correlation matrices Σ for the multinormal distribution using the three different methods (A), (B), and (C) to generate the random correlation matrices. The number of replications was 1000

d	(A)	(B)	(C)	d	(A)	(B)	(C)	d	(A)
2	100	100	100	8	14.0	38	0.005	14	0.5
3	83	96	96	9	9.1		0	15	0.2
4	65	89	78	10	5.6		0	16	0.1
5	49	79	48	11	2.9			17	0.01
6	35	65	19	12	1.7			18	0.02
7	24	52	4	13	0.9			19	0.01

Ghosh and Henderson (2002b) have developed an efficient simulation method for uniformly distributed rank correlation matrices. Thus they obtained results corresponding to our method (B) up to dimension $d = 25$. These results show that the proportion of feasible rank correlation matrices is below 1% for $d \geq 17$. Ghosh and Henderson (2002b) also give an explanation why this happens: This is due to the fact that in high dimension "most" points in a fixed set lie close to the boundaries. If we consider the unit hypercube $[0,1]^d$ its interior is the hypercube $(\epsilon/2, 1 - \epsilon/2)^d$. The ratio of the volumes of interior and the whole set is $(1 - \epsilon)^d$, which converges to 0 for increasing d. If we remember that a correlation matrix for dimension d contains $d(d-1)/2$ different correlations we have thus a good explanation why the proportion of feasible rank correlations decreases so fast although the difference between rank correlation and Pearson correlation is always below 0.0181. To fix NORTA in the case of a nonpositive definite Σ_Y Lurie and Goldberg (1998) and Ghosh and Henderson (2002a) suggest two slightly different methods to arrive at an approximate solution: They search for a matrix that is close to

Algorithm 12.11 NORTA-Multivariate

Require: Dimension d, rank correlations Σ, cdfs of all marginals $F_1, F_2, \ldots, F_d$.
Output: Random vector from NORTA distribution with desired marginals and rank correlation matrix Σ.
/* Setup */
1: Compute correlation matrix Σ_Y using $\Sigma_Y(i,j) \leftarrow 2\sin\left(\frac{\pi \Sigma(i,j)}{6}\right)$.
2: Compute Cholesky factor of Σ_Y (Algorithm 11.4, Cholesky-Decomposition).
3: **if** Σ_Y not positive definite **then** /* i.e. Algorithm 11.4 failed */
4:    **abort** (NORTA distribution not available for given Σ).
/* Generator */
5: Generate multinormal random vector $(Y_1, Y_2, \ldots, Y_d)$ with covariance matrix Σ_Y using Algorithm 11.3 (Multinormal-Cholesky).
6: **for** $i = 1$ to d **do**
7:    Find solution X_i of equation $F_i(X_i) = \Phi_N(Y_i)$ (numerically).
8: **Return** random vector $=(X_1, X_2, \ldots, X_d)$.

---

the Σ_Y and positive definite. Ghosh and Henderson (2002b) present simulation results indicating that the average maximal distance between the entries of Σ_Y and the entries of the corrected matrix is around 0.005 and seems to decrease with dimension.

As NORTA is by far the simplest solution for higher dimensions we only give the details of the multidimensional NORTA approach in Algorithm 12.11 (NORTA-Multivariate). In the case that the NORTA distribution does not work for the desired rank correlation we can use chessboard distributions or the fixing of the NORTA approach, both explained in Ghosh and Henderson (2002a). It is also possible to construct and generate from a Frechet distribution, i.e. from a mixture of maximal and minimal correlation distributions, with arbitrary (rank) correlation structure, but as this copula is not continuous this is of little practical use for most applications.

A third possibility has been found recently by Kurowicka et al. (2001), who define an elliptical multidimensional distribution. It has the property that the generation of random vectors with a given correlation matrix is as simple as for the multinormal distribution. Moreover, Kurowicka and Cooke (2001) have shown by means of a results of Bedford and Cooke (2002) that for this distribution all positive definite (rank) correlation matrices are feasible.

When we want to replace the rank correlation by Kendall's tau we have to use Transformation 12.4 instead of Transformation 12.3 to compute the correlations Σ_Y of the multinormal distribution. However, Transformation 12.3 is much closer to identity then Transformation 12.4. Thus the results of a simulation study similar to that of Table 12.7 show that the percentage of feasible Kendall's tau correlation matrices for NORTA distributions is very small; depending on the random sampling method for the correlation matrices only about 0.5 for dimension $d = 3$ and smaller than 0.01 for $d \geq 6$. These results show that using the rank correlation we have a much better chance that a random correlation matrix is feasible than when using Kendall's tau.

12.5.3 Computational Experience

The Algorithms 12.7 (Zero-Maximal-Bivariate), 12.8 (NORTU-Bivariate), and 12.9 (Chessboard-Bivariate) for generating from different two-dimensional distributions with uniform marginals and desired correlation are all short and simple. Algorithms 12.7 (Zero-Maximal-Bivariate) and 12.9 (Chessboard-Bivariate) are also fast, the latter slightly slower than inversion for the exponential distribution, the former more than 30 percent faster. Algorithm 12.8 (NORTU-Bivariate) is considerably slower (about 8 times) than inversion for the exponential distribution. This is not surprising as we have to evaluate Φ_N, the cdf of the normal distribution, twice for every generated pair.

Algorithm 12.10 (Arbitrary-Marginals) is much slower as we need in general numerical inversion for both variates of the pair we generate. The speed therefore heavily depends on the time we need to evaluate the cdf of the desired marginal distribution. For example, for gamma distributed marginals it is about 60 times slower than generating an exponential variate by inversion.

The same is also true for Algorithm 12.11 (NORTA-Multivariate). Its speed is mainly influenced by the inversion procedure. The speed is therefore also almost linear in the dimension d as the time for generating the normal vector is short compared with the numerical inversion.

12.6 Exercises

Exercise 12.1. Compute the density of the distribution generated by Algorithm 12.1 (Resampling-KDE) for the rectangular kernel and $b = 0.5$, if the sample contains the observations 1, 2, 3, and 5.

Exercise 12.2. Compute the density of the distribution generated by Algorithm 12.1 (Resampling-KDE) for the Gaussian kernel and $b = 0.5$, if the sample contains the observations 1, 2, 3, and 5.

Exercise 12.3. . Compute the feasible region in the (μ_3, μ_4)-plane for the scale mixtures where U is as in Table 12.6.
Check that the region is a subset of the feasible region for arbitrary distributions defined by $\mu_4 \geq 1 + \mu_3^2$.

Exercise 12.4. Consider a simple simulation where you want to estimate for a random pair (X, Y) the probability $X + Y < 0.1$. Compare the results of this simulation for random pairs with uniform marginals and $\rho = 0.9$ using Algorithms 12.7 (Zero-Maximal-Bivariate), 12.8 (NORTU-Bivariate), or 12.9 (Chessboard-Bivariate) to generate the random pairs. Compare the results with the density of $X + Y$ shown in Figs. 12.7, 12.8, and 12.9.

Exercise 12.5. Consider a simple simulation where you want to estimate for a random pair (X, Y) the probability $|X - Y| < 0.05$. Compare the results of this simulation for random pairs with uniform marginals and $\rho = 0.75$ using Algorithms 12.7 (Zero-Maximal-Bivariate), 12.8 (NORTU-Bivariate), or 12.9 (Chessboard-Bivariate) to generate the random pairs. Compare the results with the density of $X - Y$ shown in Figs. 12.7, 12.8, and 12.9.

Exercise 12.6. Make experiments similar to those of Exercise 12.4 and 12.5 for other values of ρ and other inequalities.

Exercise 12.7. Use Algorithm 12.10 (Arbitrary-Marginals), standard normal marginals and $\rho = 0.3, 0.5,$ and 0.7 to generate random pairs. Estimate the probability $X + Y < -3$ when using Algorithms 12.7 (Zero-Maximal-Bivariate) or 12.9 (Chessboard-Bivariate) as generator for the pairs with uniform marginals. Compare these results with the exact results for the multinormal distribution with the same ρ.

13
Time Series
(Authors Michael Hauser and Wolfgang Hörmann)

In this chapter we are concerned with times series, i.e. with the generation of sample paths of stochastic non-deterministic processes in discrete time, $\{X_t, t \in \mathbb{Z}\}$, where X_t are continuous random variates. In the first part we will focus our presentation on stationary Gaussian processes. These are most widely used in the analysis of, e.g., economic series or in signal processing.

If we want to generate time series X_t, it is clear that they can only be of finite length. A time series of fixed length d is the same as a random vector in $\mathbb{R}^d$. Theoretically we could use the generation methods of Chap. 11. But we have seen that for dimensions $d \gtrsim 10$ no universal algorithm with acceptable speed is available. So generation of longer series is generally restricted to the normal distribution. We will see in this chapter that we can generate stationary Gaussian time series up to dimension $d = 10^6$ easily.

Very important in time series analysis is the notion of stationarity. A time series is called (weakly) *stationary* if the expectation and the variance exist and are constant, and the covariance $\mathrm{Cov}(X_t, X_{t-k})$ only depends on the lag k and not on the time t. For a stationary time series we define the autocovariance and autocorrelation function as

$$\gamma_k = \mathrm{Cov}(X_t, X_{t-k}), \quad \text{and} \quad \rho_k = \frac{\gamma_k}{\gamma_0} = \frac{\mathrm{Cov}(X_t, X_{t-k})}{\mathrm{Var}(X_t)}.$$

The autocorrelation contains the main characteristics for stationary time series as the mean μ and variance σ^2 are constant and can be included easily afterwards. From ρ_k and σ we can easily find the variance-covariance matrix

$$\Sigma = \sigma^2 \begin{pmatrix} 1 & \rho_1 & \rho_2 & \rho_3 & \dots \\ \rho_1 & 1 & \rho_1 & \rho_2 & \dots \\ \rho_2 & \rho_1 & 1 & \rho_1 & \dots \\ \rho_3 & \rho_2 & \rho_1 & 1 & \dots \\ \vdots & \vdots & \vdots & \vdots & \ddots \end{pmatrix}. \qquad (13.1)$$

The variance-covariance matrix Σ has a "band" structure and is of symmetric Toeplitz matrix form.

13 Time Series (Authors Michael Hauser and Wolfgang Hörmann)

Time series textbooks rarely consider generation explicitly; a notable exception is – though an introductory book – Cryer (1986). However, the results presented under the topic prediction are applicable almost directly, see e.g. Brockwell and Davis (1991). Prediction is the projection of X_{t+1} on the closed span of the past observations, $\overline{\mathrm{sp}}\{X_t, X_{t-1}, \ldots\}$, into L^2 space. This is equivalent to taking the conditional expectation $\hat{X}_{t+1} = \mathrm{E}(X_{t+1}|X_t, X_{t-1}, \ldots)$ based on a linear model. The generation of a sample path may then be achieved by adding some white noise ϵ_t to $\hat{X}_{t+1}$: $X_{t+1} = \hat{X}_{t+1} + \epsilon_{t+1}$, provided that the finiteness of our sample path is taken adequately into account.

For conventional ARMA (autoregressive moving average) models the simplest approach is to determine initial values, e.g. by the mean, and generate a presample until it converges to the true distribution. Thus initial values for the actual generation are obtained. Similar techniques apply to ARCH and GARCH (generalized autoregressive conditional heteroscedasticity) models (see Engle, 1982; Bollerslev, 1988), which model the variance of a stationary process. These methods are, however, not exact.

More refined methods for prediction are based on the autocovariances of the true process. The Durbin-Levinson recursions (cf. Brockwell and Davis, 1991, Sect. 5.2) give $\hat{X}_{t+1}$ in terms of past values of X_t, $\{X_t, X_{t-1}, \ldots, X_1\}$, while the innovation recursions yield a representation of the form

$$\hat{X}_{t+1} = \sum_{j=1}^{t} \theta_{tj}(X_{t+1-j} - \hat{X}_{t+1-j}) = \sum_{j=1}^{t} \theta_{tj}\epsilon_{t+1-j}.$$

In Sect. 13.1 we will present the conditional distribution method (as presented in Sect. 11.1.1) for stationary Gaussian time series in matrix notation. The usage of the properties of the Toeplitz matrices is equivalent to the Durbin-Levinson algorithm. Further we develop the innovations algorithm using the properties of the Cholesky decomposition based on partitioned matrices. In addition, the spectral method of Davies and Harte (1987) is presented. So two different black-box algorithms for stationary time series are available for given (positive definite) autocorrelation functions.

The special structure of the Toeplitz matrices will be exploited extensively leading to time domain methods which are considerably faster and requiring less storage than the conventional methods for general multivariate normal distributed variates. The spectral method is particularly useful for large dimensions.

In Sect. 13.2 we will give a short overview of possibilities to generate time series with given autocorrelation. One method allows for specification of the third and fourth moments, the other for specification of the one-dimensional marginal distribution.

13.1 Stationary Gaussian Time Series

For Gaussian time series (i.e. time series where each finite section thereof is distributed as multivariate normal) the mean, variance and autocorrelation function together give a full characterization of the distribution of the time series. Contrary to the classical (linear) ARMA models, for example, fractionally integrated models (cf. Hosking, 1984 and Brockwell and Davis, 1991, p.520ff) are not representable by finite order lag polynomials. Fractionally integrated processes are characterized by (in absolute value) hyperbolically decreasing autocorrelation functions, so that far distant observations are still correlated to a non-negligible amount. During the last decade they experienced an increasing interest both from a theoretical as well as from an empirical aspect. See e.g. recently Marinucci and Robinson (2000) or Gil-Alaña and Robinson (1997). Simple recursions for the generation of the sample paths using the ARMA parameters are not feasible (reasonable approximations may be obtained by accepting very large AR or MA orders). Therefore it is important to construct a universal algorithm for general (weakly) stationary processes. There are several papers describing the generation of stationary Gaussian time series, e.g. Hosking (1984) and Marchenko and Ogorodnikov (1984a) for the time domain, and Davies and Harte (1987) for the frequency domain solution. We discuss the close relationship and compare the performance of the algorithms closely following Hauser and Hörmann (1994).

As a stationary Gaussian time series of length d with autocorrelation function ρ_k follows a d-dimensional normal distribution and has variance-covariance matrix Σ as given in (13.1) we can use Algorithm 11.3 (Multinormal-Cholesky) to generate the time series. In the setup the Cholesky decomposition of Σ is computed. The best available routines to compute the Cholesky factorization of arbitrary positive definite matrices need $d^3/3 + O(d^2)$ flops (floating point operations, i.e. additions or multiplications) and there is no faster routine available for Toeplitz matrices (cf. Golub and Van Loan, 1989). Therefore the setup of Algorithm 11.3 becomes slow even for moderate d and a lot of memory is necessary as well, at least $d^2/2 + O(d)$ floating point numbers. The generation of one time series requires d Gaussian random numbers and $d^2 + O(d)$ flops. It is interesting that it is not possible to improve Algorithm 11.3 for stationary time series in a direct way, but it is possible to find an improvement for the conditional distribution method.

13.1.1 The Conditional Distribution Method

In the following Σ will denote the correlation matrix of a stationary Gaussian time series with autocorrelations $\rho = (\rho_1, \rho_2, \ldots)'$ and $\rho_0 = 1$. Let

$$\Sigma_{t+1} = \begin{pmatrix} \Sigma_t & \Sigma'_{1t} \\ \Sigma_{1t} & \sigma_{t+1,t+1} \end{pmatrix}, \quad \text{for } t = 1, \ldots, n-1$$

where $\boldsymbol{\Sigma}_t$ denotes the upper $t \times t$ submatrix of $\boldsymbol{\Sigma}$. Then $(X_{t+1}|\mathbf{X}_t = (X_1, \ldots, X_t)')$ has expectation $\boldsymbol{\Sigma}_{1t}\boldsymbol{\Sigma}_t^{-1}\mathbf{X}_t$ and variance $\sigma_{t+1,t+1} - \boldsymbol{\Sigma}_{1t}\boldsymbol{\Sigma}_t^{-1}\boldsymbol{\Sigma}_{1t}'$. The important advantage of the conditional distribution method lies in the fact that for a symmetric Toeplitz matrix $\boldsymbol{\Sigma}$ the conditional expectation and variance of X_{t+1} can be expressed in terms of the solution $\mathbf{y}$ of the t-th order Yule-Walker equations $\boldsymbol{\Sigma}_t \mathbf{y} = -\boldsymbol{\rho}$ which can be solved recursively taking only $2d^2 + O(d)$ flops using Durbin's Algorithm (Durbin, 1960, see also Golub and Van Loan, 1989). The idea to use Durbin's Algorithm for generating stationary time series was introduced by Hosking (1984) and Marchenko and Ogorodnikov (1984a). The details are given in Algorithm 13.1 (Durbin).

To see that Algorithm 13.1 (Durbin) works we define a matrix $\mathbf{W}$ as the square matrix with ones in the secondary diagonal and zeros else. Thus $\mathbf{W} = \mathbf{W}' = \mathbf{W}^{-1}$ and $\mathbf{W}\boldsymbol{\Sigma}^{-1} = \boldsymbol{\Sigma}^{-1}\mathbf{W}$ for any symmetric Toeplitz matrix $\boldsymbol{\Sigma}$ as $\boldsymbol{\Sigma}^{-1}$ is symmetric and persymmetric (i.e. symmetric with respect to the secondary diagonal), see Golub and Van Loan (1989). We note that $\boldsymbol{\Sigma}_{1t} = (\mathbf{W}\boldsymbol{\rho})'$. Now the first two conditional moments of X_{t+1} can be expressed via the solution $\mathbf{y}$ of the Yule-Walker equations, $\boldsymbol{\Sigma}_t\mathbf{y} = -\boldsymbol{\rho}$, $\mathbf{y} = -\boldsymbol{\Sigma}_t^{-1}\boldsymbol{\rho}$. We have

$$E(X_{t+1}|\mathbf{X}_t) = \boldsymbol{\Sigma}_{1t}\boldsymbol{\Sigma}_t^{-1}\mathbf{X}_t = (\mathbf{W}\boldsymbol{\rho})'\boldsymbol{\Sigma}_t^{-1}\mathbf{X}_t$$
$$= \boldsymbol{\rho}'\boldsymbol{\Sigma}_t^{-1}\mathbf{W}\mathbf{X}_t = -(\mathbf{W}\mathbf{y})'\mathbf{X}_t = -\mathbf{y}'\mathbf{W}\mathbf{X}_t ,$$
$$\text{Var}(X_{t+1}|\mathbf{X}_t) = \sigma_{t+1,t+1} - \boldsymbol{\Sigma}_{1t}\boldsymbol{\Sigma}_t^{-1}\boldsymbol{\Sigma}_{1t}' = 1 - (\mathbf{W}\boldsymbol{\rho})'\boldsymbol{\Sigma}_t^{-1}\mathbf{W}\boldsymbol{\rho}$$
$$= 1 - \boldsymbol{\rho}'\boldsymbol{\Sigma}_t^{-1}\boldsymbol{\rho} = 1 + \boldsymbol{\rho}'\mathbf{y} .$$

One advantage of Algorithm 13.1 over Algorithm 11.3 (Multinormal-Cholesky) is that the number of flops for the setup and the generation of one time series is $4d^2 + O(d)$ instead of $d^3/3 + O(d^2)$. The second advantage are the linear storage requirements. We present a variant needing $4d + O(1)$ floating point numbers of storage, $4d^2 + O(d)$ flops for the first time series and $3d^2 + O(d)$ for additional time series.

Algorithm 13.2 (Durbin-A) is a variant that is faster at the expense of higher storage requirements. It is recommended when d is small and more than one time series with the same autocorrelations are generated. When storing the solutions $\mathbf{y}$ of the Yule-Walker equations in an array of size $d^2/2 + O(d)$ then the number of flops is reduced to $d^2 + O(d)$. It thus has the same generation speed and storage requirements as Algorithm 11.3 (Multinormal-Cholesky) but has a much faster setup.

Remark. It follows from Thm. 11.3 that Algorithm 13.1 (Durbin) can be used to multiply the Cholesky factor $\mathbf{L}$ of a Toeplitz Matrix with an arbitrary vector without computing $\mathbf{L}$ with linear storage requirements using only $4\,d^2$ flops. This is a remarkable fact as the computation of $\mathbf{L}$ itself requires at least $d^3/3$ flops.

13.1 Stationary Gaussian Time Series

Algorithm 13.1 Durbin

Require: Length of time series d, autocorrelations $\rho_1, \rho_2, \ldots, \rho_{d-1}$.
Output: Random vector (time series) of length d.
1: **for** $i = 1$ to d **do**
2:    Generate a standard normal variate ϵ_i. /* White noise */
3: Set $X_1 \leftarrow \epsilon_1$, $y_1 \leftarrow -\rho_1$, $\beta \leftarrow 1$, and $\alpha \leftarrow -\rho_1$.
4: Set $X_2 \leftarrow -X_1 y_1 + \epsilon_2 \sqrt{1 + \rho_1 y_1}$.
   /* Durbin Algorithm computes the solution y_i of the Yule-Walker equations */
5: **for** $k = 1$ to $d - 2$ **do**
6:    Set $\beta \leftarrow (1 - \alpha^2)\beta$.
7:    Set $S \leftarrow 0$.
8:    **for** $i = 1$ to k **do**
9:       Set $S \leftarrow S + \rho_{k+1-i} y_i$.
10:    Set $\alpha \leftarrow -(\rho_{k+1} + S)/\beta$.
11:    **for** $i = 1$ to $(k+1)/2$ **do**
12:       Set $h_1 \leftarrow y_i$ and $h_2 \leftarrow y_{k+1-i}$.
13:       Set $y_i \leftarrow h_1 + \alpha h_2$.
14:       Set $y_{k+1-i} \leftarrow h_2 + \alpha h_1$.
15:    Set $y_{k+1} \leftarrow \alpha$.
   /* The conditional standard deviations $\sigma_k = \sqrt{\mathrm{Var}(X_k | \mathbf{X}_{k-1})}$ in the following are computed only if the first time series for this ρ is generated. */
16:    Set $S \leftarrow 0$.
17:    **for** $i = 1$ to $k + 1$ **do**
18:       Set $S \leftarrow S + \rho_i y_i$.
19:    **if** $S \geq -1$ **then**
20:       Set $\sigma_{k+2} \leftarrow \sqrt{1 + S}$.
21:    **else**
22:       **abort** $((\rho_k)$ not positive definite$)$.
   /* Generate sample path using white noise */
23:    Set $S \leftarrow 0$.
24:    **for** $i = 1$ to $k + 1$ **do**
25:       Set $S \leftarrow S + X_i y_{k+2-i}$.
26:    Set $x_{k+2} \leftarrow -S + \epsilon_{k+2} \sigma_{k+2}$.
27: **return** time series $X_1, X_2, \ldots, X_d$.

Algorithm 13.2 Durbin-A

Require: (Short) length of time series d, autocorrelations $\rho_1, \rho_2, \ldots, \rho_{d-1}$.
Output: Random vector (time series) of length d.
   /* Same as Algorithm 13.1 but the solutions $\mathbf{y}$ of the Yule-Walker equations are stored in an array of size $d^2/2 + O(d)$. */

13.1.2 Fourier Transform Method

In many papers *Fourier transform* (FT) methods were suggested to generate stationary Gaussian time series. But the only paper we know that clearly states the correct algorithm is by Davies and Harte (1987). We need the definition of the discrete FT $\mathcal{F}(h)$ of N (possibly complex) points h_j, $j = 0, 1, \ldots, N-1$, i.e.

$$(\mathcal{F}(h))_k = \sum_{j=0}^{N-1} h_j e^{2\pi i k j / N}, \quad \text{for } k = 0, \ldots, N-1$$

where i denotes the imaginary unity $\sqrt{-1}$, and the inverse FT $\mathcal{F}^{-1}$,

$$(\mathcal{F}^{-1}(h))_k = \frac{1}{N} \sum_{j=0}^{N-1} h_j e^{-2\pi i k j / N}, \quad \text{for } k = 0, \ldots, N-1.$$

In order to reduce the computation time we make use of efficient procedures and appropriate transformations. It is well known that the discrete Fourier transform can be performed in $O(N \log_2 N)$ steps (instead of $O(N^2)$) using an algorithm called the *fast Fourier transform* (FFT). However, the standard FFT procedure works only for values of N of the form 2^k. So we define k as the smallest integer such that $N \leq N_0 = 2^k$. The corresponding d is given by $d \leq d_0 = 2^{k+1} + 1$. Thus we generate a series of length d_0 and throw away the X_j with $j \geq d$. To do this it is also necessary to elongate the autocorrelation vector up to the element $d_1 = d_0 - 1$.

We use the algorithm `drealft()` of Press, Teukolsky, Vetterling, and Flannery (1992), which computes the inverse FFT of a conjugate symmetric vector multiplied by $(N/2)$. Therefore multiplications can be saved if we compute

$$X_j = \frac{1}{2} \sum_{k=0}^{N-1} \tilde{\phi}_k \tilde{Z}_k e^{-2\pi i j k / N}, \quad \text{for } 0 \leq j < N$$

with $\tilde{\phi}_k = \phi_k / \sqrt{N/2}$ for $0 < k < N/2$ and $\tilde{\phi}_0 = \phi_0 / \sqrt{N/4}$, $\tilde{\phi}_{N/2} = \phi_{N/2}/\sqrt{N/4}$ and the $\tilde{Z}_k$ defined as the Z_k in Algorithm 13.3 (FFT) but with variance 1.

Algorithm 13.3 contains the description of a sampling algorithm based on the fast Fourier transformation. To see that this algorithm works notice that the X_k are real since we have $\phi_k Z_k = \bar{\phi}_{N-k} \bar{Z}_{N-k}$, and normal as the X_k are the sum of independent normal variables. To compute the autocorrelations of the X_k it is enough to see that we have $\mathrm{E}(Z_k \bar{Z}_k) = N$ and $\mathrm{E}(Z_k \bar{Z}_l) = 0$ for $k \neq l$. Thus

$$\mathrm{E}(X_p \bar{X}_q) = \frac{1}{N^2} \sum_{k=0}^{N-1} \sum_{m=0}^{N-1} \phi_k \phi_m e^{-2\pi i (pk-qm)/N} \mathrm{E}(Z_k \bar{Z}_m) =$$

Algorithm 13.3 FFT

Require: Length of time series d, autocorrelations $\rho_1, \rho_2, \ldots, \rho_{d-1}$.
Output: Random vector (time series) of length d.
/* Setup */
1: Store vector ρ of length $N = 2(d-1)$ of autocorrelations
$(\rho_0, \rho_1, \ldots, \rho_{N-1}) := (1, \rho_1, \ldots, \rho_{d-2}, \rho_{d-1}, \rho_{d-2}, \ldots, \rho_1)$.
/* Compute discrete Fourier transform $g = \mathcal{F}(\rho)$ of length $N = 2(d-1)$ */
2: **for** $k = 0, \ldots, N-1$ **do**
3:    Set $g_k \leftarrow \sum_{j=0}^{N-1} \rho_j e^{2\pi i k j / N}$.
4:    **if** $g_k \geq 0$ **then**
5:       Set $\phi_k \leftarrow \sqrt{g_k}$.
6:    **else** /* Method not exact for given autocorrelations */
7:       **Abort**, or set $\phi_k \leftarrow 0$ and print warning.
/* Generator */
8: Compute sequence **Z** of length N out of N iid. standard normal variates, so that z is conjugate symmetric (i.e. $z_k = \bar{z}_{N-k}$, $d-1 < k < N$) with variances $N/2$ for the real and imaginary part, and z_0, z_{d-1} real with variances N.
9: The time series **X** of length d, $\mathbf{X} = (X_0, \ldots, X_{d-1})'$,
with $X_j = \frac{1}{N} \sum_{k=0}^{N-1} \phi_k Z_k e^{-2\pi i j k / N}$.
10: **return** time series $X_0, X_1, \ldots, X_{d-1}$.

$$= \frac{1}{N} \sum_{k=0}^{N-1} \phi_k^2 e^{-2\pi i (p-q) k / N} = \rho_{p-q}$$

due to the Fourier inversion theorem and the symmetry of the autocorrelation.

The C Implementation of Algorithm 13.3 (FFT)

As the implementation of the algorithm is short but a bit tricky we include the C routine `fftrand()` given in Fig. 13.1. It requires the routine `drealft()` the double precision version of `realft()` described in Press et al. (1992, p.512ff), which calculates the FFT of a sequence of real-valued data points, and `nacr()` a generator for standard Gaussian variates described in Hörmann and Derflinger (1990). It implements Algorithm 13.3 (FFT) and generates a stationary Gaussian time series of length d in the first half of vector `data[0..2(d-1)-1]`. d-1 must be a power of 2. The autocorrelations $(1, \rho_1, \ldots, \rho_{d-1})$ must be contained in the vector `r[0..d-1]`. In order to perform the initialization of the algorithm for a new autocorrelation vector set `*setup` equal to 1. The ϕ_k calculated in the setup are stored in `r[]` as well, replacing the autocorrelation values that are no longer needed.

The setup builds up the vector ρ and computes its real FFT g which is – due to its symmetry – contained in `data[2*i]` (only g_{n_1} is stored in `data[1]`). For the case that the g_k are nonnegative, $\tilde{\phi}_k$ is computed and stored in `r[]` ($\tilde{\phi}_k = \sqrt{g_k/n_1}$ for $0 < k < n_1$ and $= \sqrt{2g_k/n_1}$ for $k = 0$ and d_1). If a g_k is

```c
void fftrand(double data[],long n,double r[],int *setup)
{   int i,n1;
    n1=n-1;
    if(*setup)
/* start of the setup*/
    {   *setup=0;
        for (i=0;i<n;i++) data[i]=r[i];
        for (i=n;i<2*n1;i++) data[i]=data[2*n1-i];
        drealft(data-1,2*n1,1);
        for(i=0;i<n1;i++) {
            if (data[2*i]<0.) {
                printf("gk negativ i=%d: %e !!!!\n",i,data[2*i]);
                r[i]=0.;
            } else r[i]=sqrt(data[2*i]/n1);
        }
        r[0]=r[0]*sqrt(2.);
        if (data[1]<0.) {
            printf("gk negativ i=%d:%e !!!!\n",n1,data[1]);
            r[n1]=0.;
        } else r[n1]=sqrt(data[1]*2./n1);
    }
/* start of the generation of the time series*/
    data[0]=r[0]*nacr();
    data[1]=r[n1]*nacr();
    for(i=2;i<2*n1;i++) data[i]=r[i/2]*nacr();
    drealft(data-1,2*n1,-1);
}
```

Fig. 13.1. C-Implementation of Algorithm 13.3 (FFT)

negative a warning is printed and the $\tilde{\phi}_k$ is set to 0, which implies that the method is not exact for the given autocorrelation.

For the generation of the time series the $\tilde{Z}_k$ are generated and multiplied by $\tilde{\phi}_k$. As $\tilde{Z}_0$ and $\tilde{Z}_{n_1}$ are real it is possible (and necessary for the call of the real FFT) to store $\tilde{\phi}_{n_1}\tilde{Z}_{n_1}$ in data[1]. Then the inverse real FFT is called.

Connection of FFT and Cholesky Algorithm

If one applies the discrete convolution theorem it is not difficult to reformulate Algorithm 13.3 (FFT). Making use of the Fourier transform we define a vector **u** so that the time series **X** can be generated as the convolution

$$\mathbf{X} = \mathbf{u} * \boldsymbol{\epsilon} \qquad \text{i.e. } x_j = \sum_{k=0}^{N-1} u_{j+k}\epsilon_k \ .$$

It is easy to see that **X** has the required autocorrelation if

$$\rho = \mathbf{u} * \mathbf{u} \quad \text{i.e.} \quad \rho_j = \sum_{k=0}^{N-1} u_{j+k} u_k = \sum_{k=0}^{N-1} u_{j-k} u_k .$$

To construct $\mathbf{u}$ by means of ρ we use the convolution theorem and get $\mathcal{F}(\rho) = \mathcal{F}(\mathbf{u}) \cdot \mathcal{F}(\mathbf{u})$ or equivalently

$$\mathbf{u} = \mathcal{F}^{-1}\left(\sqrt{\mathcal{F}(\rho)}\right) .$$

In order to write the above convolutions in matrix notation we need a $d \times N$ matrix U. As $\mathbf{u}$ is the FT of a real vector, only the first d components of $\mathbf{u}$ are necessary to construct the matrix.

$$\mathbf{U} = \begin{pmatrix} u_0, & \ldots, & u_{d-3}, u_{d-2}, u_{d-1}, & \ldots, & u_1 \\ u_1, & \ldots, & u_{d-2}, u_{d-1}, u_{d-2}, & \ldots, & u_0 \\ u_2, & \ldots, & u_{d-1}, u_{d-2}, u_{d-3}, & \ldots, & u_1 \\ \vdots & \vdots & \vdots \quad \vdots \quad \vdots & \vdots & \vdots \\ u_{d-1}, & \ldots, & u_2, \quad u_1, \quad u_0, & \ldots, & u_{d-2} \end{pmatrix} .$$

Then we have

$$\mathbf{X} = \mathbf{U}\boldsymbol{\epsilon} \quad \text{and} \quad \boldsymbol{\Sigma} = \mathbf{U}\mathbf{U}'$$

which reveals the relation to Algorithm 11.3 (Multinormal-Cholesky). The only difference is that a new decomposition of the correlation matrix $\boldsymbol{\Sigma}$ into the non-quadratic matrices $\mathbf{U}$ and $\mathbf{U}'$ is used. The advantage of that decomposition is the fact that it is a convolution and can be computed very fast. However, a vector $\boldsymbol{\epsilon}$ of length $N = 2(d-1)$ instead of d is required.

Algorithm 13.3 (FFT) uses the convolution theorem to compute $\mathbf{X}$ as

$$\mathbf{X} = \mathbf{u} * \boldsymbol{\epsilon} = \mathcal{F}^{-1}\left(\sqrt{\mathcal{F}(\rho)} \cdot \mathcal{F}(\boldsymbol{\epsilon})\right) .$$

To gain speed, $\boldsymbol{\phi} = \sqrt{\mathbf{g}} = \sqrt{\mathcal{F}(\rho)}$ is computed and stored in the setup. In addition the vector $\mathbf{z} = \mathcal{F}(\boldsymbol{\epsilon})$ is generated directly without computing a Fourier transform using the well known properties of Fourier transform of a vector of iid. Gaussian variates (cf. Step 8 of Algorithm 13.3, FFT).

One problem of Algorithm 13.3 (FFT) is that it is applicable only as long as all g_k are nonnegative. Otherwise it is impossible to find a vector $\boldsymbol{\phi}$ with the desired properties. The question for which autocorrelations Algorithm 13.3 works has been entirely neglected in the literature. It is known that as $d \to \infty$ the finite Fourier transform of the autocovariance function converges towards the spectral density of the process which must be nonnegative. This implies that the method works for d large enough if the spectral density of the process is bounded away from zero (which is the case for pure fractionally integrated processes with a nonnegative fractional integration parameter (Hosking, 1984)). Even if the spectral density has zero points our experience shows that it is not easy to find examples where the method fails for large d.

The main advantage of Algorithm 13.3 (FFT) is that it requires $O(d\log(d))$ flops for generation and setup compared with $O(d^2)$ for its competitors which makes it by far the fastest alternative if d is moderate or large.

13.1.3 Application of the Algorithms

In this section we compare Algorithms 11.3 (Multinormal-Cholesky), 13.1 (Durbin), 13.2 (Durbin-A), and 13.3 (FFT). Normal random variates were generated using algorithm NACR described in Hörmann and Derflinger (1990). All of the algorithms were coded in C under Linux.

First it is necessary to investigate for which autocorrelations ρ the different methods are applicable. In exact arithmetic Algorithms 11.3 (Multinormal-Cholesky), 13.1 (Durbin), and 13.2 (Durbin-A) work for any vector ρ that leads to a positive definite variance-covariance matrix Σ. Due to rounding errors there can occur numerical problems for certain ρ's. Considerations summarized in Golub and Van Loan (1989) imply that the stability of the Cholesky decomposition and of the Durbin algorithm are approximately the same and our experiences summarized below are in accordance with this result. For Algorithm 13.3 (FFT) the situation is different as there are cases where the algorithm does not work in precise arithmetic due to negative g_k. To get an impression how these problems can influence the applicability of the algorithms in practice we tested all four of them for the following classes of stationary processes for a variety of parameter values including the cases very close to the border of the stationary region and for d between 9 and 2049.

- MA(1) process: no problems occurred for the root of the polynomial larger than $1 + 10^{-10}$.
- AR(1) process: no problems occurred for the root of the polynomial larger than $1 + 10^{-10}$.
- pure fractionally integrated process: no problems occurred for the fractional integration parameter d between -0.5 and $0.5 - 10^{-10}$.
- AR(2) process: no problems occurred for Algorithms 11.3 (Multinormal-Cholesky), 13.1 (Durbin), and 13.2 (Durbin-A) for a lot of polynomials $(1 - \alpha_1 z - \alpha_2 z^2)$ with roots $|z| > 1 + 10^{-10}$. For Algorithm 13.3 (FFT) we found parameter values (for example $\alpha_1 = 0.5$ and $\alpha_2 = -0.995$) where negative g_k occurred. Nevertheless, for larger d (e.g. $d = 4097$) Algorithm 13.3 (FFT) worked.
- Processes with autocorrelation functions associated with polynomial spectral densities with zero points: no problems occurred for Algorithms 11.3 (Multinormal-Cholesky), 13.1 (Durbin), and 13.2 (Durbin-A). For Algorithm 13.3 (FFT) the zero point can cause problems especially if it coincides with a Fourier frequency and the spectral density is close to zero in a neighborhood.
- $\rho_k = e^{-\alpha k^2}$: Processes with this kind of correlation function are discussed by Marchenko and Ogorodnikov (1984b) as they cause problems when

α is small. In our implementation (using double precision floating point numbers) all algorithms worked for $\alpha \geq 0.1$ and all of them did not work for $\alpha \leq 0.03$. For α in between some worked and some did not. Algorithm 13.1 (Durbin) and 13.2 (Durbin-A) had of course the same behavior and were more stable than Algorithm 11.3 (Multinormal-Cholesky) which was more stable than Algorithm 13.3 (FFT). The latter fails if the spectral density of the process is very close to zero on a large interval. This is the case for small values of α.

In Marchenko and Ogorodnikov (1984b) a method for regularizing Algorithm 13.1 (Durbin) is suggested: Find an ϵ as small as possible so that Algorithm 13.1 (Durbin) using $\tilde{\Sigma} = (1-\epsilon)\Sigma + \epsilon \mathbf{I}$ instead of Σ becomes stable ($\mathbf{I}$ denotes the identity matrix). For our floating point arithmetic a value of $\epsilon = 10^{-10}$ is enough so that Algorithm 13.1 (Durbin) becomes stable for the last example with $\alpha \geq 10^{-10}$. Here the change of Σ is so small that it seems impossible that it could influence the results of a simulation.

For Algorithm 13.3 (FFT) the regularization described above is not successfull. Of course a different method could be suggested: If the negative g_k are close to zero, treat them as zero. This is simple and can remove some of the problems found in our benchmarks (especially for large d) but we do not recommend this method as it seems – in contrast to the regularization method described above – difficult to get an impression about the difference between the autocorrelation function of the regularized process and the true process.

Table 13.1. Flops and storage requirements for generation stationary time series

Algorithm	flops setup+1 time-series	flops 1 time series	memory required	normal variates
Multinormal-Cholesky	$d^3/3 + O(d^2)$	$d^2 + O(d)$	$d^2/2 + O(d)$	d
Durbin	$4d^2 + O(d)$	$3d^2 + O(d)$	$4d + O(1)$	d
Durbin-A	$4d^2 + O(d)$	$d^2 + O(d)$	$d^2/2 + O(d)$	d
FFT	$8d\log_2(2d) + O(d)$	$4d\log_2(2d) + O(d)$	$3d + O(1)$	$2(d-1)$

In order to compare the computer resources necessary for the different algorithm we present Table 13.1 that summarizes the theoretical requirements already stated in the above sections. Clearly these results have direct consequences for the speed of the compared algorithms. For large d it is clear that Algorithm 13.3 (FFT) is by far the fastest but it makes sense to compare the execution times of the algorithms for small to moderate d. Our timings (in seconds) on a PC under Linux for the setup and the generation of m time series are given in Table 13.2. Of course they depend on the computer, the compiler, and the random variate generator used. For comparsion purposes we note that the generation of 10^6 normal variates takes with our fast generator

NACR 0.8 seconds; the generation of 10^6 exponential variates with inversion takes 1.2 seconds.

Table 13.2. Execution times in seconds for m replications; 10^6 iid. normal variates take 0.80 seconds

d	10	50	100	200	500	1000	10^4	10^5	10^6
m	10000	2000	1000	500	200	100	10	1	1
Multinormal-Cholesky	0.22	0.68	1.32	3.14	22.1	117.7	*	*	*
Durbin	0.30	0.93	1.75	3.32	8.27	16.4	176	2800	*
Durbin-A	0.16	0.37	0.65	1.35	4.21	8.27	*	*	*
FFT	0.67	0.55	0.55	0.58	0.49	0.54	1.44	3.30	42.9

The results of Table 13.2 underline the strong influence of the theoretical measures of performance on the execution times. In addition we see that for $d = 10$ Algorithms 11.3 (Multinormal-Cholesky) and 13.2 (Durbin-A) have the same speed. For larger d the very slow setup makes Algorithm 11.3 (Multinormal-Cholesky) slower than Algorithm 13.2 (Durbin-A). Algorithm 13.3 (FFT) is fastest for d larger than about 100.

As general conclusion we may say that for the generation of general stationary Gaussian time series the Cholesky decomposition method should not be used unless d is small. For large d Algorithm 13.3 (FFT) is by far the fastest alternative and should be applied but problems can occur if the spectral density is close to zero in a region. For these cases Algorithm 13.1 (Durbin) and its regularization can be used as a slower but stable alternative. Algorithm 13.2 (Durbin-A) should be applied for d small to moderate if speed is of importance.

13.2 Non-Gaussian Time Series

We will discuss to three approaches to describe non-Gaussian time series. The more application oriented one is to start with the assumption that the conditional distribution, given the history of the process, is non-Gaussian. Thereby stress is laid on skewness and kurtosis contrary to specific distributions, see Davies, Spedding, and Watson (1980). As a second model we consider measurable mappings of univariate marginal distributions of the multivariate normal so that the transformed distributions, marginals of another mostly unknown multivariate distribution, are known. Here we may distinguish between the approach based on the adequate transformation of the product moment correlations (see e.g. Grigoriu, 1995; Cario and Nelson, 1996), and the one based on the Spearman's rank correlations, which seemingly has not been suggested for time series before. In both latter cases we sample in general from unknown multivariate distributions with known one-dimensional marginals and desired "auto"correlation structure. Notice, however, that for non-Gaussian time series the knowledge of marginal densities and autocorrelation function does not determine a time series uniquely. Thus we must be aware of the fact that for all generators proposed in this section, we make strong assumptions on the multivariate distribution. Thus this section could be part of Chap. 12 as well.

13.2.1 The Conditional Distribution Method

We start with the usual assumptions for ARMA models

$$\alpha(B)X_t = \beta(B)\epsilon_t$$

where $\alpha(B)$ and $\beta(B)$ are polynomials of order p and q in B with $\alpha_0 = \beta_0 = 1$. B is the backward shift operator. We assume that the roots of $\alpha(z) = 0$ and $\beta(z) = 0$ lie outside the unit circle, and that the polynomials have no roots in common. ϵ_t is white noise, i.e. the ϵ_t are iid with $E(\epsilon_t) = 0$ and $Var(\epsilon_t) = \sigma_\epsilon^2$, the location parameter zero and scale parameter constant, respectively.

If elliptically contoured or α-stable distributions are considered, the closedness with respect to linear transformations implies that we can compute the distribution of X given ϵ. On the other hand, if we allow for arbitrary skewness $\sqrt{\delta_1(\epsilon)}$ and kurtosis $\delta_2(\epsilon)$ in ϵ_t without assuming a specific distribution (see Davies et al., 1980), we may investigate the relation of the third and fourth moment of the errors with those of X.

$$\sqrt{\delta_1(X)} = \frac{E(X_t^3)}{[E(X_t^2)]^{3/2}}, \quad \delta_2(X) = \frac{E(X_t^4)}{[E(X_t^2)]^2}.$$

The relation between $\sqrt{\delta_1(\epsilon)}$ and $\sqrt{\delta_1(X)}$ may be obtained by means of impulse response coefficients, C_i, of the MA(∞) representation of the process

$$X_t = C(B)\epsilon_t = (1 + C_1 B + C_2 B^2 + \ldots)\epsilon_t,$$

$$\sqrt{\delta_1(X)} = \frac{\sum_{i=0}^{\infty} C_i^3}{[\sum_{i=0}^{\infty} C_i^2]^{3/2}} \sqrt{\delta_1(\epsilon)},$$

$$\delta_2(X) = \left(\sum_{i=0}^{\infty} C_i^4 \delta_2(\epsilon) + 6 \sum_{i=0}^{\infty}\sum_{j\neq i} C_i^2 C_j^2\right) \Big/ \left(\sum_{i=0}^{\infty} C_i^2\right)^2.$$

From this it follows that $\sqrt{\delta_1(X)} = 0$ if $\sqrt{\delta_1(\epsilon)} = 0$, and $\delta_2(\epsilon) = 3$ if and only if $\delta_2(X) = 3$, i.e. a kurtosis of 3 of ϵ is a sufficient and necessary condition for a kurtosis of 3 for X. However, there are certain parameter combinations of the αs and βs which induce a symmetric distribution in X independently of the skewness of ϵ. Further, it may even happen that the skewness of ϵ has a different sign than the skewness of X. It always holds that

$$\sqrt{\delta_1(\epsilon)} \geq \sqrt{\delta_1(X)} \quad \text{and} \quad |\delta_2(\epsilon) - 3| \geq |\delta_2(X) - 3| \,.$$

This means that the skewness and kurtosis of X are less pronounced than those of ϵ.

For the generation of a sample of X_t with given skewness, kurtosis and ARMA parameters, the impulse response coefficients have to be calculated. These are used to obtain the skewness and kurtosis of the error ϵ. Furthermore, a distribution with these properties has to be chosen, e.g. as explained in Sect. 12.4. Then, the error vector ϵ with the computed moments is generated and X_t is obtained in analogy to the recursive generation of ARMA(p,q) processes with Gaussian errors values by

$$X_t = \alpha_1 X_{t-1} + \ldots + \alpha_p X_{t-p} + \beta_1 \epsilon_{t-1} + \ldots + \beta_q \epsilon_{t-q} + \epsilon_t$$

given $X_{t-1}, \ldots, X_{t-p}$ and $\epsilon_{t-1}, \ldots, \epsilon_{t-q}$, provided that the influence of the starting values is negligible.

13.2.2 Memoryless Transformations

We consider an underlying stationary Gaussian process $\{Y_t, t \in \mathbb{Z}\}$ with mean zero, variance one, covariance function ρ_τ and marginal distribution $\Phi_N(y) = \text{Prob}(Y_t \leq y)$. This Gaussian time series is then mapped by a measurable and monotonic function g to get a non-Gaussian time series. E.g. let F be an arbitrary distribution function and let $g = F^{-1} \circ \Phi_N$. The translation process X_t is defined as

$$X_t = g(Y_t) = F^{-1} \circ \Phi_N(Y_t) \,.$$

Two strands may be distinguished: One, where the effect of the correlation structure in Y_t on the product moment (Pearson) correlations of the X_t's are investigated. This approach is called memoryless transformation by Grigoriu (1995), with respect to general covariance structures, and ARTA (autoregressive to anything) by Cario and Nelson (1996) and Deler and Nelson (2000), if properties of stationary time series are assumed. As a second possibility we suggest to consider the rank correlations induced in the X_t's. This has several advantages (see e.g. Johnson and Tenenbein, 1981) and has become quite popular for modeling multidimensional distributions related to risk analysis (see e.g. Clemen and Reilly, 1999, or Kurowicka and Cooke, 2001). An appropriate name for both approaches would be NORTA (normal to anything).

We want to stress here the simple fact that the distribution of a stationary time series is not fully specified by its autocorrelation structure and one-dimensional marginals. So both approaches implicitly include the same modeling decision; the decision to use the NORTA distribution as multidimensional distribution of the time series. For a similar approach for general (non-stationary) random vectors see Chap. 12.5.

Suppose F is a well behaved distribution function. Then

$$\text{Prob}(X_t \leq x) = \text{Prob}(F^{-1} \circ \Phi_N(Y_t) \leq x) = \text{Prob}(Y_t \leq \Phi_N^{-1}(F(x)) = F(x) \,.$$

$F(x)$ is the marginal distribution of X_t if g is a monotonically increasing function. (This and the following results will be similar for monotonically decreasing functions.) The d-dimensional density of X_t is given by

$$f_d(x_1,\ldots,x_d;t_1,\ldots,t_d) = \frac{1}{\sqrt{(2\pi)^d \det(\boldsymbol{\Sigma})}} \prod_{r=1}^d \frac{f(x_r)}{\phi_N(y_r)} \exp(-\frac{1}{2}\mathbf{y}'\boldsymbol{\Sigma}^{-1}\mathbf{y}) \ .$$

$\boldsymbol{\Sigma}$ is the variance-covariance matrix of $(Y_{t_1},\ldots,Y_{t_d})$, $y_r = \Phi_N^{-1} \circ F(x_r)$, $r = 1,\ldots,d$. $f(x) = F'(x)$ denotes the marginal density of X_t, and $\phi_N(y) = \Phi'_N(y) = (2\pi)^{-1/2} \exp(-y^2/2)$.

Specification of the Pearson Correlation Coefficients of X_t

In the following we assume that the moments up to at least order $q = 2$ exist. They are given by

$$\mathrm{E}[X_t]^q = \mathrm{E}[g(Y_t)]^q = \int_{-\infty}^{\infty} [g(y)]^q \, \phi_N(y) \, \mathrm{d}y \ .$$

The autocorrelation function of X_t, $\rho_{X,\tau} = \mathrm{E}[(X(t+\tau)-\mu_X)(X(t)-\mu_X)]/\sigma_X^2$, is obtained from

$$\mu_X^2 + \sigma_X^2 \rho_{X,\tau} = \int_{-\infty}^{\infty} \int_{-\infty}^{\infty} [g(y_1)-\mu_X][g(y_2)-\mu_X] \, \phi_N(y_1,y_2;\rho_\tau) \, \mathrm{d}y_1 \, \mathrm{d}y_2$$

in which

$$\phi_N(y_1,y_2;\rho_\tau) = \frac{1}{2\pi(1-\rho_\tau^2)^{1/2}} \exp\left(-\frac{y_1^2+y_2^2-2\rho_\tau y_1 y_2}{2(1-\rho_\tau^2)}\right)$$

denotes the joint density of the dependent standard Gaussian variables $Y(t+\tau)$ and $Y(t)$ with correlation coefficient ρ_τ. If the mapping g is strictly positive, the relationship between $\rho_{X,\tau}$ and ρ_τ can be inverted. It can be shown that $\rho_{X,\tau}$ is equal 0 (resp. 1) if and only if ρ_τ is 0 (resp. 1). However, $\rho_\tau = -1$ does not necessarily imply $\rho_{X,\tau} = -1$. If g is an odd function, $g(y) = -g(-y)$, then $\rho_{X,\tau} = -1$ follows out of $\rho_\tau = -1$. Otherwise a lower bound for $\rho_{X,\tau}$ can be given. In general

$$|\rho_{X,\tau}| \leq |\rho_\tau|$$

holds. Since the existence of the second moment of X_t is assumed and g is strictly monotone, positive definiteness of $\{\rho_Y\}$ implies also positive definiteness of $\{\rho_X\}$.

These results may be used for generation of a large class of stationary processes, X_t, with given one-dimensional marginal distribution and given autocorrelation function $\{\rho_X\}$. Numerical integration and iterative calculation of the autocorrelation function of the underlying Gaussian process $\{\rho_Y\}$ is required. A value of Y_t can then be generated by one of the methods of

Sect. 13.1 and transformed adequately to $X_t = g(Y_t)$. However, the demanding numerics are an unpleasant disadvantage.

Cario and Nelson (1996) propose a shortcut by assuming low order AR(p) models for Y_t. For AR(p) models the information of the first p values of the autocorrelation function are enough to determine the parameters of the model. Thus only the autocorrelation coefficients $\rho_{X,1}, \ldots, \rho_{X,p}$ of X_t are used to numerically compute $\rho_{Y,1}, \ldots, \rho_{Y,p}$. The computed AR($p$) model for Y_t is then used to generate Y_t. This method clearly yields an overall autocorrelation structure in X_t, which is only approximate of AR(p) form, since the $\rho_{X,j}$ for $j > p$ are not controlled for.

Specification the Spearman Rank Correlation Coefficients of X_t

More generally the translation function g may induce any distribution in X_t, even when the first moment does not exist. The process is by construction strictly stationary (cf. Kolmogorov's Theorem). The natural measure of association for general strictly stationary processes is the rank correlation coefficient. It is defined for any two random variables, X_1 and X_2, as

$$r(x_1, x_2) = \text{Cov}(F_1(x_1), F_2(x_2))$$

where F_1 and F_2 are the corresponding distribution functions. Rank correlations are invariant with respect to monotone transformations.

$$r(x_1, x_2) = r(g(y_1), g(y_2)) = r(y_1, y_2)$$

where $x_i = g(y_i)$, $i = 1, 2$. And, rank correlation matrices are also nonnegative definite. So, the rank autocorrelation function $\{r_\tau\}$ for a strictly stationary process describes a comparable dependency structure, as the product moment autocorrelation function $\{\rho_\tau\}$ for weakly stationary processes.

There is a simple relation between $r(Y_1, Y_2)$ and $\rho(Y_1, Y_2)$ for bivariate normal (Y_1, Y_2)

$$\rho(Y_1, Y_2) = 2\sin(\,r(Y_1, Y_2)\,\pi/6\,) \,. \tag{13.2}$$

So we can compute the autocorrelation vector $\rho_{Y,\tau}, \tau = 1, \ldots, d$, for the Gaussian process given the rank autocorrelations $r_{X,\tau}$. Transformation (13.2) has some nice properties: If $r = -1, 0$, or 1 then ρ results in $\rho = -1, 0$, and 1, respectively. It is an odd, strictly monotone function, so that $|\rho| > |r|$ except at -1, 0, or 1. The maximum deviation measured as $|\rho - r|$ is observed at $|r| = 0.5756$ and amounts only to 0.0181.

Using this formula we can see for very simple examples that not every positive definite autocorrelation function can be a rank autocorrelation function of a joint normal distribution. (The same observation for general multivariate distributions was made by Li and Hammond, 1975). For example, for $d = 3$ the rank autocorrelation vector $(1, 0.9, 0.63)$ implies for a Gaussian time series the autocorrelation vector $(1, 0.907981, 0.647835)$, which is not positive definite. Thus $(1, 0.9, 0.63)$ is not a feasible rank-autocorrelation for a Gaussian

13.2 Non-Gaussian Time Series

time series. There is a relation between rank autocorrelations and distributional assumptions, indicating that by means of joint normality of the Ys we cannot reproduce all possible rank correlation structures.

In order to quantify the proportion of non-feasible correlation structures for an underlying normal distribution, we randomly generate positive definite autocorrelation functions. For a similar simulation study for general random vectors see Sect. 12.5.2 and Kurowicka and Cooke (2001). We generated a sequence of length d of independent uniformly distributed random variates on the interval $(-1, 1)$ and checked for positive definiteness by the error criterion of Algorithm 11.4 (Cholesky-Decomposition). The acceptance probability for this procedure goes to zero rapidly with increasing dimension, but it is the easiest way to generate random autocorrelation vectors. The accepted vectors are considered potential rank-autocorrelation functions, $\{r_X\}$. To see if they are feasible for the multinormal distribution, we applied transformation (13.2) and computed $\rho_{Y,\tau}$; then $\rho_{Y,\tau}$ was checked for positive definiteness to see if the randomly chosen rank-autocorrelation is feasible for the joint normal distribution (i.e. Gaussian time series). The percentage of feasible rank-autocorrelation vectors observed is given in Table 13.3. According to these results the procedure is clearly not always applicable if we want to consider general autocorrelation functions.

Table 13.3. Percentage of feasible rank correlation vectors r_X for Gaussian time series. The number of replications is 1000

d	%	d	%	d	%
2	100	8	79	14	60
3	98	9	77	15	59
4	94	10	73	16	57
5	89	11	70	17	57
6	86	12	67		
7	83	13	64		

To see if problems occur for some frequently used autocorrelation patterns, we tried the autocorrelation functions as found for simple ARMA processes as rank-autocorrelation functions. Considering the AR(1) process the autocorrelation function was a feasible rank-autocorrelation for a Gaussian time series for all parameters α with $-0.999 \leq \alpha \leq 0.999$ and for all dimensions $d \leq 10^5$ tested. For MA(1) the situation is different; no problems occur only if we have $|\theta| \leq 0.765$. This is easy to understand as for $\theta \geq 0.767$, $r_{X,1}$ is so close to 0.5 that $\rho_{Y,1} > 0.5$. As $r_{X,\tau} = \rho_{Y,\tau} = 0$ for all $\tau > 1$, $\rho_{Y,\tau}$ is not positive definite if d is moderate or large. We also tried the ARMA(1,1) process. There it turned out that for the MA parameter $|\theta| \leq 0.765$ no problems occurred; for $|\theta| \geq 0.766$ problems can occur depending on the value of the AR-parameter.

For long memory (ARFIMA) processes our experiments showed that we have problems for values of $d \leq -0.3$. For $d = -0.2$ and larger values of d no problems occurred for all dimensions $d \leq 10^5$.

We close this section with Algorithm 13.4 (Non-Gaussian-Time-Series) that presents all details necessary to generate a time series with given rank-autocorrelation and arbitrary one-dimensional marginals. The algorithm generates from the NORTA distribution, i.e. it is transforming a multinormal vector (i.e. stationary Gaussian time series) component-wise to the desired marginal distribution. As discussed above the NORTA distribution does not allow for all positive definite rank-autocorrelations so Algorithm 13.4 does not work for all rank-autocorrelations either. If it is necessary to generate from such a time series we refer to Sect. 12.5.2 that gives hints to recent papers that introduce multivariate distributions that allow for all possible rank correlations.

Algorithm 13.4 Non-Gaussian-Time-Series

Require: Length of time series d, rank-autocorrelations vector r_X, cdf of one-dimensional marginal F.
Output: Random vector (time series) of length d from NORTA distribution with desired one-dimensional marginals and rank-autocorrelation vector r_X.
  /* Setup */
1: Compute the autocorrelations ρ_Y: $\rho_{Y,\tau} \leftarrow 2\sin\left(\frac{\pi r_{X,\tau}}{6}\right)$ for $\tau = 1, 2, \ldots, d-1$.
2: Run setup for Algorithm 13.3 (FFT) or 13.1 (Durbin) (necessary for Gaussian time series with autocorrelation vector ρ_Y) and check if autocorrelation vector ρ_Y is positive definite (otherwise **abort** procedure with error message).
  /* Generator */
3: Generate Gaussian time series Y_i with autocorrelation vector ρ_Y using Algorithm 13.3 (FFT) or Algorithm 13.1 (Durbin).
4: **for** $i = 1$ to d **do**
5:   Find solution X_i of equation $F_i(X_i) = \Phi_N(Y_i)$ (numerically).
  /* $\Phi_N(x)$ denotes cdf of standard normal distribution. */
6: **return** time series $(X_1, X_2, \ldots, X_d)$.

13.3 Exercises

Exercise 13.1. Spell out the details of the (approximate) recursive generation method for the AR(1), MA(1) and ARMA(1,1) processes.

Exercise 13.2. For the autocorrelation vector $\rho = (1, 0.5, 0.2)$ and the noise vector $(2, -1, 1)$ compute "by hand" the time-series generated by Algorithm 13.1 (Durbin) and by Algorithm 11.3 (Multinormal-Cholesky). Check that both result are identical.

14

Markov Chain Monte Carlo Methods

We have seen in Chapter 11 that the generation of random vectors is often not easy. The rejection based algorithms what we presented there are from a practical point of view limited to small dimensions up to at most 10. And there are lots of distributions that are even difficult to sample from in dimension three or four. A totally different approach is based on the fact that we always can easily construct a Markov chain that has the desired fixed multivariate distribution as its unique stationary distribution. Of course there is a price we have to pay for this simplicity: the dependence of the generated vectors.

The idea to use a Markov chain to simulate from random vectors is not new, but has attracted considerable attention only recently. In particular, so called *Markov chain Monte Carlo methods* (MCMC) are used extensively for the computation of difficult multidimensional integrals necessary for Bayesian statistics. An impressive number of applied and theoretical research papers on MCMC were published in the last decade together with several recent books, e.g. Gilks, Richardson, and Spiegelhalter (1996), Gamerman (1997), Chen, Shao, and Ibrahim (2000), and Evans and Swartz (2000). Also for linear optimization Markov chain algorithms are used to generate random points in and on the border of the region of feasibility. There the name *Hit-and-Run* algorithms is the most common term used, see Smith (1984) for the main early reference. The same idea is also used to compute the volume of convex bodies (see Lovasz, 1999, and the references given there). All these applications have in common that it is possible to arrive at results using directly all dependent vectors of the Markov chain.

We will not present the full theory that is also closely linked with its applications but present the main ideas of Markov chain sampling algorithms in Sect. 14.1. In Sect. 14.2 we introduce the basic idea of *perfect sampling* (Propp and Wilson, 1996), a method that allows to sample exactly from the limiting distribution of a Markov chain. In Sect. 14.3 we discuss how MCMC and perfect sampling can be used to generate independent random variates and vectors. We also introduce the first universal random variate generation algorithm based on MCMC and perfect sampling, that is currently restricted to one dimensional distributions with monotone density.

14.1 Markov Chain Sampling Algorithms

If we are given an initial vector $X_0 \in \mathbb{R}^d$ and a conditional distribution $K(X_{t+1}|X_t)$ that depends only on the current state-vector X_t we can easily generate the sequence of random vectors $X_0, X_1, X_2, \ldots$. This sequence is called a *Markov chain* and $K(X_{t+1}|X_t)$ is called the *transition kernel* of the chain. We will only consider time-homogeneous chains where the transition kernel does not depend on t. Of course it will be important for us to understand how the starting vector X_0 will affect the distribution of X_t that we denote $K^t(X_t|X_0)$. Subject to regularity conditions, $K^t(X_t|X_0)$ will converge to a unique stationary distribution, which neither depends on t nor on X_0.

In the following we describe various methods how a Markov chain with the required distribution can be generated. However, it is important to note that such generators produce dependent random variates. Moreover, the first vectors of these sequence do not follow the correct distribution and thus should be discarded. This is called the *burn-in* phase of such a Markov chain sampling algorithm. As will be seen below the required length of this burn-in (i.e. the number of discarded vectors) that guarantee convergence is difficult to estimate, which is one of the major practical problems of MCMC.

14.1.1 Metropolis-Hastings Algorithm

It is surprisingly simple and known for half a century that we can easily construct a Markov chain such that its stationary distribution is equal to our distribution of interest. The general form due to Hastings (1970) is a generalization of the method first proposed by Metropolis, Rosenbluth, Teller, and Teller (1953). Today it is commonly called *Metropolis-Hastings* algorithm. The main ingredient of the algorithm is the (multivariate) so called *proposal density* $q(x|X)$. The proposal density is not the full transition kernel as the final decision about the new state of the Markov chain is done in a rejection step. The idea is that we start with an arbitrary vector X_0. Assuming that we have generated the chain up to X_t, a proposal value $\tilde{X}$ for X_{t+1} is computed by sampling from the proposal density $q(x|X_t)$. The proposal value $\tilde{X}$ is accepted as value for X_{t+1} if an auxiliary uniform $(0,1)$ random number U is smaller than $\alpha(X_t, \tilde{X})$. So we obtain the acceptance condition

$$U \leq \alpha(X_t, \tilde{X}) = \min\left(1, \frac{f(\tilde{X})}{f(X_t)} \frac{q(X_t|\tilde{X})}{q(\tilde{X}|X_t)}\right) \tag{14.1}$$

where f denotes the (quasi-) density of the desired distribution. The formal description of this method is presented as Algorithm 14.1 (Metropolis-Hastings-Chain).

Algorithm 14.1 Metropolis-Hastings-Chain

Require: Dimension d, (quasi-) density $f(x)$;
proposal density $q(x, X)$, starting point X_0.
Output: "Infinite sequence" X_t from a Markov chain, whose stationary distribution has density f.
1: Set $t \leftarrow 0$.
2: **loop**
3:   Generate vector $\tilde{X}$ with density $q(x|X_t)$.
4:   Generate $U \sim U(0,1)$.
5:   **if** $U \leq \frac{f(\tilde{X})}{f(X_t)} \frac{q(X_t|\tilde{X})}{q(\tilde{X}|X_t)}$ **then**
6:     Set $X_{t+1} \leftarrow \tilde{X}$.
7:   **else**
8:     Set $X_{t+1} \leftarrow X_t$.
9:   Increment t.

It is remarkable that the density of the stationary distribution of the constructed Markov chain is f and that it does not depend on the choice of the proposal density. We do not deal with the mathematical background of Markov chains here but give a simple argument (following Gilks et al., 1996), why the Metropolis-Hastings algorithm works:

The transition kernel consists of a continuous part arising from acceptance and the point mass $X_{t+1} = X_t$ arising from rejection. We can write

$$K(X_{t+1}|X_t) = q(X_{t+1}|X_t)\alpha(X_t, X_{t+1}) + \mathbf{1}_{X_t}\left(1 - \int q(\tilde{X}|X_t)\alpha(X_t, \tilde{X})\,\mathrm{d}\tilde{X}\right)$$

where $\mathbf{1}$ denotes the indicator function. The definition of the acceptance probability (14.1) implies

$$f(X_t)q(X_{t+1}|X_t)\alpha(X_t|X_{t+1}) = f(X_{t+1})q(X_t|X_{t+1})\alpha(X_{t+1}|X_t)$$

which can be combined with the definition of the transition kernel above to get the detailed balance equation

$$f(X_t)K(X_{t+1}|X_t) = f(X_{t+1})K(X_t|X_{t+1}) \,.$$

Integrating both sides of this equation with respect to X_t gives

$$\int f(X_t)K(X_{t+1}|X_t)\,\mathrm{d}X_t = f(X_{t+1}) \,.$$

This shows that the stationary distribution is f but not that the Markov chain will converge to that distribution from arbitrary starting points. For discussion of convergence issues see Tierney (1994) where it is shown that under mild regularity conditions the Metropolis-Hastings algorithm converges.

MCMC is applied to obtain samples from the desired distribution f. Therefore the user hopes that the Markov chain constructed by the Metropolis-Hastings Algorithm converges fast to the correct distribution. It is also an advantage to have a chain where the correlation within the chain is small. Both aims are reached if we have a chain with good *mixing properties* (i.e. the Markov chain is moving around fast). Of course the mixing properties of the chain and thus the performance of the Metropolis-Hastings Algorithm strongly depend on the choice of the proposal density. We continue with some standard proposal densities suggested in the literature.

The Random-Walk Metropolis Algorithm (RWM)

The name comes form the fact that the proposal density for X_{t+1} is a radially symmetric distribution around X_t. This implies $q(\tilde{X}|X) = q(X|\tilde{X}) = q(|X - \tilde{X}|)$ and the acceptance probability simplifies to $\alpha(x,\tilde{x}) = \min(1, f(\tilde{x})/f(x))$. Popular choices for the radius of q include normal, Student, and uniform distributions with mean zero. Special care is necessary for choosing the scale σ of the proposal distribution. Too small values of σ will lead to a high acceptance rate but will mix slowly. Too high values of σ will lead to a small acceptance rate as $\tilde{X}$ will often fall into the far tails of the distribution. This implies that the chain will often not move, which again results in slow mixing. A good value for σ should avoid both these extremes. Roberts, Gelman, and Gilks (1997) show that for Gaussian proposal $q(.)$ and Gaussian density f the best mixing is achieved for acceptance rate 0.45 for dimension $d = 1$ and 0.23 for $d \to \infty$. For $d = 6$ the optimal acceptance rate is 0.25. Figure 14.1 shows three paths of random walk Metropolis with uniform proposal density. The scale parameter is well chosen for the top figure, too large for the middle figure, and too small for the bottom figure.

We emphasize here that RWM is not convergent if the density f is unbounded. It is easy to see that if the Markov chain is trapped in a pole of the density it remains there forever. In general a density with sharp spikes is a problem for MCMC algorithms. The regions of the spikes are sometimes called "sticky" points for the algorithm as it has problems to move away from them.

Independence Sampler

The independence sampler (Tierney, 1994) uses a proposal $q(\tilde{X}|X) = q(\tilde{X})$ that does not depend on X. Therefore the acceptance probability simplifies to

$$\alpha(X_t, \tilde{X}) = \min\left(1, \frac{f(\tilde{X})}{f(X_t)} \frac{q(X_t)}{q(\tilde{X})}\right) = \min\left(1, \frac{f(\tilde{X})}{q(\tilde{X})} \bigg/ \frac{f(X_t)}{q(X_t)}\right).$$

It should be clear that this algorithm is close to the standard rejection algorithm to generate iid. random variates (see Sect. 2.2). Instead of the hat function (or dominating distribution) $h(x)$ we have now the proposal density $q(x)$. If we write $h(x)$ for that proposal density we must not overlook that

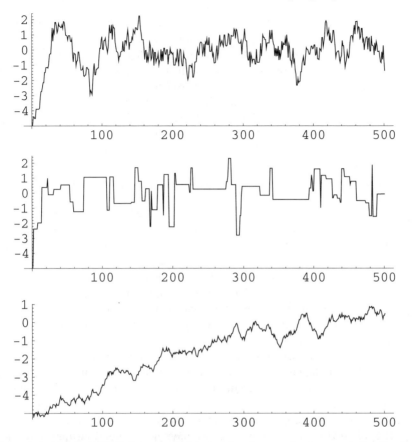

Fig. 14.1. Random Walk Metropolis chains for the standard normal distribution with starting point $X_0 = -5$ and proposal $U(-1, 1)$ (top), $U(-10, 10)$ (middle), and $U(-0.2, 0.2)$ (bottom)

$A_h = 1$ here. Thus we have to use the hat function $\max_x \left(\frac{f(x)}{h(x)}\right) h(x)$, and the acceptance condition for $\tilde{X}$ for standard rejection can be written as

$$U \max_x \left(\frac{f(x)}{h(x)}\right) \leq \frac{f(\tilde{X})}{h(\tilde{X})}.$$

If $\tilde{X}$ is rejected in a standard rejection algorithm the experiment is simply repeated until acceptance. In contrast, rejection for the independence sampler means that the Markov chain does not move and X (the old value) is taken as next value of the chain as well. But the acceptance condition for $\tilde{X}$ remains practically the same. For the independence sampler we have

$$U \left(\frac{f(X)}{q(X)}\right) \leq \frac{f(\tilde{X})}{q(\tilde{X})}.$$

This clearly implies that no rejection is possible if for the newly proposed value $\tilde{X}$ the ratio f/q is bigger than for the old X_t.

It is no surprise that the performance of the independence sampler entirely depends on the quality of the proposal density. In the case that $q(x) = f(x)$ the independence sampler directly produces iid. variates with density f. Liu (1996) gives explicit expressions for the iterated transition probabilities. It is also shown that the independence sampler is geometrically convergent with rate $1 - \inf_x(q(x)/f(x))$. Thus there will be problems with the convergence when the proposal has lower tails then the target density. Consequently the multinormal distribution may be a bad choice for $q(x)$. For that reason the multivariate t-distribution should be better when a simple general applicable choice is needed.

If $\inf_x(q(x)/f(x))$ is positive the proposal density can also be used as a hat function for a standard rejection algorithm if we can find a positive lower bound for the infimum. So for the standard rejection algorithm we have to find the bound, which gives a performance guarantee, first. Then we may sample but, if q is not a good choice, the algorithm will be slow as we need many trials until success. For the independence sampler, however, we can run the algorithm without checking if we will have geometric convergence. The algorithm will always produce a Markov chain but only after carefully assessing the properties of the generated chain we can be quite sure that the generated variates follow the desired distribution. This difference between standard rejection and the independence sampler highlights a general difference that can be observed between Markov chain algorithms and algorithms that generate exact iid. variates.

A variant of the independence sampler can be used to "correct" a standard rejection algorithm. We assume that we have a hat function $h(x)$ and know that $f(x) \leq h(x)$ is true with the exception of some small unknown regions. Then we can define an independence sampler with $q(x) = \min(f(x), h(x))$ which is called *rejection chain* in the literature. Sampling from $q(x)$ is no problem (see exercise 14.1) and it is also not difficult to see that no rejection can occur for $\tilde{X}$ as long as X_t is in the region $C = \{x | f(x) \leq h(x)\}$ (see exercise 14.2). The rejection chain can be interpreted as a "corrected" version of the standard rejection algorithm. The correction happens whenever an X with $f(X) \geq h(X)$ has been generated. In the next step this X is retained with a certain probability (depending on $f(X)$ and $h(X)$ and on the new proposed value $\tilde{X}$) as the next value of the rejection chain, thus increasing the probability for generating X by the algorithm. This is necessary as for regions with $f(X) > h(X)$ the probability to sample from X would be too small without this correction.

14.1.2 Gibbs Sampling

It is possible to reformulate the above Metropolis-Hastings algorithm such that the vector X_t is not always updated en-bloc but updated component by component. The most important special case of this single component Metropolis-Hastings Algorithm is the so called *Gibbs sampler*, which is the most popular MCMC algorithm. The Gibbs sampler has been given its name by Geman and Geman (1984) but has been used even earlier in statistical physics. The routine application to Bayesian statistics started with the papers of Gelfand and Smith (1990) and Gelfand, Hills, Racine-Poon, and Smith (1990). The idea of Gibbs sampling is that the proposal density for updating the i-th component of X_t is the full conditional distribution of f given $X_{t+1,1}, \ldots, X_{t+1,i-1}, X_{t,i+1}, \ldots, X_{t,d}$, for which we write $f(x_i|X_{t+1,1}, \ldots, X_{t+1,i-1}, X_{t,i+1}, \ldots, X_{t,d})$. For a formal description see Algorithm 14.2 (Gibbs-Chain).

Algorithm 14.2 Gibbs-Chain

Require: Dimension d, density $f(x)$; starting point $X_0 = (X_{0,1}, \ldots, X_{0,d})$.
Output: "Infinite sequence" X_t from a Markov chain, whose stationary distribution has density f.
1: Set $t \leftarrow 0$.
2: **loop**
3:   **for** $i = 1$ to d **do**
4:     Generate $\tilde{X}_i$ from the full conditional density
      $f(x_i|X_{t+1,1}, \ldots, X_{t+1,i-1}, X_{t,i+1}, \ldots, X_{t,d})$.
5:     Set $X_{t+1,i} = \tilde{X}_i$
6:   Increment t.

It is evident that for the Gibbs sampler we only need one-dimensional random variate generation algorithms. In addition no proposal density and no acceptance test is necessary. Probably these advantages led to the popularity of the Gibbs sampler. On the other hand even the one-dimensional generation problem is not trivial as in general the full conditional density changes in every step. It was this generation problem for the Gibbs sampler that motivated Gilks and Wild (1992) to develop their black-box algorithm for log-concave distributions that they called adaptive rejection sampling. That algorithm is the version of transformed density rejection with the fastest setup (see Chap. 4).

It is interesting to compare the Gibbs sampler with the conditional distribution method for generating random vectors (see Sect. 11.1.1) as these two algorithms are quite similar; both only generate from one-dimensional distributions. But for Algorithm 14.2 (Gibbs-Chain) we need starting values for the full conditional distributions which implies that the produced variates are dependent. In Algorithm 11.1 (Conditional-Distribution-Method) we get rid of the starting values by replacing the full conditional distributions by the marginal distributions. This makes the algorithm exact but intractable for most multivariate distributions, as the marginal distributions are difficult to obtain. So in this comparison we again see the typical difference between the MCMC and the iid. vector generation approach. The former allows for easier generation and tries to solve the problems afterwards when assessing the convergence. The latter wants to solve the problem right away before the generation, but this turns out to be impossible for most multivariate distributions.

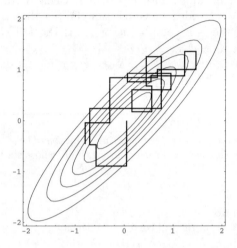

Fig. 14.2. A path of the Gibbs sampler for the normal distribution with $\rho = 0.9$

The simple structure of the Gibbs sampler has also disadvantages. It is clear from the algorithm that the Gibbs sampler can only move in directions parallel to the axis. Therefore it works well for the case that the target density $f(x)$ describes a random vector with approximately uncorrelated components. When the components are exactly independent the full conditional distributions used in the algorithm coincide with the marginals and thus the Gibbs sampler produces an iid. vector sample. But if the components of the vector have high correlations the mixing properties and the convergence speed of the sampler are poor as it cannot move in the main direction of the distribution. Figure 14.2 shows an example of a path for the multinormal distribution with $\rho = 0.9$. It is easy to construct even worse examples: If e.g. the target distribution $f(x)$ is uniform on the two squares $(-1,0) \times (-1,0)$ and $(0,1) \times (0,1)$ it is obvious that the Gibbs sampler cannot travel from one square to the other, so the Markov chain is reducible and not even ergodic.

There are several papers in the literature that deal with conditions that ensure ergodicity and geometric convergence (see e.g. Roberts and Smith, 1994). However, these do not give bounds for the convergence rate. Amit (1991) was the first who used function analytic results to prove an exact formulation of the heuristic that the Gibbs sampler will output almost iid. random vectors if the components of X are almost independent. For the bivariate normal distribution it is possible to show that the rate of convergence is ρ^2. So for many posterior distributions of Bayesian statistics it makes sense to try a (linear) reparametrization to make the correlations as small as possible. This can be done e.g. by using the results of a pilot run; but it should not be forgotten that for clearly non-normal target distributions an approximately uncorrelated parameterization need not lead to a well mixing Gibbs sampler.

14.1.3 Hit-and-Run

The reason for the mixing problems of the Gibbs sampler lie in the fact that it only samples along the coordinate axes. To get rid of this disadvantage generalizations of the Gibbs sampler, called *hit-and-run algorithms*, have been suggested that allow for updating with the full conditional distribution in random directions (Chen and Schmeiser, 1993). Special cases of that algorithm for the uniform distribution have already been suggested in the late seventies for the use in optimization problems (see Smith (1984) and references given there). The general hit-and-run algorithm has been first published in Belisle, Romeijn, and Smith (1993). The details are given in Algorithm 14.3 (Hit-and-Run-Chain).

Algorithm 14.3 Hit-and-Run-Chain

Require: Dimension d, density $f(x)$;
  distribution ν on unit sphere S^{d-1}, starting vector X_0.
Output: "Infinite sequence" X_t from a Markov chain, whose stationary distribution has density f.
1: Set $t \leftarrow 0$.
2: **loop**
3:   Generate unit length direction vector D with distribution ν.
4:   Generate $\tilde{X}$ from full conditional density along direction D, given by $f_{\text{cond}}(y) = f(X_t + y D)$.
5:   Set $X_{t+1} = \tilde{X}$
6:   Increment t.

Belisle et al. (1993) show that under fairly mild conditions on f and ν the hit-and-run algorithm converges to the correct distribution. Of course it is easiest to use the uniform distribution as distribution ν. For a target distribution with highly correlated components the mixing of the hit-and-run chain will be faster than the Gibbs sampler but still slow. Therefore Gilks, Roberts, and George (1994) have suggested "adaptive direction sampling" where the distribution ν is chosen on the basis of previously sampled points.

A practical problem of the hit-and-run algorithm is determining and sampling from the full conditional distribution along the generated direction. The situation is simplified when we restrict the algorithm to uniform target distributions. Then the full conditional distributions in arbitrary direction are all uniform. But it can still be a difficult numerical problem to find the domain of that uniform distribution.

Chen and Schmeiser (1998) explain an algorithm called *Random direction interior point* method that circumvents these problems by including a Metropolis type rejection step with a gamma proposal density. Note that the uniform assumption need not be a loss of generality as they generate vectors in $\mathbb{R}^{d+1}$ uniformly below the graph of an arbitrary density f and can then discard the last component of that vector to get vectors from the target distribution. In other words, Chen and Schmeiser are constructing a Markov chain that is moving below the graph of the target density f to generate a uniform distribution below that graph. They call this algorithm a first step toward "black-box" sampling as it is designed as a MCMC sampling algorithm that can be used for widely arbitrary target densities f. So the term black-box used there refers to MCMC sampling algorithms that produce a Markov chain, whereas we use it in this book for random variate generation algorithms that generate iid. variates or vectors.

Boender et al. (1991) present variants of hit-and-run (called *shake-and-bake algorithms*) that are designed to generate uniformly distributed points on the boundary of full dimensional bounded polyhedra.

14.2 Perfect Sampling for Markov Chain Monte Carlo

In practice the assessment of convergence of a generated Markov chain is the main problem of MCMC. So it is good news that there are recent methods for generating exactly from the stationary distribution of a Markov chain (see Propp and Wilson, 1996; Fill (1998); Wilson, 1998, maintains an annotated bibliography and link list). These methods – called *perfect sampling*, or *exact sampling* – were designed for bounded discrete state-space Markov chains. They work for many Markov chains used in statistical physics and image processing. However, there are also many discrete state-space Markov chains where perfect sampling is arbitrarily slow or impossible. We are here interested in Markov chains that have the $\mathbb{R}^d$ as state-space. In this case certain adaptions are possible to apply perfect sampling methods (see Green and Murdoch, 2000; Wilson, 2000; Murdoch, 2000; Murdoch and Meng, 2001; Mira, Moller, and Roberts, 2001) but they are still far away from a routine use. We will explain here the basic method for bounded discrete state-space. In Sect. 14.3 below we will discuss some of the adaptions necessary for continuous unbounded state-space.

14.2.1 Coupling from the Past

Coupling from the past (CFTP) is the perfect sampling algorithm introduced by Propp and Wilson (1996). We follow here closely the explanation of Wilson (2000). We assume that X_t is a Markov chain with bounded discrete state-space $\mathcal{X}$. The most important part of the CFTP algorithm is the *randomizing operation*. This is a deterministic function ϕ taking two inputs, the input state

14.2 Perfect Sampling for Markov Chain Monte Carlo

X_t and some randomness denoted as U_t (e.g. the output of a random number generator). It is assumed that the U_t are independent. The randomizing operation returns the new output state

$$X_{t+1} = \phi(X_t, U_t) \ .$$

The randomizing operation is assumed to preserve the target distribution f. Applying the randomizing operation to a given state is equivalent to running the Markov chain one step from that state. There can be many different randomizing operations consistent with a given Markov chain like there are different sampling algorithms to generate random variates from the same distribution. Natural candidates for the randomizing operation are the Metropolis-Hastings algorithm and its variants of Sect. 14.1. The properties of CFTP depend on the randomizing operation, not only on the Markov chain itself.

Now, to explain the idea and to prove the validity of the CFTP algorithm, it is convenient to assume first that we have an oracle that returns perfect random samples from distribution f. We then define an Experiment A_T.

Experiment 14.1 A_T

Require: Distribution f on bounded discrete state-space $\mathcal{X}$; randomizing operation $\phi(X_t, U_t)$, generator for randomness U_t, number of back-steps T; "oracle".
Output: Random variate with distribution f.
1: Generate the pieces of randomness $U_{-T}, U_{-T+1}, \ldots, U_{-1}$.
2: Ask oracle to draw X_{-T} from f.
3: **for** $t = -T$ to -1 **do**
4:     Compute $X_{t+1} = \phi(X_t, U_t)$ (using the U_t generated above).
5: **return** X_0.

As the distribution f is preserved by the randomizing operation, X_0 is exactly distributed according to f. In order to be less dependent on the oracle we now consider Experiment B_T. The main problem when implementing this experiment is certainly how to decide if there is only one possible value for X_0 in Step 3. If (as assumed here) the state-space is discrete and bounded we can accomplish this task by starting at time $-T$ with all possible states. Then we run all these states forward till time 0 using the same randomizing operations and the same pieces of randomness for all paths and check if they have coalesced to a single value. If this is the case this single value is X_0. As we use the same pieces of randomness in the first and in the second part of Experiment B_T it is easy to see that Experiment B_T (possibly without using the oracle) always returns precisely the same result as Experiment A_T. Therefore it is clear that Experiment B_T generates exactly from the distribution with density f.

If we want to reduce the probability to use the oracle we should pick a large value for T. But this could result in a lot of unnecessary work as it is possible that already the last few values of U_t determine X_0. So it is best to start with

Experiment 14.2 B_T

Require: Distribution f on bounded discrete state-space $\mathcal{X}$; randomizing operation $\phi(X_t, U_t)$, generator for randomness U_t, maximal number of back-steps T; "oracle".
Output: Random variate with distribution f.
1: Generate the pieces of randomness $U_{-T}, U_{-T+1}, \ldots, U_{-1}$.
2: Start from all possible states at time $-T$ and compute all possible states $X_{-T+1}, X_{-T+2}, \ldots, X_0$ using the randomizing operation $\phi(X_t, U_t)$ and the generated pieces of randomness U_t from above.
3: **if** only one possible value for X_0 **then**
4:     **return** X_0.
5: **else**
6:     Ask oracle to draw X_{-T} from f.
7:     **for** $t = -T$ to -1 **do**
8:         Compute $X_{t+1} = \phi(X_t, U_t)$ (using the U_t generated above).
9:     **return** X_0.

small values of T and to increase them later if necessary. Doing this we get Experiment C_T. It is clear that Experiment C_T produces the same output as Experiment A_T as long as the U_t and the value returned by the oracle are the same. Coupling from the past is now Experiment C_T with $T \to \infty$, see Algorithm 14.4 (CFTP). It has the important property that it never consults the oracle. From the construction it is clear that CFTP samples exactly from the distribution f.

Experiment 14.3 C_T

Require: Distribution f on bounded discrete state-space $\mathcal{X}$; randomizing operation $\phi(X_t, U_t)$, generator for randomness U_t, number of back-steps $T = 2^\tau$ ($\tau \geq 1$); "oracle".
Output: Random vector with distribution f.
1: Generate the piece of randomness U_{-1}.
2: **for** $k = 1$ to τ **do**
3:     Generate the piece of randomness $U_{-2^k}, \ldots, U_{-2^{k-1}-1}$.
4:     Start from all possible states at time -2^k and compute all possible states $X_{-2^k}, X_{-2^k+1}, \ldots, X_0$ using the randomizing operation $\phi(X_t, U_t)$ and the generated pieces of randomness U_t from above.
5:     **if** $U_{-2^k}, \ldots, U_{-1}$ uniquely determines X_0 **then**
6:         **return** X_0.
7: Ask oracle to draw X_{-T} from f.
8: **for** $t = -T$ to -1 **do**
9:     Compute $X_{t+1} = \phi(X_t, U_t)$ (using the U_t generated above).
10: **return** X_0.

14.2 Perfect Sampling for Markov Chain Monte Carlo

Algorithm 14.4 CFTP (Coupling from the past)

Require: Distribution f on bounded discrete state-space $\mathcal{X}$; generator for randomness U_t, randomizing operation $\phi(X_t, U_t)$.
Output: Random vector with probability mass function f.
1: Generate the piece of randomness U_{-1}.
2: Set $T \leftarrow 2$
3: **loop**
4:   Generate the piece of randomness $U_{-T}, \ldots, U_{-\frac{T}{2}-1}$.
5:   Start from all possible states at time $-T$ and compute all possible states $X_{-T+1}, X_{-T+2}, \ldots, X_0$ using the randomizing operation $\phi(X_t, U_t)$ and the generated pieces of randomness U_t from above.
6:   **if** $U_{-T}, \ldots, U_{-1}$ uniquely determines X_0 **then**
7:     **return** X_0.
8:   Set $T \leftarrow 2T$.

Example 14.1. We illustrate the idea of CFTP for the Markov chain with state-space $\{0, 1, 2, 3\}$ and transition probabilities $\text{Prob}(X+1|X) = \text{Prob}(X-1|X) = \text{Prob}(X|X) = 1/3$ for $X = 1, 2$, $\text{Prob}(1|0) = \text{Prob}(2|3) = 1/3$, and $\text{Prob}(0|0) = \text{Prob}(3|3) = 2/3$. We assume that U_t is -1, 0, or 1 with respective probabilities $1/3$. As randomizing operation we define $\phi(x, u) = x + u$, where we map $x + u = -1$ to 0 and $x + u = 4$ to 3 in the obvious way. Now we generate from the stationary distribution f using the sequence

$$(U_{-8}, U_{-7}, \ldots, U_{-1}) = (-1, 0, -1, 0, 1, 0, 1, 1).$$

We start with noticing that the randomizing operation U_{-1} maps $\{0, 1, 2, 3\}$ to $\{1, 2, 3\}$. So the states have not coalesced yet and we have to try again with $T = -2$,

$$\{0,1,2,3\} \xrightarrow{(1)} \{1,2,3\} \xrightarrow{(1)} \{2,3\},$$

and $T = -4$,

$$\{0,1,2,3\} \xrightarrow{(1)} \{1,2,3\} \xrightarrow{(0)} \{1,2,3\} \xrightarrow{(1)} \{2,3\} \xrightarrow{(1)} \{3\}.$$

Now the states have coalesced in $X_0 = 3$ and we output 3. To generate another variate from the stationary distribution we use the sequence:

$$(U_{-8}, U_{-7}, \ldots, U_{-1}) = (-1, 0, 1, -1, 0, -1, 1, 1).$$

Starting with times $T = -1$, -2, and -4 does not lead to coalescence. So we try $T = -8$:

$$\{0,1,2,3\} \xrightarrow{(-1)} \{0,1,2\} \xrightarrow{(0)} \{0,1,2\} \xrightarrow{(1)} \{1,2,3\} \xrightarrow{(-1)} \{0,1,2\} \xrightarrow{(0)} \{0,1,2\} \xrightarrow{(-1)} \{0,1\} \xrightarrow{(1)} \{1,2\} \xrightarrow{(1)} \{2,3\}.$$

This is also not final coalescence. We continue with generating
$(U_{-16},\ldots,U_{-9}) = (-1,1,-1,0,-1,0,0,1)$ and $T = -16$:

$$\begin{Bmatrix}0\\1\\2\\3\end{Bmatrix} \xrightarrow{(-1)} \begin{Bmatrix}0\\1\\2\end{Bmatrix} \xrightarrow{(1)} \begin{Bmatrix}1\\2\\3\end{Bmatrix} \xrightarrow{(-1)} \begin{Bmatrix}0\\1\\2\end{Bmatrix} \xrightarrow{(0)} \begin{Bmatrix}0\\1\\2\end{Bmatrix} \xrightarrow{(-1)} \begin{Bmatrix}0\\1\end{Bmatrix} \xrightarrow{(0)} \begin{Bmatrix}0\\1\end{Bmatrix} \xrightarrow{(0)} \begin{Bmatrix}0\\1\end{Bmatrix} \xrightarrow{(1)} \begin{Bmatrix}1\\2\end{Bmatrix}.$$

So $\{1,2\}$ is the state at time -8. Continuing with the sequence $(U_{-8},U_{-7},\ldots,U_{-1})$ from above we get:

$$\begin{Bmatrix}1\\2\end{Bmatrix} \xrightarrow{(-1)} \begin{Bmatrix}0\\1\end{Bmatrix} \xrightarrow{(0)} \begin{Bmatrix}0\\1\end{Bmatrix} \xrightarrow{(1)} \begin{Bmatrix}1\\2\end{Bmatrix} \xrightarrow{(-1)} \begin{Bmatrix}0\\1\end{Bmatrix} \xrightarrow{(0)} \begin{Bmatrix}0\\1\end{Bmatrix} \xrightarrow{(-1)} \{0\} \xrightarrow{(1)} \{1\} \xrightarrow{(1)} \{2\}.$$

The chain has coalesced at time $t = -2$, but we must not return X_{-2} but X_0 which is 2. Figure 14.3 illustrates this example.

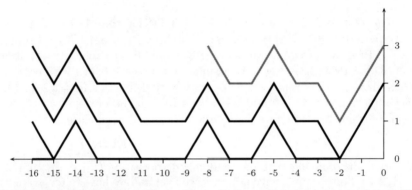

Fig. 14.3. Starting form time $T = -16$ results in a uniquely determined state X_0 (state-space $\mathcal{X} = \{0,1,2,3\}$). Starting at time $T = -8$ does not coalesce (additional gray lines) and is thus not sufficient

Looking at the above example it is easy to see how we can utilize the ordering of the state-space if we have a monotone randomizing operation with the property: $x \leq y$ implies $\phi(x,u) \leq \phi(y,u)$. It is then enough to follow only the paths of the biggest and the smallest elements in the state-space (0 and 3) in our example. Coalescence is reached when these two elements are mapped together. So for a partially ordered state-space and a monotone randomizing operation it is enough to follow the paths of only two starting values instead of all of them.

14.3 Markov Chain Monte Carlo Methods for Random Vectors

MCMC algorithms have not been designed to generate iid. random vectors. Nevertheless, they can be used for this task if we have general arguments to guarantee (approximate) convergence and independence. One way to get exact independence and convergence are perfect sampling algorithms. Unfortunately their application to generate continuous random variables is not easy, but there are possibilities to implement coupling from the past (CFTP) methods for some of the chains used in MCMC. Up to now none of these algorithms performs comparable to good standard generation methods. Nevertheless, we think that there could be future developments leading to good black-box algorithms based on CFTP. So in the first subsection we present random variate generation algorithms based on CFTP; in the second subsection we shortly discuss how to use MCMC algorithms for approximately generating random vectors.

14.3.1 Perfect Sampling Algorithms for the $\mathbb{R}^d$

Independence Sampler

As pointed out by Murdoch (2000) the independence sampler allows a comparatively simple implementation of the CFTP algorithm. The intrinsic randomness consists of a uniform random variate V to decide between acceptance and rejection and the random variate U that follows the distribution of the proposal density q. Note that U is the same as $\tilde{X}$ in Sect. 14.1. The randomizing operation is defined as

$$\phi(X, V, U) = \begin{cases} U & \text{for } V \frac{f(X)}{q(X)} \leq \frac{f(U)}{q(U)}, \\ X & \text{otherwise.} \end{cases}$$

In the case of acceptance of the independence sampler we have a good chance of coalescence to the single state U, as U does not depend on X. If we find an upper bound $b \geq \sup_x f(x)/q(x)$, we can find a condition that guarantees acceptance and does not depend on X: $V \leq \frac{f(U)}{b\,q(U)}$. Obviously this condition that guarantees coalescence is exactly the acceptance condition of the standard rejection algorithm with hat function $b\,q(x) \geq f(x)$. So the independence sampler implemented with CFTP can be seen as a reformulation of the acceptance-rejection method. Note that the new algorithm does not produce the same random variates as the standard rejection algorithm but the same number of repetitions are necessary for both algorithms.

Algorithm 14.5 (Perfect-Independence-Sampler) contains the details. They show that the independence sampler with CFTP makes it necessary to store all values generated during rejected steps as they are later used in the for-loop to generate the state X_0. So clearly standard rejection is easier to implement.

Gibbs Sampler

Wilson (2000) and Murdoch and Meng (2001) discuss how to use CFTP together with the Gibbs sampler. Both papers present solutions for certain multivariate distributions used in real-world MCMC problems. Nevertheless, these

Algorithm 14.5 Perfect-Independence-Sampler

Require: Dimension d, density $f(x)$;
  proposal density $q(x)$, bound $b \geq \sup_x f(x)/q(x)$.
Output: Random vector with density f.
1: Set $T \leftarrow 1$. /* For convenience we write positive instead of negative T. */
2: **repeat**
3:   Generate $V \sim U(0, 1)$.
4:   Generate random vector U from proposal density $q(x)$.
5:   Store V, U, and $f(U)/q(U)$ (with index T).
6:   Increment T.
7: **until** $Vb \leq f(U)/q(U)$.
8: Set $X \leftarrow U$.
9: **for** $t = T$ down to 1 **do**
10:   Read back the stored values V, U, and $f(U)/q(U)$ (with index t).
11:   **if** $Vf(X)/q(X) \leq f(U)/q(U)$ **then**
12:     Set $X \leftarrow U$.
13:   **else**
14:     Do nothing as due to rejection X is not changed.
15: **return** X.

---

algorithms seem to be of little use for random vector generation due to the poor properties of the Gibbs sampler for multivariate distributions that have highly correlated components or several modes. Especially Murdoch and Meng (2001) demonstrate this problem in their examples. If the Gibbs sampler has a poor convergence this results in a CFTP algorithm that is running very long without ever producing any output. Due to the very slowly converging Markov chain the CFTP algorithm can practically never see that all states have coalesced. This highlights a main problem of the CFTP algorithm. It can never be better than the Markov chain it is using. If this chain is not mixing well, this implies that CFTP works very slowly. From another point of view this is a virtue. Instead of delivering arbitrary vectors that do not really follow the desired distribution, CFTP is producing very little or nothing, thus indicating that the Markov chain is too slowly convergent to generate from the stationary distribution.

Perfect Slice Sampler

The key idea behind the slice sampler (Swendsen and Wang, 1987; Besag and Green, 1993), is to introduce an auxiliary variable U uniformly distributed on $(0, f(x))$. The Markov chain is then generated according to the recursion:

$$U_{t+1} \sim \text{uniform on } (0, f(X_t)) \quad \text{and} \quad X_{t+1} \sim \text{uniform on } \{x | f(x) > U_{t+1}\} .$$

The main practical problem we face in implementing the slice sampler is of course the generation of the uniform distribution over $\{x | f(x) > U_{t+1}\}$ as this set need not even be connected. Neal (2003) discusses variants of the slice sampler designed to overcome this problem. Mira et al. (2001) demonstrate that the slice sampler is well suited to perfect sampling as there exists a natural stochastic ordering with a maximal and minimal element and a "natural" possibility of coalescence; they demonstrate an example for a fixed density in dimension one and report moderate speed. The extension to multivariate distributions is difficult: Casella et al. (2002) report some success for multivariate mixture distributions, Philippe and Robert (2003) explain a perfect slice sampler algorithm for multivariate Gaussian distributions restricted to the first quadrant but only give all the details for dimension $d = 2$. We are not including the details here. The future will show if perfect slice sampling can lead to universal random variate or random vector generation algorithms.

Random Walk Metropolis (RWM) and Layered Multishift Coupling

Taking a quick glance at RWM and CFTP we could easily think that they will not work together as the probability that two paths coalesce is 0 for a continuous state-space. But there are clever ways to define the randomizing operation using for example bisection coupling (Green and Murdoch, 2000) or layered multishift coupling (Wilson, 2000) such that the continuous state-space is mapped into only a small number of different states. Unfortunately the possible rejection, which means that the chain is not moving at all, prevents us from easily applying these ideas. Green and Murdoch (2000) describe how they used bisection coupling and RWM to sample from a beta distribution but their performance report is disappointing and for random vectors the algorithms become so complicated that they seemingly were not even tried yet. We appreciate that layered multishift coupling combined with RWM and CFTP leads to entirely new random variate generation algorithms. So we try to explain the principles here although we have no practical examples yet where the new algorithms are superior to standard rejection.

Wilson (2000) introduced layered multishift coupling and applied it to the Gibbs sampler for different distributions that have full conditional distributions with invertible density. He also mentioned that it could be used for RWM but without giving any details. Layered multishift coupling is a good example of a clever coupling technique that leads to a randomizing operation that reduces the number of states dramatically compared to the straightforward randomizing operation. For the sake of simplicity we restrict our attention here to random walk with proposal distribution $U(-s, s)$. If we do not consider rejection for the moment, the standard randomizing operation can be written as

$$\tilde{\phi}(x, U) = x + U$$

where $U \sim U(-s, s)$. This randomizing operation is monotone but does not always reduce the number of states. In contrast the randomizing operation of the layered multishift coupler for the uniform distribution is defined as

$$\phi(x, U) = \left\lfloor \frac{x+s-U}{2s} \right\rfloor 2s + U$$

where again $U \sim U(-s,s)$. Figure 14.4 shows the graph of ϕ for fixed x (l.h.s.) and for fixed U (r.h.s.). For fixed values x and s, $(s-U)/(2s)$ is a $U(0,1)$ random variate. So the fractional part subtracted by the floor operation is a $U(0,1)$ random variate as well. If we ignored the floor in the definition of ϕ the expression would simplify to $x+s$. Thus by the floor operation a $U(-2s,0)$ distributed random quantity is subtracted from $x+s$ and hence $\phi(x, U^{(2)})$ is $U(x-s, x+s)$ distributed as desired. The floor operation in the definition leads to the desired property that for different values of x and a fixed value U the image is discrete. If we continuously increase x from 0 to $2s$ for fixed U the value of $\phi(x,U)$ changes only once. Therefore an interval of length l is mapped to at most $\lceil 1 + l/(2s) \rceil$ different points. Of course this is a very useful property when we try to implement CFTP for RWM. Layered multishift coupling also has the property that ϕ is monotone in the sense that $x < y$ implies $\phi(x,U) < \phi(y,U)$, which can also be observed in Fig. 14.4 (r.h.s.).

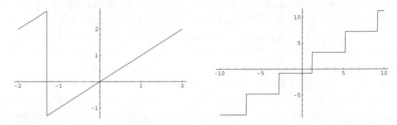

Fig. 14.4. The layered multishift coupler: $\phi(0.7, u)$ (l.h.s.) and $\phi(x, -0.8)$ (r.h.s.); both with $s = 2$

If we want to use layered multishift coupling to implement CFTP for random walk Metropolis it is necessary to include the rejection logic in the algorithm. Recall that for a proposed move from x to $\tilde{x}$ the acceptance probability is $\alpha(x, \tilde{x}) = \min(1, f(\tilde{x})/f(x))$ which means that jumps with $f(\tilde{x}) \geq f(x)$ are always accepted, jumps with $f(\tilde{x}) = 0$ are always rejected. So the standard randomizing operation for RWM including the rejection step can be written as

$$\tilde{\phi}_f(x, V, U) = \begin{cases} \tilde{x} = x + U & \text{for } V \leq \frac{f(\tilde{x})}{f(x)}, \\ x & \text{otherwise.} \end{cases}$$

Using layered multishift coupling we get

$$\phi_f(x, V, U) = \begin{cases} \tilde{x} = \lfloor \frac{x+s-U}{2s} \rfloor 2s + U & \text{for } V \leq \frac{f(\tilde{x})}{f(x)}, \\ x & \text{otherwise.} \end{cases} \quad (14.2)$$

Example 14.2. Consider the quasi-density $f(x) = 1/(1+x)^2$ on $(0, 10)$ and 0 elsewhere. The two randomizing operations $\tilde{\phi}_f$ and ϕ_f for implementing RWM

14.3 Markov Chain Monte Carlo Methods for Random Vectors

are compared in Fig. 14.5, both for $V = 0.9$ (high probability for rejection) and $U = -1.5$ (for a jump to the left). We can clearly see that the standard operation $\tilde{\phi}_f$ is not monotone. This can be easily understood as for $x < 1.5$ the jump to the left leads out of the domain and so $f(\tilde{x}) = 0$ and $\tilde{x}$ is rejected. For $x \geq 1.5$ the jump to the left remains in the interval and is always accepted as $f(\tilde{x}) > f(x)$. So $\tilde{\phi}_f$ is a non-monotone randomizing operation. In contrast to it we can see on the right hand side of Fig. 14.5 that the randomizing operation ϕ_f using the layered multishift coupler is nondecreasing for the values V and U we have chosen.

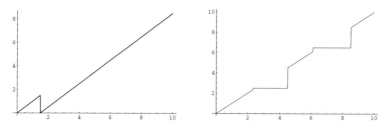

Fig. 14.5. The two randomizing operations for RWM with $f = 1/(1+x)^2$ and $s = 2$: standard operation $\tilde{\phi}_f(x, V = 0.9, U = -1.5)$ (l.h.s.) and the operation using layered multishift coupling $\phi_f(x, V = 0.9, U = -1.5)$ (r.h.s.)

The above example demonstrates that adding the rejection logic spoils the monotonicity of the standard randomizing operation $\tilde{\phi}_f$; as $\tilde{\phi}_f$ is not reducing the number of states either, the standard randomizing operation cannot be used to implement CFTP for RWM and densities on $\mathbb{R}$.

It was seemingly not yet discussed in the literature but Fig. 14.5 is a first indication that layered multishift coupling is not loosing its monotonicity as long as the density $f(x)$ is monotone. We can prove the following theorem.

Theorem 14.3. *We consider a bounded density f that is monotone on a bounded or half-open interval (b_l, b_r) and 0 outside that interval. For $b_l \leq x \leq b_r$ the randomizing operation $\phi_f(x, V, U)$ as defined in (14.2) is monotone.*

Proof. Let us first assume that there is no rejection (i.e. assume $V = 0$). We have seen above that in this case ϕ_f maps a whole interval into a single value $\tilde{x}$. We write (l, u) for the maximal interval that is mapped into $\tilde{x}$. Is it obvious from the definition of ϕ_f (and has been discussed above) that we always have $l \leq \tilde{x} \leq u$.

For the case of rejection the region of (l, u) where rejection occurs is certainly no problem for the monotonicity as rejection means that the identity is used as mapping and the identity is of course monotone.

We assume that f is monotonically non-increasing: Rejection only occurs if $V > f(\tilde{x})/f(x)$. So, due to the monotonicity, no rejection can occur for $x \geq \tilde{x}$.

Algorithm 14.6 Perfect-Random-Walk-Metropolis-Sampler

Require: Monotone bounded density $f(x)$ on (b_l, b_r);
    parameter s for the random walk proposal $U(-s, s)$ (e.g. try $s = (b_r - b_l)/4$).
Output: Random vector with density f.
1: Set $T \leftarrow 8$. /* For convenience we write positive instead of negative T. */
2: Generate and store $V_t \sim U(0,1)$ and $U_t \sim U(-s,s)$ for $t = 1, \ldots, 4$.
3: **repeat**
4:     Generate and store $V_t \sim U(0,1)$ and $U_t \sim U(-s,s)$ for $t = \frac{T}{2}+1, \ldots, T$.
5:     Set $x_l \leftarrow b_l$ and $x_u \leftarrow b_r$.
6:     **for** $t = T$ down to 1 **do**
7:         Set $\tilde{x} \leftarrow \lfloor \frac{x_l + s - U_t}{2s} \rfloor 2s + U_t$.
8:         **if** $V_t f(x_l) \leq f(\tilde{x})$ **then**
9:             Set $x_l \leftarrow \tilde{x}$.
10:        Set $\tilde{x} \leftarrow \lfloor \frac{x_u + s - U_t}{2s} \rfloor 2s + U_t$.
11:        **if** $V_t f(x_u) \leq f(\tilde{x})$ **then**
12:            Set $x_u \leftarrow \tilde{x}$.
13:     Set $T \leftarrow 2T$.
14: **until** $x_l = x_u$.
15: **return** x_l. /* which is equal to x_u */

This means that the set of all points x in (l, u) that lead to acceptance and are therefore mapped into $\tilde{x}$ form an interval that contains $(\tilde{x}, u)$ as a subset. For fixed V and U, there are subintervals (the regions of rejection) where ϕ_f is identity and subintervals where ϕ_f is mapped into the constant $\tilde{x}$. We have just seen that the latter subintervals include $\tilde{x}$ which completes the proof for monotonically non-increasing f. For monotonically non-decreasing f we can obtain the same result by the symmetry of RWM. $\square$

The above theorem shows that ϕ_f is a monotone randomizing operation for monotone densities f. Thus, to obtain a black-box algorithm for a bounded monotone density on a bounded domain, we can start the CFTP algorithm with only two states: the maximum and the minimum of the domain; we repeat to start farer from the past until we observe that the two states have coalesced to a single state at $t = 0$. Note that the acceptance condition has to be $V f(x) \leq f(\tilde{x})$.

We have used the above considerations to compile a new and simple method in Algorithm 14.6 (Perfect-Random-Walk-Metropolis-Sampler). It is interesting as it is an entirely new type of black-box algorithm. Unfortunately it does not work for unbounded densities f: if $f(b_l)$ is unbounded the Markov chain can never move away from b_l and thus is not convergent. We first tested the algorithm with the density $f(x) = 1/(1+x)^2$ on an interval $(0, b_r)$. The average necessary run-length depends strongly on the choice of s. After some experiments to find a sensible s we experienced for $b_r = 10$, $s = 4$ an average run length of about 23; for $b_r = 100$, $s = 20$ the average was 92 and for $b_r = 500$ and $s = 75$ the average was 245; for other monotone densities we observed similar results. If we compare this with the naive rejection algorithm with constant hat, that has an expected number of repetitions close to b_r for this example, the performance is really not impressive. In addition we should not forget that for one iteration we have to evaluate f twice for Algorithm 14.6 as we have to update two states but only once for standard rejection. An advantage of the new Algorithm is certainly the short setup; the choice of s is not such a big problem as the algorithm remains correct for any value of s. A sensible value for the density of interest can easily be obtained in a small simulation study, trying e.g. values of s between 0.05 and 0.3 times the length of the interval $b_r - b_l$ and comparing the average necessary run-length.

If we do not assume monotone f the randomizing operation is no longer guaranteed to be monotone around the mode. It is certainly possible to find tricks to overcome this problem but the performance for the monotone case is not a big motivation to construct an even slower algorithm for unimodal densities. It is not clear if there is a simple way to generalize RWM using layered multishift coupling to higher dimensions as it is not clear how to generalize the ordering to a half-ordering such that the monotonicity is preserved. In addition a simple consideration shows that a perhaps possible generalization of Algorithm 14.6 will become very slow in higher dimensions. To see this it is enough to consider the probability that when x is in a corner $\phi(x, u)$ will be moving away from that corner. For $d = 1$ this probability is of course 0.5. For arbitrary d we get 0.5^d which becomes very small for larger d. So it does not look too promising to find new good black-box algorithms for random vectors using RWM with CFTP.

For hit-and-run type Markov chain algorithms we do not know of any attempts to use them together with CFTP. It seems difficult to do that for general distributions but there could certainly be interesting developments for special multidimensional distributions.

14.3.2 Approximate Algorithms

How could we use the Markov chain algorithms described above to generate iid. random vectors with a difficult multivariate possibly multi-modal distribution? One natural idea to utilize approximate Markov chain algorithms is the following, suggested already by Smith (1984): Use a convergence theorem to obtain a bound for the run-length of the chain. Store the generated vectors in a big buffer and deliver them in random order to obtain approximate independence. The second part of that suggestion is easy to implement but what about the convergence bounds? For applications in algorithmic geometry and in optimization the convergence properties of random walk and hit-and-run algorithms have been studied in several recent papers (see Dyer, Frieze, and Kannan, 1991; Kannan, Lovasz, and Simonovits, 1997; Lovasz, 1999, and the references given there). In these papers it is proven that the mixing time of random walk and hit-and-run algorithms for uniform distributions over convex sets is polynomial in dimension d subject to choosing a sensible starting distribution. This sounds promising but Lovasz (1999) stresses that his results should not be used for computing the necessary number of steps. So these convergence results are of little use for us; in addition they are restricted to uniform distributions over convex sets. So for the idea to generate uniform points below the desired density and to return the projection of these points the polynomial complexity results are only applicable for concave densities. This is a very strong restriction and does not even allow for densities with infinite tails. So there seems to be no hope for a general generation procedures with complexity polynomial in the dimension.

The polynomial complexity results could also be seen as a contradiction to the comment in the previous subsection that a random walk starting in a corner has only probability 2^{-d} to move into the domain of the distribution. But this contradiction is resolved if we recall that the fast mixing results depend on a "sensible" starting value. Mixing is apparently not fast in high dimensions when we start in a corner. The same is highlighted by the oldest convergence result for hit-and-run algorithms. For the case of an uniform distribution over an open region $S \in \mathbb{R}^d$ Smith (1984) has shown that for any measurable set $A \subseteq S$

$$|\text{Prob}(X_m \in A | X_0 = x) - \text{vol}(A)/\text{vol}(S)| < (1 - \gamma/(d\,2^{d-1}))^{m-1}$$

where γ is the ratio of the d-dimensional volume of S to the d-dimensional volume of the smallest sphere containing S. This is a computable bound not depending on the starting values but computational experiments indicate that it is very pessimistic. Nevertheless, it is interesting that the bound depends on the constant γ that is the rejection constant for a rejection algorithm from the smallest sphere containing S. Also this bound indicates that for bad starting values the necessary run-length to obtain guaranteed convergence grows exponentially with the dimension.

So if f is complicated and we do not know the location of the modes, we do not have a general method to guarantee that the Markov chain is converging. The main problem for a "blind" method comes from densities with very thin spikes. Even for one-dimensional distributions no MCMC method (and no other known black-box method based on the density) has a chance to generate from a density that has poles with unknown position. Also we should not forget that all these problems become much more severe if the dimension is increasing. Therefore it is always possible to find densities for which a suggested approximate method fails badly.

Nevertheless, we mention the following three methods that could be used as approximate black-box methods if an exact algorithm is not available or much too slow.

Approximate Algorithms for Generating iid. Random Vectors

- *Random direction interior point* method of Chen and Schmeiser (1998) was designed as a black-box algorithm for Bayesian MCMC (see Sect. 14.1.3 for a very short description). If only a small fraction of the generated chain is used and the output is buffered and shoveled this method should result in a practically applicable approximate black-box algorithm to generate iid. vectors.
- To apply the Metropolis-Hastings algorithm we suggest to use a hybrid chain consisting mainly of a random walk Metropolis algorithm (with normal proposal density) but combined with an independence sampler (with t-proposal density) that is applied at all times $t \equiv 0 \bmod r$. Store the values of X before applying the independence sampler (i.e. X_t with $t \equiv -1 \bmod r$) in a large buffer and randomly output the vectors from that buffer. The necessary value of r and the necessary buffer size depends on the target density and the parameters of the chain. It is possible to compute the small lag autocorrelations of the unshoveled X_t with $t \equiv -1 \bmod r$ to check empirically if r was chosen big enough.
- It is also possible to design an "approximate rejection algorithm". For example, we can simply take a multi-t-distribution. Multiply its density with a constant k such that it is "well above" the target density for all points where you have already evaluated f and of course for all modes you know. Take a safety factor s, e.g. $s = 2$ or even $s = 10$, and use $s\,k$ to decrease the probability to cut off unknown spikes. We can compare this s used here with the r for the approximate Markov chain algorithm above. The time complexity of these two algorithms depends linearly on the design constants s and r respectively.

The complexity results cited in the last section seem to suggest that for growing dimension d the performance of Markov chain algorithms deteriorates more slowly than for rejection algorithms but these results are only valid for well behaved distributions. So it could be interesting to compare the three approximate methods for generating random vectors given above in a simulation study.

14.4 Exercises

Exercise 14.1. Assume that f and h are quasi-densities with the same domain and that a generator for h is available. Find an algorithm to sample from the density $\min(f(x), h(x))$.
Hint: Construct a rejection algorithm.

Exercise 14.2. Consider the independence sampler with proposal density $q(x) = \min(f(x), h(x))$ (which is called rejection chain). Show that $f(X_t) \leq h(X_t)$ implies that no rejection is possible when generating X_{t+1}.

Hint: Use the definition of the acceptance probability for the Metropolis-Hastings algorithm given in (14.1).

Exercise 14.3. Using the rejection chain and CFTP construct an exact universal algorithm.
Hint: The resulting algorithm is very similar to Algorithm 14.5 (Perfect-Independence-Sampler).

Exercise 14.4. Implement the combination of random walk Metropolis and independence sampler suggested above for univariate distributions with domain (0,1).

Exercise 14.5. Implement the "approximate rejection algorithm" suggested above for univariate distributions with domain $(0, 1)$.

Exercise 14.6. Compare the speed and test the independence of the generated random variates for your algorithms of Exercises 14.4 and 14.5. As quasi-density use $f(x) = 1/(x-a)$ for $0 < x < 1$ and $a > 0$; take $a = 1/i$ with $i = 1, 2, 3, \ldots$.

15
Some Simulation Examples

We hope that the reader has seen so far that random variate generation leads to interesting algorithms and is linked with sometimes difficult mathematical questions. But we should not overlook the fact that random variate generation is an applied discipline in the sense that its new developments were mainly driven by the needs of scientists and engineers using simulation. In recent years two of the most rapidly evolving areas of simulation have been Bayesian statistics and financial engineering. So we think it goes well with the aims of our book to include this final chapter, that demonstrates some very "modern" applications of random variate generation.

15.1 Financial Simulation

It is not the aim of this chapter to give an introduction to mathematical option pricing theory (see e.g. Duffie, 1996, for a well known text book) or to the practical and theoretical aspects of "value at risk" (VaR) (see e.g. Dowd, 1998). We also do not try to give a general overview of simulation techniques in financial engineering (see e.g. Staum, 2001, and the references given there). Instead we will demonstrate with several (largely arbitrarily chosen) examples how some of the random variate generation methods developed in this book can be utilized for option pricing and value at risk calculations.

15.1.1 Option Pricing

A European call option is the right to buy a particular stock for an agreed amount (called exercise price or strike price) at a specified time in the future (called expiration date). The worth of the option at expiration date is calculated by its payoff function which is for the European call option $\max(S_T - E, 0)$, where S_T denotes the price of the underlying stock at expiry T and E denotes the exercise price.

The mathematical theory of option pricing is based on stochastic differential equations thus using continuous time processes, in particular the geometric Brownian motion. For fitting these processes to real-world stock data all authors mainly consider the daily log-returns, the difference of the logarithms of the closing prices of a stock for two consecutive days, respectively. Therefore we will restrict our attention here to these daily time series. For geometric Brownian motion the daily log-returns are independent normal variates. It is therefore very simple to simulate daily stock prices following a geometric Brownian motion. The key result to obtain option prices by simulation is the "Feynman–Kac Solution" that allows to write an option price as the discounted expectation with respect to the pseudo-price process whose expected rate is exactly equal to the risk-free interest rate r (see Duffie, 1996, p. 91). (The "pseudo-price process" is a process with the same characteristics as the price process but a drift rate equal to the risk free interest rate.) Thus, to calculate an option price by simulation, we have to generate many paths of that pseudo-price process and compute the average of the discounted pay-off for all these paths. The main advantage of option pricing by simulation is its simplicity. We can simulate option prices for path dependent options (e.g. Asian options) with almost the same code as European options. The practical problem is often the slow convergence that leads to slow programs if high precision is required. (A possible solution to this problem proposed in literature are so called quasi-Monte Carlo methods which use highly uniform point sets, i.e. point sets with low discrepancy (see Sect. 8.3.3) instead of pseudo-random numbers.)

Before we can start with a first option pricing example we have to discuss how to generate a geometric random walk that has the same volatility as the underlying stock price and annual expected rate equal to the risk-free interest rate r. If we denote the yearly volatility by σ we can – assuming 252 trading days per year – easily obtain the daily volatility by $\sigma_d = \sigma/\sqrt{252}$ and the daily expectation by $r_d = r/252$. It is then natural to use the recursion

$$S_{t+1} = S_t \exp(\tilde{r}_d + \sigma_d Z)$$

where Z denotes a standard normal variate, and $\tilde{r}_d$ is some appropriate parameter chosen such that we get the required daily expected rate r_d. It can be interpreted as the "corrected drift factor" for this process. As we are transforming the normal random variate with the exponential function the price S_t is multiplied by a lognormal variate. The expectation of this lognormal variate is known to be $\tilde{r}_d + \sigma_d^2/2$. This implies that we have to chose $\tilde{r}_d = r_d - \sigma_d^2/2$ to obtain a daily expected rate of r_d. Note that this is exactly the drift rate of the corresponding continuous process obtained by Itô's Lemma. We can now generate paths of the geometric random walk (the pseudo-price process explained above) that we need for pricing options. Due to the reproduction property of the normal distribution it is not necessary to generate all steps of the random walk as we can easily compute the distribution of the stock price at the expiry date of the option. Denoting the time to expiration by τ we get

$$S_\tau = S_0 \exp(\tau \tilde{r}_d + \sigma_d \sqrt{\tau} Z) \,.$$

The discounted expectation of the payoff with respect to this distribution can be written in closed form; it is exactly the well known Black-Scholes formula. So for the assumption of normally distributed daily log-returns (i.e. for daily stock prices following a geometric random walk) no simulation is required.

However, it is well-known that the daily log-returns of most stocks do not follow a normal distribution but have considerably heavier tails. Thus many authors suggest replacing the normal distribution by heavier tailed alternatives like the t-distribution or the hyperbolic distribution, see Eberlein and Keller (1995). In that paper it is explained how, using the Esscher transform, it is possible to compute explicitly one "equivalent martingale measure". For the hyperbolic case this martingale measure is no longer unique but nevertheless it can be used to calculate option prices. Motivated by this suggestion we use an even simpler pseudo-price process with expected rate equal to the risk free interest rate by just correcting the daily drift rate. Our numerical experiments showed that the option prices calculated following the suggestions of Eberlein and Keller (1995) and our results are practically the same. Thus we will demonstrate our simple version to obtain option prices by simulation when assuming non-normal daily log-returns in the examples below.

Example 15.1 (Option prices for non-normal daily log-returns of the underlying stock).
Before we can start with the simulation we have to find the corrected drift factor $\tilde{r}_d$ that implies a daily expected rate of r_d. To do this we have to compute the daily expected rate of the random walk with drift 0, i.e. the expectation $r_{d,0} = \mathrm{E}(\exp(R))$, where R denotes the random variate of the daily log-return without drift. Computing the density of $\exp(R)$ and taking the expectation, results in

$$r_{d,0} = \int_0^\infty x \left(f_R(\log(x)) \frac{1}{x} \right) \mathrm{d}x \quad \text{and} \quad \tilde{r}_d = r_d - r_{d,0}\,, \qquad (15.1)$$

with $f_R(\cdot)$ denoting the density of R. Our numerical experiments showed that for the t-distribution and the hyperbolic distribution the corrected daily drift $\tilde{r}_d$ is very close to the value for the normal distribution, as for realistic choices of the distribution the daily return R is very close to zero. Due to the heavier tails the exact correction is a bit larger but the relative difference of the exact value and the value of the correction for the normal distribution ($\sigma_d^2/2$) was less than $1/1000$ for several realistic daily return distributions we tried. So the final results of the simulation are not influenced in a noticeable size if we use for both the t- and the hyperbolic distribution simply the formula for the normal distribution, $\tilde{r}_d = r_d - \sigma_d^2/2$. Now we can give a description of the simulation algorithm:

Require: Density f of daily log-returns R with its parameters specified such that we have zero expectation and standard deviation σ_R;
(yearly) volatility σ and the (yearly) risk-free interest rate r,
current stock price S_0, exercise price E, days to expiration τ;
number of repetitions n.

Output: 95% confidence interval for price of European call option when daily log-returns of the underlying asset have density f.

1: Set $\sigma_d \leftarrow \sigma/\sqrt{252}$. /* daily volatility */
2: Set $\tilde{r}_d = r/252 - \sigma_d^2/2$. /* approximate corrected daily drift rate */
   /* Use (15.1) for exact value */
3: Run setup for generator for density f. /* e.g. Algorithm 4.1 (TDR) */
4: **for** $i = 1$ to n **do**
5:    Set $S \leftarrow S_0$.
6:    **for** $t = 1$ to τ **do**
7:       Generate random variate R with density f.
8:       Set $S \leftarrow S \exp(\tilde{r}_d + \sigma_d R/\sigma_R)$.
9:    Set $V_i \leftarrow \exp(-r\tau/252) \max(S - E, 0)$.
10: Compute mean $\bar{V}$ and standard deviation s_V of the V_i.
11: **return** confidence interval for option price: $\bar{V} \pm 1.96\, s_V/\sqrt{n}$.

This simulation algorithm is fairly simple. The main practical problem in its application is certainly the decision about the distribution of the daily log-returns and the estimation of the distribution parameters. For our experiments we used a t-distribution with seven degrees of freedom. This distribution has standard deviation $\sqrt{7/5} = 1.183$ and will be called "our t-distribution" in the sequel. (Glasserman, Heidelberger, and Shahabuddin (2000) use degrees of freedom between four and six in their examples.) We also used the symmetric centered hyperbolic distribution (with scale parameter $\delta = 1$) which is given by the density

$$\text{hyp}_\zeta(x) = \frac{1}{2\,K_1(\zeta)} \exp\left(-\zeta\sqrt{1+x^2}\right)$$

where K_1 denotes the modified Bessel function of the third kind with index 1. We follow the presentation of Eberlein and Keller (1995), for more details and a definition of the hyperbolic distribution see Barndorff-Nielsen and Blæsild (1983). As "our hyperbolic distribution" we used the shape parameter $\zeta = 0.5$ (this corresponds to $\xi = \sqrt{2/3} \approx 0.8165$ for an alternate kind of parameterization where $\zeta = \xi^{-2} - 1$) which is in the region of the estimated ζ values reported by Eberlein and Keller (1995) and Prause (1999) for German stock data. The standard deviation of our hyperbolic distribution (with $\zeta = 0.5$) is 3.0193. For the simulation runs we have to use the standardized form of our t- and our hyperbolic distribution, respectively, i.e. where the standard deviation is 1. To have fast automatic random variate generation algorithms available is clearly an advantage for option pricing by means of Monte Carlo simulation. The universal nature of these algorithms allows us to change the distribution of the daily log-returns easily. Furthermore, the chosen distribution hardly influences the generation times which guarantees that we obtain short confidence intervals for all distributions and a given computation time when using the naive simulation method. The algorithm can generate variates from the hyperbolic or from the t-distribution (or from any other T-concave distribution) as fast as from the normal distribution with a speed that is comparable to the speed of the Box-Muller method (see Sect. 2.6). So our two distributions are only two possible examples.

Assuming a yearly volatility $\sigma = 0.2$, a risk-free interest rate $r = 0.05$ and $S_0 = 100$ for the stock price we ran the simulation for several exercise prices above and below 100, as it is easy to organize the simulation program such that it computes the confidence intervals of the option price for several different strike prices in a single run. To get an idea of the possible precision we can reach with such a simple simulation we report our observed standard deviations in Table 15.1 in the columns entitled "s-naive". The speed of the simulation is about 10^6 iterations per second on our computer which allows the generation of 10^6 paths in about 30 (10 or 1) seconds for 30 (10 or 1) days to expiry. The standard deviations report in Table 15.1 thus indicate that especially for 30 days to expiry the simulation is slow and does not provide the desired precision. The $1/\sqrt{n}$ law for the standard error implies that the simulation becomes very slow if we need confidence intervals not longer than e.g. $1/1000$.

It is possible and not difficult to speed up the above algorithm. For example we can easily save calls to exp() in the innermost loop by generating paths of the log-prices instead of the prices. On our computer this reduced the execution time by about 35 percent. As the fastest implementation is hardware and compiler dependent and the changes are not difficult, we pose them as Exercise 15.1.

To obtain a faster simulation algorithm we can try variance reduction techniques (see e.g. Law and Kelton (2000) or Staum (2001) for examples in financial engineering). Among the standard techniques control variates often lead to considerable variance reduction. As the Black-Scholes formula can be used to compute the exact result for normal-distributed daily log-returns it is an obvious idea to use the option price for this case as a control variate. This is demonstrated in the next example.

Example 15.2 (Continuation of Example 15.1 with variance reduction). To use control variates here we proceed in the following way: Together with the daily log-returns of the required distribution we generate perfectly correlated log-returns from the normal distribution using common random numbers. As result of the simulation we compute the average difference of the option price for the two different distributions of the log-returns. As the exact result for the normal case is easily computed by the Black-Scholes formula this difference is enough to compute the required option price for the non-normal case as well. The point is that we hope that the standard deviation of the difference is much smaller than the standard deviation of the option price itself. Before we report our results we give the details of the modified algorithm:

Table 15.1. European options. Simulation results for 10^6 paths with $S_0 = 100$, $\sigma = 0.2$ and $r = 0.05$; Hyperbolic distribution with $\zeta = 0.5$; t-distribution with 7 degrees of freedom; s-naive denotes the standard deviation for naive simulation (Example 15.1), s-cv the standard deviation when using control variates (Example 15.2); the units of these standard errors are 10^{-3}

days	Exercise price	Black-Scholes price	Hyperbolic distribution			t-distribution		
			price	s-naive	s-cv	price	s-naive	s-cv
1	90	10.018	10.018	1.26	0.20	10.018	1.26	0.15
	95	5.019	5.020	1.26	0.20	5.019	1.26	0.14
	100	0.513	0.469	0.78	0.14	0.489	0.77	0.11
	105	0.000	0.002	0.05	0.05	0.001	0.06	0.06
	110	0.000	0.000	0.01	0.01	0.000	0.02	0.02
10	90	10.182	10.185	3.97	0.64	10.185	3.97	0.45
	95	5.351	5.356	3.69	0.60	5.354	3.69	0.42
	100	1.689	1.674	2.46	0.47	1.678	2.46	0.34
	105	0.241	0.245	0.96	0.28	0.243	0.95	0.23
	110	0.013	0.018	0.25	0.12	0.017	0.25	0.12
30	90	10.681	10.686	6.63	1.08	10.685	6.63	0.77
	95	6.328	6.328	5.84	0.99	6.328	5.84	0.71
	100	3.051	3.044	4.41	0.84	3.046	4.41	0.60
	105	1.149	1.145	2.77	0.62	1.147	2.77	0.47
	110	0.332	0.336	1.47	0.41	0.335	1.47	0.33

Require: Density f of daily log-returns R with its parameters specified such that we have zero expectation and standard deviation σ_R; (yearly) volatility σ and the (yearly) risk-free interest rate r, current stock price S_0, exercise price E, days to expiration τ; number of repetitions n.
Output: 95% confidence interval for price of European call option when daily log-returns of the underlying asset have density f.
1: Set $\sigma_d \leftarrow \sigma/\sqrt{252}$. /* daily volatility */
2: Set $\tilde{r}_d = r/252 - \sigma_d^2/2$. /* corrected daily drift rate */
3: Run setup for generator for density f and for standard normal distribution. /* e.g. Algorithm 4.1 (TDR) */
4: **for** $i = 1$ to n **do**
5:    Set $S \leftarrow S_0$ and $\tilde{S} \leftarrow S_0$.
6:    **for** $t = 1$ to τ **do**
7:       Generate random variate R with density f and standard normal distributed random variate $\tilde{R}$. R and $\tilde{R}$ must have the highest possible correlation.
8:       Set $S \leftarrow S \exp(\tilde{r}_d + \sigma_d R/\sigma_R)$ and $\tilde{S} \leftarrow \tilde{S} \exp(\tilde{r}_d + \sigma_d \tilde{R})$.
9:    Set $D_i \leftarrow \exp(-r\tau/252) \left[(\max(S - E, 0) - \max(\tilde{S} - E, 0)\right]$.
10: Compute mean $\bar{D}$ and standard deviation s_D of the differences D_i.
11: Compute confidence interval for the difference of option price minus the Black-Scholes price: $\bar{D} \pm 1.96 \, s_D/\sqrt{n}$.
12: Compute Black-Scholes price and add it to the confidence interval.
13: **return** confidence interval.

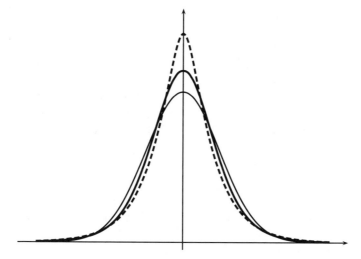

Fig. 15.1. Normal distribution (thin curve), t-distribution with 7 degrees of freedom (bold curve), and symmetric centered hyperbolic distribution with shape parameter $\zeta = 0.5$ (dashed curve); all standardized with standard deviations of 1

The computed option prices and the standard deviations for naive simulation and for the version using control variates are reported in Table 15.1. The advantage of using the control variates is obvious; disregarding the cases where the option price is very close to zero the factor of variance reduction lies between 10 and 50. The version with control variates takes less than twice the execution time of the naive simulation; so the execution times to reach the same precision are reduced by a factor between 5 and 25 which is quite impressive. The execution time to obtain all "30 days" results was less than a minute and considerably less for the "10 days" and "1 day" cases. The standard error is with one exception smaller than 10^{-3} for all presented results. So we may say that simulation can be used to obtain option pricing results for non-normal daily log-returns.

We should not overlook that this is due to the use of transformed density rejection and its correlation induction facilities. To increase the correlation of the generated hyperbolic and normal variates we have used transformed density rejection with $N = 800$ design points which result in $\varrho_{hs} \approx 1 + 10^{-6}$. (Recall that ϱ_{hs} is the ratio of the area below the hat and the area below the squeeze.) Using a smaller N (e.g. 50) results in 5 to 10 percent larger standard errors. The standard generation technique for such a simulation including common random numbers and control variates would be the inversion method. But for the hyperbolic distribution the cdf is numerically very unfriendly; so the inversion method would imply a large programming effort and a much slower code.

A last very short comment on the option prices we have calculated and presented in the three "price" columns of Table 15.1. The main message is that the Black-Scholes formula seems to be fairly robust unless the time to expiry is very short. Of course this does not come as a surprise. Comparing the results for the hyperbolic distribution ($\zeta = 0.5$) and the t-distribution (with 7 degrees of freedom) we were surprised to see that the t-distribution consistently resulted in prices closer to the Black-Scholes prices than the hyperbolic distribution, although the tails of that t-distribution are much heavier. The explanation seems to lie in the fact that in the center the t-distribution is closer to the normal distribution, see Fig. 15.1.

We restricted ourself to the case of European call options to demonstrate how option prices can be obtained by simulation and how universal random variate generation algorithms can facilitate this task. The random variate generation part does not change if we consider more complicated options as long as they depend only on daily closing prices (like most Asian options). The problem of simulating the value of a portfolio will be discussed in the next section.

15.1.2 Value at Risk

Value at Risk (VaR) (see e.g. Dowd, 1998; Jorion, 2001) has become the standard methodology to measure the risk of a random variate X, i.e. the potential loss given a level of probability α. It is defined as a quantile of X,

$$\mathrm{Prob}(X < -\mathrm{VaR}_\alpha) = \alpha \;.$$

There is a general problem in the interpretation of VaR as a measure of risk: It does not give any information about the expected size of losses in the case of very unusual events often called "crashes". This can be seen clearly by the α-probability in the definition that VaR is not including a worst case analysis. On the other hand it is probably the simplicity that made VaR a very popular concept. So we want to demonstrate how VaR can be calculated using simulation without discussing its interpretation. Note that a complementary risk measure is the expected value conditional on exceeding the quantile, that is the expected loss when X happens to be below the VaR_α. This value is also known as *tail conditional expectation* or *tail loss*.

We start with the case of a single asset. Given the shape of the distribution of the daily log-returns and an estimate for the daily volatility σ_d it is not difficult to compute the $\mathrm{VaR}_\alpha(1)$ for a single stock and a one day time horizon; we just have to compute the α-quantile, q_α, of the standardized distribution of the daily log-return to get

$$\mathrm{VaR}_\alpha(1) = S_0(1 - \exp(q_\alpha\, \sigma_d)) \;.$$

In this formula we assumed that the (corrected) drift-rate is 0. Of course we could add a non-zero drift-rate but it is difficult to estimate and, for short horizons of less than a month, it has very little influence on the VaR. Under the assumption of normal distributed log-returns the reproduction property of the normal distribution allows us to obtain a similar formula for an arbitrary horizon of τ days:

$$\text{VaR}_\alpha(\tau) = S_0(1 - \exp(\Phi_N^{-1}(\alpha)\,\sigma_d\sqrt{\tau})) \qquad (15.2)$$

where Φ_N denotes the cdf of the standard normal distribution. This formula is fairly simple but it underestimates the risk for most real world assets as the distribution of the daily log-returns is known to have heavier tails than the normal distribution. For log-returns following a hyperbolic or t-distribution and horizons of e.g. five or ten days simulation is certainly the simplest way to calculate value at risk. For one day horizon we can easily use numeric integration of the density to find the required quantile and for longer horizons like several months we can – due to the Central Limit Theorem – obtain good approximations using (15.2). In between we have to find the density of the sum of five or ten daily log-returns; this is certainly possible (using FFT and the Convolution Theorem) but requires considerably more knowledge on numerical methods than the simulation approach. We demonstrate the simplicity of the simulation approach in Example 15.3.

Example 15.3 (Value at risk for a single asset). To calculate VaR we need the distribution of the stock price at the end of the time horizon (a time τ). So we again have to simulate a large number of price paths. Thus the simulation is similar to the option pricing case but we sum up the daily log-returns. Then we use the generated stock prices S_τ to estimate the α-quantile of the distribution that can be used to calculate VaR. One computational problem is the calculation of the quantile for very large samples. This is possible without sorting or even storing the whole sample; Panny and Hatzinger (1993) describe a fast algorithm with only moderate memory requirements. The main steps of the algorithm are:

Require: Density f of daily log-returns R with its parameters specified such that we have zero expectation and standard deviation σ_R;
(yearly) volatility σ, current stock price S_0, days to expiration τ;
number of repetitions n, probability α for VaR.
Output: Estimate for $\text{VaR}_\alpha(\tau)$ of an asset when daily log-return has density f.
1: Set $\sigma_d \leftarrow \sigma/\sqrt{252}$. /* daily volatility */
2: Run setup for generator for density f. /* e.g. Algorithm 4.1 (TDR) */
3: **for** $i = 1$ to n **do**
4:     Set $S \leftarrow 0$.
5:     **for** $t = 1$ to τ **do**
6:         Generate random variate R with density f.
7:         Set $S \leftarrow S + R$.
8:     Store $R_i \leftarrow S$.
9: Compute α-quantile q_α of the R_i. /* Sort R_i and take observation $\lceil \alpha n \rceil$ */
10: **return** $\text{VaR} \leftarrow S_0(1 - \exp(q_\alpha\,\sigma_d/\sigma_R))$.

We assume yearly volatility $\sigma = 0.2$; as distributions for the daily log-return we again use "our t-distribution" with seven degrees of freedom which has the standard deviation $\sqrt{7/5} = 1.183$ and "our hyperbolic distribution" with $\zeta = 0.5$ and standard deviation 3.0193. Then we compute the required quantile for 100 samples of size 10^4 which has two advantages over taking one sample of size 10^6. We can easily sort the samples as they are not too large and we can easily compute an approximate confidence interval for the final result. In a single simulation run we can calculate estimates of the VaR for all α values we are interested in. The execution time on our computer was less than ten minutes for the 30 day horizon and considerably less for the shorter horizons. Table 15.2 contains our results.

The simulation results nicely illustrate the heavier tails and thus the higher risk of the hyperbolic and t-distributions. Notice that this is not the case for $\alpha = 5\%$ where the normal distribution leads to the highest VaR. The results for the hyperbolic and the t-distributions are close together. The differences to the normal distribution are of course much smaller for the "30 days" case where the central limit theorem implies that all VaRs are approximately the same. The simulation results also indicate that for $\alpha = 5\%$ the simple formula (15.2) is almost exact for our hyperbolic and t-distribution. But this also means that for $\alpha = 5\%$ the VaR is not reflecting all of the potential risk.

VaR for Portfolios

The VaR concept starts to become interesting with its application to portfolios. The standard method used is to assume a multinormal distribution for the log-returns of the different assets. This implies two problems: First we

Table 15.2. Value at Risk. Simulation results for $S_0 = 100$, $\sigma = 0.2$; Hyperbolic distribution with $\zeta = 0.5$ t-distribution with 7 degrees of freedom (both standardized with standard deviation 1); SE denotes the standard error we obtained for calculating the average of 100 quantile estimates from samples of size 10^4

days	α	Normal VaR	Hyperbolic VaR	SE	t-distribution VaR	SE
1	0.5%	3.193	3.904	0.003	3.658	0.003
	1%	2.888	3.193	0.002	3.142	0.003
	5%	2.051	2.028	0.001	1.999	0.001
10	0.5%	9.753	10.071	0.006	9.991	0.006
	1%	8.852	9.048	0.004	8.990	0.004
	5%	6.343	6.333	0.003	6.328	0.003
30	0.5%	16.285	16.463	0.010	16.425	0.010
	1%	14.831	14.938	0.008	14.911	0.008
	5%	10.730	10.722	0.004	10.722	0.004

have to estimate the correlation between the log-returns of all assets included. This is not a simple task especially as correlations are known to be unstable. It is not clear how long a history should be used to estimate the correlations.

The second problem is even more important: Nobody knows how (un)-realistic it is to assume multinormal log-returns. Several books and papers include remarks like the following (see Mayhew, 2002): *The hedge fund failed to recognize that the correlations between their various risk exposures would be much, much higher during an international crisis than during normal times.* Such experiences are a clear indication that for these rare-events the multinormal assumption is not even approximately correct. For risk considerations this observation is of course a big problem, as we are especially interested in the "crash" case. On the other hand, to calculate the VaR for $\alpha = 0.05$ or 0.01 the size of the crashes does not matter; it is not part of the definition of VaR anyway. We continue with the description how to use simulation to calculate the VaR for (stock) portfolios. As a matter of fact simulation is the standard (or even the only) method to answer such questions as we are concerned with integration problems in high dimensions, since a portfolio of about hundred stocks is common in practice. After having estimated all correlations and variances the simulation (for e.g. one day horizon) is very simple.

- Generate a random vector **R** of the log-returns of all assets by means of Algorithm 11.3 (Multinormal-Cholesky).
- Use vector **R** and the weights of the portfolio to compute the return of the portfolio.
- Repeat the first two steps several times and compute the desired quantiles of the simulated returns.
- Use this return-quantile and the current value of the portfolio to calculate the VaR.

The complexity of the sampling part of Algorithm 11.3 (Multinormal-Cholesky) increases quadratic with the dimension and we need (as seen in Example 15.3 above) many repetitions to get precise results. So here the speed of the simulation may become crucial in practice.

Different "copula methods" of Sect. 12.5 could be used to assess the sensitivity of the portfolio losses to the shape of the multidimensional return distribution. Discussions in that direction were intensified recently but there still seems to be no consenting opinion about which which multivariate distribution or which copula should be used. The multi-t-distribution is the choice advocated by Glasserman et al. (2000). We refer the interested reader to this paper and the references given therein for powerful variance reduction techniques necessary to calculate the VaR for large portfolios including stocks and options by simulation.

Resampling Methods

The most difficult decision before we can calculate VaR or option prices by simulation is the decision about the distribution we should use for the log-returns; the estimation of the parameters is often not a simple task either. To circumvent these problems it is often suggested to use so called "historic simulation" if we have enough (e.g. at least twelve months) of historic daily data available. It just means that instead of generating the log-returns from a theoretic distribution we simply take them randomly (with replacement) from the historic log-returns available. This is clearly a very simple procedure and it is correct if the log-returns are independent and have constant variance; an assumption that is not too unrealistic and was also (implicitly) used for all simulations we have done so far. The only parameter we have to choose in this method is the number of historic returns we should use. For VaR calculations with one day horizon this historic simulation approach has the clear disadvantage that it can only generate a limited number of returns in the left tail, those small returns present in the data. This problem can be easily solved by adding some noise to the resampled returns; this idea has been explained in Sect. 12.1. In the financial context it could be called an "improved version of historic simulation". In an empirical study Bal (2002) concludes that the variance corrected version (i.e. Algorithm 12.2, Resampling-KDEvc) is best suited for VaR calculations; this methods work well with a history of at least 100 returns whereas the shape of the noise (the kernel) and the choice of the smoothing parameter (bandwidth) has only minor influence on the results. It is surprising that this "resampling plus noise" idea that is simple and robust is seemingly not discussed at all in the computational financial literature.

To test the efficiency of the kernel smoothing approach we conducted the following simulation experiment (for the details see Bal, 2002): Generate half a year's (i.e. 122) "artifical log-returns" from a theoretic distribution; compute the VaR for these samples using resampling, resampling plus noise and the normal assumption formula (see (15.2)). Compare the mean square distance of these estimates with the exact value known from the theoretic distribution. It turned out that for all our experiments (with one day horizon) the mean square error was smaller for resampling with noise than for naive resampling. Compared with the "normal assumption", resampling with noise was only slightly worse for $\alpha = 0.1$ and $\alpha = 0.05$ when the normal distribution was used for generating the artificial returns. For the t-distribution with 7 degrees of freedom resampling with noise was slightly better, for the t-distribution with 3 degrees of freedom it performed much better than using the "normal assumption".

It is also not difficult to use this "improved version of historic simulation" for portfolios. Algorithm 12.4 (Resampling-MVKDEvc) can be used to generate the vector of log-returns. Practical problems arise for large portfolios as we cannot expect that a hundred dimensional distribution is well described by a few thousand (if available) historic return vectors. So for large portfolios it is probably better to project the problem to one dimension by calculating the historic returns of the whole portfolio. Then the one-dimensional algorithm can be used. Which method leads to better results in practice could be investigated by comparing the methods for different stock data.

The main advantage of the resampling procedures described here is certainly simplicity. A second point is their flexibility. To see how a possible future crash could influence the average value of our portfolio and the VaR we could add log-returns of, e.g. October 1987, to our historic data; and we could easily create weights to run the simulation with different crash probabilities. It is also possible to sample from the historic log-returns not uniformly but with decreasing (e.g. geometric) probabilities. Thus the most recent returns have the highest probability to be chosen which should be an advantage as the parameters and especially the volatility of the return distributions tend to be unstable. Again empirical studies are necessary to decide if these ideas are useful in practice.

Other Models in Finance

It is well known that (even without the normal assumption) the standard model of assuming iid. random variates for the daily log-returns is not correct (see Mayhew, 2002, and Joro and Na, 2002, for recent overviews). Especially the constant volatility assumption is well known to be problematic. A clear example is the phenomenon of "volatility clustering" or persistence, the tendency of extreme returns (positive and negative) to cluster together in time. The best known approach to cope with this problem are variants of the ARCH and GARCH (generalized autoregressive conditional heteroscedasticity) time series models (see Engle, 1982; Bollerslev, 1988). Long time series are required to obtain good parameter estimates. Together with the fact that the parameters tend to be unstable this creates practical problems for the parameter estimation. Having fixed these parameters the calculation of VaR for one day horizon is especially simple and does not require simulation.

Joro and Na (2002) give an overview and provide references how jump diffusion processes can be combined with geometric Brownian motion to create a more realistic continuous time process for stock prices. There the stock-price is modeled as the sum of geometric Brownian motion and a jump process; the time of the jumps follow a Poisson process whereas the size of the jumps follow a lognormal distribution. We can generate realizations of the Poisson process (possibly with non-constant rate) using the hazard rates methods described in Sect. 9.1. Then we can add up the lognormal increments coming from these jumps with the regular lognormal increments from the geometric Brownian motion to simulate discrete approximations of this process for arbitrarily short steps. So for these processes the practical problem is again the estimation of the parameters and not the generation procedure.

Another point where hazard rates can be used in finance is to describe the lifetime distribution of a company. The market's view of the hazard rate or risk of default at time T can be estimated by comparing the market value of bonds from this company with the market values of bonds from very reliable borrowers. (If the area below this hazard rate is finite this indicates that there is a positive probability that the company will not default.) After specifying the hazard rate for a company the methods of Sect. 9.1 can be applied to simulate the time of default for this company.

15.2 Bayesian Statistics

Due to the rapid development of MCMC methods (see Chap. 14) Bayesian statistics has become a research area with special interest in random variate generation methods. Therefore we think it is in place to demonstrate that some of the methods of this book are useful for Bayesian computations. We will not give an introduction into Bayesian statistics or into Bayesian modeling but we will directly describe the numerical problems and provide possible solutions for some useful standard models. For more information on Bayesian statistics and computations see e.g. Gelman, Carlin, Stern, and Rubin (1995), Gilks et al. (1996), Chen et al. (2000), or Congdon (2003).

Bayesian statistical conclusions about a parameter θ are made in term of probability statements. In opposition to classical inference parameters themselves follow a probability distribution. Assumptions or knowledge about these parameters are summarized in a so called *prior distribution* denoted by $\pi(\theta)$. Observed data X can now be used to update one's knowledge using Bayes' Theorem which states that the conditional distribution $\pi(\theta|X)$ of θ given the sample X is

$$\pi(\theta|X) \propto \pi(X|\theta)\,\pi(\theta),$$

where $\propto$ means that the two functions are proportional. Here $\pi(X|\theta)$ denotes (in a somewhat confusing manner) the probability distribution of a sample X given parameter θ. It is called the likelihood function. We will also use the notation $L(\theta|X)$, since it can be interpreted as the likelihood of the parameter θ for a fixed sample X as well. The updated density $\pi(\theta|X)$ is called the *posterior distribution* of the Bayesian model. We will write

$$\pi(\theta|X) \propto L(\theta|X)\,\pi(\theta)$$

in the following. Notice that the posterior distribution combines the information of the sample with the prior information and is the distribution of the model parameters given the observed data. To obtain e.g. a confidence interval for the parameter θ_1 we need quantiles of the marginal posterior distribution $\pi(\theta_1|X)$; to obtain a point estimate we have to calculate the median or the expectation of the marginal posterior distribution. For particular examples of Bayesian data analysis the interested reader is referred to the relevant literature (see above).

One drawback of the Bayesian approach is the fact that these posterior distributions are in general non-standard distributions. For most models of practical relevance the area below the posterior density is not simple to obtain which also implies difficulties when evaluating the integrals necessary for Bayesian inference. But it is possible to obtain the required confidence intervals and point estimates if we can generate random vectors from the posterior distribution. This is not a simple task either (especially if we have more than just a few model parameters) but it is exactly that kind of problem we have discussed throughout this book. We describe two possible solutions to this generation problem in the next two subsections.

15.2.1 Gibbs Sampling and Universal Random Variate Generation

The standard solution to this problem advocated by many practitioners and researchers is the use of the Gibbs sampler. Thanks to the freely available "WinBUGS" software (see Spiegelhalter, Thomas, Best, and Lunn, 2003; Gilks et al., 1996) Gibbs sampling experiments for a wide range of (partly very large and complex) Bayesian models can be executed easily. By means of the graphical user interface of WinBUGS models can be specified even without writing any line of programming code. The Gibbs sampler is not generating iid. vectors but a sequence of dependent vectors (see Sect. 14.1.2). For calculating quantiles and expectations from the (marginal) posterior distribution these dependent vectors could be used if the Markov chain is converging and mixing fast enough. The big practical problem is the question of how to assess the mixing and convergence properties. As this is not the topic of this book we only issue the (well known) warning here that the method is not "foolproof". Just trying different starting values and discarding the first 1000 or 10000 observations of the Markov chain does not guarantee correct results. Even looking at "trace plots" for some of the variables and calculating their autocorrelations cannot guarantee that the mixing properties of the chain are correctly assessed.

To demonstrate the use of random variate generation methods for Gibbs sampling we will consider four examples, two of them with a direct Bayesian context, the other two for demonstration purposes. As an easy start let us try to sample from the two-dimensional normal distribution using Gibbs sampling.

Example 15.4. The path of the Gibbs sampler is not influenced by the variances of the marginal distributions. Thus we assume that these variances are both one. So the only parameter remaining is the correlation ρ with $-1 < \rho < 1$ and we write for the logarithm of the joint density

$$\log f(x_1, x_2) = c_1 + \frac{-0.5}{1 - \rho^2} \left(x_1^2 - 2\rho x_1 x_2 + x_2^2 \right),$$

where c_1 denotes a normalization constant that is not important to us. For the Gibbs sampler we only need to know the full conditional distributions.

Thus for the conditional distribution of X_1 given X_2 it is enough to consider those terms containing x_1. Those only containing x_2 are not important for random variate generation and are therefore absorbed into the constant c_2:

$$\log f(x_1, x_2) = c_2 + \frac{-0.5}{1-\rho^2}\left(x_1^2 - 2\rho x_1 x_2\right).$$

We can easily see that the conditional distribution of X_1 is itself normal with mean ρx_2 and variance $(1-\rho^2)$. As the joint distribution is symmetric in x_1 and x_2 we obtain the full conditional distributions

$$(X_1|X_2 = x_2) \sim N(\rho x_2, 1-\rho^2) \quad \text{and} \quad (X_2|X_1 = x_1) \sim N(\rho x_1, 1-\rho^2),$$

where $N(a,b)$ denotes a normal distribution with mean a and variance b. Notice that we have easily derived the full conditional distributions from the density of the joint distribution, without using any special properties of the bivariate normal distribution.

Now we can directly write down the algorithm for the Gibbs sampler as it is enough to know the full conditional distributions. As both of them are normal the random variate generation problem is easily solved. We can use e.g. transformed density rejection (see Algorithm 4.1) or the Box-Muller method (Algorithm 2.11). The whole Gibbs sampling algorithm is described with the following pseudo-code.

Output: (Dependent) sequence of pairs (X_1, X_2) from bivariate normal distribution with correlation ρ.
1: Choose starting values for X_1 and X_2.
2: **loop**
3:     Generate $Z \sim N(0,1)$.
4:     $X_1 \leftarrow \rho X_2 + Z\sqrt{1-\rho^2}$.
5:     Generate $Z \sim N(0,1)$.
6:     $X_2 \leftarrow \rho X_1 + Z\sqrt{1-\rho^2}$.

We tried this Gibbs sampler for many values of ρ between -0.999 and 0.999. We expected that for large positive or negative correlation we should obtain highly autocorrelated output. It turned out that the one-step autocorrelations of X_1 and X_2 are approximately ρ^2. So we need long series to obtain usable estimates if $|\rho|$ is close to one. Of course we can easily assess the correctness of the estimates as we know that both marginal distributions are standard normal for any value of ρ. We used the starting point 0 and a short burn-in. Then we generated series of length 10^4 and recorded maximal and minimal generated values and the naive estimate for mean and standard deviation of the marginal distributions. Totally wrong results (e.g. mean $=0.3$) were only observed for $|\rho| \geq 0.995$. For $0.5 \leq |\rho| \leq 0.99$ we obtained both correct and wrong results with the mean value between one and four standard errors (i.e. between 0.01 and 0.04) away from 0.

These results demonstrate a well known property of the Gibbs sampler: For highly correlated variables convergence and mixing are slow. This problem can be easily seen from autocorrelations or a plot of the generated sequence.

And it is well known that the naive estimates for mean and variance as well as for quantiles are biased or have much larger variances than for the iid. case.

After this easy warm-up we continue with an example dealing with a posterior distribution.

Example 15.5. Let us consider the simple two-parameter Bayesian model:

$$y_i \sim N(\mu, \tau^{-1}), \text{ for } i = 1, \ldots, n, \qquad \mu \sim N(0, s_0^2), \qquad \tau \sim Ga(a_0, b_0),$$

where $Ga(a, b)$ denotes a gamma distribution with mean ab and variance ab^2. We can write the log-likelihood as

$$\log L(\mu, \tau | y_1, y_2, \ldots, y_n) = c_1 + \frac{n}{2} \log \tau - \frac{\tau}{2} \sum (y_i - \mu)^2 =$$

$$= c_1 + \frac{n}{2} \log \tau - \frac{\tau}{2} \left(\sum y_i^2 - 2\mu \sum y_i + \mu^2 n \right).$$

It is not difficult to see that for the current example the likelihood function can be interpreted as a quasi-density for the joint distribution of μ and τ and we can obtain the full conditionals of the likelihood. We get

$$\log L(\mu|\tau) = c_2 - \frac{1}{2} \left(n\tau\mu^2 - 2\tau\mu \sum y_i \right),$$

which implies that the full conditional of μ is $N(\sum y_i/n, 1/(n\tau))$. This is of course the well known result for the distribution of the sample mean. For τ, the reciprocal of the variance, we have

$$\log L(\tau|\mu) = c_3 + \frac{n}{2} \log \tau - \frac{\tau}{2} \left(\sum y_i^2 - 2\mu \sum y_i + \mu^2 n \right),$$

which implies for the full conditional of τ:

$$\tau|\mu \sim Ga\left(1 + n/2, 2/\left(\sum y_i^2 - 2\mu \sum y_i + \mu^2 n\right)\right).$$

Using the improper prior, that is constant on $\mathbb{R}$ for μ and constant on $\mathbb{R}^+$ for τ, the likelihood function we have just derived could be interpreted as a posterior distribution as well. Instead we are using the model stated at the beginning of this example that assumes conjugate priors that also lead to well known full conditionals for the posterior distribution. To see this we can write the log-posterior distribution as

$$\log \pi(\mu, \tau | y_1, y_2, ., y_n) = \log L(\mu, \tau | y_1, y_2, ., y_n) - \frac{1}{2} \left(\frac{\mu}{s_0}\right)^2 + (a_0 - 1) \log \tau - \frac{\tau}{b_0}.$$

Therefore it is simple to compute the full conditionals of the posterior distribution by taking the full conditionals of the log-likelihood and adding the logarithm of the prior distribution for that parameter. Thus we obtain

$$\pi(\mu|\tau) \sim N\left(\frac{\tau \sum y_i}{n\tau + 1/s_0^2}, \frac{1}{n\tau + 1/s_0^2}\right),$$

$$\pi(\tau|\mu) \sim Ga\left(a_0 + n/2, \frac{1}{\left(\sum y_i^2 - 2\mu \sum y_i + \mu^2 n\right)/2 + 1/b_0}\right).$$

A prior distribution is called a *flat* (or an uninformative) prior if it does not change the likelihood much. We could therefore try a large value for s_0, $a_0 = 1$, and b_0 large. For s_0 this suggestion is certainly sensible but we should not forget the interpretation of τ which is the reciprocal value of the variance. Thus a flat prior for τ favors values of the variance below one as the probability for τ to be above one is well above 0.5 for such a prior. A flat prior often suggested for τ is the gamma distribution with e.g. $a_0 = 0.001$ and $b_0 = 1000$ as it leads to a prior for τ with expectation one and a large variance. So we used these values and $s_0 = 1000$. The implementation of the Gibbs sampler is again very easy as only standard distributions are involved. For the normal distribution we can use the same algorithms as suggested in Example 15.4. For the full conditional distribution of τ it is convenient that the shape parameter of the gamma distribution does not depend on μ. So in the Gibbs sampling procedure we can always sample from the same gamma distribution and just change the scale by multiplying by the current shape parameter. As sampling procedure we can again use transformed density rejection (Algorithm 4.1, TDR) as the shape parameter is bigger than one for $n \geq 2$. The details are presented as pseudo-code:

Require: Parameters s_0, a_0, and b_0 of the prior distribution.
  A sample y_i for $i = 1, 2, \ldots, n$.
Output: (Dependent) sequence of pairs (μ, τ) from the posterior distribution.
1: Compute $\sum_{i=1}^n y_i$ and $\sum_{i=1}^n y_i^2$.
2: Choose starting values for μ and τ.
3: **loop**
4:   Generate $Z \sim N(0, 1)$.
5:   $\mu \leftarrow (\tau \sum y_i)/(n\tau + 1/s_0^2) + Z/\sqrt{n\tau + 1/s_0^2}$.
6:   Generate $G \sim Ga(a_0 + n/2, 1)$.
7:   $\tau \leftarrow G/((\sum y_i^2 - 2\mu \sum y_i + \mu^2 n)/2 + 1/b_0)$.

We tried this algorithm for a sample with $n = 5$ and all observations close to 1 which resulted in $\sum_{i=1}^n y_i = 5$ and $\sum_{i=1}^n y_i^2 = 5.1$ (i.e. sample mean 1 and sample variance $1/40$). Using $s_0 = 1000$, $a_0 = 0.001$, and $b_0 = 1000$ we ran the Gibbs sampling procedure with starting values $\mu = 1$ and $\tau = 40$, short burn-in and samplesize 10^6. The observed autocorrelation was practically zero for μ and around 0.14 for τ which shows that the sampler is mixing well. So we can estimate expectation and variance for the marginal posterior of the two parameters. For μ we obtained an expectation of 1.0001 and a standard deviation of 0.1; for τ we estimated an expectation of 39.24 and a standard deviation of 27.8. These results look sensible, the large standard deviation for τ is due the small sample size. Note that – as it is the case with real-world

Gibbs sampling applications – we have no obvious simple way to check if this simulation result is correct.

We continue with an example with real-world data found in Spiegelhalter, Thomas, Best, and Gilks (1996).

Example 15.6. Dobson (1983) discusses binary dose-response data of Bliss (1935). There the numbers of beetles killed after 5 hour exposure to carbon disulphide are recorded. The full data are contained in Table 15.3.

Table 15.3. Concentration x_i, number of beetles n_i, number of killed beetles r_i

x_i	1.6907	1.7242	1.7552	1.7842	1.8113	1.8369	1.8610	1.8839
n_i	59	60	62	56	63	59	62	60
r_i	6	13	18	28	52	53	61	60

Assuming that the observed number of deaths r_i at concentration x_i is binomial with sample size n_i and rate p_i we have for a fixed dose

$$p(r_i) \propto p_i^{r_i}(1-p_i)^{n_i-r_i} .$$

We can write the logistic model for the p_i as

$$p_i = \frac{exp(\alpha + \beta(x_i - \bar{x}))}{1 + exp(\alpha + \beta(x_i - \bar{x}))} = g(x_i, \alpha, \beta) ,$$

where $\bar{x}$ denotes the mean of the x_i (the x_i were centered to reduce the correlation between α and β). Combining these two models we obtain the log-likelihood

$$\log L(\alpha, \beta | r_1, r_2, \ldots, r_8) = c_1 + \sum_{i=1}^{8} (r_i g(x_i, \alpha, \beta) + (n_i - r_i)(1 - g(x_i, \alpha, \beta))) .$$

The full conditionals of this log-likelihood do not simplify but it is not difficult to show that they are log-concave. Assuming $N(0, s_0^2)$ prior distributions for α and β we obtain the log-posterior

$$\log \pi(\alpha, \beta | r_1, r_2, \ldots, r_8) = c_2 + \log L(\alpha, \beta | r_1, r_2, \ldots, r_8) - \frac{\alpha^2}{2s_0^2} - \frac{\beta^2}{2s_0^2} .$$

Clearly the full conditional posterior distributions for α and β are log-concave as well. So we can use a black-box algorithm for log-concave distributions to implement the Gibbs sampler. Note that for the full conditional distributions we know neither the mode nor the area below the density but it is not difficult to obtain its derivatives. As it often happens for Gibbs sampling applications we are facing a situation where the full conditional distribution is constantly changing. So we have to use a variant of transformed density rejection (TDR) (see Chap. 4) with a fast setup that does not require the mode or the area below the density. Therefore it is best to use TDR with only two starting points on either side of the mode and adaptive rejection sampling (ARS, see Sect. 4.4.5 and Gilks and Wild, 1992) to update the hat until one random variate is accepted. For our example this leads to the following implementation of the Gibbs sampler:

Require: Parameter s_0 of the prior distribution.
  Function that evaluates the log-posterior $\log \pi(\alpha, \beta)$.
Output: (Dependent) sequence of pairs (α, β) from posterior distribution.
1: Choose starting values for α and β.
2: **loop**
3:   Find design points $p_{\alpha,1}$ and $p_{\alpha,2}$ for Algorithm 4.1 (TDR) to sample from density $f_\alpha(x) = \exp(\log \pi(x, \beta))$.
   /* For our example $p_{\alpha,1} = 0$ and $p_{\alpha,2} = 1$ can be used. */
4:   Generate the random variate α with density $f_\alpha(x)$.
5:   Find design points $p_{\beta,1}$ and $p_{\beta,2}$ for Algorithm 4.1 (TDR) to sample from density $f_\beta(x) = \exp(\log \pi(\alpha, x))$.
   /* For our example $p_{\beta,1} = 30$ and $p_{\beta,2} = 45$ can be used. */
6:   Generate the random variate β with density $f_\beta(x)$.

The Gibbs sampler for this example is much slower than in the two examples above for three reasons: There are no standard distributions involved, we have to sample from a new density in every step and the evaluation of the log-posterior is really expensive. So – depending on the posterior distribution – it may be worth while to fine tune the choice of the design points by using results from previous iterations. For our example we did not try it, as we do not expect too much from this tuning here as the autocorrelations of both α and β were only about 0.1.

Our results for $s_0^2 = 1000$ and a run with 10000 repetitions were: Mean of 0.746 and standard deviation 0.138 for α and mean 34.33 and standard deviation 2.95 for β, which is very close to the results reported by Spiegelhalter et al. (1996) and Agathokleous and Dellaportas (1999).

We add a last example which is adapted from a suggestion by Neal (2003). It nicely illustrates possible problems of MCMC. For certain distributions it can happen that the Gibbs sampler (like any other MCMC sampling algorithm) mixes so badly that it never visits regions of non-negligible probability. This example stresses the importance to check the autocorrelation of the generated sequences before interpreting any results.

Example 15.7. We assume that conditional on the random variate $V \sim N(0, 1)$ we have $X \sim N(0, \exp(32V)^2)$. Of course it is very simple to sample from this two dimensional distribution directly. To test the performance of the Gibbs sampler for this distribution we calculate the logarithm of the joint density,

$$\log f(v, x) = c_1 - \frac{v^2}{2} - 32v - \frac{x^2}{2\exp(64v)}.$$

The full conditional distribution of X given V is $N(0, \exp(32V)^2)$ whereas the full conditional distribution of V has a non-standard shape but is log-concave. So we can implement the Gibbs sampler without problems using the same techniques as in Example 15.6 above. We ran this Gibbs sampler with 10^4 replications and two different starting values and obtained very similar results: the mean of V was close to 0.5 and the standard deviation close to 0.75. The minimal observed value for v in both runs was -1.43. If we compare this with the fact that V should be a standard normal variate we can see that the result of the Gibbs sampler is totally wrong, and a region with a probability larger than 0.07 was not visited at all in 10^4 repetitions. Fortunately there is an indication that this Gibbs sampler is not mixing well at all. The autocorrelation of the generated v sequence is close to 0.998. So this example is again emphasizing the fact that we should not use results from Gibbs samplers with high autocorrelations.

We hope that the above examples provided some insight into the practice of Gibbs sampling for Bayesian statistics. It is for most examples not difficult to find the full conditional distributions of the posterior for all model parameters. So the main problem is the generation of random variates from the full conditional distributions. If it is possible to see that these distributions are log-concave or T-concave it is possible to use transformed density rejection combined with adaptive rejection sampling (ARS). Exactly this solution is used in the "WinBUGS" software (see Spiegelhalter et al., 2003) for the log-concave case. After the implementation of the Gibbs sampler it is of primary importance not to forget that it is not producing iid. vectors from the posterior distribution. So all conclusions from the output must be drawn with extreme care, especially if the autocorrelations for one of the parameter-sequences is high.

What can be done as alternative to get entirely rid of these problems related to MCMC? In the next section we demonstrate that generating iid. vectors is sometimes a possible solution.

15.2.2 Vector Generation of the Posterior Distribution

We have seen in Examples 15.5 and 15.6 above how we can find the joint posterior density for all parameters. For these small models with only two parameters it is therefore not difficult to obtain a contour plot of the posterior density and to find the mode of the distribution. These informations can be used to find a vector generation algorithm to generate from the posterior distribution.

Example 15.8 (Continuation of Example 15.5). The contour plot of the log-likelihood indicates that the joint density of the pairs (μ, τ) is not orthounimodal which rules out the use of the algorithms of Sect. 11.4. It is also obvious that the contour lines do not define convex sets which makes clear that the distribution cannot be T-concave. So we can only try naive rejection from a constant hat. That is possible if we can find the mode of the density and can guess a rectangle that contains all relevant parts of the distribution. For the sample of size $n = 5$ considered above with sample mean 1 and sample variance $1/40$ using $s_0 = 1000$, $a_0 = 0.001$ and $b_0 = 1000$ for the parameters of the prior distributions it is no problem to find the mode which is at $\mu = 1$ and $\tau = 29.43$. Using the contour plot we can find the region where the log-posterior is not more than, e.g. 15, units smaller than in the mode; in our example the rectangle $(-9, 11) \times (0, 350)$. Then outside of this rectangle the density is less than 10^{-6} times the value of the mode; together with the log-concavity of the full conditional distributions this gives us some confidence that we have only chopped off negligible probabilities. We executed the naive rejection algorithm from a constant hat for the above rectangle. Of course this algorithm is quite slow (about 400 times slower than the very fast Gibbs sampler) and it took about ten seconds to generate a sample of size 10000 on our (slow) computer. But now we have an iid. sample and can draw simple conclusions. It turns out that mean and standard deviation are within one standard error from the results obtained by the Gibbs sampler. So here the vector generation approach supports the results obtained by the Gibbs sampler. Of course we do not suggest this "brute force" approach for general use, as it is too slow and the cut-off procedure could be dangerous. We just added its description here to illustrate that – for the two parameter case – there are simple alternatives to the Gibbs sampler.

If we reparameterize the posterior using the standard deviation σ instead of τ the reciprocal of the variance, it turns even out that the likelihood function is T-concave but not log-concave. This should lead to a simple generation procedure for the posterior but as the fast random vector algorithms 11.14 (TDR-Bivariate) and 11.15 (TDR-Cone) are up to now only implemented for log-concave distributions we do not give the details here.

Dellaportas had the idea to apply Algorithm 11.15 (TDR-Cone) to sample from posterior distributions of generalized linear models (GLM). Following Agathokleous and Dellaportas (1999) we demonstrate this idea for the beetles data.

Example 15.9 (Continuation of Example 15.6). In a contour plot of the posterior of the parameters α and β we can see that the contour lines form convex sets. It is also not very difficult to see from the log-posterior that it is a bivariate concave function as long as the prior distributions are log-concave themselves. The mode of the posterior is easily found and thus we can use Algorithms 11.14 (TDR-Bivariate) or 11.15 (TDR-Cone) without problems to sample iid. pairs from the posterior distribution. It turns out that for this example generating the iid. pairs is much faster than the Gibbs sampler. The results are again practically the same for both methods demonstrating the correctness of our Gibbs sampling results. We were confident that the Gibbs sampler leads to correct results here as the autocorrelations of the generated parameters were small.

The better performance of the vector generation method is due to the fact that the joint posterior distribution remains fixed. So the necessary constants can be calculated and stored in the setup and after that sampling is very fast. For the Gibbs sampler the full conditional distributions change in every step so the setup has to be calculated again and again which leads to the slow performance.

Drawing a conclusion from our examples we can say that the Gibbs sampler is for many Bayesian models (with log-concave full conditional posterior) easy to implement and we have a good chance to obtain useable results. But for our examples the same is true for the vector generation methods. Now the reader may ask herself why using the Gibbs sampler is the standard for sampling from posterior distributions in Bayesian statistics. We do not really know an answer to that question. One point is certainly that the vector generation algorithms are little known; a second that many Bayesian models (like e.g. random effect models) require the estimation of 20 or 50 parameters and not of 2 or 5. For such models the vector generation approach is probably hopeless unless we can make use of the special structure of the posterior distribution of the model. It is not certain that the Gibbs sampler works for such models but with clever reparametrization schemes there is a chance. This could be explained by the fact that the Gibbs sampler, that is only moving in the direction of the axes, is especially suited for models with conditional independence structures that freequently occur for Bayesian models with many parameters.

15.3 Exercises

Exercise 15.1. Rewrite the innermost loop of the naive option pricing simulation algorithm of Example 15.1 to reduce the number of calls to the exponential function. Measure the speed-up on your computer (it is about 35 % on ours).

Exercise 15.2. Reformulate the simulation algorithm presented in Example 15.1 that it can be used to price European put options.
Hint: European put options have payoff function $\min(E - S_T, 0)$.

Exercise 15.3. Reformulate the simulation algorithm presented in Example 15.2 that it can be used to price European put options.
Hint: European put options have payoff function $\min(E - S_T, 0)$.

Exercise 15.4. Construct a Gibbs sampler to generate pairs of the bivariate normal distribution with zero mean, unit variances and correlation ρ restricted to the first quadrant.
Hint: To generate from a one-dimensional truncated normal distribution use

Algorithm 4.1 (TDR) with its special ability to sample from truncated distributions (see Sect. 8.4.1) that is also supported by UNU.RAN.

List of Algorithms

2.1	Inversion	14
2.2	Rejection-from-Uniform-Hat	17
2.3	Rejection (Basic method)	19
2.4	Rejection-with-Inversion	19
2.5	Rejection-with-Squeeze	21
2.6	Composition	27
2.7	Composition-Rejection	27
2.8	Rejection-Immediate-Acceptance	31
2.9	RoU (Ratio-of-uniforms)	35
2.10	Almost-exact-Inversion	37
2.11	Box-Muller	38
3.1	Sequential-Search	44
3.2	Indexed-Search	46
3.3	Alias-Sample	48
3.4	Alias-Setup	48
3.5	Alias-Urn	50
4.1	TDR (Transformed-Density-Rejection)	60
4.2	TDR-Proportional-Squeeze	91
4.3	TDR-Immediate-Acceptance	92
4.4	TDR-3-Points (for $c = -1/2$)	94
4.5	TDR-T-convex	99
4.6	TDR-Mixed	100
4.7	TDR-c-Optimal	102
4.8	Adaptive-RoU (Adaptive Ratio-of-Uniforms)	107
5.1	Ahrens (Basic method)	115
5.2	Ahrens-Immediate-Acceptance	116
5.3	Ahrens-Equal-Area	122
6.1	Monotone-01 (Simple rejection for monotone density on $[0, 1]$)	128
6.2	Monotone-Bounded	128
6.3	Monotone-Unbounded	130
6.4	Simple-RoU (Simple Ratio-of-Uniforms method)	138

6.5	Simple-RoU-Mirror-Principle	139
6.6	Simple-Setup-Rejection	141
6.7	Generalized-Simple-RoU	147
6.8	Heavy-Tailed-Simple-RoU	151
7.1	Fast-Table-Inversion	163
8.1	Inversion-Truncated	184
9.1	Hazard-Rate-Inversion	195
9.2	Hazard-Rate-Composition	196
9.3	Hazard-Rate-Thinning	196
9.4	Hazard-Rate-Constant-Thinning	197
9.5	Decreasing-Hazard-Rate-Dynamic-Thinning	198
9.6	Increasing-Hazard-Rate-Dynamic-Composition	200
9.7	Convergent-Series	202
9.8	Alternating-Series	203
9.9	Fourier-Coefficients	206
9.10	Characteristic-Function-Polya	207
9.11	FVP-Density (Fejer-de la Vallee Poussin density)	208
9.12	Characteristic-Function-Smooth	209
10.1	Indexed-Search-Tail	217
10.2	Indexed-Search-2-Tails	220
10.3	Rejection-Inversion (Discrete)	222
10.4	Automatic-Rejection-Inversion (ARI)	226
10.5	Discrete-Unimodal-2ndMoment	229
10.6	Discrete-Simple-RoU (Discrete simple Ratio-of-Uniforms method)	231
10.7	Discrete-Simple-RoU-2 (Discrete simple Ratio-of-Uniforms method)	231
10.8	Generating-Function	234
10.9	Discrete-Hazard-Rate-Inversion	235
10.10	Discrete-Hazard-Rate-Sequential	236
11.1	Conditional-Distribution-Method	247
11.2	Multivariate-Rejection	248
11.3	Multinormal-Cholesky	250
11.4	Cholesky-Decomposition	251
11.5	Multi-t (Multivariate t-distribution)	252
11.6	Uniform-Sphere	253
11.7	Uniform-Ball-Rejection	254
11.8	Uniform-Ball	255
11.9	Uniform-Spacings	257
11.10	Uniform-Simplex	257
11.11	Uniform-Grid	260
11.12	Uniform-Sweep-Plane	264
11.13	TDR-mv (Multivariate transformed density rejection)	269
11.14	TDR-Bivariate	275
11.15	TDR-Cone (Multivariate transformed density rejection with cones)	279
11.16	Platymorphous	283
11.17	Monotone-Multivariate	283

11.18	Ahrens-Multivariate	285
11.19	Orthomonotone-I	287
11.20	Orthomonotone-II	289
11.21	Orthomonotone-III	291
11.22	Orthomonotone-IV	293
12.1	Resampling-KDE (Kernel density estimation)	308
12.2	Resampling-KDEvc (Kernel density estimation, variance corrected)	310
12.3	Resampling-ELK (Linear interpolation of empirical cdf)	311
12.4	Resampling-MVKDEvc	320
12.5	4-Moments-Discrete	325
12.6	4-Moments-Continuous	326
12.7	Zero-Maximal-Bivariate (Correlation mixture for uniform marginals)	331
12.8	NORTU-Bivariate	334
12.9	Chessboard-Bivariate	335
12.10	Arbitrary-Marginals	337
12.11	NORTA-Multivariate	342
13.1	Durbin	349
13.2	Durbin-A	349
13.3	FFT	351
13.4	Non-Gaussian-Time-Series	362
14.1	Metropolis-Hastings-Chain	365
14.2	Gibbs-Chain	369
14.3	Hit-and-Run-Chain	371
14.4	CFTP (Coupling from the past)	375
14.5	Perfect-Independence-Sampler	378
14.6	Perfect-Random-Walk-Metropolis-Sampler	382

References

Afflerbach, L. and W. Hörmann (1992). Nonuniform random numbers: A sensitivity analysis for transformation methods. In G. C. Pflug and U. Dieter (Eds.), *Lecture Notes in Econom. Math. Systems*, Volume 374 of *Lect. Notes Econ. Math. Syst.*, New York, pp. 135–144. Springer-Verlag. Proc. Int. Workshop Comput. Intensive Methods Simulation Optimization, Laxenburg/Austria 1990.

Afflerbach, L. and K. Wenzel (1988). Normal random numbers lying on spirals and clubs. *Statistical Papers 29*, 237–244.

Agathokleous, A. and P. Dellaportas (1999). Direct sampling from posterior densities of some common generalized linear models. Technical report, Athens University of Economics and Business, Department of Statistics.

Ahrens, J. H. (1993). Sampling from general distributions by suboptimal division of domains. *Grazer Math. Berichte 319*, 20 pp.

Ahrens, J. H. (1995). A one-table method for sampling from continuous and discrete distributions. *Computing 54*(2), 127–146.

Ahrens, J. H. and U. Dieter (1972). Computer methods for sampling from the exponential and normal distributions. *Commun. ACM 15*, 873–882.

Ahrens, J. H. and U. Dieter (1973). Extensions of forsythe's method for random sampling from the normal distribution. *Math. Comp. 27*, 927–937.

Ahrens, J. H. and U. Dieter (1982). Generating gamma variates by a modified rejection technique. *Commun. ACM 25*, 47–54.

Ahrens, J. H. and U. Dieter (1988). Efficient table-free sampling methods for the exponential, cauchy, and normal distributions. *Commun. ACM 31*, 1330–1337.

Ahrens, J. H. and U. Dieter (1989). An alias method for sampling from the normal distribution. *Computing 42*(2/3), 159–170.

Ahrens, J. H. and K. D. Kohrt (1981). Computer methods for efficient sampling from largely arbitrary statistical distributions. *Computing 26*, 19–31.

Amit, Y. (1991). On rates of convergence of stochastic relaxation for gaussian and non-gaussian distributions. *J. Mult. Anal. 38*, 82–99.

Avramidis, A. N. and J. R. Wilson (1994). A flexible method for estimating inverse distribution functions in simulation experiments. *ORSA Journal on Computing 6*, 342–355.

Bal, H. (2002). Resampling methods for simulating value at risk. M.sc. thesis, Department of Industrial Engineering, Bogazici University, Istanbul.

Banks, J., J. S. Carson, and B. L. Nelson (1999). *Discrete-Event System Simulation* (2 ed.). Upper Saddle River, New Jersey: Prentice-Hall.

Barabesi, L. (1993). Random variate generation by using the ratio-of-uniforms method. Technical Report 1–1993, Università degli Studi die Siena, Dipartimento di Metodi Quantitativi.

Barber, C. B. and H. Huhdanpaa (1995–2002). *Qhull manual*. The Geometry Center, Minneapolis MN. http://www.geom.umn.edu/software/qhull/.

Barndorff-Nielsen, O. and P. Blæsild (1983). Hyperbolic distributions. In S. Kotz and N. L. Johnson (Eds.), *Encyclopedia of Statistical Sciences*, Volume 3, pp. 700–707. New York: Wiley-Interscience.

Barton, R. R. and L. W. Schruben (2001). Resampling methods for input modeling. In B. Peters, J. Smith, D. Medeiros, and M. Rohrer (Eds.), *Proceedings of the 2001 Winter Simulation Conference*, Piscataway, New Jersey, pp. 372–378. IEEE.

Bedford, T. and R. M. Cooke (2001). Probability density decomposition for conditionally dependent random variables modeled by vines. *Annals of Mathematics and Artifical Intelligence 32*, 245–268.

Bedford, T. and R. M. Cooke (2002). Vines - a new graphical model for dependent random variables. *Annals of Statistics 30*, 1031–1068.

Belisle, C. J. P., H. E. Romeijn, and R. L. Smith (1993). Hit-and-run algorithms for generating multivariate distributions. *Mathematics of Operations Research 18*, 255–266.

Besag, J. and P. J. Green (1993). Spatial statistics and bayesian computation. *Journal of the Royal Statistical Society B 55*, 25–37.

Bieri, H. and W. Nef (1983). A sweep-plane algorithm for the volume of polyhedra represented in boolean form. *Linear Algebra Appl. 52/53*, 69–97.

Bliss, C. I. (1935). The calculation of the dosage-mortality curve. *Annals of Applied Biology 22*, 134–167.

Boender, C. G. E., R. J. Caron, J. F. McDonald, A. H. G. Rinooy Kan, H. E. Romeijn, R. L. Smith, J. Telgen, and A. C. F. Vrost (1991). Shake-and-bake algorithms for generating uniform points on the boundary of bounded polyhedra. *Operations Research 39*, 945–954.

Bollerslev, T. (1988). On the correlation structure for the generalized autoregressive conditional heteroscedastic process. *Journal of Time Series Analysis 9*, 121–131.

Box, G. E. P. and D. R. Cox (1964). An analysis of transformations. *Journal of the Royal Statistical Society, B 26*, 211–243.

Box, G. E. P. and M. E. Muller (1958). A note on the generation of random normal deviates. *Annals of Math. Stat. 29*(2), 610–611.

Bratley, P., B. L. Fox, and E. L. Schrage (1983). *A Guide to Simulation*. New York: Springer-Verlag.

Bratley, P., B. L. Fox, and E. L. Schrage (1987). *A Guide to Simulation* (2 ed.). New York: Springer-Verlag.

Brockwell, P. J. and R. A. Davis (1991). *Time Series: Theory and Methods* (2 ed.). New York: Springer.

Cario, M. C. and B. L. Nelson (1996). Time-series input processes for simulation. *Operations Research Letters 19*, 51–58.

Cario, M. C. and B. L. Nelson (1997). Modeling and generating random vectors with arbitrary marginal distributions and correlation matrix. Technical report, Northwestern University, Evanston, Illinois.

Casella, G., K. L. Mengersen, C. P. Robert, and D. M. Titterington (2002). Perfect slice samplers for mixture distributions. *Journal of the Royal Statistical Society B 64*, 777–790.

Chen, H. C. and Y. Asau (1974). On generating random variates from an empirical distribution. *AIIE Trans. 6*, 163–166.

Chen, M.-H. and B. Schmeiser (1993). Performance of the Gibbs, Hit-and-run, and Metropolis samplers. *Journal of Computational and Graphical Statistics 2*, 251–272.

Chen, M.-H. and B. Schmeiser (1998). Toward black-box sampling: a random-direction interior-point Markov chain approach. *Journal of Computational and Graphical Statistics 7*, 1–22.

Chen, M.-H., Q.-M. Shao, and J. G. Ibrahim (2000). *Monte Carlo Methods in Baysian Computation*. Springer Series in Statistics. Springer Verlag.

Chen, W. and R. L. Burford (1981). Quality evaluation of some combinations of unit uniform random generators and unit normal transformation algorithms. In R. M. Huhn, E. R. Comer, and F. O. Simons, Jr (Eds.), *Proc. 14th Annual Simulation Symp.* Florida, pp. 129–149.

Cheng, R. C. H. (1977). The generation of gamma variables with non-integral shape parameters. *Appl. Statist. 26*, 71–75.

Cheng, R. C. H. (1978). Generating beta variates with nonintegral shape parameters. *Commun. ACM 21*, 317–322.

Cheng, R. C. H. and G. M. Feast (1980). Gamma variate generators with increased shape parameter range. *Commun. ACM 23*, 389–394.

Clemen, R. T. and T. Reilly (1999). Correlations and copulas for decision and risk analysis. *Management Science 45*, 208–224.

Congdon, P. (2003). *Applied Bayesian Modelling.* New York: Wiley.

Cryer, J. (1986). *Time Series Analysis.* Boston: Duxbury Press.

Dagpunar, J. (1988). *Principles of Random Variate Generation.* Oxford, U.K.: Clarendon Oxford Science Publications.

Davies, N., T. Spedding, and W. Watson (1980). Autoregressive moving average processes with non-normal residuals. *Journal of Time Series Analysis 1*(2), 103–109.

Davies, R. B. and D. S. Harte (1987). Tests for hurst effect. *Biometrika 74*, 95–101.

Deák, I. (1978). Monte Carlo evaluation of the multidimensional normal distribution function by the ellipsoid method. *Problems Control Inform. Theory 7*, 203–212.

DeBrota, D., R. S. Dittus, S. D. Roberts, and J. R. Wilson (1989). Visual interactive fitting of bounded johnson distributions. *Simulation 52*, 199–205.

Deler, B. and B. L. Nelson (2000). Modeling and generating multivariate time series with arbitrary marginals using a vector autoregressive technique. Technical report, Northwestern University, Evanston, Illinois.

Deng, L.-Y. and R. S. Chhikara (1992). Robustness of some non-uniform random variate generators. *Statistica Neerlandia 46*(2/3), 195–207.

Derflinger, G. and W. Hörmann (2002). The optimal selection of construction points. Unpublished Manuscript.

Devroye, L. (1981). The series method for random variate generation and its application to the Kolmogorov-Smirnov distribution. *Am. J. Math. Manage Sci. 1*, 359–379.

Devroye, L. (1982). A note on approximations in random variate generation. *J. Stat. Comput. Simulation 14*, 149–158.

Devroye, L. (1984a). Methods for generating random variates with polya characteristic functions. *Stat. Probab. Lett. 2*, 257–261.

Devroye, L. (1984b). Random variate generation for unimodal and monotone densities. *Computing 32*, 43–68.

Devroye, L. (1984c). A simple algorithm for generating random variates with a log-concave density. *Computing 33*(3–4), 247–257.

Devroye, L. (1986a). *Non-Uniform Random Variate Generation*. New-York: Springer-Verlag.

Devroye, L. (1986b). Grid methods in simulation and random variate generation. *Computing 37*, 71–84.

Devroye, L. (1986c). The analysis of some algorithms for generating random variates with a given hazard rate. *Nav. Res. Logist. Q. 33*, 281–192.

Devroye, L. (1986d). An automatic method for generating random variates with a given characteristic function. *SIAM J. Appl. Math. 46*, 698–719.

Devroye, L. (1987). A simple generator for discrete log-concave distributions. *Computing 39*, 87–91.

Devroye, L. (1989). On random variate generation when only moments of fourier coefficients are known. *Math. Comput. Simulation 31*(1/2), 71–89.

Devroye, L. (1991). Algorithms for generating discrete random variables with a given generating function or a given moment sequence. *SIAM J. Sci. Stat. Comput. 12*(1), 107–126.

Devroye, L. (1997a). Random variate generation for multivariate unimodal densities. *ACM Trans. Model. Comput. Simul. 7*(4), 447–477.

Devroye, L. (1997b). Universal smoothing factor selection in density estimation, theory and practice. *Test 6*, 223–320.

Devroye, L. (2001). personal communication.

Devroye, L. and L. Györfi (1985). *Nonparametric Density Estimation: The L_1 View*. New-York: John Wiley.

Dharmadhikari, S. and L. Joagdev (1988). *Unimodality, Convexity and Applications*. New York: Academic Press.

Dieter, U. (1989). Mathematical aspects of various methods for sampling from classical distributions. In E. A. Mc Nair, K. J. Musselman, and P. Heidelberger (Eds.), *Proc. 1989 Winter Simulation Conf.*, pp. 477–483.

Dobson, A. J. (1983). *An Introduction to Statistical Modelling*. London: Chapman and Hall.

Dowd, K. (1998). *Beyond Value at Risk: the New Science of Risk Management*. New York: Wiley.

Duffie, D. (1996). *Dynamic Asset Pricing Theory* (2 ed.). Princeton, New Jersey: Princeton University Press.

Durbin, J. (1960). The fitting of time series models. *Rev. Inst. Int. Stat. 28*, 233–243.

Dyer, M. E., A. M. Frieze, and R. Kannan (1991). A random polynomial-time algorithm for approximating the volume of convex bodies. *Journal of the ACM 38*, 1–17.

Eberlein, E. and U. Keller (1995). Hyperbolic distributions in finance. *Bernoulli 1*, 281–299.

Engle, R. F. (1982). Autoregressive conditional heteroscedasticity with estimates of the variance of the united kingdom inflation. *Econometrica 50*, 987–1007.

Entacher, K. (2000). A collection of classical pseudorandom number generators with linear structures – advanced version. http://random.mat.sbg.ac.at/results/karl/server/.

Evans, M. and T. Swartz (1998). Random variable generation using concavity properties of transformed densities. *Journal of Computational and Graphical Statistics* 7(4), 514–528.

Evans, M. and T. Swartz (2000). *Approximating Integrals via Monte Carlo and Deterministic Methods*. Oxford: Oxford University Press.

Fang, K.-T. and Y. Wang (1994). *Number-theoretic Methods in Statistics*, Volume 51 of *Monographs on Statistics and Applied Probability*. London: Chapman and Hall.

Fang, K.-T., Z. Yang, and S. Kotz (2001). Generation of multivariate distributions by vertical density representation. *Statistics* 35(3), 281–293.

Feller, W. (1971). *An Introduction to Probability Theory and its Applications*, Volume 2. New York: John Wiley.

Fill, J. A. (1998). An interruptible algorithm for perfect sampling via markov chains. *The Annals of Applied Probability* 8, 131–162.

Fishman, G. S. (1996). *Monte Carlo. Concepts, algorithms, and applications*. Springer Series in Operations Research. New York: Springer.

Forsythe, G. E., M. A. Malcolm, and C. B. Moler (1977). *Computer methods for mathematical computations* (7 ed.). Prentice-Hall series in automatic computation. Englewood Cliffs, NJ: Prentice-Hall.

Gamerman, D. (1997). *Markov Chain Monte Carlo: Stochastic Simulation for Bayesian inference*. London: Chapman and Hall.

Gelfand, A. E., S. E. Hills, A. Racine-Poon, and A. F. M. Smith (1990). Illustration of bayesian inference in normal data models using gibbs sampling. *Journal of the American Statistical Association* 85, 972–985.

Gelfand, A. E. and A. F. M. Smith (1990). Sampling-based approaches to calculating marginal densities. *Journal of the American Statistical Association* 85, 398–409.

Gelman, A., J. B. Carlin, H. S. Stern, and D. B. Rubin (1995). *Bayesian data analysis*. Texts in Statistical Science. Chapman & Hall.

Geman, S. and D. Geman (1984). Stochastic relaxation, gibbs distributions and the bayesian restoration of images. *IEEE Trans. Pattn. Anal. Mach. Intel.* 6, 721–741.

Gentle, J. E. (1998). *Random Number Generation and Monte Carlo Methods*. Statistics and Computing. New York: Springer.

Ghosh, S. and S. G. Henderson (2002a). Chessboard distributions and random vectors with specified marginals and covariance matrix. *Operations Research* 50, 820–834.

Ghosh, S. and S. G. Henderson (2002b). Proerties of the norta method in higher dimensions. In E. Yücesan, C. H. Chen, J. L. Snowdon, and J. M. Charnes (Eds.), *Proceedings of the 2002 Winter Simulation Conference*, Piscataway, New Jersey. IEEE. to appear.

Gil-Alaña, L. A. and P. M. Robinson (1997). Testing of unit root and other nonstationary hypotheses in macroeconomic time series. *Journal of Econometrics* 80, 241–268.

Gilks, W. R. (1992). Derivative-free adaptive rejection sampling for Gibbs sampling. In *Bayesian Statistics 4*, pp. 641–649. Oxford, U.K.: Oxford University Press.

Gilks, W. R., S. Richardson, and D. J. Spiegelhalter (Eds.) (1996). *Markov chain Monte Carlo in practice*, London. Chapman & Hall.

Gilks, W. R., G. O. Roberts, and E. I. George (1994). Adaptive direction sampling. *The Statistician 43*, 179–189.

Gilks, W. R. and P. Wild (1992). Adaptive rejection sampling for Gibbs sampling. *Applied Statistics 41*(2), 337–348.

Glasserman, P., P. Heidelberger, and P. Shahabuddin (2000). Variance reduction techniques for value-at-risk with heavy-tailed risk factors. In J. A. Joines, R. R. Barton, K. Kang, and P. A. Fishwick (Eds.), *Proc. 2000 Winter Simulation Conference*, pp. 604–609.

Godwin, H. J. (1964). *Inequalities on Distribution Functions*. London: Charles Griffin.

Goldberg, D. (1992). What every computer scientist should know about floating-point arithmetic. *ACM Computing Surveys 23*(1), 5–48.

Golub, G. H. and C. Van Loan (1989). *Matrix Computations*. Baltimore and London: The Johns Hopkins University Press.

Goodman, J. E. and J. O'Rourke (Eds.) (1997). *Handbook of Discrete and Computational Geometry*. Boca Raton, Florida: CRC Press.

Green, P. J. and D. J. Murdoch (2000). Exact sampling for bayesian inference: towards general purpose algorithms (with discussion). In J. M. Bernardo, J. O. Berger, A. P. Dawid, and A. F. M. Smith (Eds.), *Monte Carlo Methods*, Volume 6 of *Bayesian Statistics*, Oxford, pp. 301–321. Oxford University Press.

Grigoriu, M. (1995). *Applied Non-Gaussian Processes*. Englewood Cliffs: Prentice Hall.

Grünbaum, B. (1967). *Convex Polytopes*. Interscience.

Hadwiger, H. (1968). Eine Schnittrekursion für die Eulersche Charakteristik euklidischer Polyeder. *Elem. Math. 23*(6), 121–132.

Hall, R. W. (1991). *Queueing Methods*. Prentice Hall, Englewood Cliffs, N.J.

Hastings, W. K. (1970). Monte Carlo sampling methods using Markov chains and their applications. *Biometrika 57*, 97–109.

Hauser, M. A. and W. Hörmann (1994). The generation of stationary gaussian time series. Technical report, Institut f. Statistik Wirtschaftsuniversität Wien.

Hellekalek, P. (2002). pLab – a server on the theory and practice of random number generation. http://random.mat.sbg.ac.at/.

Henderson, S. G., B. A. Chiera, and R. M. Cooke (2000). Generating "dependent" quasi-random numbers. In J. A. Joines, R. R. Barton, K. Kang, and P. A. Fishwick (Eds.), *Proc. 2000 Winter Simulation Conference*, pp. 527–536.

Henk, M., J. Richter-Gebert, and G. M. Ziegler (1997). Basic properties of convex polytopes. See Goodman and O'Rourke (1997), Chapter 13, pp. 243–270.

Hill, R. R. and C. H. Reilly (1994). Composition for multivariate random vectors. In J. D. Tew, S. Manivannan, D. A. Sadowsky, and A. F. Seila (Eds.), *Proceedings of the 1994 Winter Simulation Conference*, Piscataway, New Jersey, pp. 332–339. IEEE.

Hill, R. R. and C. H. Reilly (2000). The effects of coefficient correlation structure in two-dimensional knapsack problems on solution procedure performance. *Management Science 46*, 302–317.

Hlawka, E. and R. Mück (1972a). A transformation of equidistributed sequences. In *Appl. Number Theory numer. Analysis, Proc. Sympos. Univ. Montreal 1971*, pp. 371–388.

Hlawka, E. and R. Mück (1972b). Über eine Transformation von gleichverteilten Folgen II. *Computing 9*, 127–138.

Hörmann, W. (1993). The generation of binomial random variates. *J. Stat. Comput. Simulation* 46(1–2), 101–110.

Hörmann, W. (1994a). A note on the quality of random variates generated by the ratio of uniforms method. *ACM Trans. Model. Comput. Simul.* 4(1), 96–106.

Hörmann, W. (1994b). The quality of non-uniform random numbers. In H. Dyckhoff, U. Derings, M. Salomon, and H. C. Tijms (Eds.), *Operations Research Proceedings 1993*, Berlin, pp. 329–335. Springer Verlag.

Hörmann, W. (1995). A rejection technique for sampling from T-concave distributions. *ACM Trans. Math. Software* 21(2), 182–193.

Hörmann, W. (2000). Algorithm 802: An automatic generator for bivariate log-concave distributions. *ACM Trans. Math. Software* 26(1), 201–219.

Hörmann, W. (2002). A note on the performance of the "Ahrens algorithm". *Computing* 69, 83–89.

Hörmann, W. and O. Bayar (2000). Modeling probability distributions from data and its influence on simulation. In I. Troch and F. Breitenecker (Eds.), *Proceedings IMACS Symposium on Mathematical Modeling*, Argesim Report No. 15, pp. 429 – 435.

Hörmann, W. and G. Derflinger (1990). The acr method for generating normal random variables. *OR Spektrum* 12(3), 181–185.

Hörmann, W. and G. Derflinger (1993). A portable random number generator well suited for the rejection method. *ACM Trans. Math. Software* 19(4), 489–495.

Hörmann, W. and G. Derflinger (1994a). The transformed rejection method for generating random variables, an alternative to the ratio of uniforms method. *Commun. Stat., Simulation Comput.* 23(3), 847–860.

Hörmann, W. and G. Derflinger (1994b). Universal generators for correlation induction. In R. Dutter and W. Grossmann (Eds.), *Compstat, Proceedings in Computational Statistics*, Heidelberg, pp. 52–57. Physica-Verlag.

Hörmann, W. and G. Derflinger (1996). Rejection-inversion to generate variates from monotone discrete distributions. *ACM Trans. Model. Comput. Simul.* 6(3), 169–184.

Hörmann, W. and G. Derflinger (1997). An automatic generator for a large class of unimodal discrete distributions. In A. R. Kaylan and A. Lehmann (Eds.), *ESM 97*, pp. 139–144.

Hörmann, W. and G. Derflinger (2002). Fast generation of order statistics. *ACM Trans. Model. Comput. Simul.* 12(2), 83–93.

Hörmann, W. and J. Leydold (2000). Automatic random variate generation for simulation input. In J. A. Joines, R. Barton, P. Fishwick, and K. Kang (Eds.), *Proceedings of the 2000 Winter Simulation Conference*, pp. 675–682.

Hörmann, W. and J. Leydold (2003). Continuous random variate generation by fast numerical inversion. *ACM Trans. Model. Comput. Simul.* 13(4). to appear.

Hosking, J. R. M. (1984). Modeling persistence in hydrological time series using fractional differencing. *Water Resources Research* 20, 1898–1908.

IMSL (1994–2003). *IMSL Fortran Library User's Guide, STAT/LIBRARY Volume 2.* Visual Numerics, Inc. Version 5.0, http://www.vni.com/books/docs/imsl/StatV2.pdf.

Johnson, M. E. (1987). *Multivariate Statistical Simulation*. Wiley Series in Probability and Mathematical Statistics. Applied Probability and Statistics. New York: John Wiley & Sons.

Johnson, M. E. and A. Tenenbein (1981). A bivariate distribution family with specified marginals. *J. Amer. Statist. Assoc. 76*, 198–201.

Johnson, N. L., S. Kotz, and N. Balakrishnan (1994). *Continuous Univariate Distributions* (2nd ed.), Volume 1. New York: John Wiley & Sons, Inc.

Johnson, N. L., S. Kotz, and N. Balakrishnan (1995). *Continuous Univariate Distributions* (2nd ed.), Volume 2. New York: John Wiley & Sons, Inc.

Jones, M. C. and A. D. Lunn (1996). Transformations and random variate generation: Generalised ratio-of-uniforms methods. *J. Stat. Comput. Simulation 55*(1–2), 49–55.

Jorion, P. (2001). *Value at Risk* (2nd ed.). McGraw-Hill.

Joro, T. and P. Na (2002). Credit risk modeling for catastrophic events. In E. Yücesan, C. H. Chen, J. L. Snowdon, and J. M. Charnes (Eds.), *Proceedings of the 2002 Winter Simulation Conference*, Piscataway, New Jersey. IEEE. 1511–1514.

Kannan, R., L. Lovasz, and M. Simonovits (1997). Random walks and an $o^*(n^5)$ volume algorithm. *Random Structures and Algorithms 11*, 1–50.

Kemp, C. D. and S. Loukas (1978). The computer generation of bivariate discrete random variables. *J. R. Stat. Soc., Ser. A 141*, 513–519.

Kinderman, A. J. and J. F. Monahan (1977). Computer generation of random variables using the ratio of uniform deviates. *ACM Trans. Math. Software 3*(3), 257–260.

Kinderman, A. J. and J. G. Ramage (1976). Computer generation of normal random variables. *J. Am. Stat. Assoc. 71*(356), 893–898.

Knuth, D. E. (1998). *The Art of Computer Programming. Vol. 2: Seminumerical Algorithms* (3rd ed.). Addison-Wesley.

Kotz, S., K.-T. Fang, and J.-J. Liang (1997). On multivariate vertical density representation and its application to random number generation. *Statistics 30*(2), 163–180.

Kronmal, R. A. and A. V. Peterson (1979). On the alias method for generating random variables from a discrete distribution. *Amer. Statist. 33*(4), 214–218.

Kruskal, W. (1958). Ordinal measures of association. *Journal of the American Statistical Association 53*, 814–861.

Kurowicka, D. and R. M. Cooke (2001). Conditional, partial and rank correlation for the elliptical copula: Dependence modelling in uncertainty analysis. In E. Zio, M. Demichela, and N. Piccinini (Eds.), *European Safety and Reliability Conference ESREL 2001, Torino, Italy, 16–20 Sptember 2001*, Volume 3 of *Safety and Reliability*, pp. 1795–1802.

Kurowicka, D., J. Misiewicz, and R. M. Cooke (2001). Elliptical copulae. In G. I. Schuëller and P. D. Spanos (Eds.), *Monte Carlo Simulation*, Rotterdam, pp. 209–214. Balkema.

Law, A. M. and W. D. Kelton (2000). *Simulation Modeling and Analysis* (3 ed.). McGraw-Hill.

Lawrence, J. (1991). Polytope volume computation. *Math. Comput. 57*(195), 259–271.

L'Ecuyer, P. (1998). Random number generation. In J. Banks (Ed.), *Handbook of Simulation*, Chapter 4, pp. 93–137. Wiley.

L'Ecuyer, P. (1999). Good parameters and implementations for combined multiple recursive random number generators. *Operations Research 47*(1), 159–164.

L'Ecuyer, P. and R. Simard (2002). TestU01: A software library in ANSI C for empirical testing of random number generation. http://www.iro.umontreal.ca/~simardr/.

Lee, C. W. (1997). Subdivisions and triangulations of polytopes. See Goodman and O'Rourke (1997), Chapter 14, pp. 271–290.

Leobacher, G. and F. Pillichshammer (2002). A method for approximate inversion of the hyperbolic cdf. *Computing* 69(4), 291–303.

Lewis, P. A. W. and G. S. Shedler (1979). Simulation of nonhomogeneous Poisson processes by thinning. *Naval Res. Logist. Quart.* 26, 403–413.

Leydold, J. (1998). A rejection technique for sampling from log-concave multivariate distributions. *ACM Trans. Model. Comput. Simul.* 8(3), 254–280.

Leydold, J. (2000a). Automatic sampling with the ratio-of-uniforms method. *ACM Trans. Math. Software* 26(1), 78–98.

Leydold, J. (2000b). A note on transformed density rejection. *Computing* 65(2), 187–192.

Leydold, J. (2001). A simple universal generator for continuous and discrete univariate T-concave distributions. *ACM Trans. Math. Software* 27(1), 66–82.

Leydold, J. (2003). Short universal generators via generalized ratio-of-uniforms method. *Mathematics of Computation* 72(243), 1453–1471.

Leydold, J., G. Derflinger, G. Tirler, and W. Hörmann (2003). An automatic code generator for nonuniform random variate generation. *Mathematics and Computers in Simulation* 62(3–6), 405–412.

Leydold, J. and W. Hörmann (1998). A sweep-plane algorithm for generating random tuples in simple polytopes. *Mathematics of Computation* 67(224), 1617–1635.

Leydold, J., W. Hörmann, E. Janka, and G. Tirler (2002). *UNU.RAN – A Library for Non-Uniform Universal Random Variate Generation*. A-1090 Wien, Austria: Institut für Statistik, WU Wien. available at http://statistik.wu-wien.ac.at/unuran/.

Leydold, J., E. Janka, and W. Hörmann (2002). Variants of transformed density rejection and correlation induction. In K.-T. Fang, F. J. Hickernell, and H. Niederreiter (Eds.), *Monte Carlo and Quasi-Monte Carlo Methods 2000*, Heidelberg, pp. 345–356. Springer-Verlag.

Leydold, J., H. Leeb, and W. Hörmann (2000). Higher dimensional properties of non-uniform pseudo-random variates. In H. Niederreiter and J. Spanier (Eds.), *Monte Carlo and Quasi-Monte Carlo Methods 1998*, Berlin, Heidelberg, pp. 341–355. Springer-Verlag.

Li, S. T. and J. L. Hammond (1975). Generation of pseudo-random numbers with specified univariate distributions and correlation coefficients. *IEEE Transactions on Systems, Man, and Cybernetics* 5, 557–560.

Liu, J. S. (1996). Metropolized independent sampling with comparisons to rejection sampling and importance sampling. *Stat. and Comput* 6, 113–119.

Loukas, S. (1984). Simple methods for computer generation of bivariate beta random variables. *J. Statist. Comput. Simulation* 20, 145–152.

Loukas, S. and C. D. Kemp (1983). On computer sampling from trivariate and multivariate discrete distributions. *J. Stat. Comput. Simulation* 17, 113–123.

Loukas, S. and C. D. Kemp (1986). The computer generation of bivariate binomial and negative binomial random variables. *Commun. Stat., Simulation Comput.* 15, 15–25.

Lovasz, L. (1999). Hit-and-run mixes fast. *Mathematical Programming, Series A 86*, 443–461.

Lurie, P. M. and M. S. Goldberg (1998). An approximate method for sampling correlated random variables from partially-specified distributions. *Management Science 44*, 203–218.

Marchenko, A. S. and V. A. Ogorodnikov (1984a). The modelling of very long stationary gaussian sequences with an arbitrary correlation function. *USSR Comput. Maths. Math. Phys. 24*, 141–144.

Marchenko, A. S. and V. A. Ogorodnikov (1984b). The modelling of very long stationary Gaussian sequences with an arbitrary correlation function. *Zh. Vychisl. Mat. i Mat. Fiz. 24*, 1514–1519. in Russian.

Mardia, K. V. (1970). A translation family of bivariate distributions and frechet's bounds. *Sankhya, Series A 32*, 119–122.

Marinucci, D. and P. M. Robinson (2000). Weak convergence of multivariate fractional processes. *Stochastic Processes and their Applications 86*, 103–120.

Marsaglia, G. (1963). Generating discrete random variables in a computer. *Commun. ACM 6*, 37–38.

Marsaglia, G. (1984). The exact–approximation method for generating random variables in a computer. *Journal of the American Statistical Association 79*, 218–221.

Marsaglia, G. (1985). A current view of random number generators. In L. Billard (Ed.), *Computer Science and Statistics: The Interface*, pp. 3–10. Amsterdam: Elsevier Science Publishers B.V.

Marsaglia, G. (1996). DIEHARD: A battery of tests of randomness. http://stat.fsu.edu/~geo/diehard.html.

Marsaglia, G., M. D. MacLaren, and T. A. Bray (1964). A fast procedure for generating normal random variables. *Commun. ACM 7*, 4–10.

Marsaglia, G. and W. W. Tsang (1984). A fast, easily implemented method for sampling from decreasing or symmetric unimodal density functions. *SIAM J. Sci. Statist. Comput. 5*, 349–359.

Matsumoto, M. and T. Nishimura (1998). Mersenne twister: a 623-dimensionally equidistributed uniform pseudo-random number generator. *ACM Trans. Model. Comput. Simul. 8*(1), 3–30.

Mayhew, S. (2002). Security price dynamics and simulation in financial engineering. In E. Yücesan, C. H. Chen, J. L. Snowdon, and J. M. Charnes (Eds.), *Proceedings of the 2002 Winter Simulation Conference*, Piscataway, New Jersey. IEEE. 1568–1574.

McMullen, P. (1970). The maximum number of faces of a convex polytope. *Mathematika 17*, 179–184.

Metropolis, N., A. W. Rosenbluth, A. H. Teller, and E. Teller (1953). Equations of state calculations by fast computing machines. *Journal of Chemical Physics 21*, 1087–1091.

Meuwissen, A. H. M. and T. J. Bedford (1997). Minimal informative distributions with given rank correlation for use in uncertainty analysis. *Journal of Statistical Computation and Simulation 57*, 143–175.

Mira, A., J. Moller, and G. O. Roberts (2001). Perfect slice samplers. *Journal of the Royal Statistical Society B 63*, 593 – 606.

Monahan, J. F. (1985). Accuracy in random number generation. *Math. Comp. 45*(172), 559–568.

Murdoch, D. J. (2000). Exact sampling for bayesian inference: unbounded state spaces. In N. Madras (Ed.), *Monte Carlo Methods*, Volume 26 of *Fields Institute Communications*, pp. 111–121. American Mathematical Society.

Murdoch, D. J. and X.-L. Meng (2001). Towards perfect sampling for bayesian mixture priors. In *ISBA 2000, Proceedings*, pp. ?? ISBA and Eurostat.

Neal, R. M. (2003). Slice sampling. *Annals of Statistics 31*, 705–767.

Neave, H. R. (1973). On using the Box-Müller transformation with multiplicative congruential pseudo-random number generators. *Appl. Statist. 22*(1), 92–97.

Nelson, B. L. and M. Yamnitsky (1998). Input modeling tools for complex problems. In D. J. Medeiros, E. F. Watson, J. S. Carson, and M. S. Manivannan (Eds.), *Proceedings of the 1998 Winter Simulation Conference*, pp. 105–112. Piscataway, New Jersey: Institute of Electrical and Electronics Engineers.

Neumann, J. v. (1951). Various techniques used in connection with random digits. In A. S. Householder et al. (Eds.), *The Monte Carlo Method*, Number 12 in Nat. Bur. Standards Appl. Math. Ser., pp. 36–38.

Niederreiter, H. (1992). *Random number generation and quasi-Monte Carlo methods*, Volume 63 of *SIAM CBMS-NSF Regional Conference Series in Applied Mathematics*. Philadelphia: SIAM.

NIST. Random number generation and testing. http://csrc.nist.gov/rng/.

Overton, M. L. (2001). *Numerical Computing with IEEE Floating Point Arithmetic*. Philadelphia: SIAM.

Özgül, E. (2002). The generation of random variates with a given hazard rate. M.sc. thesis, Department of Industrial Engineering, Bogazici University, Istanbul.

Pang, W. K., Z. H. Yang, S. H. Hou, and M. D. Troutt (2001). Further results on multivariate vertical density representation and an application to random vector generation. *Statistics 35*(4), 463–477.

Panny, W. and R. Hatzinger (1993). Single and twin-heaps as natural data structures for percentile point simulation algorithms. *Statistics and Computing 3*, 163–170.

Park, S. K. and K. W. Miller (1988). Random number generators: good ones are hard to find. *Commun. ACM 31*, 1192–1201.

Peterson, A. V. and R. A. Kronmal (1982). On mixture methods for the computer generation of random variables. *Amer. Statist. 36*(3), 184–191.

Philippe, A. and C. P. Robert (2003). Perfect simulation of positive gaussian distributions. *Statistics and Computing 13*, 179–186.

Poirion, F. (2001). Simulation of random vectors with given moments. In G. I. Schuëller and P. D. Spanos (Eds.), *Monte Carlo Simulation*, Rotterdam, pp. 145–150. Balkema.

Prause, K. (1999). *The Generalized Hyperbolic Model: Estimation, Financial Derivatives, and Risk Measures*. Ph.D. dissertation, University of Freiburg 1999. available at http://www.freidok.uni-freiburg.de/volltexte/15.

Prekopa, A. (1973). On logarithmic concave measures and functions. *Acta Scientiarium Mathematicarum Hungarica 34*, 335–343.

Press, W. H., S. A. Teukolsky, W. T. Vetterling, and B. P. Flannery (1992). *Numerical Recipes in C* (2 ed.). Cambridge University Press.

Propp, J. G. and D. B. Wilson (1996). Exact sampling with coupled markov chains and applications to statistical mechanics. *Random Structures and Algorithms 9*, 223–252.

Ripley, B. D. (1987). *Stochastic Simulation*. New York: John Wiley & Sons.

Roberts, G. O., A. Gelman, and W. R. Gilks (1997). Weak convergence and optimal scaling of random walk Metropolis algorithms. *Annals of Applied Probability 7*(1), 110–120.

Roberts, G. O. and A. F. M. Smith (1994). Simple conditions for the convergence of the gibbs sampler and hasting-metropolis algorithms. *Stoch. Proc. Appl. 49*, 207–216.

Rubin, P. A. (1984). Generating random points in a polytope. *Communications in Statistics – Simulation and Computation 13*, 375–396.

Rubinstein, R. Y. (1982). Generating random vectors uniformly distributed inside and on the surface of different regions. *Eur. J. Oper. Res. 10*, 205–209.

Schmeiser, B. W. and V. Kachitvichyanukul (1990). Non-inverse correlation induction: guidelines for algorithm development. *J. Comput. Appl. Math. 31*, 173–180.

Schmeiser, B. W. and R. Lal (1980). Squeeze methods for generating gamma variates. *J. Amer. Statist. Assoc. 75*, 679–682.

Schmetterer, L. (1994). Private Communication at a Conference in Linz, Austria.

Shanker, A. and W. D. Kelton (1991). Empirical input distributions: an alternative to standard input distributions in simulation modeling. In B. Nelson, M. Clark, and W. Kelton (Eds.), *Proceedings of the 1991 Winter Simulation Conference*, Piscataway, New Jersey, pp. 975–985. IEEE.

Shanthikumar, J. G. (1985). Discrete random variate generation using uniformization. *Eur. J. Oper. Res. 21*, 387–398.

Shao, J. and D. Tu (1995). *The Jacknife and Bootstrap*. New York: Springer-Verlag.

Shepp, L. (1962). Symmetric random walk. *Transactions of the American Mathematical Society 104*, 144–153.

Silverman, B. W. (1986). *Density Estimation for Statistics and Data Analysis*. London: Chapman and Hall.

Sivazlian, B. D. (1981). On a multivariate extension of the gamma and beta distributions. *SIAM Journal on Applied Mathematics 41*, 205–209.

Smith, R. L. (1984). Efficient monte carlo procedures for generating points uniformly distributed over bounded regions. *Operations Research 32*, 1296–1308.

Spiegelhalter, D. J., A. Thomas, N. G. Best, and W. R. Gilks (1996). *BUGS Examples Volume 2, Version 0.5*. MRC Biostatistics Unit. available at http://www.mrc-bsu.cam.ac.uk/bugs/.

Spiegelhalter, D. J., A. Thomas, N. G. Best, and D. Lunn (2003). *WinBUGS Version 1.4 User Manual*. MRC Biostatistics Unit. available at http://www.mrc-bsu.cam.ac.uk/bugs/.

Stadlober, E. (1989a). Ratio of uniforms as a convenient method for sampling from classical discrete distributions. In E. A. MacNair, K. J. Musselman, and P. Heidelberger (Eds.), *Proc. 1989 Winter Simulation Conf.*, pp. 484–489.

Stadlober, E. (1989b). *Sampling from Poisson, binomial and hypergeometric distributions: Ratio of uniforms as a simple and fast alternative*. Number 303 in Bericht der Mathematisch-Statistischen Sektion. Forschungsgesellschaft Joanneum-Graz.

Stadlober, E. and H. Zechner (1999). The patchwork rejection technique for sampling from unimodal distributions. *ACM Trans. Model. Comput. Simul. 9*(1), 59–80.

Staum, J. (2001). Simulation in financial engineering. In B. Peters, J. Smith, D. Medeiros, and M. Rohrer (Eds.), *Proceedings of the 2001 Winter Simulation Conference*, Piscataway, New Jersey, pp. 123–133. IEEE.

Ştefănescu, S. (1985). Algorithms for computer generation of uniformly distributed points in a simplex in R^n. *Stud. Cercet. Mat. 37*, 83–93. in Romanian. English summary.

Stefănescu, S. and I. Văduva (1987). On computer generation of random vectors by transformations of uniformly distributed vectors. *Computing 39*, 141–153.

Stefănescu, S. V. (1990). Generating uniformly distributed points in a hypersphere from $\mathbb{R}^{2p}$. *Stud. Cercet. Mat. 42*(5/6), 465–469.

Stefănescu, S. V. (1994). On some reduction-extension procedures for generating uniform random points in a hypersphere. *Syst. Anal. Modelling Simulation 16*(2), 153–162.

Swain, J. J., S. Venkatraman, and J. R. Wilson (1988). Least squares estimation of distribution functions in johnson's translation system. *Journal of Statistical Computation and Simulation 29*, 271–297.

Swendsen, R. and J. Wang (1987). Non-universal critical dynamics in monte carlo simulation. *Physical Review Letters 58*, 86–88.

Taylor, M. S. and J. R. Thompson (1986). A data based algorithm for the generation of random vectors. *Comput. Statist. Data Anal. 4*, 93–101.

Tezuka, S. (1995). *Uniform Random Numbers: Theory and Practice*. Kluwer Academic Publishers.

Tierney, L. (1994). Markov chains for exploring posterior distributions. *Annals of Statistics 22*, 1701–1728.

Tirler, G., P. Dalgaard, W. Hörmann, and J. Leydold (2003). An note on the Kinderman-Ramage Gaussian variate generator. Technical report, Insitute of Statistics, Vienna University of Economics and Business Administration.

Todd, M. J. (1976). *The Computation of Fixed Points and Applications*, Volume 124 of *Lecture Notes in Economics and Mathematical Systems*. Berlin: Springer.

Troutt, M. D. (1991). A theorem on the density of the density ordinate and an alternative interpretation of the box-muller method. *Math. Operationsforsch. Statist. Ser. Statist. 22*(2), 463–466.

Troutt, M. D. (1993). Vertical density representation and a further remark on the box-muller method. *Statistics 24*(1), 81–83.

Wagner, M. A. F. and J. R. Wilson (1996). Using univariate bezier distributions to model simulation input processes. *IIE Transactions 28*, 699–711.

Wakefield, J. C., A. E. Gelfand, and A. F. M. Smith (1991). Efficient generation of random variates via the ratio-of-uniforms method. *Statist. Comput. 1*(2), 129–133.

Walker, A. J. (1974). New fast method for generating discrete random numbers with arbitrary frequency distributions. *Electron. Lett. 10*, 127–128.

Walker, A. J. (1977). An efficient method for generating discrete random variables with general distributions. *ACM Trans. Math. Software 3*, 253–256.

Wallace, C. S. (1976). Transformed rejection generators for gamma and normal pseudo-random variables. *Australian Computer Journal 8*, 103–105.

Wand, M. P. and M. C. Jones (1995). *Kernel Smoothing*. London: Chapman and Hall.

Wilson, D. B. (1998). Annotated bibliography of perfectly random sampling with markov chains. In D. Aldous and J. Propp (Eds.), *Microsurveys in Discrete Probability*, Volume 41 of *Series in Discrete Mathematics and Theoretical Computer Science*, pp. 209–220. American Mathematical Society. Updated versions to appear at http://www.dbwilson.com/exact/.

Wilson, D. B. (2000). Layered multishift coupling for use in perfect sampling algorithms (with a primer on cftp). In N. Madras (Ed.), *Monte Carlo Methods*, Volume 26 of *Fields Institute Communications*, pp. 141–176. American Mathematical Society.

Zaman, A. (1996). Generation of random numbers from an arbitrary unimodal density by cutting corners. unpublished manuscript.

Ziegler, G. M. (1995). *Lectures on Polytopes*, Volume 152 of *Graduate Texts in Mathematics*. New York: Springer-Verlag.

Author index

Afflerbach, L. 175
Agathokleous, A. 266, 406, 408
Ahrens, J. H. 5, 6, 37, 41, 65, 113, 115, 118, 119, 121, 160, 161
Amit, Y. 370
Asau, Y. 5, 45, 215
Avramidis, A. N. 306

Bal, H. 319, 398
Balakrishnan, N. 69
Banks, J. 311, 313
Barabesi, L. 103
Barber, C. B. 258
Barndorff-Nielsen, O. 390
Barton, R. R. 307
Bayar, O. 307
Bedford, T. J. 335, 340–342
Belisle, C. J. P. 371
Besag, J. 378
Best, N. G. 401, 405–407
Bieri, H. 260
Blæsild, P. 390
Bliss, C. I. 405
Boender, C. G. E. 372
Bollerslev, T. 346, 399
Box, G. E. P. 5, 38, 61, 174, 176
Bratley, P. 15, 156, 185, 306, 311, 312
Bray, T. A. 5, 113
Brockwell, P. J. 346, 347
Burford, R. L. 176, 177

Cario, M. C. 330, 336, 356, 358, 360
Carlin, J. B. 400
Caron, R. J. 372
Carson, J. S. 311, 313
Casella, G. 379
Chen, H. C. 5, 45, 215
Chen, M.-H. 363, 371, 372, 385, 400
Chen, W. 176, 177
Cheng, R. C. H. 5, 183
Chhikara, R. S. 174
Chiera, B. A. 180
Clemen, R. T. 328, 358
Congdon, P. 400
Cooke, R. M. 180, 330, 336, 340–342, 358, 361
Cox, D. R. 61
Cryer, J. 346

Dagpunar, J. 5, 13, 105, 216
Dalgaard, P. 174
Davies, N. 356, 357
Davies, R. B. 346, 347, 350
Davis, R. A. 346, 347
Deák, I. 253, 254
DeBrota, D. 322
Deler, B. 358
Dellaportas, P. 266, 406, 408
Deng, L.-Y. 174
Derflinger, G. 5, 7, 37, 40, 51, 68, 70, 71, 73, 78, 79, 169, 175, 176, 185, 186, 221, 224, 225, 227, 351, 354
Devroye, L. 3, 5, 6, 11, 13, 18, 23, 36, 44, 51, 63, 65, 75, 77, 103, 113, 116, 119, 126, 133–135, 138, 148, 152, 156, 174, 175, 182, 188, 193, 194, 196–198, 201, 203, 205–208, 227–229, 232–235, 245, 253, 255,

256, 259, 265, 280–282, 284, 297, 306, 308, 323, 328, 332, 340
Dharmadhikari, S.   280
Dieter, U.   5, 37, 41, 104, 113
Dittus, R. S.   322
Dobson, A. J.   405
Dowd, K.   387, 394
Duffie, D.   387, 388
Durbin, J.   348
Dyer, M. E.   383

Eberlein, E.   389, 390
Engle, R. F.   346, 399
Entacher, K.   4
Evans, M.   89, 97, 100, 266, 363

Fang, K.-T.   180, 249
Feast, G. M.   5
Feller, W.   281
Fill, J. A.   372
Fishman, G. S.   13
Flannery, B. P.   350, 351
Forsythe, G. E.   277
Fox, B. L.   15, 156, 185, 306, 311, 312
Frieze, A. M.   383

Gamerman, D.   363
Gelfand, A. E.   35, 142, 249, 369
Gelman, A.   366, 400
Geman, D.   369
Geman, S.   369
Gentle, J. E.   13
George, E. I.   371
Ghosh, S.   330, 333, 340–342
Gil-Alaña, L. A.   347
Gilks, W. R.   6, 55, 61, 81, 82, 89, 366, 369, 371, 405, 406
Glasserman, P.   390, 397
Godwin, H. J.   327
Goldberg, D.   182
Goldberg, M. S.   328, 341
Golub, G. H.   347, 348, 354
Green, P. J.   372, 378, 379
Grigoriu, M.   356, 358
Grünbaum, B.   258, 261
Györfi, L.   306

Hadwiger, H.   260
Hall, R. W.   316

Hammond, J. L.   330, 340, 360
Harte, D. S.   346, 347, 350
Hastings, W. K.   364
Hatzinger, R.   395
Hauser, M. A.   251, 347
Heidelberger, P.   390, 397
Hellekalek, P.   4, 174
Henderson, S. G.   180, 330, 333, 340–342
Henk, M.   258
Hill, R. R.   328, 330, 340
Hills, S. E.   369
Hlawka, E.   180, 181
Hörmann, W.   4–7, 37, 40, 51, 55, 56, 66, 68, 70, 71, 73, 75, 78, 79, 84, 93, 120, 123, 161, 163, 166, 167, 169, 174–177, 180, 185, 186, 221, 224, 225, 227, 245, 251, 260, 263, 265, 268, 269, 275, 297, 307, 347, 351, 354
Hosking, J. R. M.   347, 348, 353
Hou, S. H.   249
Huhdanpaa, H.   258

Ibrahim, J. G.   363, 400
IMSL   174

Janka, E.   4, 84, 166, 167, 185, 186, 275
Joagdev, L.   280
Johnson, M. E.   245, 336, 358
Johnson, N. L.   69
Jones, M. C.   37, 308
Jorion, P.   394
Joro, T.   399

Kachitvichyanukul, V.   185, 186
Kannan, R.   383
Keller, U.   389, 390
Kelton, W. D.   306, 311, 313, 316, 391
Kemp, C. D.   296–298
Kinderman, A. J.   5, 27, 33, 103, 106, 174, 176, 177
Knuth, D. E.   174, 178
Kohrt, K. D.   160, 161
Kotz, S.   69, 249
Kronmal, R. A.   47, 49, 215
Kruskal, W.   333
Kurowicka, D.   330, 336, 340–342, 358, 361

Lal, R.   5
Law, A. M.   306, 311, 313, 391
Lawrence, J.   260
L'Ecuyer, P.   4, 8, 174
Lee, C. W.   258
Leeb, H.   177, 180
Leobacher, G.   181
Lewis, P. A. W.   194
Leydold, J.   4, 6, 7, 80, 84, 91, 104, 105, 107, 136, 137, 141, 143, 146, 147, 149, 150, 161, 163, 166, 167, 169, 174, 177, 180, 185, 186, 229, 230, 245, 260, 263, 265, 268, 269, 275, 276, 279, 307
Li, S. T.   330, 340, 360
Liang, J.-J.   249
Liu, J. S.   368
Loukas, S.   296–298
Lovasz, L.   363, 383, 384
Lunn, A. D.   37
Lunn, D.   401, 407
Lurie, P. M.   328, 341

MacLaren, M. D.   5, 113
Malcolm, M. A.   277
Marchenko, A. S.   347, 348, 354, 355
Mardia, K. V.   330
Marinucci, D.   347
Marsaglia, G.   5, 36, 113, 123, 174, 215
Matsumoto, M.   4, 178
Mayhew, S.   397, 399
McDonald, J. F.   372
McMullen, P.   258
Meng, X.-L.   372, 377, 378
Mengersen, K. L.   379
Metropolis, N.   364
Meuwissen, A. H. M.   335
Miller, K. W.   177
Mira, A.   372, 379
Misiewicz, J.   336, 342
Moler, C. B.   277
Moller, J.   372, 379
Monahan, J. F.   33, 103, 106, 174, 176, 184
Mück, R.   180, 181
Muller, M. E.   5, 38, 174, 176
Murdoch, D. J.   372, 377–379

Na, P.   399

Neal, R. M.   378, 406
Neave, H. R.   174
Nef, W.   260
Nelson, B. L.   311, 313, 322, 330, 336, 356, 358, 360
Neumann, J. v.   4, 16
Niederreiter, H.   174, 180, 181
Nishimura, T.   4, 178

Ogorodnikov, V. A.   347, 348, 354, 355
Overton, M. L.   182
Özgül, E.   198, 199

Pang, W. K.   249
Panny, W.   395
Park, S. K.   177
Peterson, A. V.   47, 49, 215
Philippe, A.   379
Pillichshammer, F.   181
Poirion, F.   327
Prause, K.   390
Prekopa, A.   68, 263, 268, 300
Press, W. H.   350, 351
Propp, J. G.   363, 372

Racine-Poon, A.   369
Ramage, J. G.   5, 27, 174, 176, 177
Reilly, C. H.   328, 330, 340
Reilly, T.   328, 358
Richter-Gebert, J.   258
Rinooy Kan, A. H. G.   372
Ripley, B. D.   251
Robert, C. P.   379
Roberts, G. O.   366, 370–372, 379
Roberts, S. D.   322
Robinson, P. M.   347
Romeijn, H. E.   371, 372
Rosenbluth, A. W.   364
Rubin, D. B.   400
Rubin, P. A.   258
Rubinstein, R. Y.   253–255

Schmeiser, B. W.   5, 185, 186, 371, 372, 385
Schmetterer, L.   66
Schrage, E. L.   15, 156, 185, 306, 311, 312
Schruben, L. W.   307
Shahabuddin, P.   390, 397
Shanker, A.   316

Shanthikumar, J. G.   235, 236, 241
Shao, J.   307
Shao, Q.-M.   363, 400
Shedler, G. S.   194
Shepp, L.   281
Silverman, B. W.   306, 308, 309, 319
Simard, R.   174
Simonovits, M.   383
Sivazlian, B. D.   284
Smith, A. F. M.   35, 142, 249, 369, 370
Smith, R. L.   363, 371, 372, 383, 384
Spedding, T.   356, 357
Spiegelhalter, D. J.   401, 405–407
Stadlober, E.   27, 34, 104, 229
Staum, J.   387, 391
Ştefănescu, S. V.   142, 249, 255
Stern, H. S.   400
Swain, J. J.   306
Swartz, T.   89, 97, 100, 266, 363
Swendsen, R.   378

Taylor, M. S.   319, 321
Telgen, J.   372
Teller, A. H.   364
Teller, E.   364
Tenenbein, A.   336, 358
Teukolsky, S. A.   350, 351
Tezuka, S.   175
Thomas, A.   401, 405–407
Thompson, J. R.   319, 321
Tierney, L.   365, 366
Tirler, G.   4, 7, 166, 167, 169, 174, 275

Titterington, D. M.   379
Todd, M. J.   278, 279
Troutt, M. D.   249
Tsang, W. W.   113, 123
Tu, D.   307

Văduva, I.   142, 249
Van Loan, C.   347, 348, 354
Venkatraman, S.   306
Vetterling, W. T.   350, 351
Vrost, A. C. F.   372

Wagner, M. A. F.   306, 322
Wakefield, J. C.   35, 142, 249
Walker, A. J.   5, 47, 215
Wallace, C. S.   36, 37
Wand, M. P.   308
Wang, J.   378
Wang, Y.   180
Watson, W.   356, 357
Wenzel, K.   175
Wild, P.   6, 55, 61, 81, 82, 89, 369, 406
Wilson, D. B.   363, 372, 377, 379
Wilson, J. R.   306, 322

Yamnitsky, M.   322
Yang, Z.   249
Yang, Z. H.   249

Zaman, A.   119
Zechner, H.   27
Ziegler, G. M.   258

Selected Notation

x	variable
$\mathbb{R}$	real numbers
$\mathbb{R}^+$	positive real numbers $(x > 0)$
$\mathbb{N}_0$	integers $0, 1, 2, \ldots$
$[a, b]$	closed interval, $x\colon a \leq x \leq b$
(a, b)	open interval, $\{x\colon a < x < b\}$
$\mathbf{1}_{(a,b)}$	indicator function of the interval (a, b)
$\mathbb{R}^d$	real space of dimension d
$\mathbf{x}$	vector
$\mathbf{x} = (x_1, \ldots, x_d)$	vector $\mathbf{x} \in \mathbb{R}^d$ with components $x_1, \ldots, x_d$
$\|\mathbf{x}\|$	Euclidean norm of vektor $\mathbf{x}$, i.e. $\sqrt{x_1^2 + \ldots + x_d^2}$
$\langle \mathbf{x}, \mathbf{y} \rangle$	scalar product of vectors $\mathbf{x}$ and $\mathbf{y}$, i.e. $\sqrt{x_1 y_1 + \ldots + x_d y_d}$
$\mathbf{A}$	matrix
$O(\cdot)$	Landau symbol; in abuse of language $f(x) = O(g(x))$ means that there exists a constant $k > 0$ such that $f(x) \leq k\, g(x)$ for sufficiently large x
$\Gamma(\cdot)$	gamma function
$\text{Prob}(A)$	probability of event A
X	random variate
$\text{E}(X)$	expectation of random variate X
$\text{Var}(X)$	variance of random variate X
σ_X	standard deviation of random variate X, i.e. $\sqrt{\text{Var} X}$
μ_r	r-th moment of a random variate (p. 129)
$\text{Cov}(X, Y)$	covariance between X and Y
$\rho(X, Y)$	correlation between X and Y (p. 329)
$r(X, Y)$	rank correlation between X and Y (p. 329)
ρ_k	autocorrelation of a time series (p. 345)
$U(0, 1)$	uniform $(0, 1)$ distribution
$U(a, b)$	uniform distribution on interval (a, b)
$X \sim U(0, 1)$	random variate with law $U(0, 1)$

Selected Notation

$N(\mu, \sigma^2)$	normal distribution with mean μ and variance σ^2
$\Phi_N(\cdot)$	cdf of the standard normal distribution
$\mathcal{G}_f$	region between the graph of the (quasi-) density f and the x-axis, i.e. the set $\{(x,y): 0 < y \leq f(x)\}$ (p. 17)
$[b_l, b_r]$	domain with left boundary b_l and right boundary b_r for (quasi-) density (p. 17)
A_f	area below quasi-density f, i.e. $\int_{\mathbb{R}} f(x)\,dx$ or $\int_{\mathbb{R}^d} f(\mathbf{x})\,d\mathbf{x}$ (p. 21)
$\check{A}_f$	upper bound for area below quasi-density f, i.e., $\check{A}_f \geq A_f$ (p. 127)
A_h, A_s	resp. areas below hat and squeeze (p. 21)
A_{h-s}	area between hat h and squeeze s, i.e. $A_h - A_s$ (p. 22)
ϱ_{hs}	ratio between area below hat h and area below squeeze s, i.e. A_h/A_s (p. 21)
α	rejection constant; given by the ratio A_h/A_f between area below hat h and area below density f (p. 21)
$\mathrm{E}(I)$	expected number of iterations until an algorithm terminates; for rejection algorithms $E(I) = \alpha$
$\mathrm{E}(\#f)$	expected number of evaluations of density f necessary to generate one random variate (p. 21)
$\mathrm{E}(\#U)$	expected number of uniform random numbers necessary to generate one random variate (p. 21)
Ψ	transformation used for almost-exact inversion (p. 35)
$T(\cdot)$	monotone increasing differentiable transformation
T_c	familiy of transformations: $T_c(x) = \mathrm{sign}(c)\,(x^c)$ and $T_0(x) = \log(x)$ (p. 61)
$\tilde{f}(x)$	transformed density, i.e. $T(f(x))$ (p. 56)
$F_T(x)$	antiderivative of inverse transformation, i.e. $\int_{-\infty}^{x} T^{-1}(t)\,dt$ (p. 59)
$\mathrm{lc}_f(x)$	local concavity of f in x (p. 66)
N	number of construction points (p. 56)
p_i	construction points, for $i = 1, 2, \ldots, N$ (p. 56)
b_i	interval borders for TDR, for $i = 0, 1 \ldots, N$ (p. 57)
A_i	area below hat in interval (b_{i-1}, b_i), i.e. $\int_{b_{i-1}}^{b_i} h_i(x)\,dx$ (p. 58)
$\mathcal{A}_f$	region of acceptance for ratio-of-uniforms method, i.e., $\{(u,v): 0 < v \leq \sqrt{f(u/v)}\}$ (p. 136)
$\mathcal{P}^e$	Enveloping polygon of automatic RoU method (p. 105)
$\mathcal{P}^s$	Squeeze polygon of automatic RoU method (p. 105)
ϵ_U	maximal allowed absolute error in u-direction (p. 157)
$f_i(u)$	full conditional density of $f(\mathbf{x})$ given that $x_j = 0$ for $j \neq i$, i.e. $f(0, \ldots, 0, u, 0, \ldots, 0)$ with u in the i-th position (p. 286)

Subject Index and Glossary

acceptance condition
  for almost-exact inversion 35
  for independence sampler 367
  for Metropolis-Hastings algorithm 364
  for ratio-of-uniforms method 34
  for rejection method 19
  for rejection-inversion 222
  for thinning method 196
acceptance probability $(\frac{1}{\alpha})$ 21, 22
acceptance region
  for RoU $(\mathcal{A}_f)$ 33
acceptance-rejection method *see* rejection method
adaptive rejection sampling (ARS) 81–84, 118–119, 269
Ahrens method 114
  basic properties 189
  for truncated distributions 185
  multivariate 284–286
  performance of 116
  timing results 170
alias method 47–49, 215
alias-urn method 49–50
almost-exact inversion 35–37
  acceptance condition 35
alternating series method 203–204
antiderivative of inverse transformation (F_T) 59
antithetic variates 185, 330
application programming interface (API) 166

ARCH: autoregressive conditional heteroscedasticity 346
area
  below density (A_f) 21, 31
  below hat (A_h) 21, 31
  below squeeze (A_s) 21, 31
  between hat and squeeze (A_{h-s}) 22
ARFIMA *see* long memory processes
ARMA: autoregressive moving average 346
ARS *see* adaptive rejection sampling
ARTA: auto-regressive to anything 358
autocorrelation (ρ_k) 345
  of Gibbs sampler output 402
automatic generator *see* universal generator
automatic ratio-of-uniforms method 103–108
  basic properties 188
  construction points 108
  enveloping polygon 105
  relation to TDR 104
  timing results 170

Bayesian statistics 400
Bernoulli distribution 323
beta distribution 69
  timing for 170
beta-prime distribution 69
Bézier distribution 322
binomial distribution 232
  timing for 237
bisection method 157

black-box generator *see* universal generator
block monotone 281
Box-Muller method 38, 174
Brownian motion 388
burn-in 364
Burr distribution 69

Cauchy distribution 15, 32, 69
  defining a transformation 95
  timing for 170
cdf: cumulative distribution function 14
CFTP: coupling from the past 372
characteristic function 206–210
  bound for corresponding density 208
  Polya 207
chessboard distributions 333
Cholesky decomposition 250, 348, 356
code generator 7, 168
coding function 297
common random numbers 185, 391
composition method 26–27
  for hazard rate 195
  multivariate 249
composition-rejection 27
conditional distribution 246, 296
conditional distribution method 246–248, 347–348, 357
  discrete 295–297
cone 262, 276
conjugate density 36
construction points
  adaptive rejection sampling (ARS) 81–84, 118–119, 269
  derandomized adaptive rejection sampling (DARS) 84–85, 118–119, 276, 285
  equal area rule 119–121
  equiangular 80–81, 108
  equidistant 116–117
  for Ahrens method 116–121
    optimal 117–118
  for multivariate TDR 269
  for multivariate TDR with cone regions 277
  for TDR 69
    asymptotically optimal 78–80

    convergence 85
    equiangular 80–81
    one optimal 70–74, 204
    résumé 86
    three optimal 75–77
    three simple 77
    two optimal 74–75
    for TPR 223
control variates 391
convergent series method 201–203
convex polytope 255
convex unimodal 265, 280
copula 329, 331, 397
correlation 328, 329
  Kendall's tau 329
  Pearson 329
  rank 329
correlation induction 185–186, 393
coupling from the past (CFTP) 372–376
cumulative hazard rate 193
curse of dimensionality 246, 265, 270

DARS *see* derandomized adaptive rejection sampling
decreasing hazard rate 197
derandomized adaptive rejection sampling (DARS) 84–85, 118–119, 276, 285
design points *see* construction points
difference of order r 232
discrepancy 180
discrete hazard rate 235–236
discretization error 183
distribution object 167
domain $[b_l, b_r]$: region where (e.g. density) function is defined
dominating function 18
Durbin's algorithm 348
dynamic composition 198

$E(X)$: expectation of random variate X
empirical distribution 306–311
Epanechnikov kernel 309
equal area rule for construction points 119–121
equiangular construction points 80–81, 108

equidistant construction points 116–117
error in u-direction (ϵ_U) 157
European call option 387
exact sampling 372
exact-approximation method 36
exponential distribution 15, 69, 169
 timing for 170
exponential power distribution 69

face lattice 258
factorial moment 233
failure rate 193
fast Fourier transform (FFT) 350
feasible correlation matrices 341
Fejer-de la Vallee Poussin density 207
finance 387
floating point numbers 182
flops: floating point operations, i.e. additions or multiplications 347
Fourier coefficients 204–205
Fourier transform (FT) 350–351
Frechet distribution 331
full conditional distribution 246, 402

gamma(a): gamma distribution with shape parameter a and scale parameter 1
gamma distribution 5, 69, 403
 timing for 170
GARCH: generalized autoregressive conditional heteroscedasticity 346
Gaussian distribution *see* normal distribution
general position 255
generalized inverse Gaussian distribution 69
generalized ratio-of-uniforms method 142
 equivalence with TDR 143
generalizing a sample *see* empirical distribution
generalizing a vector-sample 319
generating function 232–233
generator object 166, 167
geometric Brownian motion 388
geometric distribution 217, 232
Gibbs sampler 368–371, 377, 401

for posterior distribution 404, 405
Gram's relation 262
grid method 113, 259–260
guide table method *see* indexed search
Gumbel distribution 69

hash table method *see* indexed search
hat distribution: distribution with density proportional to the hat function 18, 58
hat function: dominating function (or upper bound) for (quasi-) density
hat function 18, 57, 64
 characteristic point 71
 finding a good 22–23
 stair-case shaped 25
 universal
  for T_c-concave density 144
  for $T_{-1/2}$-concave density 140
  for heavy-tailed T_c-concave distribution 149
  for log-concave density 135, 148
hazard rate 193
 composition method for 195
 discrete 235–236
 dynamic composition 198
 inversion method for 194–195
 thinning method for 196–198
heavy-tailed T_c-concave distribution 149
Hermite interpolation 161
histogram estimate 312
hit-and-run algorithm 371–372
hyperbolic distribution 69, 390
 timing for 170
hyperbolic secant distribution 15, 39
hypergeometric distribution
 timing for 237

IEEE 754-1985 floating point standard 182
iid.: independent and identically distributed
immediate acceptance 29–32
 for Ahrens method 116
 for automatic ratio-of-uniforms method 106
 for TDR 91

increasing hazard rate   198
independence sampler   366–368, 377
indexed search   45–46
   for right tail   216–218
   for two tails   218–219
   for unbounded domain   216–219
intersection point   57
inverse cdf (F^{-1})
   definition   14
   examples in closed form   15
inversion   13–14
   and floating point arithmetic   183
   basic properties   190
   by Hermite interpolation   160–163
     error bound   162
   by indexed search   45–46
   by indexed search for unbounded domain   216–219
   by iterated linear interpolation   159–160
   by linear interpolation   159
   by sequential search   43–44, 296, 298
   discrete   43–44
   for hazard rate   194–195
   numerical   155–158
   numerical with tables   158
   quality   178
   timing results   170
inversion-rejection   156

Kendall's tau   329
kernel   307
kernel density estimation   307–310
   multivariate   319–321
Khinchine's theorem   324

Laplace distribution   39
layered multishift coupling   379–383
LCG: linear congruential generator   175
likelihood function   400
linear interpolation of the cdf   159
linear interpolation of the empirical cdf   310–311
Lipschitz density   134
local concavity ($\mathrm{lc}_f(x)$)   66, 101, 149
log-concave   56, 65
   discrete   221, 228–230
   full conditionals   405
   posterior   408
log-concave density   135, 287, 292
   universal hat for   135
log-concave distribution   131, 148
   multivariate   265
log-log-concave   103
logarithmic distribution
   timing for   237
logistic distribution   15
logistic model   405
lognormal distribution   69, 388
long memory processes   361

marginal distribution   246, 295, 328
marginal execution time: average time required to generate one random variate (without taking the setup into account)   5
Markov chain   364
Markov Chain Monte Carlo   363
MCMC: Markov Chain Monte Carlo   363
Metropolis-Hastings algorithm   364–366, 385
mirror principle   138
mixing property   366
modeling   305
moment problem   327
moments   323
   bound for   132
   log-concave distribution   131
   matching the first four   323
monotone density   126
   block monotone   281
   bound for   126
   orthomonotone   280
Morgenstern distribution   332
multi-t-distribution   252
multinomial distribution   296–297
multinormal distribution   250–252
multivariate density with compact support   289–291
multivariate distribution   245, 407
multivariate table method   284–286

$N(\mu, \sigma^2)$: normal distribution with mean μ and variance σ^2
naive resampling   307
Neave effect   175

negative binomial distribution 232
Newton method 156, 160
non-Gaussian time series 356
normal copula 333
normal distribution 5, 69
   almost-exact-inversion for 36
   as kernel 309
   Bayesian parameter estimation for 403
   Box-Muller method 38, 174
   Kinderman-Ramage generator 174
   multivariate 250–252
   of log-returns 388
   of multivariate log-returns 397
   rejection method for 19
   timing for 170
   two-dimensional 267, 370, 401
normal mixture 40
   bivariate 331
NORTA: normal to anything 330, 336, 340–343, 358
NORTU: normal to uniform 333
number of evaluations of the density 21
number of iterations till acceptance 21
numerical inversion 155

object-oriented design 166
option
   European call 387
   price simulation 389
order statistics
   generation of 186
   log-concave 68
   T-concave 68
   timing for 170
orthomonotone 280, 299
orthounimodal 280

parameter object 167
Pareto distribution 15
patchwork method 27
Pearson correlation 329, 359
Pearson VI distribution 69
perfect sampling 372
   for R^d 377
Perks distribution 69
Planck distribution 69
platymorphous distribution 282

pmf: probability mass function 43, 215
Poisson distribution 232
   timing for 237
Poisson process 194
Polya characteristic function 207
polytope 258
portfolio 396
posterior distribution 400
   vector algorithm for 408
prior distribution 400
PRNG: pseudo-random number generator 173
Prob(A): probability of event A
probability generating function 232
probability mass function 43, 215
probability vector 43, 215
product density 286–287
product moment correlation see Pearson correlation
proposal density 364
pseudo-price process 388
pseudo-random numbers
   streams of 8

quality of generated streams 173
quasi-density: any multiple of a density function, i.e., a nonnegative integrable function 16
quasi-probability mass function (quasi-pmf): any multiple of a pmf, i.e. a nonnegative summable sequence 215
queuing theory 316

radially symmetric distribution 253
random direction interior point method 372, 385
random-walk Metropolis (RWM) 366, 379–383
randomizing operation 372
   monotone 376
rank correlation 329, 360
ratio between area below hat and area below squeeze (ϱ_{hs}) 21
ratio-of-uniforms method 32–35
   acceptance region ($\mathcal{A}_f$) 33
   basic properties 189
   minimal bounding rectangle 34
   multivariate 249

440 Subject Index and Glossary

quality 175
timing results 170
with short setup 135–138
recycling uniform random numbers 28, 46
region below density f ($\mathcal{G}_f$) 16, 18
regula falsi 156, 160
rejection constant (α) 21
rejection method 16–20
  acceptance condition 19
  acceptance probability ($\frac{1}{\alpha}$) 21, 22
  acceptance region for density f ($\mathcal{G}_f$) 16, 18
  based on global inequality 125, 281
  discrete 50–51, 219
  for monotone densities with bounded domain 126–128
  for monotone densities with unbounded domain 129–133
  multivariate 248–249
  multivariate discrete 299
  on unit cube 282–284
  using conditional densities 286–294
  using Fourier coefficients 204–205
  using series 201–204
  using staircase-shaped hat *see* Ahrens method
  with short setup for $T_{-1/2}$-concave distributions 141
rejection sampler 368
rejection-inversion 51, 221–222
  transformed probability rejection 223–227
relative generation time: average generation time divided by marginal generation time for one exponential random variate using inversion 11, 169
resampling 307–310, 398
RNG: random number generator
Robin Hood algorithm 47–48
RoU: ratio-of-uniforms 32
RWM *see* random-walk Metropolis

sampling part: second part of an algorithm for generating random variates 5
scale invariant 60
scale mixture 324–327

scatter plot 178
sequential search 43–44
series method 201
  alternating 203–204, 208
  convergent 201–203, 205
setup: first part of an automatic algorithm to compute necessary constants 5
setup time: time required for the setup 5
shake-and-bake algorithm 372
simple cone 262, 276
simple polytope 260
simplex 255
slice sampler 378
smoothed bootstrap 307
smoothing parameter 307
Snedecor's F distribution 69
Spearman's rho *see* rank correlation
squeeze: lower bound for density
squeeze 20–21
  for Ahrens method 114
  for automatic ratio-of-uniforms method 105
  for multivariate TDR 271
  for series method 201
  for TDR 56, 57
    figure 64
  for TPR 225
star-unimodal 281
stationary Gaussian time series 347
stationary time series 345
string API 168
strip method 113
  horizontal strips 123
support: closure of region where function is non-zero
sweep-plane technique
  for log-concave distribution 266
  for uniform distribution 260–264
symmetric multivariate density 287–289

T-concave 264
  definition 56
  discrete 219, 228
  distributions 63–69
T-concave distribution
  multivariate 265

T-convex 97
t-distribution   69, 390
  multivariate   252, 397
tangent   57
TDR   *see* transformed density rejection
thinning method   196–198
  acceptance condition   196
time series   345
timing
  for continuous distributions   169–171
  for discrete distributions   237
  for Gaussian timeseries   355
  for multivariate distributions   294
Toeplitz matrix   345
TPR   *see* transformed probability rejection
transformation T_c 59–61, 223
transformation theorem   41
transformed density rejection   55–59
  basic properties   188
  for T-convex distributions   97
  for bivariate distributions   271–276
  for log-concave multivariate distributions   277–279
  for multivariate distributions   266–271, 276–279
  for non-T-concave distributions   99
  for truncated distributions   185
  generalized   103
  immediate acceptance   91–92
  multivariate   264
  non-constant c 100–103
  other transformations   95
  properties of transformation   57
  proportional squeeze   89–90
  quality   178, 182
  three design points   92–93, 147
  timing results   170
  without derivatives   88–89
transformed probability rejection (TPR)   216, 219–221
transformed rejection   37
transition kernel   364
triangular distribution   15, 37
triangulation   105, 258, 278
truncated distribution   184

$U(0,1)$: uniform distribution on the interval $(0,1)$
uniform distribution   15
uniform in simplex   255–257
uniform in the sphere   254–255
uniform on the sphere   253–254
uniform spacings   256
uniform vectors   252
unimodal density
  convex unimodal   280
  discrete with given moment   227–228
  orthounimodal   280
  star-unimodal   281
unit sphere   253
universal generator   5
UNU.RAN   166–168

value at risk   394
  portfolio   396
VaR: value at risk   394
variance reduction   391
vertical density representation   249

Weibull distribution   15, 69
window width   308

Yule-Walker equations   348

zero-maximal correlation mixture   330
ziggurat method   123

Printing: Saladruck Berlin
Binding: Stürtz AG, Würzburg